A. Togni, H. Grützmacher
Catalyc Heterofunctionalization

Other Titles of Interest

A. Togni, R. L. Halterman

Metallocenes

2 Volumes

XXXXII. 790 pages with 461 figures and
62 tables
1998
Hardcover
ISBN 3-527-29539-9

B. Cornils, W. A. Herrmann

Applied Homogeneous Catalysis with Organometallic Compounds

2 Volumes

XXXVI. 1246 pages with 1000 figures
and 100 tables
1996
Hardcover
ISBN 3-527-29286-1
Softcover
ISBN 3-527-29594-1

B. Cornils, W. A. Herrmann, R. Schlögl,
C.-H. Wong

Catalysis from A to Z

XVIII. 640 pages with more than
300 figures and 14 tables
2000
Hardcover
ISBN 3-527-29855-X

M. Beller, C. Bolm

Transition Metals for Organic Synthesis

2 Volumes

LVIII. 1062 pages with 733 figures and
75 tables
1998
Hardcover
ISBN 3-527-29501-1

D. E. De Vos, I. F. J. Vankelecom,
P. A. Jacobs

Chiral Catalyst Immobilization and Recycling

XX. 320 pages with 199 figures and
45 tables
2000
Hardcover
ISBN 3-527-29952-1

R. A. Sheldon, H. van Bekkum

Fine Chemicals through Heterogeneous Catalysis

XXV. 611 pages with 182 figures and
94 tables
2000
Hardcover
ISBN 3-527-29951-3

Catalytic Heterofunctionalization

From Hydroanimation to Hydrozirconation

Edited by
Antonio Togni, Hansjörg Grützmacher

WILEY-VCH

Weinheim – New York – Chichester – Brisbane – Singapore – Toronto

The Editors of this Volume

Prof. Dr. Antonio Togni
Department of Chemistry
Swiss Federal Institute of Technology
ETH-Hönggerberg
CH-8093 Zürich
Switzerland

Prof. Dr. Hansjörg Grützmacher
Department of Chemistry
Swiss Federal Institute of Technology
ETH-Hönggerberg
CH-8093 Zürich
Switzerland

■ This book was carefully produced. Nevertheless, authors, editors and publisher do not warrant the information contained therein to be free of errors. Readers are advised to keep in mind that statements, data, illustrations, procedural details or other items may inadvertently be inaccurate.

First Edition 2001

Library of Congress Card No.:
applied for

British Library Cataloguing-in-Publication Data:
A catalogue record for this book is available from the British Library.

Die Deutsche Bibliothek – CIP Cataloguing-in-Publication-Data:
A catalogue record for this publications is available from Die Deutsche Bibliothek.

© Wiley-VCH Verlag GmbH, Weinheim; 2001

Printed in the Federal Republic of Germany.

Printed on chlorine-free paper.

Composition Datascan GmbH, Darmstadt
Printing Strauss Offsetdruck GmbH, Mörlenbach
Bookbinding J. Schäffer GmbH & Co. KG, Grünstadt

ISBN 3-527-30234-4 ac✻

Contents

Preface

Finding molecules which are able to catalyze the reaction between others is an important contribution of molecular chemists to increase the efficiency of chemical reactions whereby our daily life based on consumption of chemicals is shifted closer to an ecologically and economically tolerable equilibrium with our environment. Processes, where large amounts of energy are consumed - mostly in order to overcome the activation barrier of a reaction – will disturb significantly and irreversibly our living conditions. Considering the fact that only a small part of the world population lives under acceptable conditions, it would be cynical to call for a reduction of industrial production and development. On the contrary, the production of fine chemicals for any pharmaceutical and agricultural use must increase immensely.

Meanwhile, synthetic organic chemistry has reached a level where probably for any molecule composed of the elements carbon, hydrogen, nitrogen, and oxygen (to name only the most relevant elements of functionalized organic molecules) a suitable synthesis can be found via a retro-synthetic approach using the fund of known reaction principles [1]. However, depending on the complexity of the target molecule (which will increase with our understanding of the interaction of molecular entities with its surroundings) these syntheses correspond actually to reaction schemes including a multitude of single reaction steps. The thermodynamic parameters for any of these steps are given. Also the costs of a reaction calculated per atom (i.e. carbon, hydrogen, nitrogen, oxygen, etc.) are almost fixed by the prices of the basic chemicals on the world-market. Making a reaction sequence shorter and inevitable reaction steps faster can reach the aim of increasing the productivity while keeping energy consumption on a tolerable level. For example, the – especially stereospecific – synthesis of alcohols or amines requires often a lengthy multistep synthesis by which suitable functionalized intermediates are formed. Clearly, the direct stereospecific addition of water or an amine to a prochiral C=C function would be the ultimate response to this problem. However, this addition is connected with a very high activation barrier and without a catalyst (which in its most elegant form may also intervene to control the stereochemistry of the addition process) this reaction is ineffective and useless.

This book is divided in eight chapters and each of them is devoted to the state-of-the art of the homogeneously catalysed addition of E-H or E-E' heteroelement bonds* to unsaturated substrates with C=C, C≡C and C=X functions (X = O, S). Em-

---

* By this term we understand bonds between heteroelements in the sense of classical organic chemistry, i.e. bonds which do not include the element carbon.

phasis is not only given to highlight achievements, which have been made in each domain, but also to clearly show the limitations. Reactions including the addition of dihydrogen (hydrogenations) or C-H bonds (C-H activation) are not considered and the reader is referred to recent monographs covering these topics [2,3]. The ordering of the chapters follows simply the ordering by which the heteroelement is positioned in the periodic table. The addition of reagents containing main group elements is treated first and the one of transition metal containing reagents last. Hence the first chapter discusses the catalyzed addition of boron reagents and the last one gives an overview about hydrozirconation reactions.

A catalyzed hydroboration reaction has first been described in 1985. In the beginning, the advantage of this reaction was seen in the use of cheap boron reagents which are easy to handle but little reactive in the non-catalyzed reaction. The considerable progress, concerning the development of the boron reagents and the catalysts employed in these types of reactions, is traced by N. Miyaura in the first chapter. Nowadays a wide variety of catalyst types based on complexes of Ti, Zr, Sm, Ru, Rh, Ir, Ni, Pd, and different boranes are employed which allow to control the stereoselectivity and specificity of the borane addition to a manyfold of substrates containing C=C, C≡C, and C=X multiple bonds. More recently the addition of B-B, B-Si, B-Ge, and B-Sn bonds to unsaturated substrates attracted attention. These reactions are generally catalyzed by Pd(0) Pt(0) or Rh(I) complexes. They allow the elegant syntheses of highly functionalized products in few steps. Furthermore, the catalyzed cross-coupling reaction of diboranes, $R_2B\text{-}BR_2$, with organic halides opened a straightforward route to aryl and allyl boranes which themselves are valuable intermediates.

In the second chapter, homogeneously catalyzed hydroalumination reactions of alkenes and alkynes are surveyed. Although alanes are more reactive than boranes and many hydroaluminations proceed indeed without a catalyst (especially those of alkynes), metallocene chlorides, such as Cp_2TiCl_2 or Cp_2ZrCl_2, nickel or cobalt salts, or palladium(II) complexes not only accelerate the reaction but also influence the stereochemistry of the addition reaction. Apart from (enantioselective) hydroaluminations of carbon-carbon multiple bonds and allyl ether cleavages, the reader will learn about highly selective reductive ring opening reactions, which were invented in the group of M. Lautens who is, with M. Dahlmann, the author of this chapter. This reaction is another good example for the short and elegant synthesis of complex molecules by a novel approach using a catalytic heterofunctionalization as the key step.

The transition metal-catalyzed hydrosilylation belongs to the „old-timers" of catalytic heterofunctionalizations and numerous applications have been established. Therefore, J. Tang and T. Hayashi concentrate in the third chapter on the progress made in enantioselective hydrosilylations. Frequently, precursor complexes with platinum and rhodium as active centres and a chiral phosphine as ligand are employed in these reactions. However, recently also palladium complexes carrying a monodendate axial-chiral phosphine were introduced as highly efficient catalysts for enantioselective hydrosilylations. Furthermore, new lanthanide and group 3 metallocene complexes were found to be active complementing the established list of d^0 metal hydrosilylation catalysts, i.e. titanocenes and zirconocenes. Notable progress has also been made in the asymmetric syntheses of functionalzsed carbocycles by hydrosilylation of suitable dienes. Catalyzed by chiral palladium(II) oxazoline or rhodium(I) bisphosphine complexes, C-C, C-Si and C-H bonds are stereoselectively formed within one catalytic cycle making the efficiency of catalytic heterofunctionalizations evident.

The fourth chapter gives a comprehensive review about catalyzed hydroaminations of carbon carbon multiple bond systems from the beginning of this century to the state-of-the-art today. As was mentioned above, the direct - and whenever possible stereoselective - addition of amines to unsaturated hydrocarbons is one of the shortest routes to produce (chiral) amines. Provided that a catalyst of sufficient activity and stability can be found, this heterofunctionalization reaction could compete with classical substitution chemistry and is of high industrial interest. As the authors J. J. Brunet and D. Neibecker show in their contribution, almost any transition metal salt has been subjected to this reaction and numerous reaction conditions were tested. However, although considerable progress has been made and enantioselectivites of 95% could be reached, all catalytic systems known to date suffer from low activity (TOF < 500 h^{-1}) or/and low stability. The most effective systems are represented by some iridium phosphine or cyclopentadienyl samarium complexes.

The discussion of the activation of bonds containing a group 15 element is continued in chapter five. D.K. Wicht and D.S. Glueck discuss the addition of phosphines, R_2P-H, phosphites, $(RO)_2P(=O)H$, and phosphine oxides $R_2P(=O)H$ to unsaturated substrates. Although the addition of P-H bonds can be sometimes achieved directly, the transition metal-catalyzed reaction is usually faster and may proceed with a different stereochemistry. As in hydrosilylations, palladium and platinum complexes are frequently employed as catalyst precursors for P-H additions to unsaturated hydrocarbons, but (chiral) lanthanide complexes were used with great success for the (enantioselective) addition to heteropolar double bond systems, such as aldehydes and imines whereby pharmaceutically valuable α-hydroxy or α-amino phosphonates were obtained efficiently.

In the sixth chapter the activation of O-H bonds of water, alcohols and carboxylic acids, and their addition to multiple bonds is reported. Since the formally oxidative addition of ROH gives rise to hydrido(hydroxo) complexes, [MH(OR)Ln] which are postulated as intermediates in many important reactions (water gas shift reaction, Wacker-chemistry, catalytic transfer hydrogenations etc.) the authors of this chapter, K. Tani and Y. Kataoka, begin their discussion with an overview about the synthesis and isolation of such species. Many of them contain Ru, Os, Rh, Ir, Pd, or Pt and complexes with these metals appear also to be the most active catalysts. Their stoichiometric reactions, as well as the progress made in catalytic hydrations, hydroalcoxylations, and hydrocarboxylations of triple bond systems, i.e. nitriles and alkynes, is reviewed. However, as in catalytic hydroaminations the „holy grail", the addition of O-H bonds across non-activated C=C double bonds under mild conditions has not been achieved yet.

H. Kuniyasu continues the discussion of the activation of group 16 element bonds with an overview on S(Se)-X additions to unsaturated substrates. For some time, it was believed that sulfur compounds „poison" systematically transition metal complexes by forming very robust metal sulfides. However, as it is shown in this seventh chapter, a wide variety of thiols, disulfides, diselenides, silyl and germyl sulfides and selenides, and thioboranes can be successfully added to carbon carbon multiple bonds, especially alkynes, with the aid of metal catalysts. Frequently, the „ubiquitous" metal complexes used in homogeneous catalysis like the phosphine complexes of palladium, platinum, and rhodium can be used to afford a wide range of chalcogenato compounds. Also cobalt, nickel, and ruthenium complexes, and some Lewis-acids were studied as catalysts.

A chapter written by A. Igau reviewing hydrozirconations concludes this book. As was demonstrated in recent years, the addition of the Schwartz reagent, $[Cp_2ZrHCl]_n$, to unsaturated substrates containing C=C, C≡C, C=N, C=P, and C=O entities allowed the synthesis of a wide range of highly functionalized zirconium derivatives which proved to be valuable intermediates in organic synthesis. Since the primary products of the hydrozirconation reaction contain a highly polar zirconium(δ+) X(δ-) bond (X = C, N, O, etc), they can be easily transformed further by substitution reactions with halides or insertion reactions of another equivalent of an unsaturated substrate into the Zr-X bond. Although catalytic hydrozirconations are just being discovered and most of the reactions described in this chapter are stoichiometric, the reader will find many useful applications of this type of heterofunctionalization.

For some of the reactions described in this book, rather precise and detailed ideas about the reaction mechanism exist. However, for many catalytic reactions, the mechanistic understanding is very poor and further experimental studies are certainly needed. Calculations proved to be a highly valuable tool to gain a more precise picture of the reaction pathways. However, mostly only model systems can be studied due to the complexity of the problem. Anyway, it is the firm believe of the authors that for any reaction with an activation barrier a suitable catalyst can be found. This book shall give an insight into what has been achieved in this area concerning the synthesis of heterofunctionalized organic molecules. It is the hope of all contributors that future retro-synthetic schemes will include the catalytic approaches outlined in this book.

[1] D. Seebach, *Angew. Chem.* **1990**, *102*, 13; *Angew. Chem. Int. Ed. Engl.* **1990**, *29*, 1320.
[2] P. A. Chaloner, M. A. Esteruelas, F. Joó, L. A. Oro, *Homogeneous Hydrogenation*, Kluwer Academic Publishers, Dordrecht, 1994.
[3] J. A. Davies, P. L. Watson, J. F. Liebman, A. Greenberg, *Selective Hydrocarbon Activation*, VCH-Wiley, Weinheim, 1990.

Zürich, July 2001   H. Grützmacher   A. Togni

List of Contributors

Dr. Jean-Jacques Brunet
Laboratoire de Chimie de Coordination
du CNRS
UPR 8241
205, route de Narbonne
31077 Toulouse Cedex 4
France

Marc Dahlmann
Department of Chemistry
University of Toronto
Toronto, Ontario
Canada M5S 3H6

Prof. Dr. David S. Glueck
6128 Burke Laboratory
Department of Chemistry
Dartmouth College
Hanover
New Hampshire 03755
USA

Prof. Dr. Tamio Hayashi
Kyoto University
Faculty of Science
Department of Chemistry
Sakyo
Kyoto 606-01
Japan

Dr. Alain Igau
Laboratoire de Chimie de Coordination
Du CNRS
UPR 8241
205, route de Narbonne
31077 Toulouse Cedex 04
France

Prof. Dr. Hitoshi Kuniyasu
Department of Applied Chemistry
Osaka University
Suita
Osaka 565-0871
Japan

Prof. Dr. Mark Lautens
Department of Chemistry
University of Toronto
Toronto, Ontario
Canada M5S 3H6

Prof. Dr. Norio Miyaura
Division of Molecular Chemistry
Graduate School of Engineering
Hokkaido University
Sapporo 060-8628
Japan

Dr. Denis Neibecker
Laboratoire de Chimie de Coordination
du CNRS
UPR 8241
205, route de Narbonne
31077 Toulouse Cedex 4
France

Prof. Dr. Kazuhide Tani
Department of Chemistry
Graduate School of Engineering Science
Osaka University, Toyonaka
Osaka 560-8531
Japan

Juan Tang
Kyoto University
Faculty of Science
Department of Chemistry
Sakyo
Kyoto 606-01
Japan

Denyce K. Wicht
6128 Burke Laboratory
Department of Chemistry
Dartmouth College
Hanover
New Hampshire 03755
USA

1
Hydroboration, Diboration, Silylboration, and Stannylboration

Norio Miyaura

1.1
Introduction

In this work, particular attention will be given to the synthesis of organoboron compounds *via* the metal-catalyzed addition and coupling reactions of H–B, B–B, B–Si, and B–Sn reagents [1, 2]. The classical methods for the synthesis of organoboron compounds are based on the reaction of trialkyl borates with Grignard or lithium reagents (transmetalation) or the addition of H–B reagents to alkenes or alkynes (uncatalyzed hydroboration) [3]. Although these methods are now most common and convenient for large-scale preparations, the metal-catalyzed reactions are advantageous in terms of efficiency and selectivity of the transformations. Hydroboration of alkenes and alkynes is one of the most studied of reactions in the synthesis of organoboron compounds and their application to organic synthesis. However, catalyzed hydroboration did not attract much attention until Männig and Nöth in 1985 [4] reported that a Wilkinson complex [RhCl(PPh$_3$)$_3$] catalyzes the addition of catecholborane to alkenes or alkynes. Although the transition metal complexes significantly accelerate the slow reaction of (dialkoxy)boranes, the catalyzed hydroboration is a more interesting strategy to realize the different chemo-, regio-, diastereo-, and enantioselectivities, relative to the uncatalyzed reaction, because the catalyzed reaction can change the metal-hydride species which interacts with the unsaturated C–C bond. The addition of diboron tetrahalides B$_2$X$_4$ (X=F, Cl, Br) to unsaturated hydrocarbons (diboration), first discovered by Schlesinger in 1954 [5, 6], is an attractive and straightforward method to introduce boryl groups into organic molecules, but the synthetic use has been severely limited because of the instability and limited availability of the reagents. Although tetra(alkoxo)diboron dramatically enhances the stability of the B–B species at the expense of reactivity for organic nucleophiles, the B–B compounds oxidatively add to low-valent transition metals with the B–B bond cleavage, thus allowing the catalyzed transfer of boron to unsaturated organic substrates. The metal-catalyzed addition of B–B, B–Si, or B–Sn reagents to alkenes or alkynes provides a new class of boron compounds including heterofunctionalized alkyl-, alkenyl-, and allylboronates. The cross-coupling reaction of metal-boryl reagents is an alternative to the transmetalation method and perhaps a more convenient and direct protocol for the synthesis of organoboron compounds from organic halides and other electrophiles.

Much attention has recently been focused on organoboronic acids and their esters because of their practical usefulness for synthetic organic reactions including asymmetric synthesis, combinatorial synthesis, and polymer synthesis [1, 3, 7–9], molecular recognition such as host-guest compounds [10], and neutron capture therapy in treatment of malignant melanoma and brain tumor [11]. New synthetic procedures reviewed in this article will serve to find further applications of organoboron compounds.

1.2
Metal-Catalyzed Hydroboration

1.2.1
Hydroboration of Alkenes and Alkynes

Most studies of catalyzed hydroboration have employed catecholborane 1 (HBcat) [12] because of its high reactivity for various transition metal catalysts (Scheme 1-1). However, pinacolborane 2 (HBpin) [13] has recently been found to be an excellent alternative because it is a more stable, easily stored and prepared hydroboration reagent. The high stability of the resulting products (pinacol esters of alkyl- or 1-alkenylboronates) to moisture and chromatography is also very convenient for organic chemists. Other borane reagents including 4,4,6-trimethyl-1,3,2-dioxaborinane (3) [14], oxazaborolidines (4) [15] , benzo-1,3,2-diazaborolane (5) [16], and borazine (8) [17] may also be used, but the scope of these reagents remains to be explored.

There is no systematic study of the effect of borane reagents, and the best choice would be highly dependent on the catalysts and substrates. A series of di(alkoxy)boranes have recently been synthesized and subjected to hydroboration of cyclopentene at ambient temperature in the presence of RhCl(PPh₃)₃ (Scheme 1-2) [18]. The

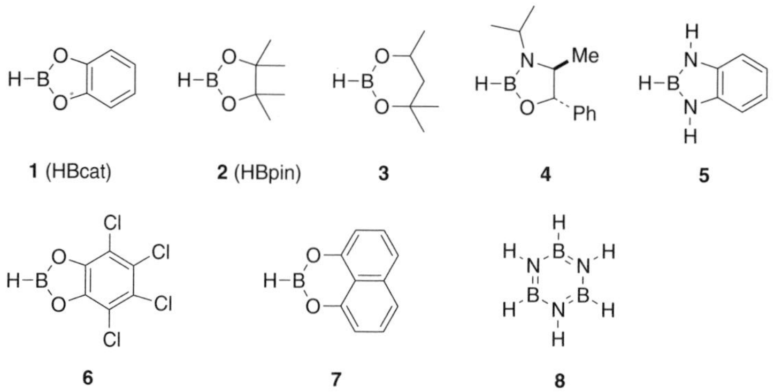

Scheme 1-1   Borane Reagents for Catalyzed Hydroboration

$$\text{[cyclopentene]} + \underset{\text{(2 equivs)}}{\text{HB(OR)}_2} \xrightarrow[\text{CDCl}_3/\text{r.t.}]{\text{RhCl(PPh}_3)_3 \ (< 1 \text{ mol\%)}} \text{[cyclopentyl]}-\text{B(OR)}_2$$

borane	time for >90% conversion
7	4 min
6	30 min
1	90 min
HB(OCH$_2$Ph)$_2$	no reaction

Scheme 1-2   Effect of Borane Reagents

superiority of more Lewis-acidic boranes is suggested because acyclic dialkoxybo-
ranes do not participate in the catalytic cycle and tetrachlorocatecholborane (**6**) reacts adequately faster than catecholborane. However, the less acidic six-membered
borane **7** is, unexpectedly, the best reagent for the rhodium-catalyzed hydroboration.

Hydroboration of styrene derivatives has been extensively studied, and perhaps
these are the best substrates to consider in a discussion of the efficiency and selectivity of the catalysts (Table 1-1). A neutral rhodium-phosphine complex

Tab. 1-1   Catalysts for Hydroboration of Vinylarenes

$$\text{ArCH=CH}_2 \xrightarrow[\text{2. H}_2\text{O}_2/\text{OH}^-]{\text{1. HBX}_2/\text{Rh catalyst/THF/r.t.}} \underset{\mathbf{9}}{\text{Ar}\overset{\text{OH}}{\underset{\text{CH}_3}{\diagdown}}} + \underset{\mathbf{10}}{\text{Ar}\diagdown\diagup\text{OH}}$$

Ar=Ph or 4-MePh

Entry	Borane	Catalyst	Yield/%	9	10	Ref.
1	HBcat	RhCl(PPh$_3$)$_3$ (argon)	80	>99	<1	[19, 20]
2		RhCl(PPh$_3$)$_3$ (air)	–	24	76	[19, 20]
3		[RhCl(cod)]$_2$	45	20	80	[20, 22]
4		[RhCl(cod)]$_2$/4PPh$_3$	90	98	2	[19, 20, 23]
5		[RhCl(cod)]$_2$/2dppe	50	34	66	[22]
6		[RhCl(cod)]$_2$/2dppb	83	45	55	[22]
7		[RhCl(cod)]$_2$/2dppf	83	10	90	[22]
8		Rh(η^3-2-Me-allyl) (dppb)	–	>99	<1	[24]
9		[Rh(cod)$_2$]BF$_4$/2PPh$_3$	93	99	1	[22, 25]
10		[Rh(cod)$_2$]BF$_4$/dppb	99	99	1	[22, 25]
11		[Cp*IrCl$_2$]$_2$	–	0	100	[28]
12		[Ir(cod)(Py)(PCy$_3$)]OTf	–	22	78	[28]
13		RuCl$_2$(PPh$_3$)$_4$	–	0	100	[29]
14		RuCl$_2$(PPh$_3$)$_2$(MeOH)	–	1	99	[29]
15		Cp$_2$TiMe$_2$ (in benzene)	89	0	100	[30]
16		Cp*$_2$Sm(THF) (in benzene)	89	0	100	[31]
17	HBpin	RhCl(PPh$_3$)$_3$ (CH$_2$Cl$_2$)	99[a]	41	59	[33]
18		Rh(CO)(PPh$_3$)$_2$Cl (CH$_2$Cl$_2$)	99	1	99	[33]
19		CpNiCl(PPh$_3$) (CH$_2$Cl$_2$)	99	1	99	[33]
20		[Ir(cod)Cl]$_2$/2dppp (CH$_2$Cl$_2$)	97	<1	>99	[34]

a) PhCH=CHB(OR)$_2$ (15%) was also accompanied.

$$\text{RhCl(PPh}_3)_3 \xrightarrow{\text{O}_2} \text{RhCl(O}_2)(\text{PPh}_3)_3 \longrightarrow \begin{array}{l} [\text{Rh}(\mu\text{-Cl})(\text{PPh}_3)_2]_2 \\ + [\text{RhCl(O}_2)(\text{PPh}_3)_2]_2 \\ + \text{Ph}_3\text{P=O} \end{array}$$

Scheme 1-3   Air Oxidation of Wilkinson Catalyst

$RhCl(PPh_3)_3$ is the most studied catalyst for hydroboration of alkenes, but the complex is unfortunately highly sensitive to air. Thus, handling the catalyst under argon or air results in different regioselectivity (entries 1 and 2) [19, 20]. The changes in regioselectivity resulted from lowering the triphenylphosphine-to-rhodium ratio *via* oxidation of phosphine to the oxide (Scheme 1-3) [21]. Thus, the *in situ* preparation of the catalyst from $[RhCl(cod)]_2$ and a limited amount of phosphine (entries 3–7) [19–23] or the use of an air stable π-allylrhodium complex (entry 8) [24] is a convenient alternative, but the use of a large excess of the ligand should be avoided because of its higher coordination ability to the metal than that of alkenes. An addition of phosphine to $[Rh(cod)_2]BF_4$ generates a highly active species to catalyze hydroboration even at temperatures lower than 0°C (entries 9 and 10) [22, 25]. The regiochemical preference giving terminal (**10**) or internal products (**9**) depending on the ligand and the valence state of the metal has not yet been well understood. The high internal selectivity of vinylarenes is accounted by a contribution of a π-benzylrhodium species [22]; however, the catalyzed reaction commonly exhibits high internal selectivity for alkenes having an electron-withdrawing group such as vinylarenes, fluoroalkenes [26], and α,β-unsaturated esters or amides [27], and the cationic rhodium catalysts would further increase the internal selectivity. The iridium(I) [28] and ruthenium(II) or (III) [29] complexes analogously catalyze hydroboration with catecholborane (entries 11–14). Although the neutral phosphine complexes reveal a high terminal selectivity, the scope of these catalysts has not yet been studied in detail. The cyclopentadienyl complexes such as Cp_2TiMe_2 [30] and $Cp^*_2Sm(THF)$ [31] are excellent catalysts for the addition of boron to the terminal carbon (entries 15 and 16). Such high terminal selectivity can be accounted for by steric hindrance of the cyclopentadienyl ligand, since the Cp* complex exhibits higher terminal selectivity than that of the Cp ligand and SmI_3 [32] results in a mixture of both isomers. The steric effect of borane reagents also plays an important role in selectivity. Pinacolborane selectively adds to the terminal carbon because of its bulkiness (entries 17–20), which is in sharp contrast to the internal addition of catecholborane according to the electronic effect of the phenyl group. Although $RhCl(PPh_3)_3$ results in a complex mixture for styrene including regioisomers (**9, 10**) and a dehydrogenative coupling product $PhCH=CHBpin$ (entry 17), other Rh(I), Ni(II), and Ir(I) catalysts reveal high terminal selectivity (entries 18–20) [33–34].

Various metal complexes catalyze the addition of catecholborane and pinacolborane to aliphatic terminal alkenes (Table 1-2). Neither the borane reagents nor the catalysts alter the high terminal selectivity, but a titanium catalyst does (entry 3). Although Cp_2TiMe_2 [30] exhibits high terminal selectivity for vinylarenes, aliphatic alkenes afford appreciable amounts of internal products, whereas an analogous $Cp^*_2Sm(THF)$ [31] allows selective addition of catecholborane to the terminal car-

Tab. 1-2 Catalysts for Hydroboration of Aliphatic 1-Alkenes

$$RCH=CH_2 \xrightarrow[\text{2. } H_2O_2/OH^-]{\text{1. } HBX_2/Rh \text{ catalyst/r.t.}} \underset{\underset{11}{CH_3}}{R \underset{|}{\overset{\quad OH}{\diagup}}} + \underset{12}{R \diagdown\diagup OH}$$

Entry	Alkene R=	Borane	Catalyst (solvent)	Yield/%	11	12	Ref.
1	C_4H_9	HBcat	$RhCl(PPh_3)_3$ (THF)	–	1	99	[35]
2			$[Rh(nbd)(dppb)]BF_4$ (THF)	–	10	90	[35]
3			Cp_2TiMe_2 (benzene)	85	34	66	[30]
4			Cp^*_2Sm(THF)	78	1	>99	[31]
5	C_8H_{17}		SmI_3 (3 h, THF)	8	25	75	[32]
6			SmI_3 (18 h, THF)	98	2	98	[32]
7	C_6H_{13}	HBpin	$RhCl(PPh_3)_3$ (CH_2Cl_2)	99	0	100	[36]
8			Cp_2ZrHCl (CH_2Cl_2)	70	0	100	[36]
9			$[Ir(cod)Cl]_2/2dppm$ (CH_2Cl_2)	89	<1	>99	[34]
10	$PhOCH_2$		$RhCl(PPh_3)_3$ (CH_2Cl_2)	90	0	100	[33]
11			$[Ir(cod)Cl]_2/2dppm$ (CH_2Cl_2)	89	<1	>99	[34]
12	$CH_3C(O)O$		$RhCl(PPh_3)_3$ (CH_2Cl_2)	84	0	100	[33]

bon (entry 4), which is due to differences in the metal hydride species participating in the insertion of alkene (see Section 1.2.2). SmI_3 exhibits a unique regioselectivity depending on the reaction time (entries 5 and 6) [32]. The reaction initially yields a mixture of internal and terminal product, but the catalyst slowly isomerizes secondary alkylboronate to the primary one on prolongation of reaction time, thus suggesting reversible formation of the C–B bond. On the other hand, the catalysts do not alter the high terminal selectivity of pinacolborane (entries 7–12) [33, 34, 36].

The differences in the steric effect between catecholborane and pinacolborane, and the valence effect between a cationic or neutral rhodium complex reverse the regioselectivity for fluoroalkenes (Scheme 1-4) [26]. The reaction affords one of two possible isomers with excellent regioselectivity by selecting borane and the catalyst appropriately, whereas the uncatalyzed reaction of 9-BBN or Sia_2BH failed to yield the hydroboration products because of the low nucleophilicity of fluoroalkenes. The regiochemical preference is consistent with the selectivity that is observed in the hydroboration of styrene. Thus, the internal products are selectively obtained when using a cationic rhodium and small catecholborane while bulky pinacolborane yields terminal products in the presence of a neutral rhodium catalyst.

The isomerization of internal alkenes to terminal ones before hydrometalation or the isomerization during hydrometalation results in the formation of terminal prod-

$$\underset{\text{internal >97\%}}{\underset{R_F}{\overset{CH_3}{\diagup}}OH} \xleftarrow[\text{2. } H_2O_2/OH^-]{\text{1. HBcat} \atop /[Rh(cod)(dppb)]BF_4} R_F\text{-}CH=CH_2 \xrightarrow[\text{2. } H_2O_2/OH^-]{\text{1. HBpin} \atop /RhCl(PPh_3)_3} \underset{\text{terminal >92\%}}{R_F \diagdown\diagup OH}$$

$$R_F=CF_3, C_4F_9, C_6F_{13}$$

Scheme 1-4 Regioselectivity of Fluoralkenes

Tab. 1-3   Isomerization to the Terminal Carbon

$$(E)\text{-}C_3H_7CH\!=\!CHC_3H_7 \quad \xrightarrow[\text{2. } H_2O_2/OH^-]{\text{1. } HBX_2/\text{catalyst}} \quad \text{1-octanol + 2-octanol + 3-octanol + 4-octanol}$$

Entry	Borane	Catalyst	Yield/%	1-ol	2-ol	3-ol	4-ol	Ref.
1	HBcat	RhCl(PPh$_3$)$_3$ (THF)	–	0	0	0	100	[20]
2		[Rh(nbd)(dppb)]BF$_4$ (THF)	–	4	2	7	87	[20]
3	HBpin	Rh(CO)(PPh$_3$)$_2$Cl (CH$_2$Cl$_2$)	94	3	0	0	97	[33]
4		CpNiCl(PPh$_3$) (CH$_2$Cl$_2$)	97	1	0	0	99	[33]
5		RhCl(PPh$_3$)$_3$ (CH$_2$Cl$_2$)	92	100	0	0	0	[33, 36]
6		[Ir(cod)Cl]$_2$/2dppp (CH$_2$Cl$_2$)	77	100	0	0	0	[34]

ucts for internal alkenes. For example, the hydroboration-oxidation of 4-octene yields terminal or internal alcohols depending on the boranes and catalysts employed (Table 1-3). The cationic rhodium and the iridium complexes are more prone to isomerize the boron atom to the terminal carbon than the neutral rhodium complexes (entries 1 and 2) [20]. A bulky pinacolborane has a strong tendency to isomerize to the terminal carbon [33, 36]. Thus, all selectivities shown in Table 1-3 illustrate the superiority of pinacolborane for the synthesis of terminal boron compounds. The bulkiness of the pinacolato group may have the effect of accelerating β-hydride elimination and slowing down the C–B bond formation from R-Rh(III)-Bpin intermediate so that the rhodium can migrate to the terminal carbon *via* an addition-elimination sequence of the H-Rh(III)-Bpin species. An uncatalyzed sequence of hydroboration/isomerization at elevated temperature is an alternative to synthesizing terminal alcohols from internal alkenes or a mixture of terminal and internal ones [37].

The catalyzed hydroboration of alkynes with catecholborane or pinacolborane affords (*E*)-1-alkenylboron compounds at room temperature (Table 1-4). The RhCl(PPh$_3$)$_3$-catalyzed reaction of phenylacetylene yields a complex mixture of two regioisomers of alkenylboronates (**13** and **15**), two hydrogenation products of **13** and **15**, and a trace of a diboration product (entry 1) [19]. The nickel- [33, 39] or palladium-phosphine complexes [40] and Cp$_2$Ti(CO)2 [30] are good catalysts for catecholborane giving selectivity comparable to that of the uncatalyzed reaction (entries 2, 7, 8, and 11–13), and Cp$_2$ZrHCl [38], Rh(CO)(PPh$_3$)$_2$Cl [33] or CpNiCl(PPh$_3$) [33] for pinacolborane (entries 4–6 and 9–10). The *cis*- and *anti*-Markovnikov addition to terminal alkynes may have no significant advantage over the uncatalyzed reaction since the same compounds can be reliably synthesized by the uncatalyzed reaction at slightly elevated temperature. However, the differences in the metal hydride species between the catalyzed and uncatalyzed reactions often alter the chemo-, regio-, and stereoselectivity. For example, the catalysts reverse the regioselectivity in the hydroboration of 1-phenyl-1-propyne (entries 7 and 8). The Cp$_2$Ti(CO)$_2$ [30] prefers the addition of boron to the carbon adjacent to phenyl according to its electronic effect, and steric hindrance of the phosphine ligand of NiCl$_2$(dppe) [39] forces the addition to the β-carbon. The uncatalyzed hydroboration of thioalkynes with di-

Tab. 1-4 Catalysts for Hydroboration of Alkynes

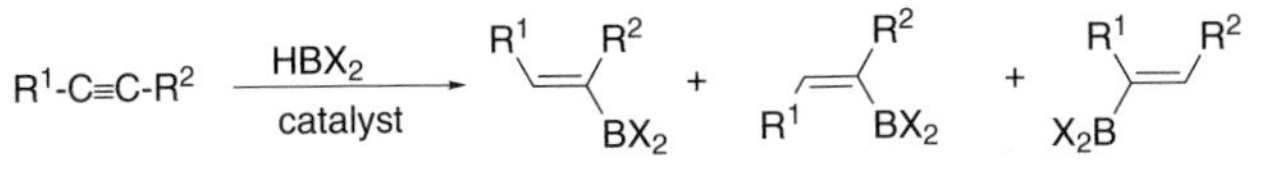

$$R^1-C\equiv C-R^2 \xrightarrow[\text{catalyst}]{HBX_2}$$

						13	**14**	**15**	
Entry	R¹=	R²=	Borane	Catalyst (solvent)	Yield/%	13	14	15	Ref.
1	Ph	H	HBcat	RhCl(PPh₃)₃	—a)	60	0	40	[19]
2	p-Tol	H		Cp₂Ti(CO)₂	96	100	0	0	[30]
3	Ph	H	HBpin	Cp₂ZrHCl	75	98	1	1	[38]
4	Ph	H		RhCl(PPh₃)₃	99	48	0	52	[33]
5	Ph	H		Rh(CO)(PPh₃)₂Cl	99	98	0	2	[33]
6	Ph	H		CpNiCl(PPh₃)	99	98	0	2	[33]
7	Ph	Me	HBcat	Cp₂Ti(CO)₂	89	33	0	67	[30]
8	Ph	Me		NiCl₂(dppe)	98	67	0	33	[39]
9	C₄H₉	H	HBpin	Rh(CO)(PPh₃)₂Cl	99	99	0	1	[33]
10	C₃H₇	C₃H₇		RhCl(PPh₃)₃	99	99	0	–	[33]
11	EtS	Me	HBcat	NiCl₂(dppe)	99	>99	0	<1	[39]
12	PhS	H		NiCl₂(dppp)	93	>99	0	<1	[39]
13	MeSe	H		Pd(PPh₃)₄	74	>97	–	–	[40]
14	Ph	H		[Rh(cod)Cl]₂/4PⁱPr₃	60b)	1	99	0	[44]
15	C₈H₁₇	H		[Rh(cod)Cl]₂/4PⁱPr₃	86b)	1	99	0	[44]

a) 1-Phenylethylborare and 2-phenylethylborate were also produced.
b) The reaction was carried out at room temperature in CH₂Cl₂ in
   the presence of Et₃N (1 equiv) and excess of alkyne (1.2.equivs).

cyclohexylborane yields **15** because the α-carbon adjacent to the alkylthio group is more nucleophilic than the β-carbon [41]. However, the nickel- or palladium-phosphine complexes allow a complete reversal of the regiochemical preference, resulting in a selective formation of β-alkylthio-1-alkenylboronates (entries 11 and 12) [39]. The (Z)-1-alkenylboronates have been synthesized by a two-step method based on the intramolecular S$_N$2-type substitution of 1-halo-1-alkenylboronates with metal hydrides [42] or the *cis*-hydrogenation of 1-alkynylborates [43]. The rhodium(I)- or iridium(I)-ⁱPr₃P complex has recently been found to catalyze the *trans*-hydroboration of terminal alkynes directly yielding *cis*-1-alkenylboron compounds (entries 14–15)

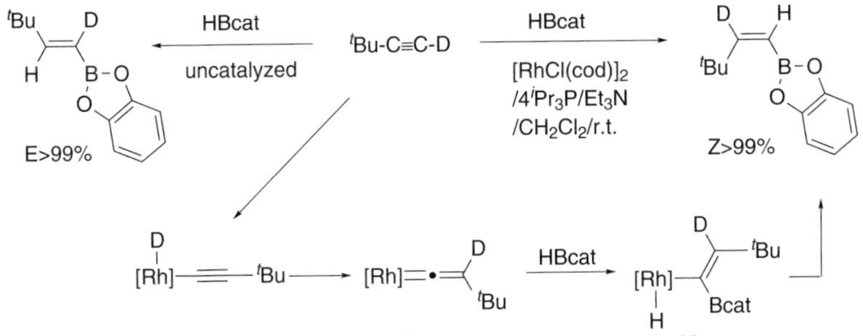

Scheme 1-5 *Cis*- and *Trans*-Hydroboration of Terminal Alkynes

[44]. The dominant factors reversing the conventional *cis*-hydroboration to the *trans*-hydroboration are the use of alkyne in excess of catecholborane or pinacolborane and the presence of more than 1 equiv. of Et$_3$N. The β-hydrogen in the *cis*-product unexpectedly does not derive from the borane reagents because a deuterium label at the terminal carbon selectively migrates to the β-carbon (Scheme 1-5). A vinylidene complex (**17**) [45] generated by the oxidative addition of the terminal C–H bond to the catalyst is proposed as a key intermediate of the formal *trans*-hydroboration.

The catalyzed hydroboration of conjugate dienes, 1,2-dienes (allenes), and enynes proceeding though a π-allylmetal intermediate realizes very different regioselectivities relative to the uncatalyzed reactions. The palladium-catalyzed hydroboration of conjugate 1,3-dienes yields allylboronates *via* an oxidative addition-insertion-reductive elimination process (Scheme 1-6) [46, 47]. The *cis*-addition of the H–B bond to a diene coordinated to palladium(0) affords *cis*-allylboronate (Z>99%), and the selective migration of a hydrogen to the unsubstituted double bond gives a single regioisomer for asymmetric dienes, though analogous reaction of 1,3-pentadiene [46] or 1-phenyl-1,3-butadiene [47] fails to yield allylboronates. The rhodium complex gives a complex mixture for alicyclic dienes, but Rh$_4$(CO)$_{12}$ is recognized to be the best catalyst for 1,4-hydroboration of 1,3-cyclohexadiene (92%).

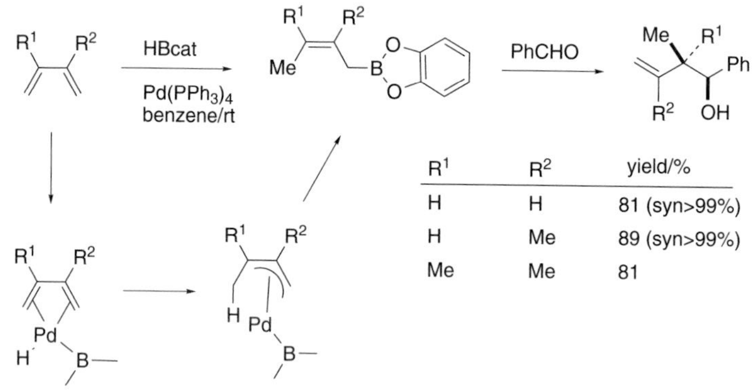

R^1	R^2	yield/%
H	H	81 (syn>99%)
H	Me	89 (syn>99%)
Me	Me	81

Scheme 1-6   Hydroboration of 1,3-Dienes

The uncatalyzed hydroboration of allenes suffers from the formation of a mixture of monohydroboration and dihydroboration products or a mixture of four possible regioisomers (**19–22**) [48]; however, the phosphine ligand on the platinum(0) catalyst controls the regio- and stereoselectivity so as to provide **21** or **22** for alkoxyallenes, and **20** or **22** for aliphatic and aromatic allenes (Scheme 1-7) [49]. The addition to aliphatic and aromatic allenes with Pt(dba)$_2$/2P^{t}Bu$_3$ occurs at the internal double bond to selectively provide **20**, whereas the electronic effect of MeO or TBSO play a major role in influencing the course of the reaction as evidenced by the preferential formation of the terminal *cis*-isomer (**21**) by way of attack from the less-hindered side of the allenes (Z>84–91%). Thus, a platinum(0)/2^{t}Bu$_3$P complex affords the internal products (**20**) or the terminal *anti*-Markovnikov products (**21**) depending on the electron-donating property of the substituents. On the other hand, a very

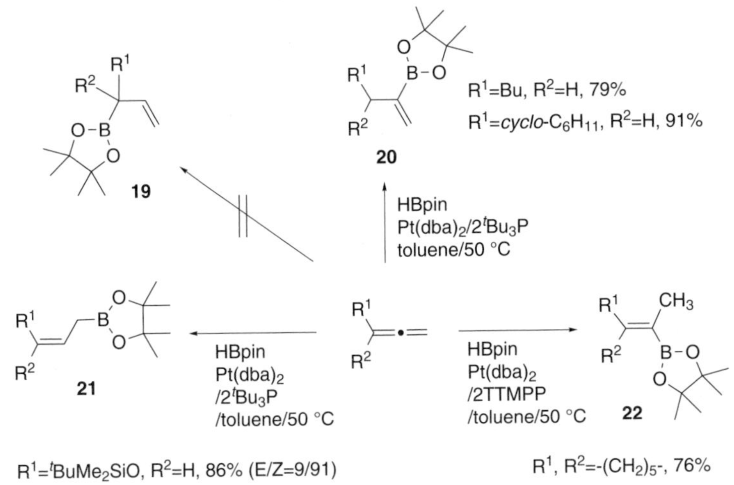

Scheme 1-7 Hydroboration of Terminal Allenes

bulky and basic tris(2,4,6-trimethoxy-phenyl)phosphine (TTMPP) exhibits a characteristic effect which dramatically changes the regioselectivity to the Markovnikov addition (**22**) for the representative terminal allenes.

The hydroboration of enynes yields either of 1,4-addition and 1,2-addition products, the ratio of which dramatically changes with the phosphine ligand as well as the molar ratio of the ligand to the palladium (Scheme 1-8) [46–51]. (*E*)-1,3-Dienylboronate (**24**) is selectively obtained in the presence of a chelating bisphosphine such as dppf and dppe. On the other hand, a combination of Pd$_2$(dba)$_3$ with Ph$_2$PC$_6$F$_5$ (1–2 equiv. per palladium) yields allenylboronate (**23**) as the major product. Thus, a double coordination of two C–C unsaturated bonds of enyne to a coordinate unsaturated catalyst affords 1,4-addition product. On the other hand, a monocoordination of an acetylenic triple bond to a rhodium(I)/bisphosphine complex leads to **24**. Thus, asymmetric hydroboration of 1-buten-3-yne giving (R)-allenylboronate with 61% *ee* is carried out by using a chiral monophosphine (S)-(–)-MeO-MOP (MeO-MOP=2-diphenylphosphino-2′-methoxy-1,1′-binaphthyl) [52].

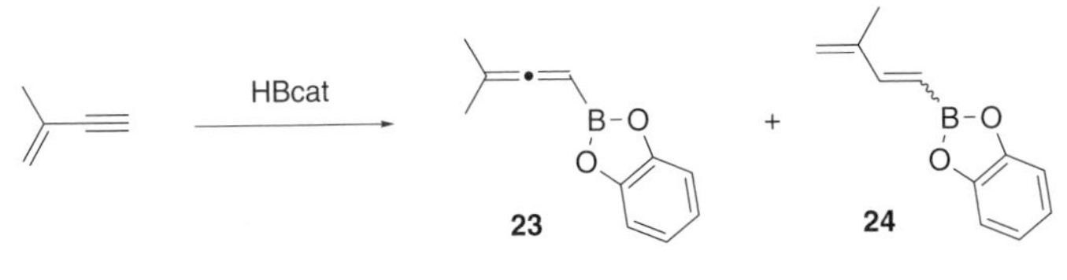

	23	**24**	
Pd(PPh$_3$)$_4$ [46, 50]	45%	30%	(E/Z=15/85)
Pd$_2$(dba)$_3$/2Ph$_2$PC$_6$F$_5$ [50]	61%	12%	(Z>99%)
Pd$_2$(dba)$_3$/dppf [50]	0%	89%	(E>99%)
NiCl$_2$(dppe) [51]	0%	>95%	(E>99%)

Scheme 1-8 Hydroboration of Enynes

Tab. 1-5 Dehydrogenative Coupling giving 1-Alkenylboronates

$$RCH=CH_2 \xrightarrow[\text{catalyst}]{HBX_2} \overset{R}{\underset{BX_2}{\diagdown\!=\!\diagup}} + RCH_2CH_3$$

Alkene	Borane	Catalyst (solvent)	Yield/%	Ref.
4-MeOPhCH=CH$_2$	4	[RhCl(alkene)$_2$]$_2$ (toluene)	98	[15]
4-ClPhCH=CH$_2$	4	[RhCl(alkene)$_2$]$_2$ (toluene)	99	[15]
PhCH=CH$_2$	2 (HBpin)	[RhCl(cod)]$_2$ (toluene)	81	[53]
4-MeO$_2$CPhCH=CH$_2$	2 (HBpin)	[RhCl(cod)]$_2$ (toluene)	80	[53]
Ph(Me)C=CH$_2$	1 (HBcat)	RhCl(PPh$_3$)$_3$/10PPh$_3$ (THF)	70[a]	[54]
CH$_2$=CH$_2$	5	Cp*$_2$Ti(η^2-CH$_2$CH$_2$)	58	[55, 56]

a) CH$_3$CH(Ph)CH$_2$Bcat (10%) and Ph(Me)CHCH(Bcat)$_2$ (18%) was accompanied.

The dehydrogenative coupling of borane is a very attractive method since the re-action directly yields (*E*)-1-alkenylboronates from alkenes (Table 1-5). However, the reaction can be limitedly applied to the synthesis of styrylboron derivatives [15, 53–56]. The reported procedures recommend a combination of vinylarene, a steri-cally hindered borane such as oxazaborolidine **4** or pinacolborane **2**, and a phos-phine-free rhodium(I) catalyst for achieving selective coupling. The reaction re-quires more than two equivalents of vinylarene because H$_2$, generated by β-hydride elimination, hydrogenates a molar amount of vinylarene, as is discussed in the mechanistic section. Cp*$_2$Ti exceptionally catalyzes the borylation of ethylene, but the scope of the reaction for other inactivated alkenes has not yet been explored [55].

1.2.2
Catalytic Cycles

Catalyzed hydroboration often results in a complex mixture of products derived not only from catalyzed hydroboration but also uncatalyzed hydroboration and hydro-genation of alkenes, because the reaction of RhCl(PPh3)3 with catecholborane (HB-cat) yields various borane and rhodium species (Scheme 1-9) [19]. The oxidative ad-dition of HBcat to RhCl(PPh3)3 affords a coordinate unsaturated borylrhodium complex (**25**) [57], which is believed to be an active species of the catalyzed hydrobo-ration. However, further oxidative addition of the borane to **25** generates H2 and a diborylrhodium complex (**26**) [58]. The diboryl complex **26** will then undergo dibo-ration or reductive monoborylation of alkenes (see Section 1.3.2), and dihydrogen thus generated will hydrogenate a part of the alkenes. Thus, the reaction is often accompanied by small amounts of RCH(Bcat)CH2(Bcat), RC(Bcat)=CH2, RCH=CH(Bcat), and RCH2CH3, along with the desired hydroboration product. On the other hand, the degradation of catecholborane makes the reaction more complex when the catalyzed reaction is very slow. The phosphine eliminated from the cata-lyst reacts with catecholborane to yield H3B·L and B$_2$cat$_3$ (cat=O$_2$C$_6$H$_4$) (**29**) [59]. Al-though the borane/phosphine complex thus generated fortunately does not hydrob-

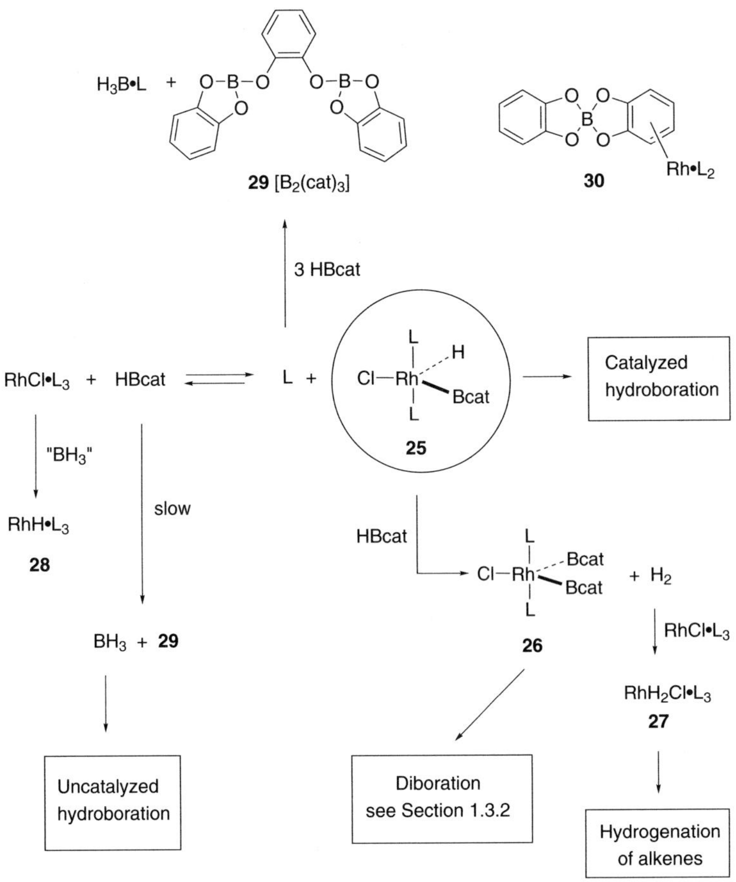

Scheme 1-9 Reaction of RhCl(PPh$_3$)$_3$ with Cartecholborane (HBcat)

orate alkenes, **29** may contribute to the production of other rhodium species such as **30** [24], which has high catalyst activity comparable to that of **25**. The reaction often suffers from the competitive uncatalyzed hydroboration with BH$_3$, and the formation of such products does not reflect the true selectivity of the catalyzed reactions [19, 60, 61]. The degradation of catecholborane to BH$_3$ and B$_2$cat$_3$ **29** is, in general, very slow at room temperature; however, the reaction of BH$_3$ will compete with the catalyzed hydroboration when the reaction is very slow because of decomposition of the catalyst or low catalyst loading or activity. Although there are many probable processes leading to side reactions, catecholborane undergoes clean hydroboration when the catalyst is selected appropriately.

One most important observation for the mechanistic discussion is the oxidative addition/insertion/reductive elimination processes of the iridium complex (**31**) (Scheme 1-10) [62]. The oxidative addition of catecholborane yields an octahedral iridium-boryl complex (**32**) which allows the *anti*-Markovnikov insertion of alkyne into the H–Ir bond giving a 1-alkenyliridium(III) intermediate (**34**). The electron-

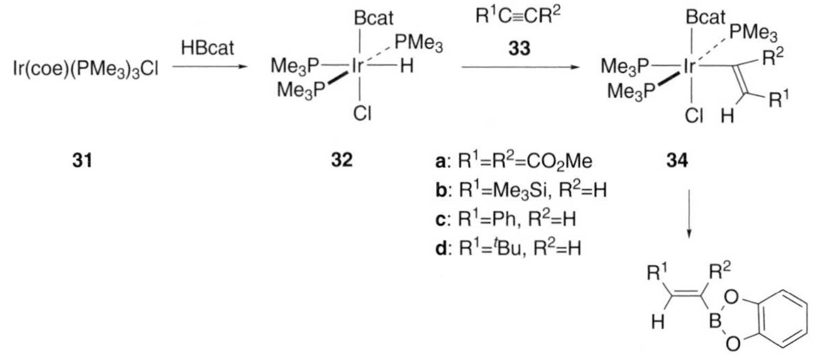

Scheme 1-10 Oxidative Addition/Insertion/Reductive Elimination Process

withdrawing groups as in **33a** strongly stabilize **34** while **34b–d** are led to reductive elimination to give hydroboration products. However, these intermediates are well characterized in solution *via* ^{1}H NMR spectroscopy. The observed regio- and stereoselectivities are in good agreement with the catalyzed hydroboration of alkynes with a rhodium, palladium, or nickel complex. It is also reasonable to assume the same catalytic cycle for the hydroboration of alkenes by Group 9 or 10 metal catalysts.

A catalytic cycle proposed for the metal-phosphine complexes involves the oxidative addition of borane to a low-valent metal yielding a boryl complex (**35**), the coordination of alkene to the vacant orbital of the metal or by displacing a phosphine ligand (**35 → 36**) leads to the insertion of the double bond into the M-H bond (**36 → 37**) and finally the reductive elimination to afford a hydroboration product (Scheme 1-11) [1]. A variety of transition metal-boryl complexes have been synthesized *via* oxidative addition of the B–H bond to low-valent metals to investigate their role in cat-

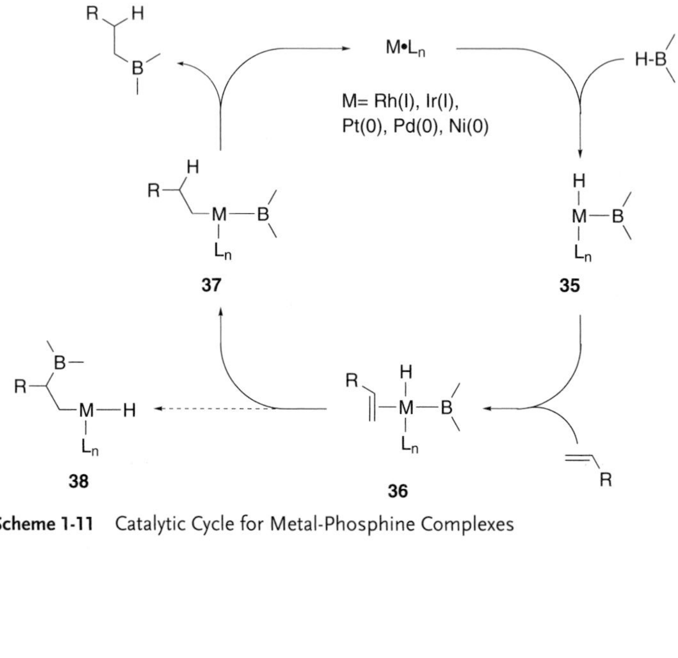

Scheme 1-11 Catalytic Cycle for Metal-Phosphine Complexes

alyzed hydroboration as well as other catalyzed reactions of boron compounds [2]. The reaction of catecholborane (HBcat) with RhCl(PPh$_3$)$_3$ yields a complex mixture including RhHCl(PPh$_3$)$_2$(Bcat) (**25**), but the related rhodium dimer [Rh(PPh$_3$)$_2$(μ-Cl)]$_2$ affords a pure **25** [57], as is shown in Scheme 1-9. IrCl(CO)(PPh$_2$)$_2$ and catecholborane similarly yield *trans*-IrHCl(CO)(PPh$_3$)$_2$(Bcat) [28], which also has high catalyst activity for hydroboration. Perhaps the most important step for determining the regio- and stereoselectivity is the insertion of alkene into the M–H bond, but relevant information is very scarce because the next reductive elimination is too fast to observe the intermediate (**37**). However, the high terminal regioselectivity commonly observed in terminal alkenes suggests the first migration of H to the β-carbon (**36** → **37**), similarly to that of alkyne to the H–Ir bond [62]. An *ab initio* investigation of the rhodium(I)-catalyzed hydroboration of C$_2$H$_4$ with BH$_3$ supports the above cycle, suggesting the existence of several 14-electron complexes as the reaction intermediates and the first migration of H from **36**, rather than BH$_2$, to the double bond [63, 64].

A catalytic cycle proceeding through an insertion of alkene into Rh–B bond (**36** → **38**) is also a possible process proposed by another *ab initio* molecular orbital calculation [65]. The mechanism does not fit the experimental results observed in the major products; however, a competitive formation of vinylboronates *via* an insertion-β-hydride elimination sequence suggests the existence of the Rh–B bond insertion process. The reductive coupling of borane is not a major reaction in the rhodium-phosphine complexes and predominates in the phosphine-free rhodium complexes, as illustrated in Table 1-5. A cycle proposed for the phosphine-free catalysts is quite

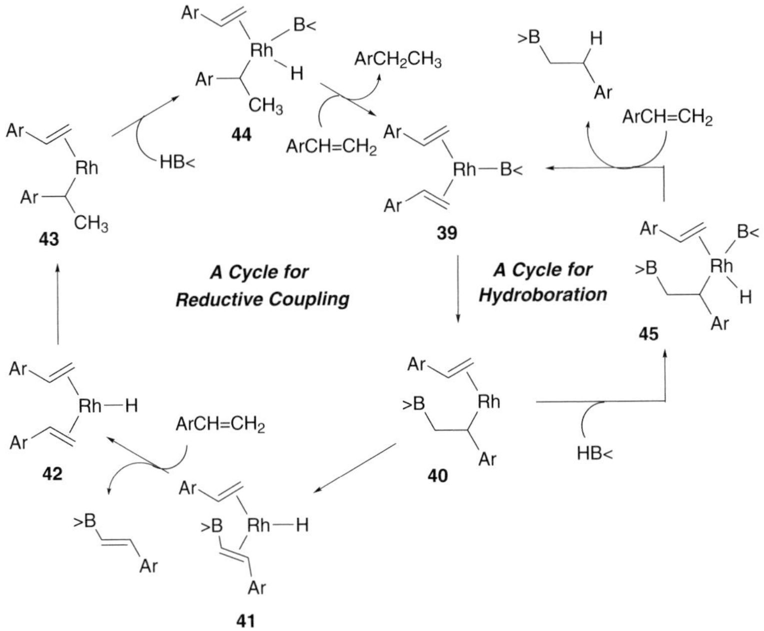

Scheme 1-12 Catalytic Cycle for Phosphine-Free Rh(I) Complexes

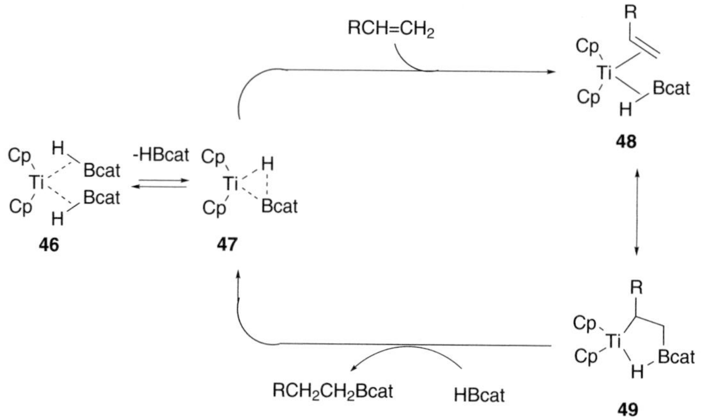

Scheme 1-13 Catalytic Cycle for Cp$_2$Ti

different from that of the phosphine-based catalysts (Scheme 1-12) [15]. The absence of phosphine allows smooth coordination of alkenes to the rhodium(I) center, and the species participating in the insertion step is Rh(I)–H (**42**) or Rh(I)–B (**39**) complex rather than rhodium(III) complex as in **36**. The regiochemical preference yielding a stable benzylrhodium species in the insertion of vinylarenes (**40, 43, 45**) is the same as that of rhodium/phosphine catalysts; however, the first migration of the boryl group (**39 → 40**) reverses the regioselectivity between the phosphine-free catalysts and the rhodium/phosphine catalysts (see entries 1–3 in Table 1-1). A common intermediate (**40**) undergoes two competing reactions; the β-hydride elimination (**40 → 41**) leads to the reductive coupling and the next hydrogenation process of alkene, while the oxidative addition of a further molecule of borane leads to the hydroboration of vinylarenes (**40 → 45**). The β-hydride elimination will predominate in relatively bulky boranes such as pinacolborane and oxazaborolidine (**4**), but catecholborane yields a mixture of two products because of its high reactivity for oxidative addition.

The cyclopentadienyl (Cp) complexes of Ti, Nb, Ta, Sm, and La catalyze the hydroboration by quite different mechanisms from that of Group 9 and 10 metals. The titanocene-catalyzed reaction involves initial dissociation of a coordinated borane (**46**) to generate a monoborane intermediate (**47**) (Scheme 1-13) [30]. The coordination of alkene leads to the alkene-borane complex stabilized by a B-H interaction (**48, 49**). The coordination of a further molecule of borane then yields a hydroboration product with regeneration of **47**. A kinetic study revealed that the borane behaves as not only a reagent but also a ligand on the resting state of the catalyst (**46**) [56, 66]. The steric interaction between R and a catecholate group in **48**, which is larger than that of R with the Cp ligands, induces the formation of a terminal boron product. The electron-withdrawing aryl group also stabilizes the transition state **48**, but the donation group destabilizes it to result in reduced regioselectivity for aliphatic 1-alkenes.

The proposed mechanism for the metallocene complexes of Group 5 metals and lanthanides involves the coordination of alkene to the metal hydride (**51**), followed

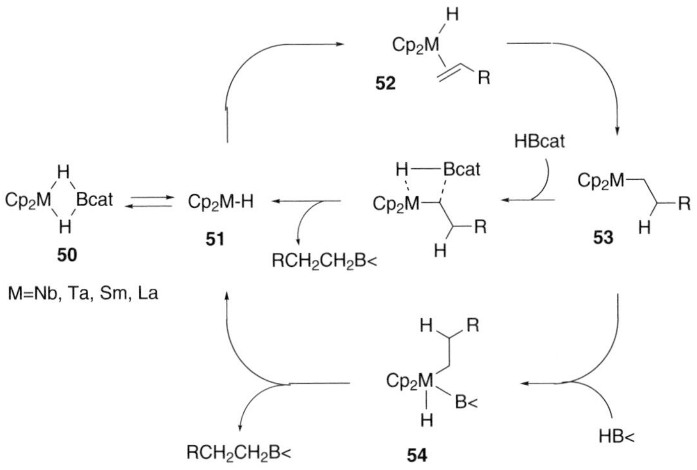

Scheme 1-14 Catalytic Cycle for Lantanide and Group 5 Metals

by insertion *via* migration of H to the β-carbon (**53**) (Scheme 1-14) [67]. The product will be then provided *via* the direct σ-bond metathesis between HBcat and **53** or an oxidative addition/reductive elimination sequence (**53 → 54 → 51**). These processes are studied in detail for the Group 5 metals such as Nb and Ta [67–69]; however, an *ab initio* calculation [70] for the hydroboration of C_2H_4 with $HB(OH)_2$ supports an analogous cycle for the Cp_2Sm catalyst. Due to the high oxophilicity of the samarium atom, the last σ-bond metathesis (**53 → 51**) yields an acid/base complex between $EtB(OH)_2$ and Cp_2SmH. The rate-determining step is the dissociation of $Cp_2Sm–H$ coordinated to an oxygen in $EtB(OH)_2$. The coordination of alkene to **51** avoiding a steric interaction between the Cp ligand and R (**52**) leads to a terminal addition product. Thus, $Cp*_2Sm(THF)$ exhibits indeed high terminal selectivity for both aromatic and aliphatic 1-alkenes.

1.2.3
Synthetic Applications

The unique chemo-, regio-, enantio-, and diastereoselectivities observed in catalyzed hydroboration have realized various synthetic applications. The rate of hydroboration with $RhCl(PPh_3)_3$ highly depends on the substituents of alkenes (Scheme 1-15). For terminal alkenes, the reaction completes within minutes at room temperature, which is adequately faster than that of disubstituted alkene or reduction of ketone carbonyl group. More highly substituted alkenes are less reactive, and the reaction often requires a higher catalyst loading and an excess of a borane reagent to ensure that the reaction proceeds to completion at a convenient rate. Tetrasubstituted alkenes are, in general, highly inert to the catalyzed hydroboration. The order of reactivity (terminal > terminal disubstituted > internal disubstituted > trisubstituted alkenes) [35] allows selective hydroboration at the terminal double bond in preference to other double bonds (**55** [35], **56** [71], **57** [72]) or ketone carbonyl groups (**58**

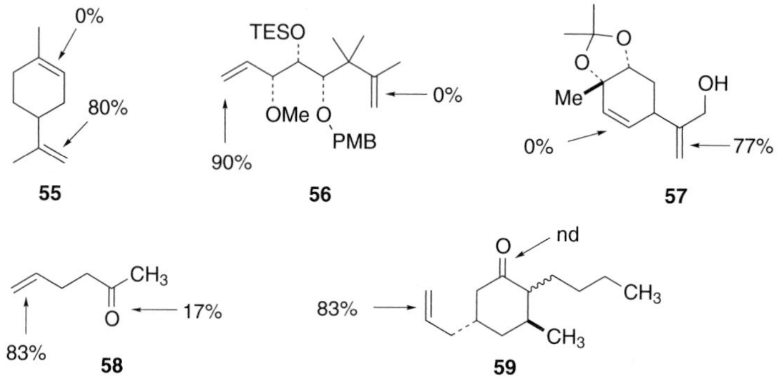

Scheme 1-15 Hydroboration at the Terminal Double Bond

[4], **59** [73]). The chemoselectivity of catecholborane can be in the order of RCH=O, $R_2S=O > CH_2=CH– > R_2C=O$, oxirane $> RCO_2R > C≡N$ [12].

The hydroboration of *exo*-cyclic alkenes affords stereochemically complemental products between the catalyzed and uncatalyzed reaction (Scheme 1-16). The hy-

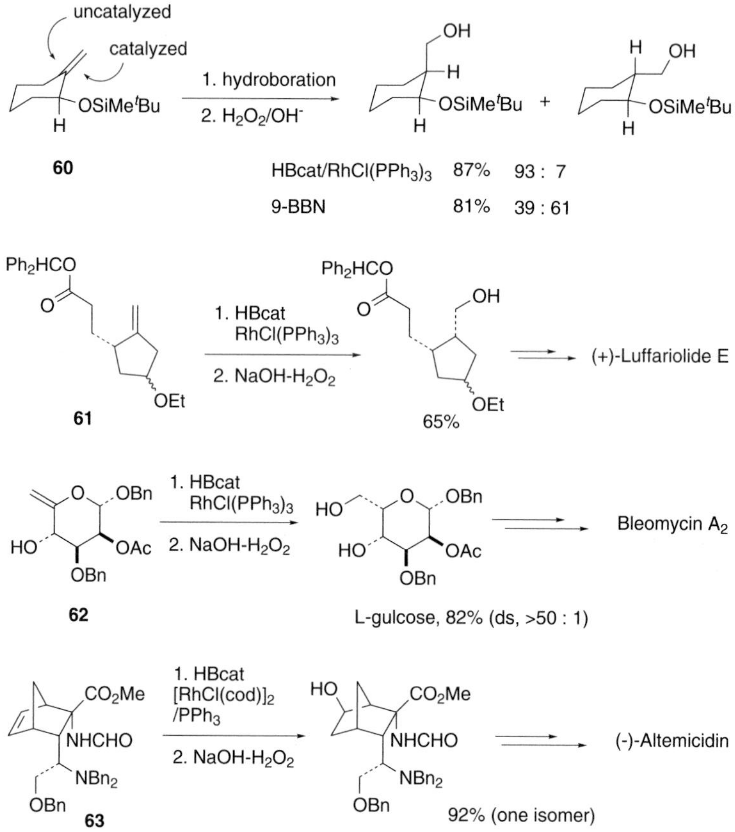

Scheme 1-16 Regio- and Stereoselectivity of Cyclic Alkenes

droboration-oxidation of **60** [35] with 9-BBN yields two alcohols with the *trans*-isomer predominating in a ratio of 39/61. The reaction contrasts highly with the catalyzed hydroboration yielding the *cis*-alcohol with 93% selectivity by the addition from the face opposite to the allylic substituent. Thus, the catalyzed reaction is more sensitive to the steric effect than the electronic effect, whereas 9-BBN prefers to attack from the more electron-rich face of the double bond (stereoelectronic effect) [74, 75]. High *cis*-stereoselectivity observed in *exo*-cyclic alkenes succeeds in various natural product syntheses. An intermediate for the synthesis of (+)-luffariolide is obtained by the RhCl(PPh₃)₃-catalyzed hydroboration of *exo*-methylene cyclopentane (**61**) [76] with catecholborane. The rhodium(I)-catalyzed hydroboration-oxidation of the *exo*-methylene dihydropyran (**62**) [77] furnishes l-gulose with diastereoselectivity of >50:1, which is a key intermediate required for the synthesis of bleomycin A2. A small allylic alcohol protecting group such as CH₂Ph or the free alcohol itself is found to favor the attack from the β-face. The rhodium-catalyzed hydroboration of the norbornene (**63**) [78] takes place with striking regioselectivity to give 6-*exo*-norborneol, which is led to the naturally occurring sulfonamide altemicidin. The remote substituent effect is not well understood, but the selectivity is highly depend-

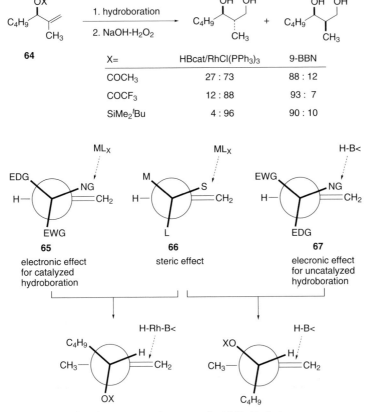

Scheme 1-17 Preferred Reactive Conformations for Allylic Alcohols

ent on the *endo* substituents, which will result in increased polarization of the neighboring double bond [79].

The diastereoselective hydroboration of acyclic alkenes is one of the most successful results achieved by catalyzed hydroboration (Scheme 1-17) [35, 80, 81]. The rhodium-catalyzed hydroboration-oxidation of allylic alcohol **64** [81] yields the *syn*-diol, whereas the uncatalyzed reaction of 9-BBN preferentially produces the *anti*-diol. The origin of diastereofacial selectivity arises from differences in the mechanism of π-complexation between the transition metals (**65**) and the main metals (**67**), together with the steric effect of the substituents (**66**) [81]. The π-complexation of transition metals can be stabilized by the increased back-donation from a filled d-orbital of the metal to a π*-orbital of alkene, which is enhanced when the best σ-acceptor adopts the anti position (**65**) and the smallest group (S) is in the inside crowded position (**66**). Thus, the diastereoselection is optimal if OX is a good σ-acceptor and is large relative to the other substituents on the asymmetric center. Thus, the CF₃COO group increases the selectivity more than CH₃COO since the former is a better electron-withdrawing group (EGW) and the selectivity is significantly increased in a bulky protecting group such as SiMe₂^{t}Bu. On the other hand, the uncatalyzed hydroboration predominantly leads to the *anti*-diols when the electron-donating group (EDG) occupies the *anti* position (**67**) because this electronic effect enhances the interaction of the empty p orbital of boron and the filled p-orbital of alkene [82].

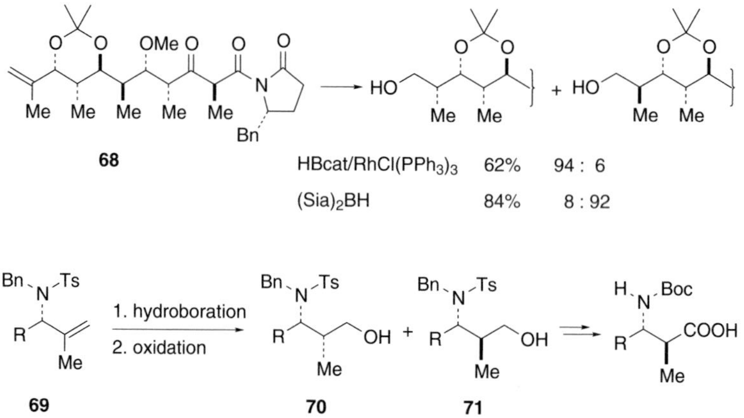

R=	HBcat/[Rh]^a 70 : 71	uncatalyzed 70 : 71	
PhCH₂	95 : 5	93 : 7	(9-BBN)
		6 : 94	(BH₃)
Me₂CHCH₂	86 : 14	5 : 95	(BH₃)
^{t}BuO₂CCH₂	----	2 : 98	(BH₃)

^{a}[Rh]= [Rh(cod)Cl]₂/4PPh₃

Scheme 1-18 Diastereoselectivity of Acyclic Alkenes

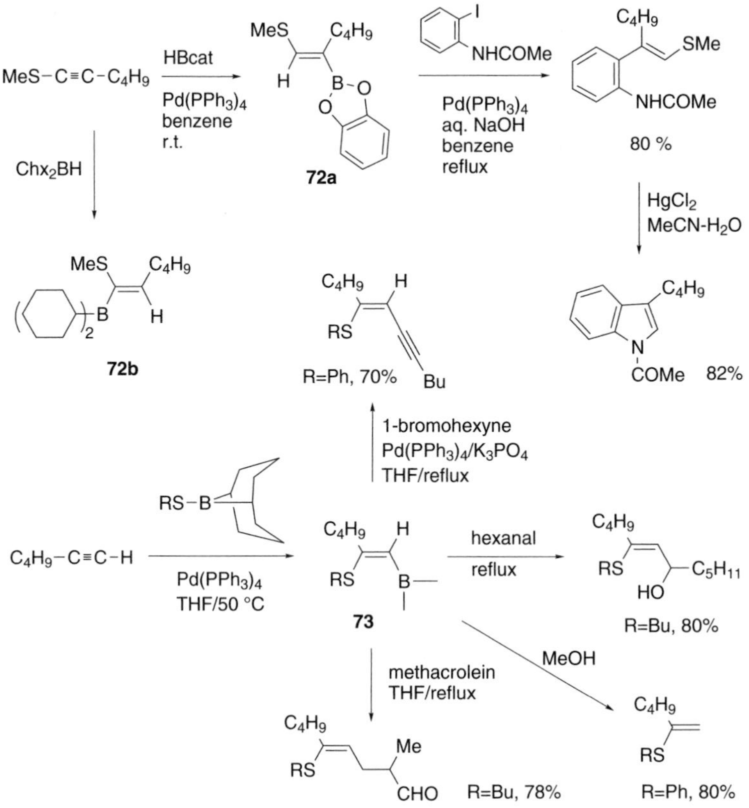

Scheme 1-19 Catalyzed Hydroboration and Thioboration

The accessibility of both diasteroisomers from a single alkene is greatly advantageous for organic syntheses, which is demonstrated in various syntheses of natural products (Scheme 1-18). The catalyzed hydroboration-oxidation of **68** [83] yields a diastereomeric alcohol with 94% selectivity, which is used as an intermediate for the synthesis of the polyether antibiotic, Ionomycin A. Another isomer required for the stereochemical assignments is easily obtained by changing the hydroboration method to the uncatalyzed reaction of disiamylborane (Sia$_2$BH). The effect of the TsNH group is strongly enhanced by *N*-benzylation to give the TsNBn group, since the bulkiness of the allylic substituent serves to increase the *syn*-diastereoselectivity (**70**) [84], as is discussed in Scheme 1-17. Although 9-BBN unexpectedly results in the same diastereoselection to that of the catalyzed reaction, the borane-tetrahydro-furane complex selectively yields *anti*-amino alcohols (**71**) [85], which are a versatile intermediate for the synthesis of optically active β-amino acids *via* the RuCl$_3$-catalyzed oxidation of alcohols with NaIO$_4$.

Hydroboration of alkenes or alkynes followed by cross-coupling with organic electrophiles provides a straightforward method for the carbon-carbon bond formation (Scheme 1-19). The hydroboration of thioalkynes with catecholborane in the presence of a nickel or palladium catalyst yields β-(alkylthio)-1-alkenylboronates (**72a**)

[39], and the uncatalyzed reaction of (dicyclohexyl)borane (Chx$_2$BH) provides α-(alkylthio) derivatives (**72b**) [41]. Since the vinylic sulfide is synthetically equivalent to a carbonyl compound, the palladium-catalyzed cross-coupling of **72a** with aromatic halide having an *ortho*-acetylamino or a protected hydroxy group provides valuable precursors for heterocyclic compounds [86]. For example, a three-step procedure in the same flask affords indole derivatives. An alternative method for the synthesis of β-(alkylthio)-1-alkenylboranes is thioboration of terminal alkyne with 9-RS-9-BBN (9-BBN=9-borabicyclo[3.3.1]nonane) [87]. A selective *cis*-addition catalyzed by a palladium(0) complex affords 9-[(Z)-2-RS-1-alkenyl]-9-BBN (**73**), which exhibits exceptionally high reactivity on protonolysis with methanol [87], cross-coupling reaction with organic halides [87], and nucleophilic addition to carbonyl compounds [88]. The ready availability of these boron reagents by catalytic hydroboration and thioboration may offer a flexible and reliable route to such stereodefined alkenyl sulfides in combination with numerous reactions reported in organoboron compounds.

The conjugate reduction of α,β-unsaturated ketones with catecholborane proceeds without the assistance of any catalyst, but the addition to the corresponding imides, esters, and amides [27] can be accelerated in the presence of RhCl(PPh$_3$)$_3$ (Scheme 1-20). A stereoselective reduction of *N*-acylimidate **74** [89] to the desired *N*-acyl aminal for the synthesis of a vesicant isolated from the beetle (Pederin) is carried out by using the catalyzed addition of catecholborane. A catalytic cycle analogous to the catalyzed hydroboration of alkenes is proposed to account for the stereochemistry of the reduction. On the other hand, the reduction of β-ionone generates a single (Z)-enolate without a catalyst, which undergoes the aldol reaction with acetaldehyde giving a *syn*-product with good selectivity [27]. The (Z)-boron enolate will also be anticipated in the rhodium-catalyzed conjugate reduction.

Hydroboration of alkenes with chiral hydroboration reagents such as diisopinocamphenylborane succeeds in various syntheses of chiral organoboron com-

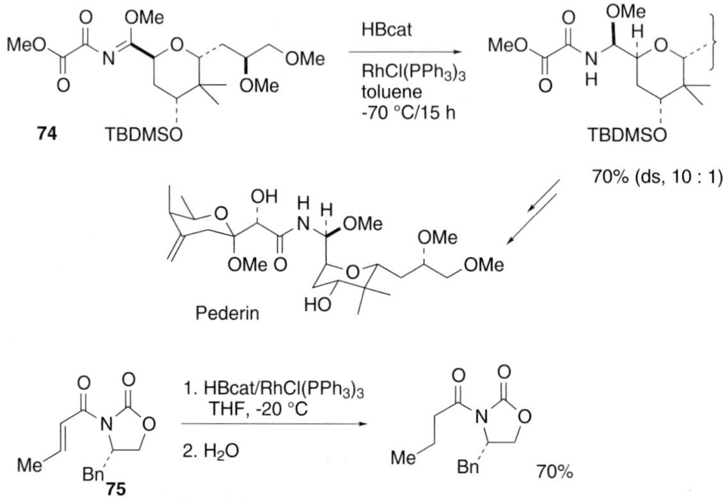

Scheme 1-20 Conjugate Reduction

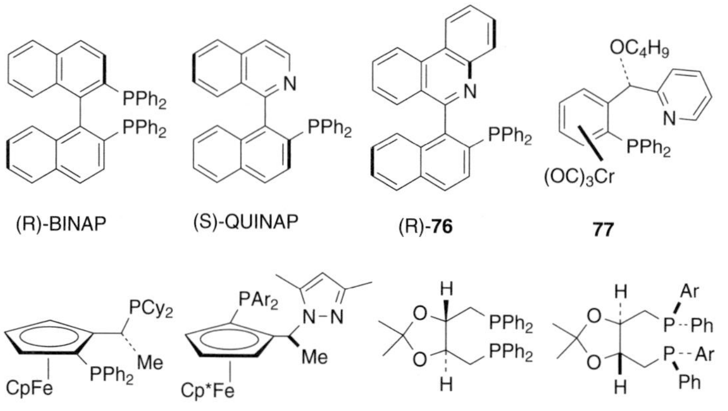

(R)-BINAP (S)-QUINAP (R)-76 77

(R)(S)-78 (S)(R)-79 (S,S)-DIOP (Sp,Sp)-80

Ar= Ph, 4-CF$_3$C$_6$H$_4$ Ar= 2-MeOC$_6$H$_4$

Scheme 1-21 Chiral Ligands

pounds [90]. However, an asymmetric catalyst is one of the most exciting features of catalyzed hydroboration since chiral phosphine ligands are the chiral auxiliaries most extensively studied for transition metal-catalyzed asymmetric reactions of alkenes (Scheme 1-21) [1b, e]. These ligands were originally designed for asymmetric hydrogenation and hydroformylation, but they also work well in catalyzed hydroboration because both reactions involve a quite analogous molecular recognition mechanism for alkenes.

Asymmetric hydroboration has been extensively studied, but the representative results are briefly discussed here because the reaction has been recently reviewed in detail [1e] (Table 1-6). Various chiral ligands including bisphosphines and aminophosphines succeed in achieving high enantioselectivities exceeding 90% *ee* for styrene [23, 25, 91–94]. The enantioselectivity is strongly dependent on the reaction temperature, higher selectivity being observed at lower temperature. Thus, the cationic rhodium complexes are recognized to be an efficient catalyst achieving high activity for styrene derivatives at –78°C and high regioselectivity giving benzyl alcohol derivatives [91, 95]. Chiral aminophosphines such as (S)-(–)-QUINAP [1-(2-diphenylphosphino-1-naphthyl)isoquinoline] and **76** exceptionally afford excellent enantioselectivity for various vinylarenes even at room temperature. On the other hand, these catalysts are less effective for aliphatic alkenes [95–97].

The carbon-carbon bond-forming reactions extensively studied in organoboron chemistry now allow further transformation of the chiral carbon thus constructed on the boron atom (Scheme 1-22) [98, 99]. The one-carbon homologation of chiral alkylboronates has been recently demonstrated to occur with complete retention of configuration. The alkylation of chiral borates with halomethyllithiums yields alcohols, aldehydes, and carboxylic acids with one-carbon homologation. The hydroboration is carried out by catecholborane because of its high reactivity even at –78°C, but its conversion into the pinacol ester is more convenient for the next isolation and homologation. The reaction involves the first formation of an ate-complex, [ClCH$_2$BR*(OR)$_2$]Li, followed by intramolecular S$_N$2 substitution of Cl with R*.

Tab. 1-6 Asymmetric Hydroboration

Alkene	Catalyst/temp	Yield/%	% EE	Ref.
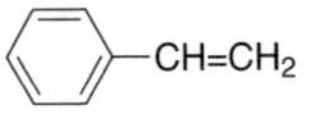 —CH=CH₂	[Rh(cod)₂]BF₄/(R)-BINAP/-78°C	91	96 (R)	[25]
	1/2[RhCl(cod)₂]/(S)-BINAP/-78°C	63	92 (S)	[23]
	[Rh(cod)/(S)-QUINAP]OTf/rt	71	91 (S)	[91]
	[Rh(nbd)₂]BF₄/(R)(S)-**78**/-78°C	65	92 (R)	[92]
	[Rh(cod)₂]BF₄/**77**/0°C	86	82 (R)	[93]
	[Rh(cod)₂]BF₄/(S)(R)-**79**/20°C	76	94 (R)	[94]
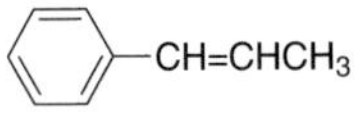 —CH=CHCH₃	[Rh(cod)/(S)-QUINAP]OTf/rt	64	95 (S)	[91]
	[Rh(cod)/(R)-**76**]OTf/rt	60	91 (R)	[91]
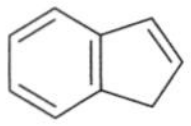	[Rh(cod)/(S)-QUINAP]OTf/rt	58	91 (S)	[91]
	1/2[Rh(cod)₂]/(R,R)-DIOP/-30°C	–	74 (S)	[95]
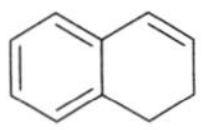	[Rh(cod)/(R)-**76**]OTf/rt	69	84 (R)	[91]
	1/2[RhCl(cod)₂]/(S_p,S_p)-**80**/-25°C	–	84 (R)	[95]
	1/2[RhCl(cod)₂]/(R)-BINAP/-25°C	–	65 (R)	[96]
	1/2[Rh(cod)₂]/(R,R)-DIOP/-25°C	–	60 (R)	[96, 97]
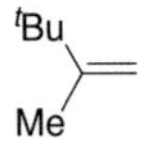 ᵗBu⧵═ / Me	1/2[RhCl(cod)₂]/(R,R)-DIOP/-5°C	–	69 (R)	[96]

1.3
Metal-Catalyzed Diboration, Silylboration, and Stannylboration

1.3.1
B–B, B–Si, B–Ge, and B–Sn Reagents

Various B–B, B–Si, and B–Sn compounds are now available for the metal-catalyzed borylation of alkenes and alkynes (Scheme 1-23). A direct synthesis of tetra(alkoxy)diborons by the Wurtz coupling of (RO)₂BCl is unsuccessful, but tetra(amido)diborons are available in large quantity by the corresponding reaction of (Me₂N)₂BX (X=Cl, Br) with sodium or potassium [100–102]. Thus, tetra(alkoxo)diborons (**81** [101], **82** [101, 102], **83** [103, 104], **84** [104] **85** [105] and **86** [106]) are synthesized by the reaction of (Me₂N)₂BB(NMe₂)₂ with an appropriate alcohol or diol in the presence of 4 equiv. of HCl. Two (alkoxo)diborons, **82** and **83**, are now commercially available because reactions of these reagents have been extensively studied because of their high reactivity for transition metals and high stability in air. On the other hand, the reaction of (R₂N)₂BX with a silyl- or stannyllithium provides B–Si (**87**) [107] or B–Sn (**90**) [108] reagents. The reaction of **87** with pinacol [109] or, more conveniently, the reaction of PhMe₂SiLi or PhMe₂GeLi with *i*-PrOBpin yields **88** or **89** [110].

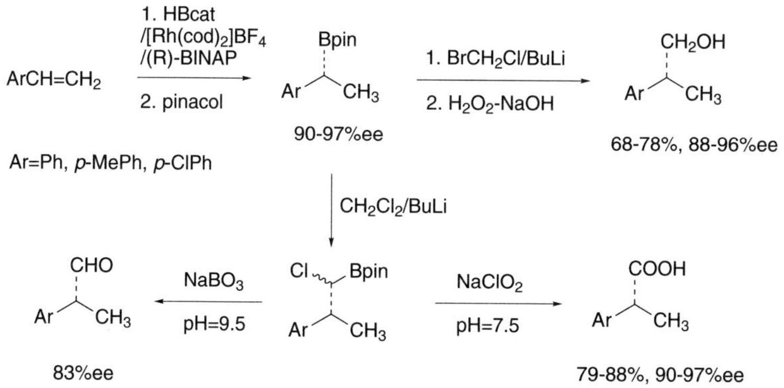

Scheme 1-22 HOmologation of Chiral Boronic Esters

1.3.2
Addition to Alkynes and Alkenes

The diboration of alkynes is efficiently catalyzed by a platinum(0) complex such as $Pt(PPh_3)_4$, $Pt(C_2H_4)(PPh_3)_2$ or $Pt(CO)_2(PPh_3)_2$ (3 mol%) yielding *cis*-1,2-dibory-lalkenes in high yields [111–113] (Scheme 1-24), though platinum(II), palladium, and rhodium(I) complexes are ineffective. Platinum(0) complexes analogously catalyze the addition of other (alkoxo)diborons, **81**, **83** and **84**, but the reaction of $(Me_2N)_2BB(NM_2)_2$ is very slow even at 120°C because of its slow oxidative addition to the platinum(0) complex. There are no large differences in the yields and the reaction rates between internal and terminal alkynes and the reaction is available with various functional groups. The carbon-carbon double bond, chloro, epoxy, ester, cyano, and ketone carbonyl groups remain intact during the addition to the C-C triple bonds [111]. The first synthesis of tetraborylethylene $(catB)_2C=C(Bcat)_2$ is

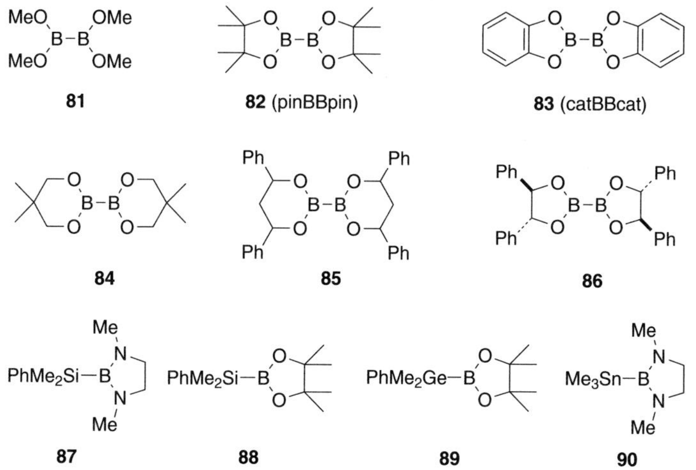

Scheme 1-23 Borylation Reagents

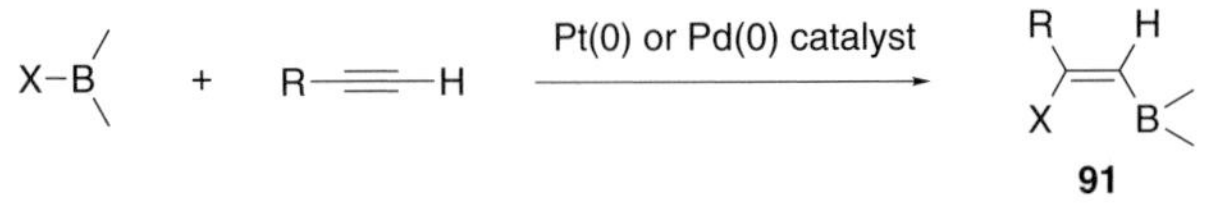

Scheme 1-24 Addition to Alkynes

Entry	X-B<	R	Catalyst	Yield/%	Ref.
1	(MeO)$_2$BB(OMe)$_2$	n-C$_6$H$_{13}$	Pt(PPh$_3$)$_4$/80°C	89	[111, 112]
2	catBBcat (83)	n-C$_6$H$_{13}$	Pt(PPh$_3$)$_4$/80°C	>95	[112, 113]
3	pinBBpin (82)	n-C$_6$H$_{13}$	Pt(PPh$_3$)$_4$/80°C	92	[111, 112]
4		Ph	Pt(PPh$_3$)$_4$/80°C	79	[111, 112]
5		N≡C(CH$_2$)$_3$	Pt(PPh$_3$)$_4$/80°C	79	[112]
6		MeO$_2$C(CH$_2$)$_4$	Pt(PPh$_3$)$_4$/80°C	89	[112]
7	(Me$_2$N)$_2$BB(NMe$_2$)$_2$	n-C$_6$H$_{13}$	Pt(PPh$_3$)$_4$/120°C	7	[112]
8	R$_3$SiB(NR$_2$)$_2$ (87)	n-C$_6$H$_{13}$	Pd$_2$(dba)$_3$/$_4$etpo/80°C[a]	92	[109, 114]
9	R$_3$SiBpin (88)	n-C$_6$H$_{13}$	Pd(OAc)$_2$/15RNC/110°C[b]	92	[109, 115]
10	Me$_3$SnB(NR$_2$)$_2$ (90)	n-C$_6$H$_{13}$	Pd(PPh$_3$)$_4$/rt	83	[116]
11		Ph	Pd(PPh$_3$)$_4$/rt	73	[116]

a) etpo=P(OCH$_2$)$_3$CCH$_2$CH$_3$.
b) RNC='BuCH$_2$CMe$_2$NC

achieved by addition of catBBcat (83) to catBC≡CBcat with Pt(PPh$_3$)$_4$ in refluxing toluene [113]. Both amido and pinacol derivatives of B–Si and B–Sn compounds similarly add to terminal and internal alkynes in the presence of a palladium catalyst [114–116]. The addition of B–Sn reagent (90) takes place even at room temperature, whereas the reaction of B–Si reagents (87, 88) only proceeds at over 100°C, presumably because of the slow oxidative addition of the B–Si bond to the palladium(0). Both reactions stereo- and regioselectively provide cis-products via addition of silicone or stannane to the internal carbon and boron to the terminal carbon.

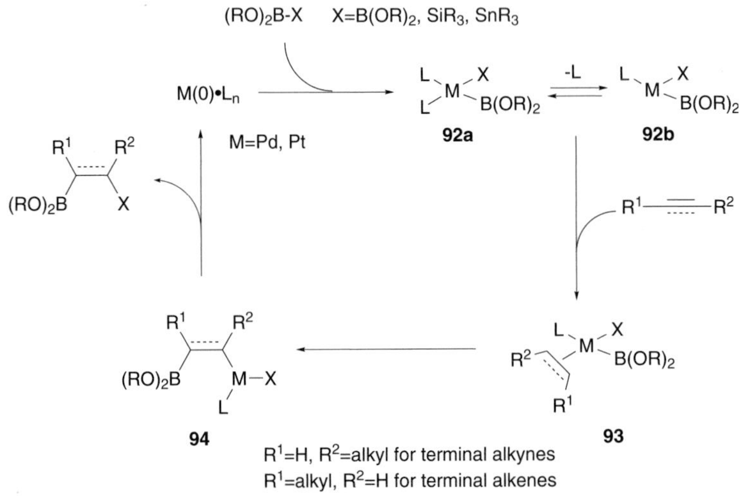

Scheme 1-25 Catalytic Cycle for Addition to Alkenes and Alkynes

The addition proceeds through (a) oxidative addition of the B–X bond to a low-valent metal (M=Pd, Pt) giving a cis-B–M–X complex (92), (b) migratory insertion of alkene or alkyne into the B–M bond (93 → 94), and finally (c) reductive elimination to give an addition product (Scheme 1-25) [111, 112, 117]. Monitoring of the reaction mixture of diboron 82 and Pt(PPh₃)₄ by multinuclear NMR spectroscopy reveals the formation of a new Pt(II) species which shows reasonable thermal stability to allow isolation and is characteristic of phosphines cis-coordinated to a platinum(II) center. A single crystal of (PPh₃)₂Pt(Bpin)₂ consists of a distorted square-planar coordination geometry for the Pt atom containing two cis boryl and phosphine ligands where the significant bond angles are P(1)–Pt–P(2) (102.65°) and B(1)–Pt–B(2) (75.3°) [111]. Treatment of this complex with 1-alkyne yields 1,2-bis(boryl)alkene as the sole product, which is consistent with the proposed catalytic cycle. The catalyst activity significantly decreases in the presence of added PPh₃ suggesting that phosphine dissociation (92a → 92b) is a critical step in the catalytic cycle [117]. A theoretical study of the diboration of alkenes and alkynes also supports the oxidative addition-insertion process [118–120].

Although the mechanism of silylboration or stannylboration is relatively unexplored, the reactions can be rationalized by the same catalytic cycle shown in Scheme 1-25. The addition of a B–Sn compound to palladium(0) yields indeed an analogous (η²-Me₂PCH₂CH₂PMe₂)Pd(SnMe₃)[B(NR₂)₂] which exhibits high catalyst activity for stannylboration [116]. The high reactivity of B–X compounds for the oxidative addition has been recently demonstrated by theoretical calculation [121, 122]; thus, the reaction of (HO)₂B–XH₃ (X= Si, Ge, Sn) to M(PH₃)₂ (M= Pd, Pt) occurs with a very small barrier or no barrier for X=Si, Ge, Sn. The bond energy decreases in the order of B–Si > B–Ge > B–Sn, M–B > M–Si > M–Ge > M–Sn, and Pt–B > Pd–B. The high M–B bond energy over that of other elements is due to the π-back-donating interaction between the M dₚ and B(OH)₂ pₚ orbitals. However, another factor rather than the bond energies is significant in the next insertion step since all experimental results suggest the first migration of the boryl group to a coordinated alkene or alkyne (93 → 94).

The reaction of silylboranes 88 or germylborane 89 with a large excess of terminal or internal alkynes in the presence of a free-phosphine Ni-catalyst in situ generated

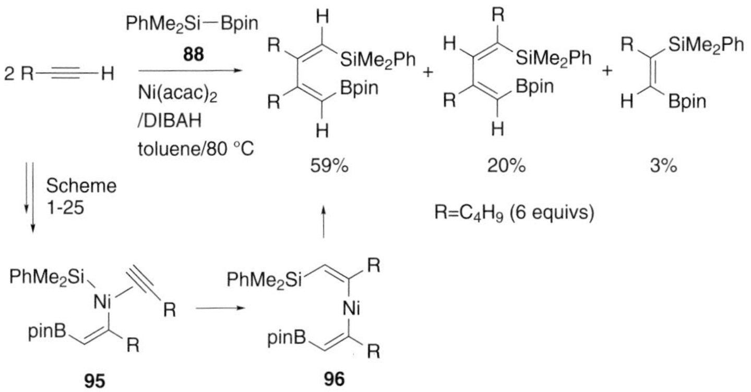

Scheme 1-26 Silaborative Dimerization of Alkynes

from Ni(acac)$_2$ and *i*-Bu$_2$AlH affords dimerization products in moderate to high yields (Scheme 1-26) [123]. The dimerization giving a Z,Z-isomer as a major product with head-to-head coupling of terminal alkyne is best explained by the stepwise double insertion into the B–Ni and then the Si–Ni bond yielding a divinyl nickel intermediate (**96**). The low regioselectivity in the later insertion into the Si–Ni bond results in a mixture of isomers. The use of a phosphine-free catalyst is critical for selective dimerization since the coordinate unsaturated metal allows the multiple coordination of alkyne.

The palladium-catalyzed stannylboration (**90**) [124] or silylboration (**87**) [109, 114] succeeds in the intramolecular carbocyclization of diynes and enynes (Scheme 1-27). It is interesting that a very strained four-membered cyclization of hexa-1,5-diyne proceeds without any difficulties, similarly to five- or six-membered cyclization. The boryl group is selectively introduced into the more reactive C≡CH rather than C=C for enynes and into the terminal C≡CH rather than the internal C≡CR for diynes, again suggesting a mechanism proceeding through the first insertion into the Pd–B bond in preference to the Pd–Sn or Pd–Si bond.

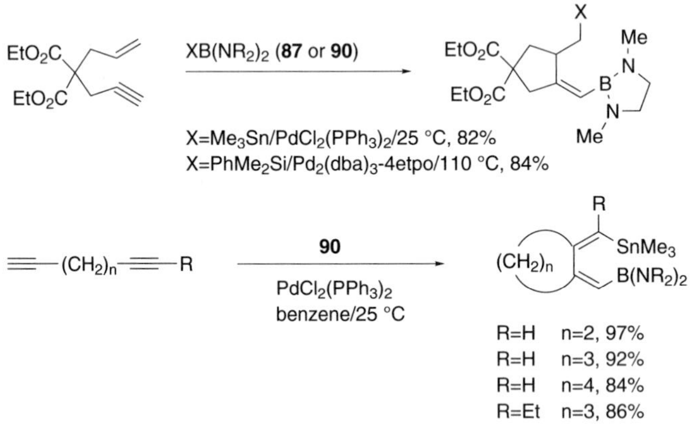

Scheme 1-27 Carbocyclization of Diynes and Enynes

The transformation of 1,2-bis(boryl)-1-alkenes *via* a cross-coupling reaction provides a method for regio- and stereoselective synthesis of 1-alkenylboranes because they have potential reactivity difference between two C–B bonds [125] (Scheme 1-28). Although the reaction is often accompanied with a double-coupling product at both C–B bonds (5-10%), high terminal-selectivity of over 99% can be readily achieved in the coupling with aryl, 1-alkenyl, benzyl, and cinnamyl halides. Under basic conditions, the palladium-catalyzed cross-coupling at the C–B bond is adequately faster than at the C–Si and C–Sn bond. Thus, a silylboration-cross coupling [115] or silylboration/conjugate addition [109, 126] sequence provides a stereoselective method for the synthesis of alkenylsilanes, presumably also alkenylstannanes. These twostep procedures are synthetically equivalent to carbo-heterofunctionalization of alkynes.

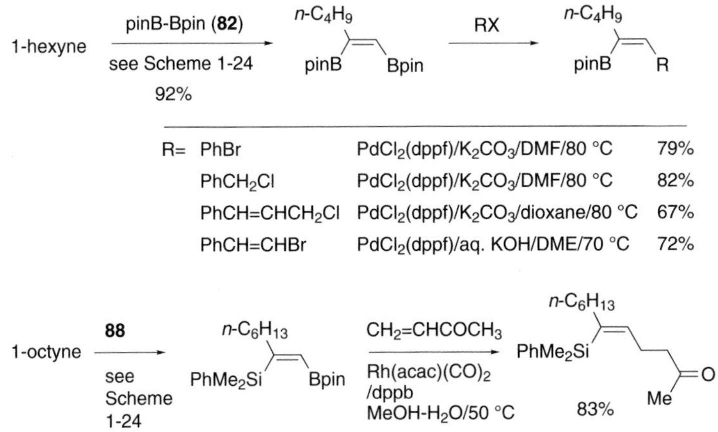

R=			
PhBr	$PdCl_2$(dppf)/K_2CO_3/DMF/80 °C	79%
$PhCH_2Cl$	$PdCl_2$(dppf)/K_2CO_3/DMF/80 °C	82%
$PhCH{=}CHCH_2Cl$	$PdCl_2$(dppf)/K_2CO_3/dioxane/80 °C	67%
PhCH=CHBr	$PdCl_2$(dppf)/aq. KOH/DME/70 °C	72%

Scheme 1-28 Diboration or Silylboration-Coupling Sequence

The utility of the stepwise, double-coupling procedure is demonstrated in the parallel synthesis of Tamoxifen derivatives on solid support [127] (Scheme 1-29). 1-Alkenylboronates thus obtained by a diboration-cross coupling sequence are further coupled with *p*-silyliodobenzene supported on polymer resin. Using this strategy, each position about the ethylene core is modified by the appropriate choice of alkyne, aryl halide, and cleavage conditions for the synthesis of a library of Tamoxifen derivatives.

The phosphine-based platinum(0) catalysts do not catalyze the diboration of alkenes because of the high coordination ability of phosphine over the alkene double bond, but platinum(0) complexes without a phosphine ligand such as $Pt(dba)_2$ [128] and $Pt(cod)_2$ [129] are an excellent catalyst allowing the alkene insertion into the B–Pt bond under mild conditions (Scheme 1-30). The diboration of aliphatic and aromatic terminal alkenes takes place smoothly at 50°C or even at room temperature. The reaction is significantly slow for disubstituted alkenes and cyclic alkenes, but cyclic alkenes having an internal strain afford *cis*-diboration products in high

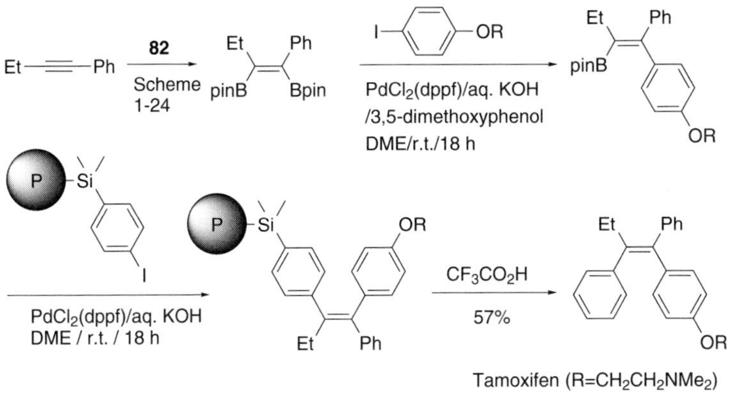

Scheme 1-29 Solid-Phase Synthesis of Tamoxifen

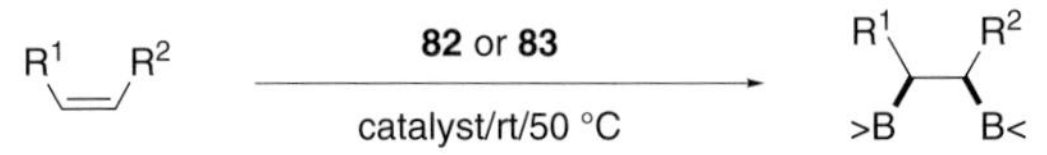

(RO)₂BB(OR)₂	Alkene	Catalyst	Yield/%	Ref.
pinB-Bpin	1-decene	Pt(dba)₂/50°C	82	[128]
(82)	styrene	Pt(dba)₂/50°C	86	[128]
	cyclopentene	Pt(dba)₂/50°C	85	[128]
	cyclohexene	Pt(dba)₂/50°C	0	[128]
	norbornene	Pt(dba)₂/50°C	85	[128]
catB-Bcat	1-hexene	Pt(cod)₂rt	95	[129]
(83)	norbornene	Pt(cod)₂rt	93	[129]
	4-MeOPhCH=CH₂	AuCl(PEt₃)-2dcpe/rt	high	[130]
	styrene	[(acac)Rh(dppm)]B₂cat₃/rt	87	[131]
	PhCH=CHPh	[(acac)Rh(dppm)]B₂cat₃/rt	92-99	[131]

dcpe=ethane-1,2-diylbis(dicyclohexylphosphane)
B₂cat₃=See, **29** in scheme 1-9

Scheme 1-30 Diboration of Alkenes

yields. A phosphine-gold(I) complex [130] and a zwitterionic rhodium(I) complex [131] analogously catalyze the diboration of styrene derivatives.

Silyl(pinacol)borane (**88**) also adds to terminal alkenes in the presence of a coordinate unsaturated platinum complex (Scheme 1-31) [132]. The reaction selectively provides 1,2-adducts (**97**) for vinylarenes, but aliphatic alkenes are accompanied by some 1,1-adducts (**98**). The formation of two products can be rationalized by the mechanism proceeding through the insertion of alkene into the B–Pt bond giving **99** or **100**. The reductive elimination of **97** occurs very smoothly, but a fast β-hydride elimination from the secondary alkyl-platinum species (**100**) leads to isomerization to the terminal carbon.

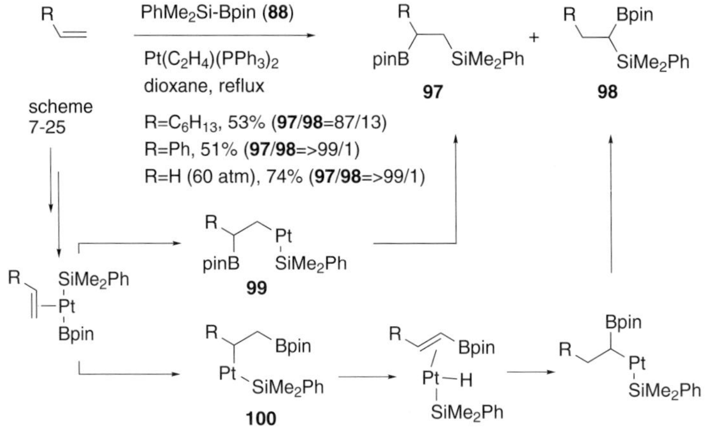

Scheme 1-31 Silylboration of Alkenes

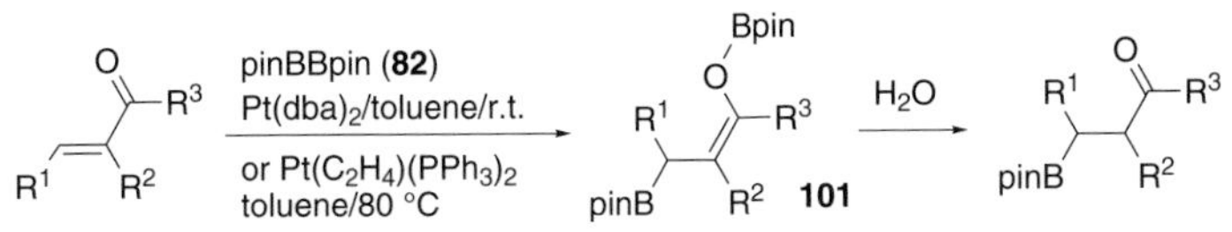

R¹	R²	R³	yield/%	Ref.
Ph	H	CH$_3$	high	[133]
H	CH$_3$	CH$_3$	64	[134]
CH$_3$	H	CH$_3$	72	[134]
Ph	H	Ph	high	[133]
2-cyclopentenone			88	[134]
2-cyclohexenone			61	[134]

Scheme 1-32 Diboration of α,β-Unsaturated Ketones

The diboration of enones with **82** or **83** yields 1,4-addition products (**101**) in the presence of a platinum(0) catalyst such as Pt(C$_2$H$_4$)(PPh$_3$)$_2$ [133] at 80°C or Pt(dba)$_2$ [134] at room temperature (Scheme 1-32). The reaction catalyzed by Pt(C$_2$H$_4$)(PPh$_3$)$_2$ affords a single isomer which is assumed to be the Z-enolate [133]. The hydrolysis of **101** with water gives β-borylketones in high yields, the conversion of which is synthetically equivalent to the conjugate 1,4-addition of a boryl anion to enones. The diboration of α,β-unsaturated esters and nitriles affords similar products.

Methylenecyclopropane and its derivatives are of interest as the substrate for the transition metal-catalyzed addition reactions because of their high and unique reactivities originating from the highly strained structure. The platinum-catalyzed reaction of diboron (**82**) proceeds through the proximal bond cleavage of the cyclopropane ring (Scheme 1-33) [135]. The catalytic cycle involving the insertion of methylenecyclopropane into the B–Pt bond of **92**, followed by the rearrangement to a homoallylplatinum(II) species gives a ring-opening product. The selective formation of the *cis*-isomers for bicyclic methylenecyclopropane suggests a four-centered cyclic transition state for the ring-opening rearrangement. The results also provide information on the insertion mechanism of **92**, i.e., the addition of the Pt–B bond to terminal alkene gives a primary alkyl-platinum intermediate, the selectivity being similar to that for the silylboration of alkenes [132].

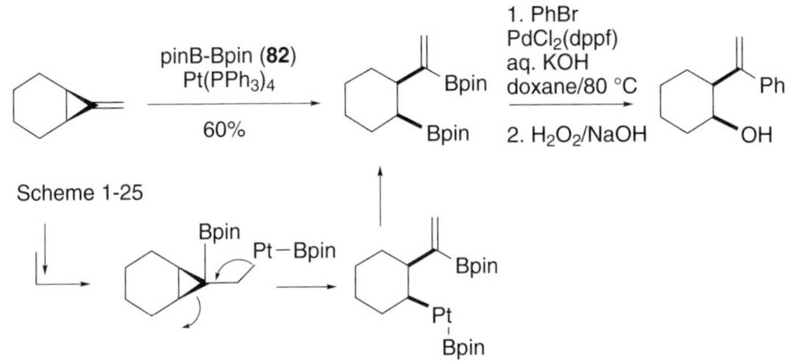

Scheme 1-25

Scheme 1-33 Diboration of Methylenecyclopropanes

1.3.3
Addition to 1,2-Dienes (Allenes) and 1,3-Dienes

The addition to 1,3-dienes affords a new class of allylboron compounds (Scheme 1-34). The Pt(PPh$_3$)$_4$-catalyzed diboration stereoselectively yields *cis*-1,4-addition products **(102)** [106, 136] for the representative aliphatic and alicyclic dienes. The mechanism involves an *S*-*cis* coordination of a diene to (PPh$_3$)$_2$Pt(Bpin)$_2$ **(92)** by displacing two phosphines, insertion of the less-substituted double bond into the B–Pt bond giving *anti*-π-allyl(boryl)platinum(II), and finally reductive elimination of **102**. Interestingly, a phosphine-free platinum(0) catalyst Pt(dba)$_2$ (dba=dibenzalacetone) allows dimerization of the diene before the addition of diboron. Isoprene provides a 1,8-addition product **(103)** [136] having (*E,E*) configuration and a symmetrical structure derived from the head-to-head coupling. Such dimerization *via* the stepwise insertion of two isoprenes into B–Rh bonds is efficiently catalyzed by a phosphine-free platinum complex since the generation of a coordinate unsaturated π-allylplatinum(II) species allows further coordination of the diene and its insertion into the Pt–B bond yielding a di-π-allylplatinum(II) species. Addition of silylborane **(88)** [137] and stannylborane **(90)** [138] to 1,3-dienes with a nickel(0) or palladium(0) complex affords analogous 1,4-addition products in high yields. The silylboration is unfortu-

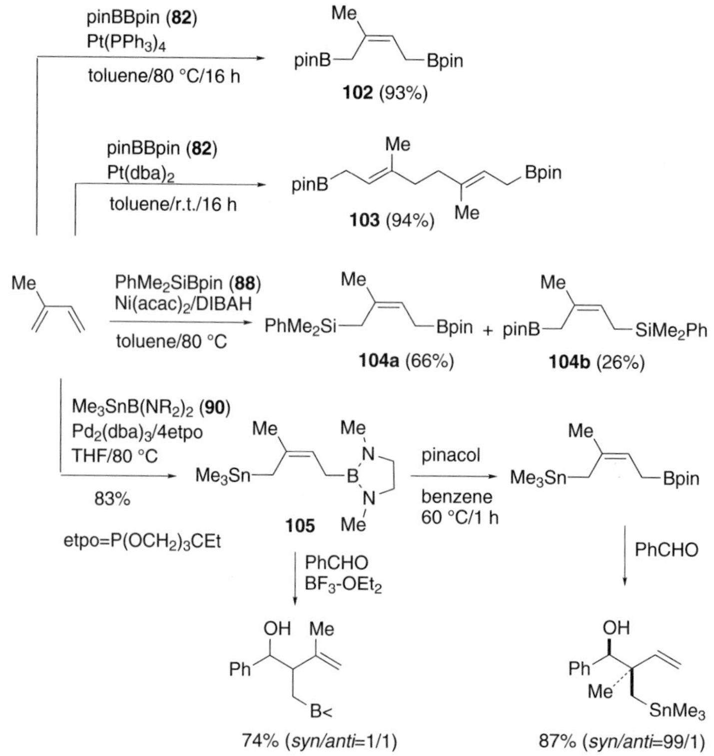

Scheme 1-34 Diboration, Silaboration and Stannylboration of 1,3-Dienes

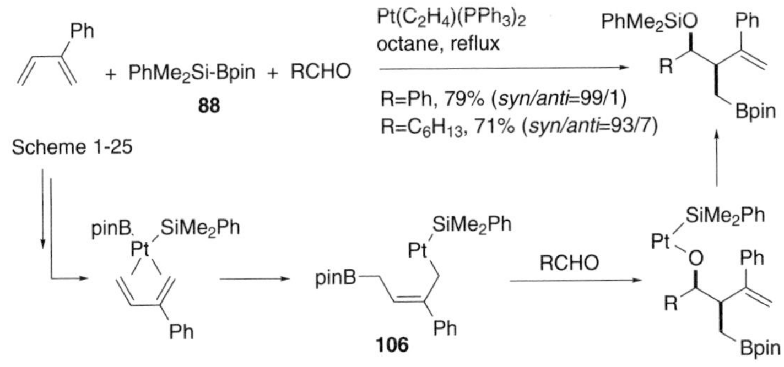

Scheme 1-25

Scheme 1-35   Allylation of Aldehydes

nately not sufficiently regioselective and suffers from stereochemical isomerization, but the stannylboration yields a single product by selective migration of the boryl group to the less-substituted double bond. The acid-catalyzed allylstannation of aldehydes results in a mixture of diastereoisomers, but uncatalyzed allylboration is highly selective to yield *syn*-adducts [138].

Although the platinum(0)-catalyzed silylboration of 1,3-diene suffers from the formation of regioisomers or stereoisomers, analogous reaction in the presence of aldehydes diastereoselectively yields allylation products (Scheme 1-35) [139]. Allylsilanes do not undergo allylmethalation of carbonyl compounds without assistance of a base or acid (CsF, Bu_4HF, or BF_3), but it is interesting that the allylation smoothly occurs at the C–Si bond and the C–B bond remains intact during the reaction. A probable species active for allylmethalation is allylplatinum(II) complex **106** generated by the migratory insertion of the boryl group to the less-substituted double bond. The electron-withdrawing silyl ligand may increase the Lewis acidity of platinum(II), thus allowing coordination and the next insertion of the carbonyl group. The proposed mechanism, involving a nucleophilic π-allyl transition metal species, is also reported in the nickel-catalyzed hydrosilylation of a diene having a carbonyl group [140].

The diboration and silylboration of 1,2-dienes (allenes) affords another series of allylboron compounds (Scheme 1-36) [141–143]. The addition of diboron **82** has a strong tendency to occur at the internal double bond, but steric hindrance in both the allenes and the phosphine ligands forces the addition towards the terminal double bond. Thus, the reaction of monosubstituted allenes with a less bulky PPh_3 preferentially produces the internal adducts **107a**, and the bulky and electron-donating $P(Cy)_3$ (Cy=cyclohexyl) affords the terminal adducts **107b** for 1,1-disubstituted allenes [141]. The reaction selectively yields (*E*)-**107b** for 1,2-heptadiene ($R^1=C_4H_9$, $R^2=H$) by way of addition from the less-hindered side of the terminal double bond. The palladium-catalyzed silylboration [142, 143] of allenes exhibits a much stronger tendency to occur at the internal double bond. However, it is interesting that addition to 1,1-dimethylallene reveals completely opposite selectivity between the palladium(0) and platinum(0) catalysts. The mechanism has not yet been elucidated, but one probable speculation, based on observation that the boryl group always adds to

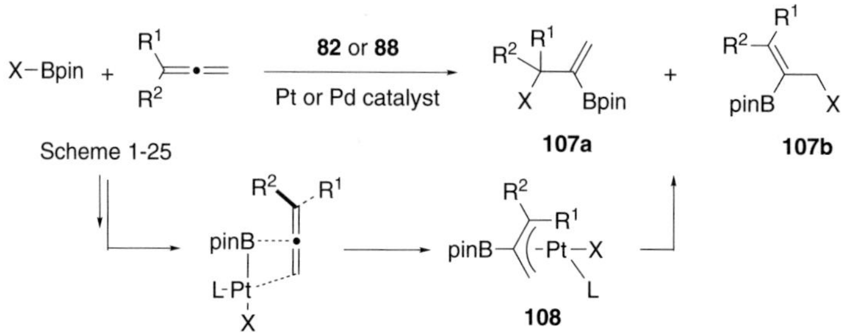

Scheme 1-25

Scheme 1-36   Diboration and Silylboration of Allenes

X=	R^1	R^2	Catalyst	Yield/% (107a/107b)	Ref.
pinB	n-C$_4$H$_9$	H	Pt(PPh$_3$)$_4$/80°C	97 (94/6)	[141]
	Ph	H	Pt(PPh$_3$)$_4$/80°C	94 (71/29)	[141]
	Me	Me	Pt(dba)$_2$-Cy$_3$P/50°C	99 (2/98)	[141]
	MeO	H	Pt(dba)$_2$-Cy$_3$P/50°C	85 (0/100)	[141]
PhMe$_2$Si	Me	H	Pd$_2$(dba)$_3$-4etpo/80°C	98 (100/0)	[142]
	Ph	H	Pd(acac)$_2$-2,6-xylylNC/120°C	95 (86/14)	[143]
	Me	Me	Pd$_2$(dba)$_3$-4Ph$_3$P/80°C	90 (100/0)	[142]
	Me	Me	Pt(C$_2$H$_4$)(PPh$_3$)$_4$/80°C	94 (0/100)	[142]
	MeO	H	Pd$_2$(dba)$_3$-4etpo/80°C	93 (100/0)	[142]

the central carbon, is the insertion of allene into the M–B (M=Pt, Pd) bond giving a π-allylmetal intermediate (108) which undergoes reductive elimination at the internal or terminal carbon depending on the effect of substituents, ligands, or metals.

1.4
Metal-Catalyzed Cross-Coupling Reaction

The best method for the synthesis of organoboron compounds from organic halides is based on the reaction of trialkyl borates with Grignard or lithium reagents. However, the cross-coupling reaction of diboron or pinacolborane solves the difficulties associated with the use of Mg/Li compounds. The transition metal-catalyzed cross-coupling reaction of disilanes [144] and distannanes [145] is an elegant method for the synthesis of organosilicone and organotin compounds directly from organic electrophiles, but the lack of suitable boron nucleophiles has limited this protocol for boron compounds. However, (pinacolato)diboron 82 acts as the boron-nucleophile in the presence of a base for the palladium-catalyzed cross-coupling reaction of organic electrophiles [146, 147]. More recently, pinacolborane (Me$_4$C$_2$O$_2$)BH has also been found to couple with aryl and vinyl electrophiles in the presence of a palladium catalyst and triethylamine [148]. Both reactions now provide a simple and direct method for the borylation of organic halides and triflates.

1.4.1
Coupling with Aryl and Vinyl Halides and Triflates

The cross-coupling reaction of diborons with aryl halides and triflates proceeds in the presence of a base and a palladium catalyst (Scheme 1-37) [146, 147]. The use of $PdCl_2$(dppf) and KOAc in DMSO or DMF is found to provide the best conditions for aryl iodides and bromides because a stronger base, such as K_3PO_4 and K_2CO_3, prompts the further coupling of arylboronates with haloarenes, resulting in contamination of a substantial amount of biaryls (36–60% yields) [146]. The coupling with aryl chlorides is very slow because of their slow oxidative addition to palladium(0). Bulky and electron-donating trialkylphosphines such as P(t-Bu)$_3$ and PCy$_3$ (Cy=cyclohexyl) are better ligands than triarylphosphines for such substrates. A combination of Pd(dba)$_2$ (3 mol%)/PCy$_3$ (7.2 mol%)/KOAc(1.5 equiv.)/pinBBpin (**82**) (1.1 equiv.) in dioxane at 80°C affords good yields of arylboronates, e.g., 88% for 4-chloronitrobenzene, 87% for methyl 4-chlorobenzoate, and 76% for 4-chlorotouene [149]. The coupling of triflates is best carried out in dioxane in the presence of an additional dppf ligand [147]. Arylboronic acid synthesis using Grignard or lithium reagents needs the protection of functional groups sensitive to these reagents, but the coupling reaction is available with various functional groups, e.g., CO_2Me, COMe, NO_2 and CN.

	ArX (Ar=)	ArX (X=)	yield/%
For halides: PdCl$_2$(dppf)/AcOK	p-NCC$_6$H$_4$	Br	76
/DMSO/80 °C		OTf	75
For triflates: PdCl$_2$(dppf)-dppf	p-CH$_3$COC$_6$H$_4$	Br	80
/AcOK/dioxane/80 °C		OTf	92
	p-MeOC$_6$H$_4$	I	82
		OTf	93
	p-BrC$_6$H$_4$	I	86
	2,4,6-Me$_3$C$_6$H$_2$	I	71
	3-bromoquinoline		84

Scheme 1-37 Borylation of Aryl Halides and Triflates

Pinacolborane is a unique boron nucleophile for analogous coupling with aryl iodides, bromides, or triflates (Scheme 1-38) [148]. It is interesting that ester, ketone, and nitro groups remain intact during the coupling reaction of pinacolborane at 80°C, though the reaction competes with the hydrogenation of aryl halides (10–20%). The reaction is accelerated in the presence of an electron-donating group, while a withdrawing group slows down the rate of coupling, the electronic effect of which is the reverse of the cross-coupling reactions of diborons and other organoboronic acids.

The reaction proceeds through the formation of an (acetoxo)palladium(II) species (**110**) prior to transmetalation with diboron (Scheme 1-39). The presence of a base

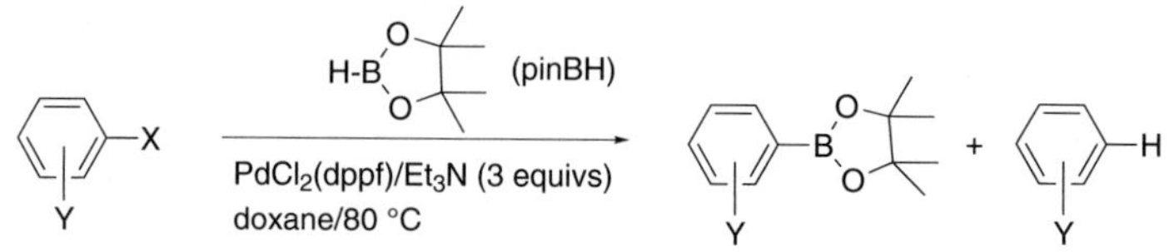

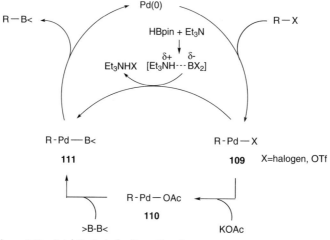

Scheme 1-38 Borylation of Aryl Halides and Triflates

ArX (Ar=)	ArX (X=)	yield/%	yield/%
p-MeOC$_6$H$_4$	I	77	16
	Br	77	11
	OTf	81	12
p-ClCOC$_6$H$_4$	I	83	nd
p-EtO$_2$CC$_6$H$_4$	I	79	17
	Br	43	20
	OTf	77	21
p-NO$_2$C$_6$H$_4$	I	84	nd
p-Me$_2$NC$_6$H$_4$	I	79	nd
2-iodothiophene		80	nd

such as KOAc is critical for the coupling reaction of diborons, suggesting a trans-metalation mechanism occurring from Ar-Pd-OAc generated by displacement of X (**109**) with an acetate anion. There is no reaction between diboron (**82**) and trans-PhPd(Br)(PPh$_3$)$_2$ (**109**), but the palladium acetate complex (**110**) [150] indeed reacts with diboron at room temperature yielding arylboronate [146]. The high reactivity of the oxo-palladium complexes for the transmetalation with organoboron compounds can be attributed to both the high reactivity of the Pd–O bond, which consists of a soft acid and a hard base combination, and the high oxophilicity of the boron center. On the other hand, the presence of Et$_3$N plays an important role in not only pre-venting the production of Ar–H but also facilitating the B–C bond formation in the

Scheme 1-39 Catalytic Cycle for Cross-Coupling

coupling reaction of pinacolborane. The transmetalation involves the displacement of Pd–X with a weakly nucleophilic boryl anion species generated by the reaction of Et$_3$N and HBpin [148b] or σ-bond metathesis between H–Pd–Bpin and ArX leading to Ar–Pd–Bpin (**111**).

Both Pd(PPh$_3$)$_4$ and PdCl$_2$(dppf) catalyze the cross-coupling reaction well, but Pd(PPh$_3$)$_4$ often suffers from the formation of phenylboronate (PhBpin) derived from the coupling with phosphine-bound aryls [146]. The contamination by such an unexpected by-product is observed in the reaction of electron-rich haloarenes such as (4-methoxy)- and (4-dimethylamino)bromobenzene, though they are in negligibly small amount for electron-deficient haloarenes. The aryl exchange [ArPd(X)(PPh$_3$)$_2$ → PhPd(X)(PPh$_3$)(PPh$_2$Ar)] occurs before transmetalation [151]; thus, the slow transmetalation due to steric and electronic reasons will result in enhancing the coupling of phosphine-bound aryls.

(4-Boronophenyl)alanine (BPA) is a practical boron compound which is clinically used for the treatment of not only malignant melanoma but also of brain tumors, in neutron capture therapy (Scheme 1-40) [105, 152, 153]. Although (pinacolato)di-boron (**82**) is an excellent reagent to afford the corresponding boronate in 88% yield, it strongly resists the hydrolysis to arylboronic acids. Alternatively, the 1,3-diphenyl-propanediol ester (**85**) is more convenient to deprotect the diol moiety by catalytic hydrogenolysis [105].

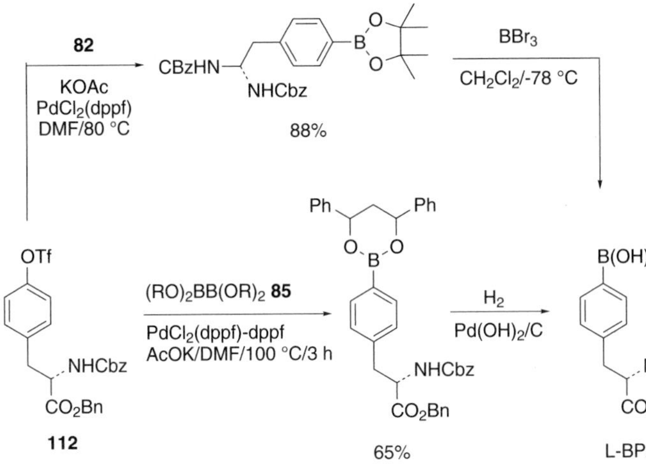

Scheme 1-40 Synthesis of L-BPA

The direct preparation of arylboronic esters from aryl halides or triflates now allows a one-pot, two-step procedure for the synthesis of unsymmetrical biaryls (Scheme 1-41) [147]. The synthesis of biaryls is readily carried out in the same flask when the first coupling of the triflate with diboron **82** is followed by the next reaction with another triflate. The synthesis of naturally occurring biflavanoids and the coupling of *N*-(phenylfluorenyl)amino carbonyl compounds to polymeric supports are reported [154].

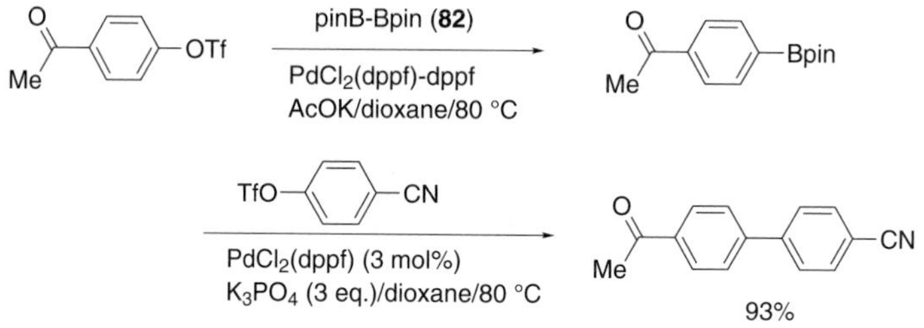

Scheme 1-41 One-Pot Synthesis of Biaryls

The protocol offers a direct and efficient method for the synthesis of the boronic ester in the solid phase, which hitherto met with little success using classical methodology (Scheme 1-42). A solid-phase boronate (**113** [155], **114** [156]) is quantitatively obtained by treating a polymer-bound iodoarene with the diboron (**82**). The subsequent coupling with haloarenes furnishes various biaryls. The robot synthesis or the parallel synthesis on the surface of resin is the topic of further accounts of the research.

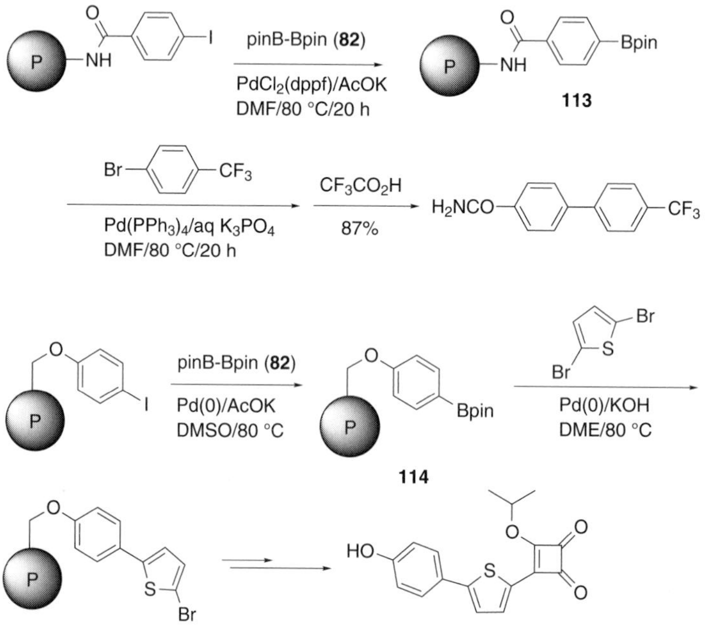

Scheme 1-42 Solid-Phase Synthesis

The synthesis of 1-alkenylboronic acids from 1-alkenylmagnesiums or -lithiums suffers from difficulty in retaining the stereochemistry of 1-alkenyl halides, but the palladium-catalyzed coupling reaction of diboron **82** with 1-alkenyl halides or triflates directly provides 1-alkenylboronic esters (Scheme 1-43) [157, 158]. Although the reaction conditions applied to the aryl coupling resulted in the formation of an

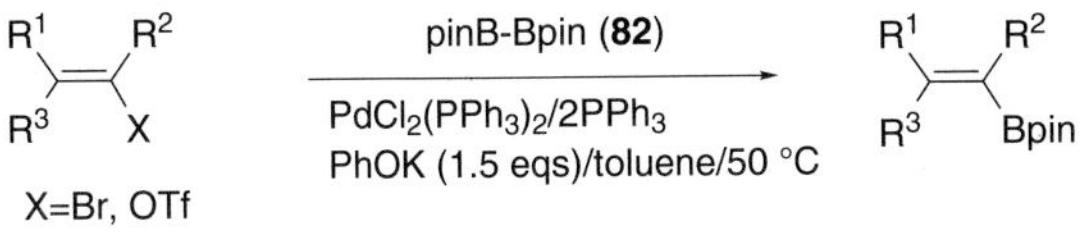

R^1	R^2	R^3	X	yield/%
H	n-C$_8$H$_{17}$	H	Br	92
H	Ph	H	Br	88
H	NC(CH$_2$)$_3$	H	Br	85
H	H	n-C$_8$H$_{17}$	Br	74
	- (CH$_2$)$_4$ -	H	Br	99
	- (CH$_2$)$_4$ -	H	OTf	88

Scheme 1-43 Borylation of Vinyl Halides and Triflates

inseparable mixture of several products including the Heck coupling, the homocoupling of haloalkenes along with the desired coupling, the selective synthesis of 1-alkenylboronates is achieved in the presence of a palladium-triphenylphosphine complex and solid PhOK in toluene [157]. The reaction affords cyclic and acyclic vinylboronic esters possessing various functional groups, which are not available by conventional hydroboration of alkynes or by a transmetalation method.

1.4.2
Coupling to Allyl Halides and Acetates

Allylboron compounds are valuable reagents in organic synthesis, addition of which to the carbon-oxygen or the carbon-nitrogen double bond diastereoselectively provides homoallylic alcohols or amines *via* a chair-like, six-membered cyclic transition state. Allylboron compounds have been synthesized by the transmetalation of allyllithiums or allylmagnesiums to the boron, but the palladium-catalyzed cross-coupling reaction of diboron **82** with allyl acetates or chlorides regio- and stereoselectively yields allylboronic esters (Scheme 1-44) [159, 160]. The reaction with allyl acetates [159] occurs without the assistance of a base because their oxidative addition directly yields an (acetoxo)palladium(II) species corresponding to **110**, but the presence of AcOK is critical in the coupling with allyl chlorides for producing π-allylpalladium acetate intermediate [160]. The coupling at the less hindered terminal carbon and the formation of thermally stable (E)-allylboronates *via* the isomerization of *anti*-π-allyl-palladium to the more stable syn-complex are observed for various allyl electrophiles. Thus, the (E)- and (Z)-cinnamyl acetate, chloride, and its secondary alcohol derivative all afford an (E)-cinnamylboronate. The borylation of prenyl acetate and chloride results in low yields, but the corresponding tertiary derivative can be used for the same purpose because the boron atom exclusively bonds to its primary carbon.

Intramolecular addition of allylmetal reagents to carbonyl substrates is a powerful tool for the synthesis of cyclic homoallyl alcohols with high regio- and stereoselectivity, but the corresponding reaction of allylboranes has not been well developed mainly because of the lack of a general method for the synthesis of allylboronates having a carbonyl group. The palladium(0)-catalyzed cross-coupling reaction of di-

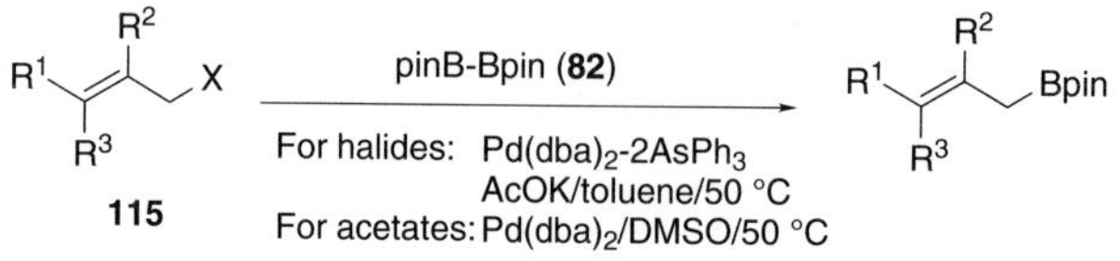

R¹	R²	R³	X	time/h	yield/%
Ph	H	H	Cl	10	70
Ph	H	H	OAc	26	73
H	Ph	H	Cl	5	71
H	Ph	H	OAc	16	89
Me	H	Me	Cl	5	78
Me	H	Me	OAc	60	24
3-chlorocyclohexene				5	70
3-acetoxycyclohexene				60	4

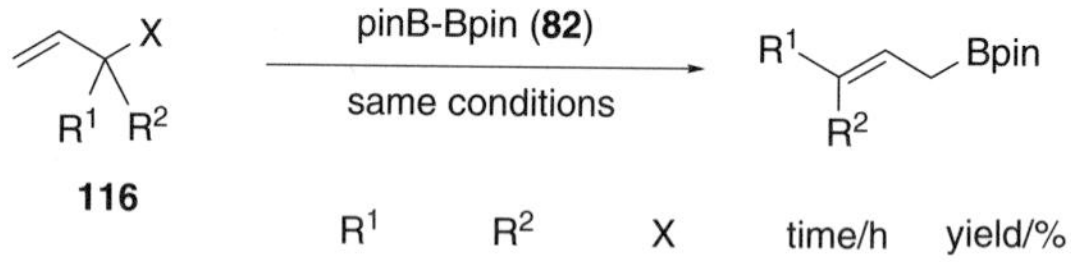

R¹	R²	X	time/h	yield/%
Ph	H	Cl	10	64
Ph	H	OAc	16	83
Me	Me	OAc	16	83

Scheme 1-44 Borylation of Allylic Halides and Acetates

boron **82** provides an efficient and convenient access to variously functionalized allyboranes under neutral conditions (Scheme 1-45) [161]. A variety of 5-5, 6-5, and 7-5 *cis*-fused exomethylene cyclopentanols are synthesized from β-ketoesters (**117**) *via* a cross-coupling/intramolecular allylboration sequence. A diastereoselective cyclization observed in **119** may suggest the intramolecular allylboration via a chair-like, six-membered transition state.

The coupling with benzyl halides is carried out under conditions similar to those used for allyl chlorides [162]. The selective coupling is achieved by an electron-rich

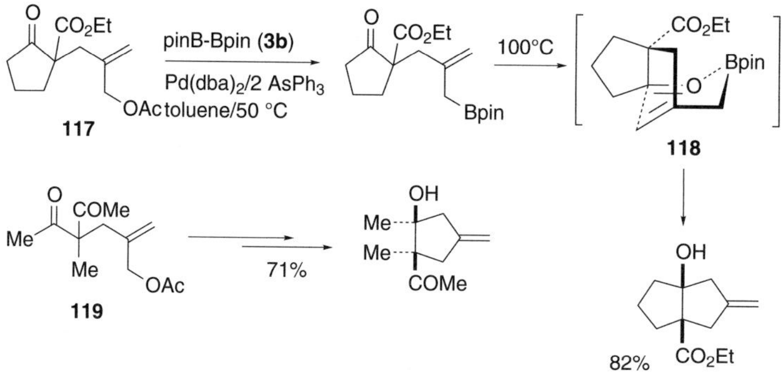

Scheme 1-45 Intramolecular Allylboration

ligand such as palladium(0)-tris(*p*-methoxyphenyl)phosphine complex. On the other hand, the slow reaction with Pd(PPh₃)₄ resulted in a mixture of benzylboronate and benzyl acetate.

1.5
Conclusions

The metal-catalyzed addition and coupling reaction of organoboron reagents including H–B, B–B, Si–B, and Sn–B compounds provide expeditious access to numerous polyfunctionalized organoboron compounds. The applications of organoboron compounds range from asymmetric synthesis and the preparation of biologically active molecules and new materials to combinatorial chemistry. It is hoped that further innovative synthetic chemistry in diverse areas, stimulated by the metal-catalyzed methodology, will appear in the future.

Abbreviations

cod	1,5-cyclooctadiene	etpo	P(OCH2)3CCH2CH3
coe	cyclooctene	TTMPP	tris(2,4,6-trimethoxyphenyl)phosphine,
nbd	norbornadiene	dppf	1,1′-bis(diphenylphosphino)ferrocene
dba	dibenzalacetone	dppm	$Ph_2PCH_2PPh_2$
Cy	cyclohexyl	dppe	$Ph_2P(CH_2)_2PPh_2$
Py	pyridine	dppp	$Ph_2P(CH_2)_3PPh_2$
Cp	η^5-C₅H₅	dppb	$Ph_2P(CH_2)_4PPh_2$
Cp*	η^5-Me₅C₅		

References

1 Reviews for catalyzed hydroboration, see: (a) L. Delox, M. Srebnik, *Chem. Rev.* **1993**, *93*, 763-784. (b) K. Burgess, M. J. Ohmeyer, *Chem. Rev.* **1991**, *91*, 1179. (c) I. Beletskaya, A. Pelter, *Tetrahedron*, **1997**, *53*, 4957-5026. (d) G. C. Fu in *Transition Metals for Organic Synthesis*, M. Beller, C. Bolm (eds.), Wiley-VCH, Weinheim, **1998**, *Vol. 2*, 141-146. (e) T. Hayashi in *Comprehensive Asymmetric Synthesis*, E. N. Jacobsen, A. Pfaltz, H. Yamamoto (eds.), Springer, Berlin, **1999**, *Vol. I*, 351-366.

2 Reviews for diboration, silylboration, and stannylboration, see: (a) H. Wadepohl, *Angew. Chem. Int. Ed. Engl.* **1997**, *36*, 2441-2444. (b) H. Braunschweig, *Angew. Chem. Int. Ed. Engl.* **1998**, *37*, 1786-1801. (c) T. B. Marder, N. Norman, *Topics in Catalysis*, **1998**, *5*, 63-73. (d) G. J. Irvine, M. J. Lesley, T. B. Marder, N. C. Norman, C. R. Rice, E. G. Robins, W. R. Roper, G. R. Whittel, L. J. Wright, *Chem. Rev.* **1998**, *98*, 2685-2722. (e) L-B. Han, M. Tanaka, *Chem. Commun.* **1999**, 395-402. (f) T. Ishiyama, N. Miyaura, *J. Synthetic Organic Chemistry, Japan,* **1999**, *57*, 503-511.

3 Reviews for uncatalyzed hydroboration: (a) T. Onak, *Organoborane Chemistry*, Academic, New York, **1975**. (b) B. M. Mikhailov, Tu. N. Bubnov, *Organoboron Compounds in Organic Synthesis*, Harwood Academic Pub., Amsterdam, **1983**. (c) A. Pelter, K. Smith, H. C. Brown, *Borane Reagents*, Academic, New-York, **1988**. (d) D. S. Matteson, *Stereodirected Synthesis with Organoboranes*, Springer, Berlin, **1995**.

4 D. Männig, H. Nöth, *Angew. Chem. Int. Ed. Engl.* **1985**, *24*, 878-879.

5 G. Urry, J. Kerrigan, T. D. Parsons, H. I. Schlesinger, *J. Am. Chem. Soc.* **1954**, *76*, 5299-5301; C. A. Ceron, A, Finch, J. Frey, J. Kerringan, T. Parsons, G. Urry, H. I. Schlesinger, *Ibid.* **1959**, *81*, 6368-6371; M. Zeldin, T. Wartik, *Ibid.* **1966**, *88*, 1336-1338; T. D. Coyle, J. J. Ritter, *Ibid.* **1967**, *89*, 5739-5741; M. Zeldin, A. R. Gatti, T. Wartik, *Ibid.* **1997**, *89*, 4216-4218.

6 A review for uncatalyzed diboration, see: T. D. Coyle, J. J. Ritter in *Advances in Organometallic Chemistry*, Academic, New York, **1972**, *Vol. 10*, 237-272.

7 N. Miyaura, A. Suzuki, *Chem. Rev.* **1995**, *95*, 2457-2483.

8 A. Suzuki in *Metal-catalyzed Cross-Coupling Reactions*, F. Diederich, P. J. Stang (eds.), Wiley-VCH, Weinheim, **1998**, 49-97.

9 N. Miyaura in *Advances in Metal-Organic Chemistry*, L. S. Liebeskind (ed.), JAI, Stamford, **1998**, *Vol. 6*, 187-243.

10 R. E. London, S. A. Gabel, *J. Am. Chem. Soc.* **1994**, *116*, 2562-2569; K. R. A. S. Sandanayake, K. Nakashima, S. Shinkai, *J. Chem. Soc., Chem. Commun.* **1994**, 1621-1622 K. R. A. S. Sandanayake, S. Shinkai, *Ibid.* **1994**, 1083-1084; G. Wuff, S. Krieger, B. Kuhneweg, A. Steigel, *J. Am. Chem. Soc.* **1994**, *116*, 409-410; M. T. Reetz, J. Rudolph, K. Töllner, A. Deege, R. Goddard, *Ibid.* **1994**, *116*, 11588-11589.

11 M. F. Howthorne, *Angew. Chem. Int. Ed. Engl.* **1993**, *32*, 950. A. H. Soloway,

Chem. Rev. **1998**, *98*, 1515-1582.

12 H. C. Brown, S. K. Gupta, J. Am.
Chem. Soc. **1971**, *93*, 1816-1819; C. F.
Lane, G. W. Kabalka, *Tetrahedron*,
1976, *32*, 981-990.

13 C. E. Tucker, J. Davidson, P.
Knochel, *J. Org. Chem.* **1992**, *57*, 3482-
3485.

14 W. G. Woods, P. L. Strong, *J. Am.
Chem. Soc.* **1966**, *88*, 4667-4671.

15 J. M. Brown, G. C. Lloyd-Jones, *J.
Chem. Soc. Chem. Commun.* **1992**, 710-
712; J. M. Brown, G. C. Lloyd-Jones, *J.
Am. Chem. Soc.* **1994**, *116*, 866-878.

16 H. R. Molales, H. Tlahuext, F. Santi-
esteban, R. Contreras, *Spectrochim.
Acta* **1984**, *40A*, 855-862; C. Camacho,
M. A. Paz-Sandoval, R. Contreras,
Polyhedron **1989**, *5*, 1723-1732.

17 A. T. Lynch, L. G. Sneddon, *J. Am.
Chem. Soc.* **1987**, *109*, 5867-5868; P. J.
Fazen, L. G. Sneddon, *Organometallics*
1994, *13*, 2867-2877; A. T. Lynch, L. G.
Sneddon, *J. Am. Chem. Soc.* **1989**, *111*,
6201-6209.

18 A. Lang, H. Nöth, M. Thomann-Al-
bach, *Chem. Ber.* **1997**, *130*, 363-369.

19 K. Burgess, W. A. van der Donk, S. A.
Westcott, T. B. Marder, R. T. Baker, J.
C. Calabrese, *J. Am. Chem. Soc.* **1992**,
114, 9350-9359.

20 D. A. Evans, G. C. Fu, B. A. Anderson,
J. Am. Chem. Soc. **1992**, *114*, 6679-6685.

21 M. J. Bennett, P. B. Donaldson, *Inorg.
Chem.* **1997**, *16*, 1581-1594; M. T. Atlay,
L. R. Gahan, K. Kite, K. Moss, G.
Read, *J. Mol. Catal.* **1980**, *7*, 31.

22 T. Hayashi, Y. Matsumoto, Y. Ito,
Tetrahedron: Asymm. **1991**, *2*, 601-612.

23 J. Zhang, B. Lou, G. Guo, L. Dai, *J.
Org. Chem.* **1991**, *56*, 1670-1672.

24 S. A. Westcott, H. P. Blom, T. B.
Marder, R. T. Baker, *J. Am. Chem. Soc.*
1992, *114*, 8863-8869.

25 T. Hayashi, Y. Matsumoto, Y. Ito, *J.
Am. Chem. Soc.* **1989**, *111*, 3426-3428.

26 P. V. Ramachandran, M. P. Jennings,
H. C. Brown, *Org. Lett.* **1999**, *1*, 1399-
1402.

27 D. A. Evans, G. C. Fu, *J. Org. Chem.*
1990, *55*, 5678-5680.

28 S. A. Westcott, T. B. Marder, R. T.
Baker, J. C. Calabrese, *Can. J. Chem.*
1993, *71*, 930-936.

29 K. Burgess, M. Jaspars,
Organometallics **1993**, *12*, 4197-4200.

30 (a) X. He, J. F. Hartwig, *J. Am. Chem.
Soc.* **1996**, *118*, 1696-1702; (b) J. F.
Hartwig, C. N. Muhoro,
Organometallics **2000**, *19*, 30-38.

31 (a) K. N. Harrison, T. J. Marks, *J. Am.
Chem. Soc.* **1992**, *114*, 9220-9221. (b) E.
A. Bijpost, R. Duchateau, J. H.
Teuben, *J. Mol. Catal. A* **1995**, *95*, 121-
128.

32 D. A. Evans, A. R. Muci, R. Stürmer,
J. Org. Chem. **1993**, *58*, 5307-5309.

33 S. Pereira, M. Srebnik, *Tetrahedron
Lett.* **1996**, *37*, 3283-3286.

34 T. Umemoto, Y. Yamamoto, N. Miyaura, unpublished results.

35 D. A. Evans, G. C. Fu, A. H. Hoveyda,
J. Am. Chem. Soc. **1992**, *114*, 6671-
6679; D. A. Evans, G. C. Fu, A. H.
Hoveyda, *J. Am. Chem. Soc.* **1988**, *110*,
6917-6918.

36 S. Pereira, M. Srebnik, *J. Am. Chem.
Soc.* **1996**, *118*, 909-910.

37 H. C. Brown, G. Zweifel, *J. Am.
Chem. Soc.* **1996**, *88*, 1433; H. C.
Brown, J. B. Campbell, *Aldrichimica
Acta* **1981**, *14*, 3.

38 S. Pereira, M. Srebnik,
Organometallics **1995**, *14*, 3127-3128.

39 I. D. Gridnev, N. Miyaura, A. Suzuki,
Organometallics, **1993**, *12*, 589-592.

40 D.-Y. Yang, X. Huang, *J. Chem. Research (S)* **1997**, 62-63.

41 M. Hoshi, Y. Masuda, A. Arase, *Bull.
Chem. Soc. Jpn.* **1990**, *63*, 447-452.

42 (a) E. Negishi, R. M. Williams, G.
Lew, T. Yoshida, *J. Organomet. Chem.*
1975, *92*, C4. (b) J. B. Jr. Campbell, G.
A. Molander, *J. Organomet. Chem.*
1978, *156*, 71. (c) H. C. Brown, T. Imai,
Organometallics **1984**, *3*, 1392-1395.

43 (a) M. Srebnik, N. G. Bhat, H. C.
Brown, *Tetrahedron Lett.* **1988**, *29*,
2635-2638. (b) L. Deloux, M. Srebnik,
J. Org. Chem. **1994**, *59*, 6871-6873.

44 T. Ohmura, Y. Yamamoto, N.
Miyaura, *J. Am. Chem. Soc.* **2000**, in
press.

45 (a) J. WOLF, H. WERNER, O. SERHADLI, M. L. ZIEGLER, *Angew. Chem. Int. Ed. Engl.* **1983**, *22*, 414-416. (b) F. J. G. ALONSO, A. HÖHN, J. WOLF, H. OTTO, H. WERNER, *Angew. Chem. Int. Ed. Engl.* **1985**, *24*, 406-408.

46 M. SATO, Y. NOMOTO, N. MIYAURA, A. SUZUKI, *Tetrahedron Lett.* **1989**, *30*, 3789-3792.

47 Y. MATSUMOTO, T. HAYASHI, *Tetrahedron Lett.* **1991**, *32*, 3387-3390.

48 D. S. SETHI, G. C. JOSHI, D. DEVAPRABHAKARA, *Can. J. Chem.*, **1969**, *47*, 1083-1086; R. H. FISH, *J. Am. Chem. Soc.*, **1968**, *90*, 4435-4439; H. C. BROWN, R. LIOTTA, G. W. KRAMER, *J. Am. Chem. Soc.*, **1979**, *101*, 2966-2970.

49 Y. YAMAMOTO, R. FIJIKAWA, A. YAMADA, N. MIYAURA, *Chem. Lett.* **1999**, 1069-1070.

50 Y. MATSUMOTO, M. NAITO, T. HAYASHI, *Organometallics* **1992**, *11*, 2732-2734.

51 M. ZAIDLEWICZ, J. MELLER, *Collect. Czech. Chem. Commun.* **1999**, *64*, 1049-1056.

52 Y. MATSUMOTO, M. NAITO, Y. UOZUMI, T. HAYASHI, *J. Chem. Soc. Chem. Commun.* **1993**, 1468-1469.

53 M. MURATA, S. WATANABE, Y. MASUDA, *Tetrahedron Lett.* **1999**, *40*, 2585-2588.

54 S. A. WESTCOTT, T. B. MARDER, R. T. BAKER, *Organometallics* **1993**, *12*, 975-979.

55 D. H. MOTRY, A. G. BRAZIL, M. R. SMITH III, *J. Am. Chem. Soc.* **1997**, *119*, 2743-2744.

56 D. H. MOTRY, M. R. SMITH III, *J. Am. Chem. Soc.* **1995**, *117*, 6615-6616.

57 (a) S. A. WESTCOTT, N. J. TAYLOR, T. B. MARDER, R. T. BAKER, N. J. JONES, J. C. CALABRESE, *J. Chem. Soc., Chem. Commun.* **1991**, 304-305. (b) H. KONO, K. ITO, Y. NAGAI, *Chem. Lett.* **1975**, 1095-1096.

58 R. T. BAKER, J. C. CALABRESE, S. A. WESTCOTT, P. NGUYEN, T. B. MARDER, *J. Am. Chem. Soc.* **1993**, *115*, 4367-4368.

59 S. A. WESTCOTT, H. P. BLOM, T. B. MARDER, R. T. BAKER, J. C. CALABRESE, *Inorg. Chem.* **1993**, *32*, 2175-2182.

60 K. BURGESS, M. JASPARS, *Tetrahedron Lett.* **1993**, *34*, 6813-6816.

61 K. BURGESS, W. A. VAN DER DONK, *Organometallics* **1994**, *13*, 3616-3620.

62 J. R. KNORR, J. S. MEROLA, *Organometallics* **1990**, *9*, 3008-3010. The opposite regioselectivity is reported due to misprint: a private communication from Merola.

63 A. E. DORIGO, P. VON R. SCHLEYER, *Angew. Chem. Int. Ed. Engl.* **1995**, *34*, 115-118.

64 P. R. RABLEN, J. F. HARTWIG, S. P. NOLAN, *J. Am. Chem. Soc.* **1994**, *116*, 4121-4122.

65 D. G. MUSAEV, A. M. MEBEL, K. MOROKUMA, *J. Am. Chem. Soc.* **1994**, *116*, 10693-10702.

66 P. BINGER, F. SANDMEYER, C. KRÜGER, J. KUHNIGK, R. GODDARD, G. ERKER, *Angew. Chem. Int. Ed. Engl.* **1994**, *33*, 197-198.

67 D. R. LANTERO, D. L. WARD, M. R. SMITH III, *J. Am. Chem. Soc.* **1997**, *119*, 9699-9708.

68 D. R. LANTERO, D. H. MOTRY, D. L. WARD, M. R. SMITH III, *J. Am. Chem. Soc.* **1994**, *116*, 10811-10812.

69 J. F. HARTWIG, S. R. DE GALA, *J. Am. Chem. Soc.* **1994**, *116*, 3661-3662.

70 S. A. KULKARNI, N. KOGA, *J. Mol. Struct. (Theochem)* **1999**, *461-462*, 297-310.

71 D. A. EVANS, J. R. GAGE, *J. Org. Chem.* **1992**, *57*, 1958-1961.

72 S. HATAKEYAMA, M. KAWAMURA, S. SAKAMOTO, H. IRIE, *Tetrahedron Lett.* **1994**, *35*, 7993-7996.

73 J. COSSY, S. BOUZBOUZ, *Tetrahedron Lett.* **1996**, *37*, 5091-5094.

74 Y. NOMOTO, N. MIYAURA, A. SUZUKI, *Synlett* **1992**, 727-729. X-L. HOU, Q-C. XIE, L-X. DAI, *J. Chem. Research (S)* **1997**, 436-437.

75 K. BURGESS, W. A. VAN DER DONK, M. B. JARSTFER, M. J. OHLMEYER, *J. Am. Chem. Soc.* **1991**, *113*, 6139-6144.

76 G. HAREAU-VITTINI, P. J. KOCIENSKI, *Synlett* **1995**, 893-894.

77 D. L. BOGER, T. HONDA, *J. Am. Chem. Soc.* **1994**, *116*, 5647-5656.

78 A. S. KENDE, K. LIU, K. M. J. BRANDS, *J. Am. Chem. Soc.* **1995**, *117*, 10597-10598.

79 K. M. J. BRANDS, A. S. KENDE, *Tetrahedron Lett.* **1992**, *33*, 5887-5890.

80 K. BURGESS, M. J. OHLMEYER, *Tetrahedron Lett.* **1989**, *30*, 395-398.

81 K. Burgess, J. Cassidy, M. J. Ohlmeyer, *J. Org. Chem.* **1991**, *56*, 1020-1027.

82 K. N. Houk, N. G. Rondan, Y. Wu, J. T. Metz, M. N. Paddon-Row, *Tetrahedron* **1984**, *40*, 2257-2274.

83 D. A. Evans, G. S. Sheppard, *J. Org. Chem.* **1990**, *55*, 5192-5194.

84 K. Burgess, M. J. Ohlmeyer, *J. Org. Chem.* **1991**, *56*, 1027-1036.

85 K. Burgess, L. T. Liu, B. Pal, *J. Org. Chem.* **1993**, *58*, 4758-4763.

86 I. D. Gridnev, N. Miyaura, A. Suzuki, *J. Org. Chem.* **1993**, *58*, 5351-5354.

87 T. Ishiyama, K. Nishijima, N. Miyaura, S. Suzuki, *J. Am. Chem. Soc.* **1993**, *115*, 7219-7225.

88 S. Machida, T. Ishiyama, N. Miyaura, unpublished results.

89 P. Kocienski, K. Jarowicki, S. Marczak, *Synthesis* **1991**, 1191-1200.

90 M. Srebnik, P. V. Ramachandran, *Aldchimica Acta* **1987**, *20*, 9 and reference 1.

91 J. M. Valk, G. A. Whitlock, T. P. Layzell, J. M. Brown, *Tetrahedron: Asymm.* **1995**, *6*, 2593-2596; J. M. Brown, D. I. Hulmes, T. P. Layzell, *J. Chem. Soc., Chem. Commun.* **1993**, 1673-1674.

92 A. Togni, C. Breutel, A. Schnyder, F. Spindler, H. Landert, A. Tijani, *J. Am. Chem. Soc.* **1994**, *116*, 4062-4066.

93 S. U. Son, H.-Y. Jang, J. W. Han, I. S. Lee, Y. K. Chung, *Tetrahedron: Asymm.* **1999**, *10*, 347-354.

94 H. C. L. Abbenhuis, U. Burckhardt, V. Gramlich, A. Martelletti, J. Spencer, I. Steiner, A. Togni, *Organometallics* **1996**, *15*, 1614-1621; A. Schnyder, L. Hintermann, A. Togni, *Angew. Chem. Int. Ed. Engl.* **1995**, *34*, 931-933; A. Schnyder, A. Togni, U. Wiesli, *Organometallics* **1997**, *16*, 255-260.

95 K. Burgess, M. J. Ohlmeyer, K. H. Whitmire, *Organometallics* **1992**, *11*, 3588-3600.

96 K. Burgess, W. A. van der Donk, M. J. Ohlmeyer, *Tetrahedron: Asymm.* **1991**, *2*, 613-621.

97 K. Burgess, M. J. Ohlmeyer, *J. Org. Chem.* **1988**, *53*, 5178-5179.

98 A. C. Chen, L. Ren, C. M. Crudden, *Chem. Commun.* **1999**, 611-612.

99 A. Chen, L. Ren, C. M. Crudden, *J. Org. Chem.* **1999**, *64*, 9704-9710.

100 H. Nöth, *Z. Naturforsch.* **1984**, *39b*, 1463.

101 R. J. Brotherton, A. L. McCloskey, L. L. Petterson, H. Steinberg, *J. Am. Chem. Soc.* **1960**, *82*, 6242-6245. R. J. Brotherton, A. L. McCloskey, J. L. Boone, H. M. Manasevit, *Ibid.* **1960**, *82*, 6245-6248.

102 T. Ishiyama, M. Murata, T. Ahiko, N. Miyaura, *Organic Syntheses* **1999**, *77*, 176-181.

103 C. N. Welch, S. G. Shore, *Inorg. Chem.* **1968**, *7*, 225-230.

104 F. J. Lawlor, N. C. Norman, N. L. Pickett, E. G. Robins, P. Nguyen, G. Lesley, T. B. Marder, J. A. Ashmore, J. C. Green, *Inorg. Chem.* **1998**, *37*, 5282-5288.

105 H. Nakamura, M. Fujiwara, Y. Yamamoto, *J. Org. Chem.* **1998**, *63*, 7529-7530.

106 W. Clegg, T. R. F. Johann, T. B. Marder, N. C. Norman, A. G. Orpen, T. M. Peakman, M. J. Quayle, C. R. Rice, A. J. Scott, *J. Chem. Soc., Dalton Trans.* **1998**, 1431-1438.

107 (a) D. Seyferth, H. P. Kögler, *J. Inorg. Nucl. Chem.* **1960**, *15*, 99. (b) A. H. Cowley, H. H. Sisler, G. E. Ryschewitsch, *J. Am. Chem. Soc.* **1960**, *82*, 501-502. (c) H. Nöth, G. Höllerer, *Chem. Ber.* **1966**, *99*, 2197-2205. (d) W. Biffar, H. Nöth, R. Schwerthöffer, *Liebigs Ann. Chem.* **1981**, 2067-2080. (e) J. D. Buynak, B. Geng, *Organometallics* **1995**, *14*, 3112-3115.

108 K. Niedenzu, E. F. Rothgery, *Synth. Inorg. Met.-Org. Chem.* **1972**, *2*, 1.

109 M. Suginome, T. Matsuda, H. Nakamura, Y. Ito, *Tetrahedron* **1999**, *55*, 8787-8800.

110 The reaction of PhMe$_2$SiLi and HBpin (2 equiv.) or *i*-PrOBpin (2 equiv.) in THF at 0°C yields PhMe$_2$SiBpin in 74% yield. A same reaction of PhMe$_2$GeLi and *i*-PrOBpin (2 equiv.) gives PhMe$_2$GeBpin in 81% yield. See, The 76th Annual Meeting of Japan Chemical Society, Vol. 2, 1B634.

111 (a) T. Ishiyama, N. Matsuda, N. Miyaura, A. Suzuki, *J. Am. Chem. Soc.* **1993**,
115, 11018-11019. (b) T. Ishiyama, N.
Matsuda, M. Murata, F. Ozawa, A.
Suzuki, N. Miyaura, *Organometallics*
1996, *15*, 713-720.

112 (a) G. Lesley, P. Nguyen, N. J. Taylor,
T. B. Marder, A. J. Scott, W. Clegg, N.
C. Norman, *Organometallics* **1996**, *15*,
5137-5154. (b) P. Nguyen, G. Lesley, N.
J. Taylor, T. B. Marder, *Inorg. Chem.*
1994, *33*, 4623-4624. (c) C. Dai, G.
Stringer, T. B. Marder, R. T. Baker, A.
J. Scott, W. Clegg, N. C. Norman,
Can. J. Chem. **1996**, *74*, 2026-2031. (d)
T. B. Marder, N. C. Norman, C. R.
Rice, E. G. Robins, *J. Chem. Soc.,
Chem. Commun.* **1997**, 53-54. (e) C. Dai,
G. Stringer, T. B. Marder, A. J. Scott,
W. Clegg, N. C. Norman, *Inorg. Chem.*
1997, *36*, 272-273. (f) W. Clegg, F. J.
Lawlor, G. Lesley, T. B. Marder, N. C.
Norman, A. G. Orpen, M. J. Quayle,
C. R. Rice, A. J. Scott, F. E. S. Souza,
J. Organomet. Chem. **1999**, *550*, 183-192.

113 A. Maderna, H. Pritzkow, W. Siebert,
Angew. Chem. Int. Ed. Engl. **1996**, *35*,
1501-1502.

114 S. Onozawa, Y. Hatanaka, M. Tanaka,
J. Chem. Soc., Chem. Commun. **1997**,
1229-1230.

115 M. Suginome, H. Nakamura, Y. Ito, *J.
Chem. Soc., Chem. Commun.* **1996**,
2777-2778.

116 S. Onozawa, Y. Hatanaka, T. Sakaku-
ra, S. Shimada, M. Tanaka,
Organometallics **1996**, *15*, 5450-5452.

117 (a) C. N. Iverson, M. R. Smith III, *J.
Am. Chem. Soc.* **1995**, *117*, 4403-4406;
(b) C. N. Iverson, M. R. Smith III,
Organometallics **1996**, *15*, 5155-5165.

118 Q. Cui, D. G. Musaev, K. Morokuma,
Organometallics **1997**, *16*, 1355-1364.

119 Q. Cui, D. G. Musaev, K. Morokuma,
Organometallics **1998**, *17*, 742-751.

120 S. Sakaki, T. Kikuno, *Inorg. Chem.*
1997, *36*, 226-229.

121 S. Sakaki, S. Kai, M. Sugimoto,
Organometallics **1999**, *18*, 4825-4837.

122 B. Biswas, M. Sugimoto, S. Sakaki,
Organometallics **1999**, *18*, 4015-4026.

123 M. Suginome, T. Matsuda, Y. Ito,
Organometallics **1998**, *17*, 5233-5235.

124 S. Onozawa, Y. Hatanaka, N. Choi,
M. Tanaka, *Organometallics* **1997**, *16*,
5389-5391.

125 T. Ishiyama, M. Yamamoto, N. Miyaura, *Chem. Lett.* **1996**, 1117-1118.

126 The conjugate 1,4-addition of 1-alkenyl-
and arylboronic acids to enones, see:
Sakai, M.; Hayashi, H.; Miyaura, N.
Organometallics **1997**, *16*, 4229-4231;
Takaya, Y.; Ogasawara, M.; Hayashi,
T.; Sakai, M.; Miyaura, N. *J. Am.
Chem. Soc.*, **1998**, *120*, 5579-5580.

127 (a) S. D. Brown, R. W. Armstrong, *J.
Am. Chem. Soc.* **1996**, *118*, 6331-6332;
(b) S. D. Brown, R. W. Armstrong, *J.
Org. Chem.* **1997**, *62*, 7076-7077.

128 (a) T. Ishiyama, M. Yamamoto, N.
Miyaura, *Chem. Commun.* **1997**, 689-
690; (b) T. B. Marder, N. C. Normant,
C. R. Rice, *Tetrahedron Lett.* **1998**, *39*,
155-158.

129 C. N. Iverson, M. R. Smith III,
Organometallics **1997**, *16*, 2757-2759.

130 R. T. Baker, P. Nguyen, T. B. Marder,
S. A. Westcott, *Angew. Chem. Int. Ed.
Engl.* **1995**, *34*, 1336-1338.

131 C. Dai, E. G. Robins, A. J. Scott, W.
Clegg, D. S. Yufit, J. A. K. Howard,
T. B. Marder, *Chem. Commun.* **1998**,
1983-1984.

132 M. Suginome, H. Nakamura, Y. Ito,
Angew. Chem. Int. Ed. Engl. **1997**, *36*,
2516-2518.

133 Y. G. Lawson, M. J. G. Lesley, T. B.
Marder, N. C. Norman, C. R. Rice,
Chem. Commun. **1997**, 2051-2052.

134 H. Shima, T. Ishiyama, N. Miyaura,
unpublished results.

135 T. Ishiyama, S. Momota, and N.
Miyaura, *Synlett* **1999**, 1790-1791.

136 T. Ishiyama, M. Yamamoto, N. Miyaura, *Chem. Commun.* **1996**, 2073-2074.

137 M. Suginome, T. Matsuda, T. Yoshi-
moto, Y. Ito, *Org. Lett.* **1999**, *1*, 1567-
1569.

138 S. Onozawa, Y. Hatanaka, M. Tanaka,
Tetrahedron Lett. **1998**, *39*, 9043-9046.

139 M. Suginome, H. Nakamura, T. Matsuda, Y. Ito, *J. Am. Chem. Soc.* **1998**,
120, 4248-4249.

140 Y. Sato, M. Takimoto, K. Hayashi, T.
Katsuhara, K. Takagi, M. Mori, *J.
Am. Chem. Soc.* **1994**, *116*, 9771-9772.

141 T. Ishiyama, T. Kitano, N. Miyaura, *Tetrahedron Lett.* **1998**, *39*, 2357-2360.

142 S. Onozawa, Y. Hatanaka, M. Tanaka, *J. Chem. Soc., Chem. Commun.* **1999**, 1863-1864.

143 M. Suginome, Y. Ohmori, Y. Ito, *Synlett* **1999**, 1567-1568.

144 K. A. Horn, *Chem. Rev.* **1995**, *95*, 1317-1350; M. Murata, K. Suzuki, S. Watanabe, Y. Masuda, *J. Org. Chem.* **1997**, *62*, 8569-8571.

145 M. Kosugi, K. Shimazu, A. Ohtani, T. Migita, *Chem. Lett.* **1981**, 829-830; T. N. Mitchell, *Synthesis* **1992**, 803-815.

146 T. Ishiyama, M. Murata, N. Miyaura, *J. Org. Chem.* **1995**, *60*, 7508-7510.

147 T. Ishiyama, Y. Itoh, T. Kitano, N. Miyaura, *Tetrahedron. Lett.* **1997**, *38*, 3447-3450.

148 (a) M. Murata, S. Watanabe, Y. Masuda, *J. Org. Chem.* **1997**, *62*, 6458-6459. (b) M. Murata, T. Oyama, S. Watanabe, Y. Masuda, *J. Org. Chem.* **2000**, *65*, 164-168.

149 T. Ishiyama, K. Ishida, N. Miyaura, unpublished results.

150 C. Amatore, A. Jutand, M. A. M'Barki, *Organometallics* **1995**, *14*, 1818-1826.

151 (a) K-C. Kong, C-H. Cheng, *J. Am. Chem. Soc.* **1991**, *113*, 6313-6315; (b) D. K. Morita, J. K. Stille, J. R. Norton, *J. Am. Chem. Soc.* **1995**, *117*, 8576-8581; (c) reference 9, pp. 210-216.

152 (a) C. Malan, C. Morin, *J. Org. Chem.* **1998**, *63*, 8019-8020. (b) Other methods for the deprotection of a pinacol moiety, see: $HN(CH_2CH_2OH)_2$, M. E. Jung, T. I. Lazarova, *J. Org. Chem.* **1999**, *64*, 2976-2977; $NaIO_4/NH_4OAc$, S. J. Coutts, J. Adams, D. Krolikowski, R. Snow, *Tetrahedron Lett.* **1994**, *37*, 5109-5112.

153 (a) F. Firooznia, C. Gude, K. Chan, N. Marcopulos, Y. Satoh, *Tetrahedron Lett.* **1999**, *40*, 213-216; (b) M. E. Jung, T. I. Lazarova, *J. Org. Chem.* **1999**, *64*, 2976-2977.

154 (a) F. Gosselin, J. V. Betsbrugge, M. Hatatam, W. Lubell, *J. Org. Chem.* **1999**, *64*, 2486-2493. (b) D. E. Zembower, H. Zhang, *J. Org. Chem.* **1998**, *63*, 9300-9305.

155 (a) S. R. Piettre, S. Baltzer, *Tetrahedron Lett.* **1997**, *38*, 1197-1200. (b) A. Giroux, Y. Han, P. Prasit, *Tetrahedron Lett.* **1997**, *38*, 3841-3844.

156 P. A. Tempest, R. W. Armstrong, *J. Am. Chem. Soc.* **1997**, *119*, 7607-7608.

157 K. Takahashi, J. Takagi, T. Ishiyama, N. Miyaura, *Chem. Lett.* **2000**, 126-127.

158 (a) S. M. Marcuccio, M. Rodopoulos, H. Weigold, 10th International Conference on Boron Chemistry, Durham, England, July **1999**, PB-35; (b) M. Murata, T. Oyama, S. Watanabe, and Y. Masuda, 76th Annual Meeting of Chemical Society of Japan, March **1999**, 2B712.

159 T. Ishiyama, T.-a. Ahiko, N. Miyaura, *Tetrahedron Lett.* **1996**, *38*, 6889-6892.

160 S. Momota, T.-a. Ahiko, N. Miyaura, unpublished results.

161 T.-a. Ahiko, T. Ishiyama, N. Miyaura, *Chem. Lett.* **1997**, 811-812.

162 Z. Oohashi, T.-a. Ahiko, T. Ishiyama, N. Miyaura, unpublished results.

2
Metal-Catalyzed Hydroalumination Reactions

Marc Dahlmann, Mark Lautens

2.1
Introduction

The addition of aluminum hydrides to non-polar carbon-carbon multiple bonds is a valuable route to organoaluminum compounds (hydroalumination) [1–3]. During the course of the reaction a new Al–C and C–H bond is formed at the expense of an Al–H and a C–C π-bond, making this process favorable from a thermodynamic point of view. In fact, many hydroalumination reactions occur at elevated temperature without the necessity of a catalyst. However, the use of specific catalyst systems leads to a pronounced enhancement of the reaction rate and allows the catalyzed hydroalumination reaction to proceed under much milder conditions than those necessary for the un-catalyzed processes [4, 5]. In some cases, hydroalumination only occurs in the presence of a catalyst. In addition, hydroaluminations of unsymmetrically substituted alkenes and alkynes can exhibit an altered regioselectivity if a transition metal catalyst is present. Tuning of the steric and electronic properties of the transition metal catalyst by choosing appropriate ligands can also have a direct influence on the regio- and stereochemistry of hydroalumination. The use of optically active ligands opens the way for a direct control of the configuration of newly formed stereogenic centers.

In contrast to the related organoboranes, which are mostly used in the addition to non-polar carbon-carbon multiple bonds, aluminum hydrides have found their widest use in organic synthesis in the addition reaction to polar carbon-carbon and carbon-heteroatom multiple bonds including carbonyl, nitrile and imino groups as well as their α,β-unsaturated analogs. Although these reduction reactions are also sometimes referred as hydroalumination reactions in the literature, they are outside the scope of this review.

2.2
Organoaluminum Compounds as Intermediates in Organic Synthesis

Because of the polarity of the carbon-aluminum bond, organoaluminum compounds are valuable intermediates in organic synthesis, which exhibit higher reactivity than the organometallic products of more common hydrometallation reactions, especially organoboron, tin and silicon compounds. The Al–C bond can be cleaved by a number of electrophilic reagents [1, 6]. For example, protonolysis leads to the hydrocarbons [7a], deuterolysis to the respective deuterated analogs. Halogenation with Br_2 or $I2$ yields halogenated compounds [7b], oxidation with O_2 gives alcohols [7c], and reaction with SO_2 and SO_3 leads to sulfenic and sulfonic acids, respectively [7d]. Carbon-carbon bond formation can be achieved by carbonation [7e] or alkylation with reactive alkyl halides [7f], epoxides [7g] and carbonyl compounds [7h]. The reactivity profile of the organoaluminum compounds can further be altered by transmetallation reactions to various main group and transition metals [7i]. In general, Al–C bonds on sp^2 centers are much more reactive in these reactions than those on sp^3 carbons.

2.3
Survey of Catalyst Systems and Catalysis Mechanism

2.3.1
General Considerations

Three general classes of aluminum hydride sources have found application for the hydroalumination of unsaturated carbon-carbon bonds: (1) complex aluminates, $MAlR_{4-n}H_n$, (2) neutral hydridoalanes, $R_{3-n}AlH_n$, and (3) neutral alanes, which do not contain an Al–H bond but instead bear an alkyl group with one or more β-hydrogens. The most commonly used representatives of each class are $LiAlH_4$, $^iBu_2AlH$ and $^iBu_3Al$, which are all commercially available. Transition metal catalyst systems exist for each aluminum hydride source, making the choice of a specific aluminum reagent/catalyst system largely dependent on the compatibility of the functional groups present in the substrate with the aluminum compound. Special applications have been developed for some reagent systems.

In order to get reliable information about the yield of organoaluminum compounds formed in a transition metal-catalyzed hydroalumination reaction it is essential to derivatize the organometallic products by quenching the reaction mixture with electrophiles like D_2O, O_2 or halogens. It is often observed that hydrogenation

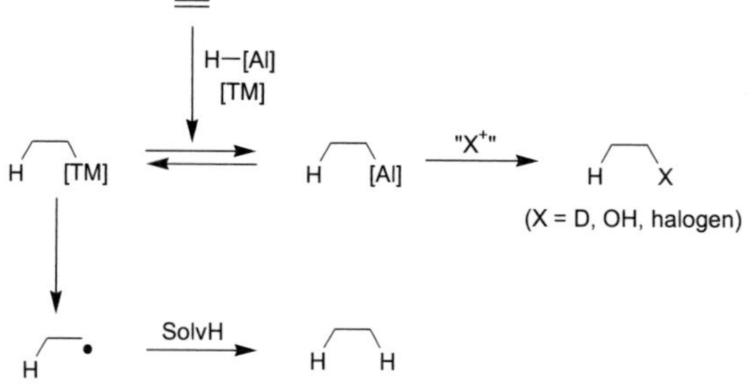

Scheme 2-1

of the carbon-carbon π-bond is a significant side reaction. The hydrogenated by-products are formed by homolysis of the aluminum-carbon or transition metal(TM)-carbon bond and subsequent abstraction of a hydrogen radical from a solvent molecule (Scheme 2-1) [8, 9]. Stabilization of the transition metal by a suitable ligand leads to a more stable metal-carbon bond and thus to higher yields of organoaluminum compounds [8]. Since organoalanes are stable towards homolysis of the Al–C bond at the low temperatures usually required, cleavage of this bond is likely to be catalyzed by the transition metal complex.

2.3.2
Uranium Catalysts

The reduction of terminal and internal *cis*-alkenes with $LiAlH_4$ in the presence of catalytic amounts of UCl_3 or UCl4 has been reported by Folcher [10, 11]. Linear 1-alkenes are reduced to the respective alkanes in quantitative yields by reacting with $LiAlH_4$ and 1 mol% of UCl_4 and subsequent hydrolysis. Workup of the reaction mixtures with D_2O, I_2 or formaldehyde indicated the formation of aluminum alkyls as the primary products. The catalytically active species in these reactions was spectroscopically identified as a U(III) compound, which is generated *in situ* by reduction of UCl_4 by $LiAlH_4$. Further investigations imply that a uranium aluminohydride such as $U(AlH_4)_3$ is formed. Although detailed mechanistic studies were not performed it is believed that coordination of the alkene to the uranium center in the bimetallic active species facilitates the insertion into the aluminum hydride bond.

2.3.3
Titanium and Zirconium Catalysts

Early attempts by Asinger to enlarge the scope of hydroalumination by the use of transition metal catalysts included the conversion of mixtures of isomeric linear alkenes into linear alcohols by hydroalumination with $^iBu_3Al$ or $^iBu_2AlH$ at temperatures as high as 110°C and subsequent oxidation of the formed organoaluminum compounds [12]. Simple transition metal salts were used as catalysts, including titanium(IV) and zirconium(IV) chlorides and oxochlorides. The role of the transition metal in these reactions is likely limited to the isomerization of internal alkenes to terminal ones since no catalyst is required for the hydroalumination of a terminal alkene under these reaction conditions.

In 1976, Sato reported the hydroalumination of terminal alkenes with $LiAlH_4$ in the presence of $ZrCl_4$ [13]. For example, 1-hexene was quantitatively converted into n-hexane at room temperature after hydrolytic workup, whereas no reaction occurred in the absence of a catalyst. Halogenation of the reaction mixtures indicated that these reactions in fact proceed through organoaluminum intermediates. Later, $TiCl_4$ was found to be an even more active catalyst [14, 15].

The hydroalumination of terminal alkenes by $LiAlH_4$ catalyzed by titanocene and zirconocene dichloride, Cp_2TiCl_2 and Cp_2ZrCl_2, respectively, has been reported by Isagawa [16] and Sato [14]. Again, the titanium compound proved to be more active

$$Al + H_2 + 2 R_2NH \xrightarrow[\text{benzene}]{\Delta,\ 2000\ psi} HAl(NR_2)_2 + 3/2\ H_2$$

Scheme 2-2

than its zirconium counterpart. A trimetallic complex $Cp_2Ti(AlH_3)_2$ was suggested to be the catalytically active species [17].

Group 4 metallocene dichlorides also catalyze the hydroalumination of olefins with neutral alanes as aluminum hydride sources. Ashby studied the reactivity of various alanes of the general formula R_2AlH towards the hydroalumination of alkenes using Cp_2TiCl_2 as a catalyst [18, 19]. While dialkoxyalanes (R = O^{t}Bu, O^{i}Pr, OMe) exhibited poor reactivity in the hydroalumination of 1-octene, dichloroalane (R = Cl) proved to be much more reactive. The highest yields were obtained using diamidoalanes, with reactivities increasing in the order R = NEt$_2$ ≈ NiPr$_2$ < N(SiMe$_3$)$_2$. The latter alanes can be manufactured by reaction of the corresponding amines with hydrogen and aluminum (Scheme 2-2) [20].

In the course of studies on the zirconocene dichloride-catalyzed carboalumination of alkenes and alkynes, Negishi and coworkers noticed that the desired reaction was accompanied by significant amounts of hydroalumination if a trialkylalane R_3Al was used in which the R groups contain sp^3 bound β-hydrogen atoms [21]. The latter reaction occured exclusively when iBu3Al was used as the aluminum source. Thus, reaction of 1-octene with 2 equiv. of ^{i}Bu$_3$Al and 10 mol% of Cp_2ZrCl_2 in 1,2-dichloroethane for 6 h at 0°C gave, after protonolysis of the reaction mixture, octane in quantitative yield. The authors note that no reaction occurs under identical conditions if Cp_2ZrCl_2 is omitted. The use of ^{i}Bu$_2$AlH as the reducing agent also did not yield any detectable amount of the alkane. It was therefore concluded that ^{i}Bu$_2$AlH is not an intermediate in the catalytic hydroalumination reaction. ^{1}H NMR spectroscopic analysis of the reaction mixture, however, showed the intermediacy of $Cp_2Zr(Cl)$n-octyl. A mechanistic proposal was derived in which Cp_2ZrCl_2 is alkylated by ^{i}Bu$_3$Al, followed by β–hydrogen elimination to form the zirconocene hydridochloride $Cp_2Zr(H)Cl$ (Scheme 2-3), which in turn is known to be able to hydrozirconate alkenes and alkynes [22]. Subsequent transmetallation releases the organoalane and regenerates Cp_2ZrCl_2.

$TiCl_4$ and $ZrCl_4$ were found to be less effective catalysts than Cp_2ZrCl_2, and Cp_2TiCl_2 shows no catalytic activity under similar conditions [23]. However, the hy-

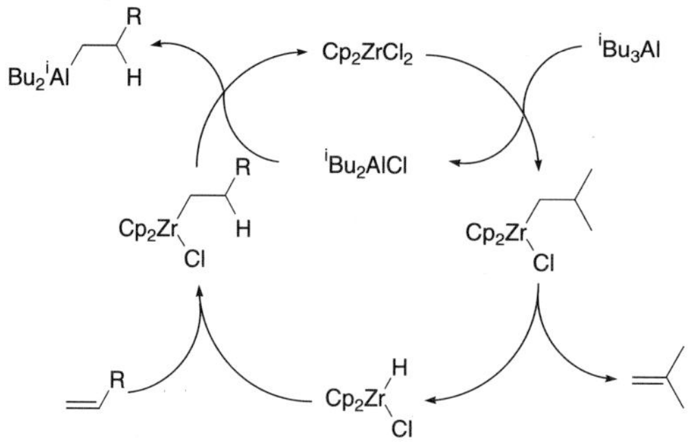

Scheme 2-3

droalumination of terminal alkenes with Et$_3$Al was reported to be efficiently catalyzed by Cp$_2$TiCl$_2$ at 30–35°C when the reaction is carried out in a hydrocarbon medium like hexane, cyclohexane, benzene or toluene [24].

2.3.4
Nickel Catalysts

In 1954, Ziegler and coworkers observed that the course of the reaction of ethene with trialkylalanes was drastically altered by the presence of traces of nickel salts [25]. Instead of low molecular weight polyethylene, the only product was 1-butene. Obviously, the transition metal strongly supports the displacement reaction of the alkyl group bonded to the aluminum by ethylene, a reaction which can be formally described as transfer of a hydridoalane.

In 1968, Eisch and Foxton showed that nickel(II) salts enhance the rate of ^{i}Bu$_2$AlH addition to internal alkynes by a factor of ca. 1000 compared to the process in the absence of a catalyst [26]. Similar catalytic activity of nickel compounds was found for the addition of aluminum hydrides to alkenes.

The mechanism of catalysis by nickel compounds has been a matter of debate for many years and is to date still not clearly understood. While Ziegler originally ascribed the catalytic activity to colloidal dispersed elemental nickel [27], studies by Giacomelli and Pino led to the conclusion that the active species is a nickel(II) complex which serves to activate the alkene for the Al–H addition [28]. Eisch and Foxton suggested that nickel hydride species like R–Ni–H or R$_2$Al–Ni–H play a key role in the catalytic reaction [26]. Based on extensive investigations into the chemistry of well-defined nickel(0) complexes, Wilke and coworkers suggested that the catalytically active species contains nickel in an oxidation state of 0 [29, 30]. Furthermore, they could show that nickel(II) salts are rapidly reduced to elemental nickel or nickel(0) complexes by aluminum alkyls, thereby disproving the mechanistic proposal by Pino. Wilke precluded the participation of nickel(II) species, claiming that compounds like H–Ni–X will rapidly undergo reductive elimination to form nickel(0). Studies on the reactions of tris(alkene)nickel(0) complexes with aluminum alkyls led to the following observations:

(i)   Tris(ethylene)nickel seems to form an adduct with Et$_3$Al which leads to a significant stabilization of the nickel complex. Moreover, if perdeuterated ethylene is used as a ligand for nickel, its exchange with the aluminum alkyl groups can be observed by the formation of deuterated Et$_3$Al.

(ii) While norbornene does not undergo any reaction with ^{n}Bu$_3$Al, a catalytic amount of tris(norbornene)nickel leads to a rapid hydroalumination of the C=C bond. The aluminum is added to the exo-face of the norbornene, which is also occupied by the nickel in tris(norbornene)nickel as confirmed by X-ray crystallography.

Based on these observations the authors propose the following mechanism for the nickel-catalyzed hydroalumination (Scheme 2-4): During the catalysis process

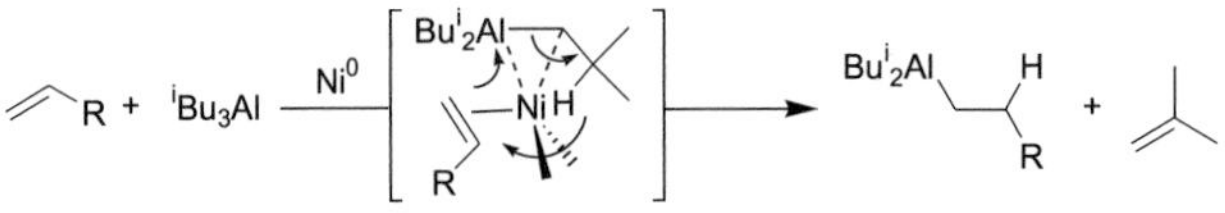

Scheme 2-4

nickel rests in an oxidation state of 0. If nickel(II) salts are used for the reaction they are reduced to nickel(0) by the trialkylaluminum compound.

Nickel(0) reacts with the olefin to form a nickel(0)-olefin complex, which can also coordinate the alkyl aluminum compound *via* a multicenter bond between the nickel, the aluminum and the α carbon atom of the trialkylaluminum. In a concerted reaction the aluminum and the hydride are transferred to the olefin. In this mechanistic hypothesis the nickel thus mostly serves as a template to bring the olefin and the aluminum compound into close proximity. No free Al–H or Ni–H species is ever formed in the course of the reaction. The adduct of an amine-stabilized dimethylaluminum hydride and (cyclododecatriene)nickel, whose structure was determined by X-ray crystallography, was considered to serve as a model for this type of mechanism since it shows the hydride bridging the aluminum and alkene-coordinated nickel center [31].

Eisch proposed a modified version of this mechanistic interpretation [32]. In a series of experiments it was found that the regiochemistry of Al–H addition to unsymmetrical alkenes and alkynes differs significantly according to whether the hydroalumination is carried out in the absence or in the presence of a nickel catalyst (Scheme 2-5). Thus, the reaction of 1,1-dimethylindene (1a) with $^iBu_2AlH$ gave exclusively 2, the product of an aluminum addition in the position distal to the methyl substituents, as confirmed by deuterolysis of the intermediate organoaluminum intermediate. In contrast, in the presence of $Ni(acac)_2$ as a catalyst precursor, hydroalumination of 1,1-dimethylindene-2-d_1 (1b) yielded a product mixture consisting of 65% of trans-1,1-dimethylindane-2,3-d_2 (3), 26% of 1,1-dimethylindane-2,2-d_2 (4) and 9% 1,1-dimethylindane-2-d_1 (5).

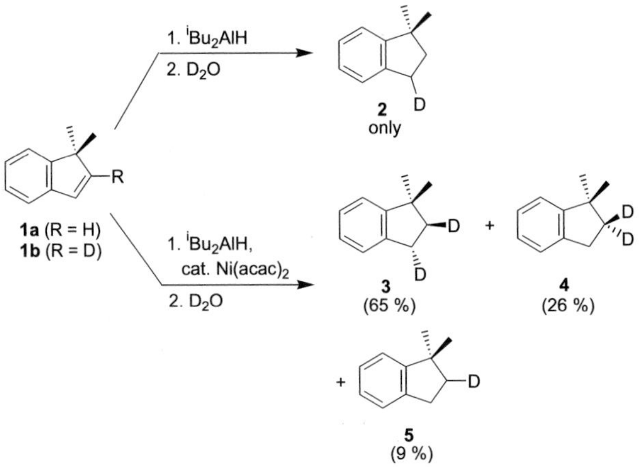

Scheme 2-5

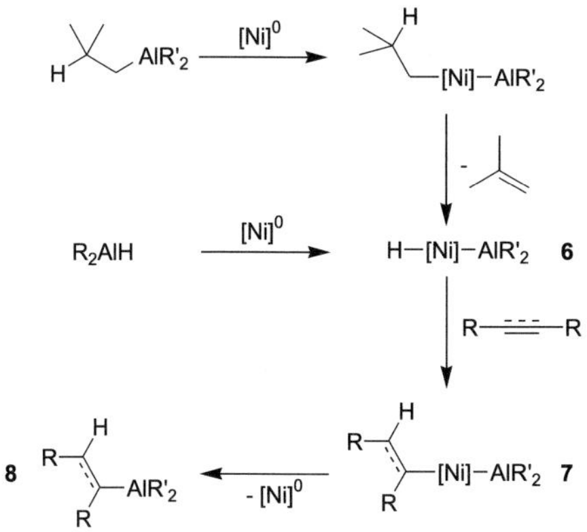

Scheme 2-6

These results confirm that the nickel-catalyzed hydroalumination occurs *via* a *syn* addition of Al–H to the carbon-carbon double bond in analogy to the thermal process [1]. However, in the presence of nickel both **3** and **4** are produced. Similar results were obtained in the hydroalumination of 1-phenylpropyne. These observations led the authors to the conclusion that the active hydrometallating agent in the nickel-catalyzed reactions must possess polar and steric requirements very different from those exhibited by di- or trialkyl alanes. Based on the analogy between nickel-catalyzed hydroalumination and hydrogen transfer of dihydroaromatic compounds, a mechanism was suggested in which an aluminum nickel hydride complex **6** is the key intermediate. Hydride **6** can either be generated by the oxidative addition of a dialkyl aluminum hydride at the nickel(0) catalyst (path a in Scheme 2-6) or by oxidative addition of a trialkyl aluminum reagent with subsequent β-hydride elimination (path b). Coordination of the alkene and subsequent insertion into the Ni–H bond leads to the alkyl nickel complex **7**, which can undergo reductive elimination to release the organoaluminum compound **8** and regenerate the nickel(0) catalyst species.

The formation of $R_2Al–Ni–H$ as the active hydrometallating agent may account for the lower degree of regioselectivity in the catalyzed hydroalumination compared to the thermal process. The lower polarity of the Ni–H bond compared to the Al–H bond as well as the lower steric congestion of the aluminum-nickel fragment compared to the bulky dialkylaluminum moiety may explain the regiochemical outcome. Increasing the steric bulk of the nickel center by coordination of PPh_3 results in a drastic increase of the regioselectivity [33].

More recent studies by Eisch and Wilke support this mechanistic proposal [34]. The interaction of nickel(0) complexes with $^iBu_3Al$ and $^iBu_2AlH$ was studied by NMR spectroscopy as was the reactivity of the resulting complexes. The most direct evidence for an oxidative insertion of Ni(0) into an Al–H bond was provided by the ^1H NMR spectrum of the product of the reaction of $^iBu_2AlH$ with $Ni(PEt_3)_4$, which exhibits a signal at –12.33 ppm, i.e. in a region typical for Ni–H signals. Less unequivocal ^1H and 13C NMR spectroscopic evidence was found for the insertion of Ni(0) into the Al–C bond of $^iBu_3Al$, resulting in only small shifts of the resonances of the ^iBu groups.

2.3.5
Other Transition Metal Catalysts

Extensive studies of the application of first row transition metal chlorides as catalysts for the hydroalumination of carbon-carbon multiple bonds using different aluminum hydride sources were carried out by Ashby and coworkers [35–37]. Reduction of terminal alkenes with LiAlH$_4$ was found to be efficiently catalyzed by CoCl$_2$
and NiCl$_2$ to yield the respective alkanes in high yields upon aqueous workup [35,
36]. However, deuterolytic workup led to only <50% deuterium incorporation, indicating that the organoaluminum intermediates undergo C–Al to C–H reactions.
Similar results were obtained using alkynes as substrates. Reduction of 1-octene
and cis-2-hexene by (^{i}Pr$_2$N)$_2$AlH is also efficiently catalyzed by CoCl$_2$ and NiCl$_2$,
Ni(acac)$_2$ as well as various nickel phosphine complexes, but again poor deuterium
incorporation (<10%) was observed upon D$_2$O workup [37].

Hydrogen transfer hydroalumination of terminal alkenes using ^{i}Bu$_3$Al was
achieved using various late transition metal catalysts containing cobalt, rhodium,
nickel, palladium, and platinum, as recently published by Negishi and coworkers
[38]. Among the different complexes tested, (PPh$_3$)$_2$PdCl$_2$ proved to be the most efficient catalyst (see Table 2-1). Significantly lower yields of the hydroalumination
product were obtained using ^{i}Bu$_3$Al in the presence of preformed Pd(0) complexes
such as (PPh$_3$)$_4$Pd as catalysts or using the (PPh$_3$)$_2$PdCl$_2$ catalyst system and aluminum hydride sources such as ^{i}Bu$_2$AlH or LiAlH$_4$, which are capable of reducing
Pd(II) complexes to Pd(0) complexes. These findings indicate that the active species
contains Pd(II) rather than Pd(0). The presence of chlorine ligands seems to be crucial for the catalytic activity, as shown by the failure of Pd(OAc)$_2$ to act as a catalyst.
The authors speculate that bimetallic activation involving Al–Cl–Pd bonds plays an
important role in the catalytic cycle.

(η^3-Cyclooctenyl)(cyclooctadiene)cobalt was also reported to catalyze the hydroalumination of terminal alkenes by ^{i}Bu$_3$Al [30]. It should be noted that organoboranes
catalyze the addition of Cl$_2$AlH to alkenes [39–41] and allenes [42, 43].

Catalyst	Yield 1-iododecane
(PPh$_3$)$_2$PdCl$_2$	90%
Li$_2$PdCl$_4$	78–86%
K$_2$PtCl$_4$	86%
(PPh$_3$)$_2$NiCl$_2$	65%
(PPh$_3$)$_3$CoCl	76%
(PPh$_3$)$_3$RhCl	79%
(PPh$_3$)$_4$Pd	25%
Pd(OAc)$_2$	no reaction

a) Conditions: 1.1 equivalents of
^{i}Bu$_3$Al, 2.5-5 mol-% catalyst,
CH$_2$Cl$_2$, 25°C

Tab. 2-1 Hydroalumination-iodinolysis of 1-decene
with ^{i}Bus$_3$Al in the presence of late transition metal
compexes[a]

2.4
Hydroalumination of Functional Groups

2.4.1
Alkenes

The reactivity of carbon-carbon double bonds towards hydroalumination catalyzed by transition metal compounds largely depends on steric factors. In general, terminal monosubstituted bonds tend to be the most reactive and thus can be hydroaluminated under very mild conditions (0°C to room temperature). Substituents in the 2-position of a terminal alkene significantly diminish the rate of hydroalumination. Internal disubstituted double bonds also require higher reaction temperatures and prolonged reaction times. *Cis*-alkenes react faster than their *trans*-substituted counterparts. If the 1,2-disubstituted alkene is part of a strained ring system, as in norbornene or norbornadiene, milder conditions can be applied. Trisubstituted double bonds usually give the respective organoaluminum intermediates only in moderate to poor yields; tetrasubstituted alkenes are essentially unreactive.

Using these differences in reactivity it is often possible to selectively hydroaluminate one of several double bonds present in a substrate. Some representative examples are given in Scheme 2-7 [14, 44, 45]. A high degree of selectivity in the final example is obtained by attaching a bulky $^iBu_2Al$ group to the alcohol functionality and thereby deactivating the terminal allylic C=C bond.

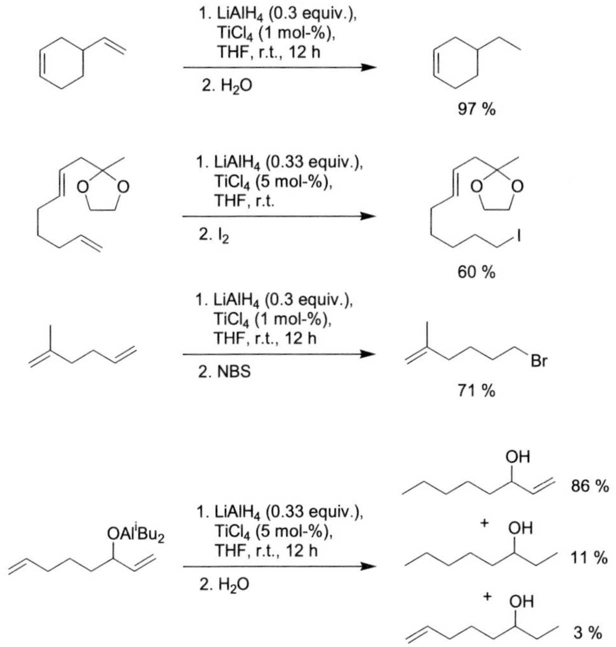

Scheme 2-7

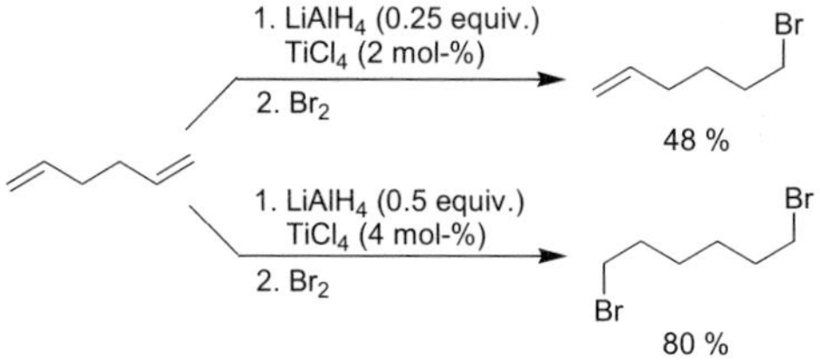

Scheme 2-8

LiAlH$_4$ is one of the most commonly used hydroalumination reagents for the re-action of terminal alkenes in the presence of TiCl$_4$, ZrCl$_4$, Cp$_2$TiCl$_2$ or Cp$_2$ZrCl$_2$ as catalysts (Scheme 2-8) [13–16]. The reactions proceed smoothly at room tempera-ture in etheral solvents. Titanium catalysts exhibit slightly higher activities than their zirconium counterparts. However, the latter were found to be better suited for allylic ethers as substrates [46]. Terminal alkenes with substituents in the 2-position react somewhat more slowly, internal alkenes require long reaction times to form, under isomerization, n-alkylaluminum intermediates [15]. α,ω-Dienes were report-ed to be selectively mono- or dihydroaluminated by reacting with $^1/_4$ and $^1/_2$ molar equiv. of LiAlH$_4$, respectively. The first transformation is of particular interest since cyclization occurs with other reagent systems [47].

The use of LiAlH$_4$ has several advantages over other hydroalumination reagents such as dialkylalanes or trialkylalanes. Because one mol of LiAlH$_4$ can react with up to four equivalents of the alkene, the yield of products per mol of aluminum is usu-ally high. In addition, the alkylalanate complexes formed contain no other organic ligand which might complicate further transformations and workup procedures, as is observed in some cases where the above-mentioned aluminum sources are used [1]. A number of transformations for the intermediate lithium alkylalanates have been developed, allowing simple, selective one-pot functionalizations at terminal C=C bonds (Scheme 2-9) [48–54].

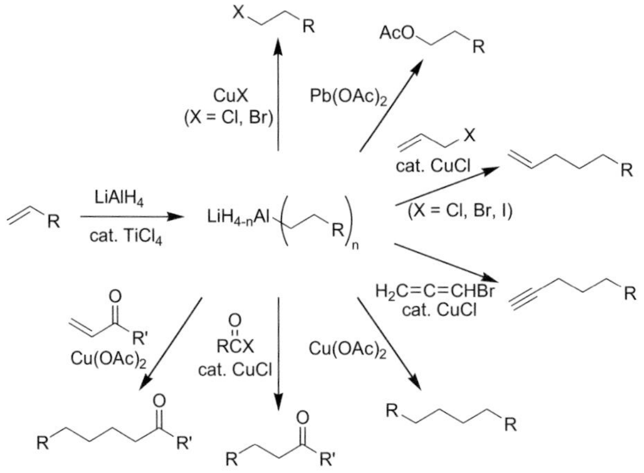

Scheme 2-9

Lithium tetraalkylalanates obtained by $TiCl_4$-catalyzed hydroalumination of terminal olefins with $LiAlH_4$ were stereospecifically added to the keto group of phenylglyoxylic acid (–)-menthyl ester yielding α-substituted mandelic acid esters in diasteriomeric excesses of 70% [55]. Cp_2TiCl_2-catalyzed hydroalumination of terminal C=C bonds with $LiAlH_4$ and subsequent reaction of the lithium tetraalkylalanates with optically active 1-benzyloxy-2,2,2-trifluoroethyl tosylate results in the nucleophilic displacement of the tosylate with inversion of configuration and 80–100% chirality transfer [56].

Other complex aluminum hydrides such as $NaAlH_4$, $LiAlMe_3H$, $NaAlMe_3H$, $NaAl(OCH_2CH_2OCH_3)_2H_2$ (vitride), $LiAlH_2(NR_2)_2$ and $NaAlH_2(NR_2)_2$ were also applied in a similar way as hydroaluminating reagents [57].

Bis(diamino)alanes $(R_2N)_2AlH$ were used for the hydroalumination of terminal and internal alkenes [18, 19]. $TiCl_4$ and Cp_2TiCl_2 are suitable catalysts for these reactions, whereas Cp_2ZrCl_2 exhibits low catalytic activity. The hydroaluminations are carried out in benzene or THF solution at elevated temperatures (60°C). Internal linear *cis-* and *trans*-alkenes are converted into *n*-alkylalanes *via* an isomerization process. Cycloalkenes give only moderate yields; tri- and tetrasubstituted double bonds are inert. Hydroalumination of conjugated dienes like butadiene and 1,3-hexadiene proceeds with only poor selectivity. The structure of the hydroalumination product of 1,5-hexadiene depends on the solvent used. While in benzene cyclization is observed, the reaction carried out in THF yields linear products (Scheme 2-10).

The addition of dialkylaluminum hydrides R_2AlH and trialkylalanes R_3Al, in which the group R bears a hydrogen in the β-position to the aluminum, to alkenes can be catalyzed by zirconium compounds such as $ZrCl_4$, chlorozirconium alkoxides and Cp_2ZrCl_2. $^iBu_2AlH$ reacts with terminal monosubstituted double bonds and norbornene at 20°C in the presence of $ZrCl_4$ [58]. Terminal alkenes with substituents in the 2-position require higher reaction temperatures (60°C). For cycloalkenes, chlorozirconium alkoxides, which can be generated by reaction of $ZrCl_4$ with an alcohol, tend to be the best catalysts [59]. The reaction rate depends on the ring size and decreases in the order: $C_5 > C_6 > C_7 > C_{12} > C_8$. $^iBu_2AlH$ adds to norbornene at temperatures as low as 0°C, while trisubstituted double bonds like that in 1-methylcyclohexene do not react.

$^iBu_3Al$ is capable of hydroaluminating terminal alkenes, internal alkenes and cycloalkenes at 0°C in the presence of 10 mol% of Cp_2ZrCl_2 in 1,2-dichloroethane [21]. Quenching the reaction mixture with D_2O, I_2 or O_2 revealed that formation of organoaluminum intermediates had occured in high yield. The reaction of cycloalkenes and internal alkenes requires prolonged reaction times, but no informa-

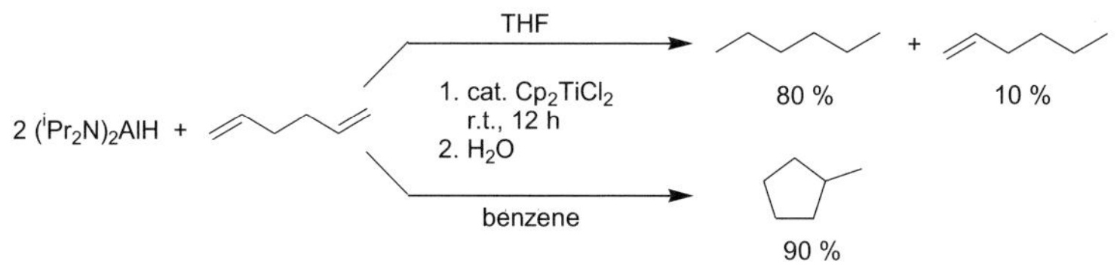

Scheme 2-10

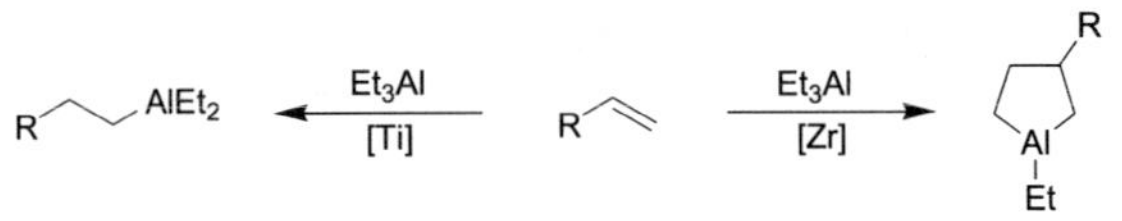

Scheme 2-11

tion concerning the isomerization to yield *n*-alkylaluminum intermediates has been published in the latter case. Under the mild reaction conditions, functional groups such as SPh, OH and Br are tolerated. For example, allyl phenyl sulfide was converted to a 97:3 mixture of 3- and 2-deutero-*n*-propyl phenyl sulfide in 77% combined yield *via* hydroalumination-deuterolysis, the amount of benzenethiol formed *via* cleavage of the C–S bond being only 12%. In contrast, application of the LiAlH$_4$/TiCl$_4$ reagent system led to extensive C–S cleavage (53%) and a low yield (17%) of the desired sulfide.

Hydroalumination of terminal alkenes using Et$_3$Al as the hydride source must be carried out with titanium catalysts [24], since zirconium compounds lead to the formation of alumacyclopentanes [60, 61] (Scheme 2-11) and carbometallated products [62]. Suitable substrates for hydroalumination include styrene, allylnaphthalene and vinylsilanes. Only one of the ethyl groups in Et$_3$Al takes part in these reactions, allowing the synthesis of diethylalkylalanes, which are difficult to obtain by other methods.

The hydroalumination of alkenes with ^{i}Bu$_2$AlCl catalyzed by Cp$_2$ZrCl$_2$ produces higher dialkylaluminum chlorides, which cannot be prepared by non-catalytic hydroalumination (Scheme 2-12) [63–65]. Terminal alkenes, internal linear alkenes and cycloalkenes can serve as substrates at reaction temperatures increasing in this order. 1,5-Dienes react to give cyclized products.

The hydroalumination of terminal alkenes and dienes using ^{i}Bu$_3$Al as aluminum source can also be catalyzed by late transition metal complexes such as (PPh$_3$)$_2$PdCl$_2$ [38]. The reactions are carried out at room temperature, the use of dichloromethane as a solvent is crucial because it probably regenerates the catalytically active Pd(II) species by reoxidizing Pd(0) complexes formed in side reactions. Internal C=C bonds are not hydroaluminated under these conditions. Functional groups such as Cl, Br, and PhS in a position remote to the terminal C=C bond are tolerated, however, certain substrates such as 4-bromo-1-butene, 11-iodo-1-undecene, allyl phenyl ether, allyl benzyl ether, and (3*E*)-1,3-decadiene did not provide the desired hydroa-

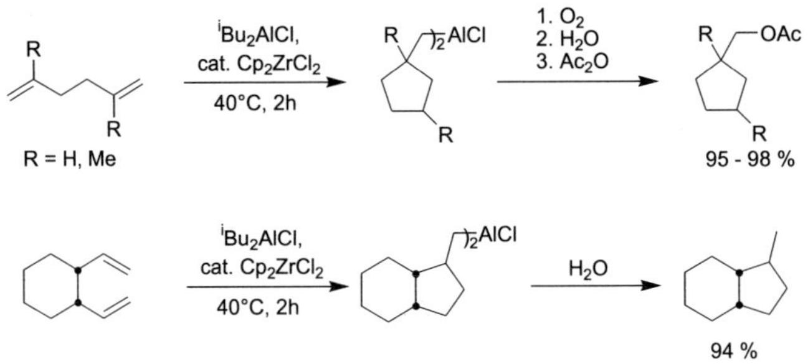

Scheme 2-12

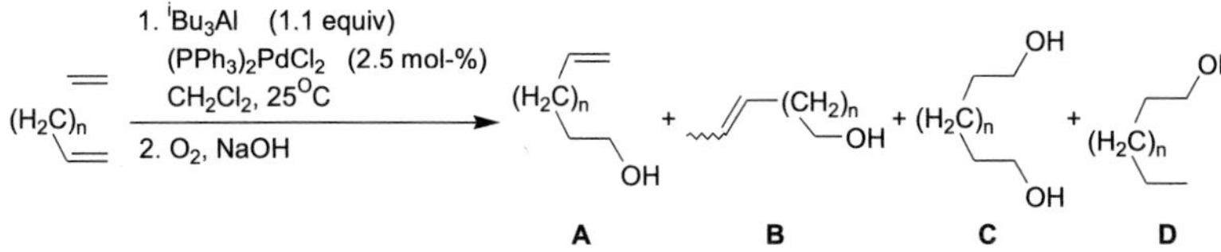

Reaction scheme:

1. ^{i}Bu$_3$Al (1.1 equiv)
   (PPh$_3$)$_2$PdCl$_2$ (2.5 mol-%)
   CH$_2$Cl$_2$, 25°C
2. O$_2$, NaOH

(H$_2$C)$_n$ → A + B + C + D

Entry	n of diene	Unreacted diene/%	Product yield/%			
			A	B	C	D
1	3	0	32	12	10	22
2	4	0	20	23	14	21
3	5	0	12	24	14	25
4	6	0	11	24	16	27
5	10	0	0	26	19	54

Tab. 2-2 Hydroalumination of α,ω-dienes with ^{i}Bu$_3$Al catalyzed by (PPh$_3$)$_2$PdCl$_2$

lumination product. While one equivalent of ^{i}Bu$_3$Al per terminal C=C bond is consumed in the case of mono-olefins, both C=C bonds in α,ω-dienes can undergo hydroalumination under the same reaction conditions. While 1,5-hexadiene undergoes a hydrometallation-cyclic carbometallation process to form a cyclopentyl-carbinylalane, longer α,ω-dienes yield a mixture of mono- and dihydroaluminated product **A** and **C**, respectively, along with double bond-migrated internal alkene **B** and saturated alcohol **D** (Table 2-2). The product ratio highly depends on the chain length of the α,ω-diene. No cyclic carbometallation is observed in these cases.

Nickel catalysts promote the hydroalumination of alkenes using trialkylalanes R$_3$Al and dialkylalanes such as ^{i}Bu$_2$AlH as the aluminum hydride sources [9, 29, 30, 33]. However, exhaustive studies of the range of substrates capable of hydroalumination with these reagents has not been carried out. Linear terminal alkenes like 1-octene react quantitatively with ^{i}Bu$_3$Al at 0°C within 1–2 h in the presence of catalytic amounts of Ni(COD)$_2$ [30]. Internal double bonds are inert under these conditions, whereas with 1,5-hexadiene cyclization occurs.

1,1-Dimethylindene derivatives were hydroaluminated with ^{i}Bu$_2$AlH using Ni(acac)$_2$ as the nickel(0) precursor (Scheme 2-13) [9]. Hydroalumination of the trisubstituted double bond in **9** was also achieved, although more forcing conditions

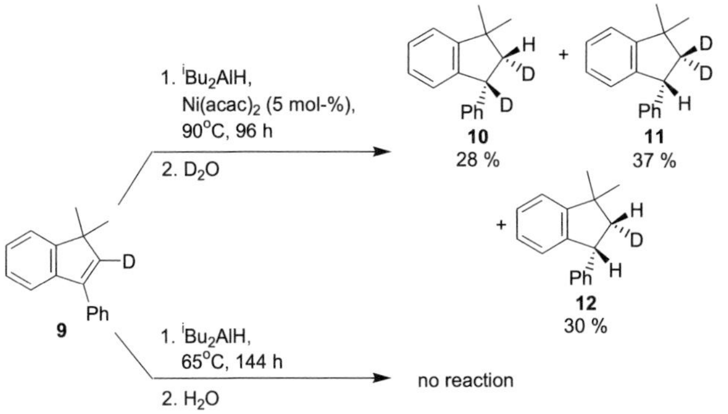

1. ^{i}Bu$_2$AlH, Ni(acac)$_2$ (5 mol-%), 90°C, 96 h
2. D$_2$O

10 28 % + **11** 37 % + **12** 30 %

1. ^{i}Bu$_2$AlH, 65°C, 144 h
2. H$_2$O

no reaction

Scheme 2-13

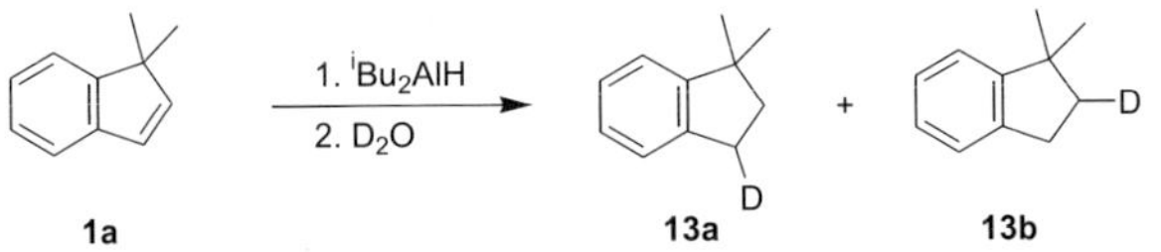

	1a	13a	:	13b
10 mol-% Ni(acac)$_2$		4	:	1
10 mol-% Ni(acac)$_2$ 10 mol-% PPh$_3$		99	:	1

Scheme 2-14

were required. However, deuterolytic quenching of the reaction mixture revealed the presence of **12** as a major by-product, suggesting that the C–Al bond underwent a hydrogen atom abstraction. In the absence of a nickel catalyst no reaction occurred.

An advantage of nickel catalysts over other metal systems is that the properties of the active species can easily be tuned by the addition of suitable ligands. For example, the presence of PPh_3 was shown to have a direct influence on the regiochemistry of hydroalumination of 1,1-dimethylindene **1a** [33]. While the reaction of $^{i}Bu_2AlH$ with 1a gave a 4:1 mixture of regioisomeric products **13a/13b** after deuterolytic workup, the same reaction carried out in the presence of PPh_3 yielded **13a** and **13b** in a ratio of >99:1 (Scheme 2-14).

Nickel-catalyzed hydroalumination has found some specific applications in organic synthesis (see below).

2.4.1.1 Applications in Organic Synthesis

Reductive ring-opening reactions A very useful application of a transition metal-catalyzed hydroalumination process was developed by Lautens and coworkers for the synthesis of cyclohexenols and cycloheptenols bearing multiple stereogenic centers. Rigid oxabicyclo[3.2.1] and -[2.2.1] templates, which can be synthesized and modified by a number of highly stereoselective reactions [66], undergo a formal S_N2' displacement of the bridging oxygen atom by a hydride equivalent (Scheme 2-15). Diisobutylaluminum hydride was found to be the best hydrometallating agent for this transformation [67, 68]. Thus, treatment of **14** with $^{i}Bu_2AlH$ in refluxing hexanes afforded the ring-opened product **16** in 50% yield.

The reaction may proceed *via* the hydroalumination of the olefin moiety to give **15**, which can undergo β-alkoxy elimination to yield, upon workup, the alcohol **16**. How-

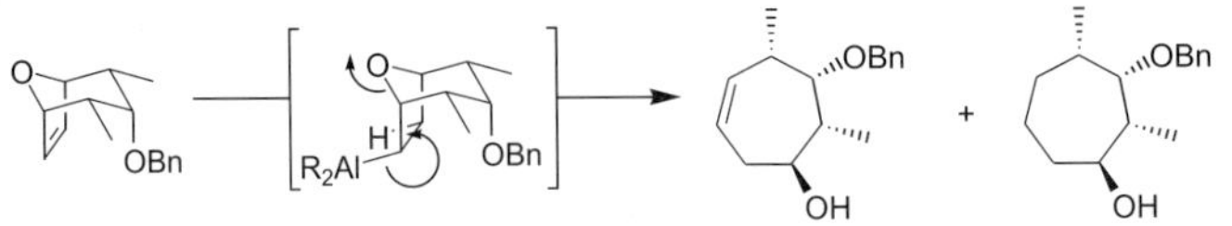

	14	15	16	17
7.0 equiv. $^{i}Bu_2AlH$, hexanes, reflux, 6.5 h:			50 %	27 %
1.1 equiv. $^{i}Bu_2AlH$, 10 mol-% Ni(COD)$_2$, toluene, r.t., 30 min., then $^{i}Bu_2AlCl$, reflux, 2 h:			76 %	<5 %

Scheme 2-15

Tab. 2-3 Reductive ring opening of unsymmetric oxabicyclic substrates[a]

Entry	Substrate	Products		Yields (ratio a : b)
1	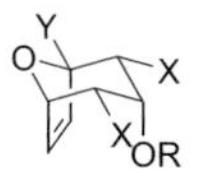 18, X=Y=Me, R=H[b]	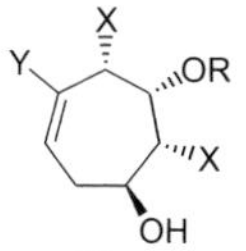 19a	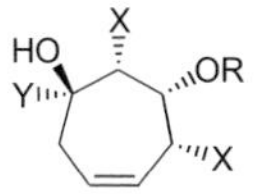 19b	75% (1:1.3)
2	18[c]	19a	19b	85% (9.5:1)
3	18[d]	19a	19b	75% (98:2)
4	18[e]	19a	19b	82% (300:1)
5	20, X=Y=R=Me	21a	21b	80% (380:1)
6	22, X=Y=Me, R=TBDMS	23a	23b	76% (>49:1)
7	24, X=R=H, Y=Me	25a	25b	72% (167:1)
8	26, X=Me, Y=Ph, R=H	27a	27b	93% (>49:1)
9	28, X=Me, Y=TMS, R=H	29a	29b	no reaction
10	30, X=Me, Y='Bu, R=H	31a	31b	no reaction
11	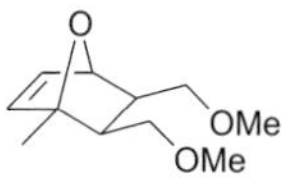 32	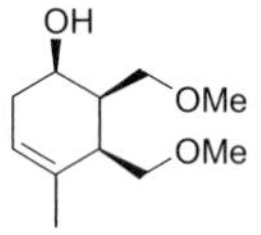 33a	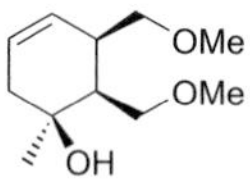 33b	78% (39:1)

a) Reaction conditions unless otherwise stated: 10 mol-% Ni(COD)$_2$,
   20 mol-% dppd, 1.2 – 3.0 equiv. od ^{i}Bu$_2$AlH
b) ^{i}Bu$_2$AlH
c) ^{i}Bu$_2$AlH + Ni(COD)$_2$
d) ^{i}Bu2AlH + Ni(COD)$_2$ + PPh$_3$ (4.0 equiv.)

ever, overreduction of **16** leading to **17** was observed to a significant extent. This side reaction could be suppressed by the use of Ni(COD)$_2$ as a catalyst [69]. The hydroalumination of **14** can be performed under much milder conditions and the subsequent alkoxy elimination step is facilitated by the addition of a Lewis acid (Scheme 2-15). Cycloheptenol **16** is obtained in 76% yield, along with traces of **17** (<5%).

The addition of phosphine ligands has a large influence on the regiochemistry in the ring-opening reaction of unsymmetrical oxabicyclic substrates bearing substituents at the bridgehead positions. While the uncatalyzed reaction of oxabicyclic alcohol **18** with ^{i}Bu$_2$AlH yielded a product mixture with a slight preference for the tertiary alcohol **19b**, in the presence of Ni(COD)$_2$ the primary alcohol **19a** was the main product of the reaction (see Table 2-3). The regioselectivity favouring **19a** was further increased by the addition of PPh$_3$ as a ligand for the catalytically active Ni species. The highest regioselectivity (**19a/19b** 250:1) and yield (82%) was achieved using bis(diphenylphosphino)butane (dppb). To date, dppb is the most effective ligand in regioselective ring opening of unsymmetrical oxabicyclic compounds.

The presence of a free hydroxyl group is not a requirement for high selectivity, as demonstrated by the reductive ring opening reaction of the methyl and TBDMS

(TBDMS = tert-butyldimethylsilyl) ether **20** and **22**, respectively. The reactivity of the oxabicyclic substrates is determined by the size of the bridgehead substituent: While the phenyl-substituted compound **26** undergoes a highly regioselective catalytic ring-opening reaction, the trimethylsilyl and *tert*-butyl-substituted oxabicycles **28** and **30**, repectively, did not react under the chosen reaction conditions. Oxabicyclic [2.2.1] substrates also undergo reductive ring opening to yield trisubstituted cyclohexenols in high yields and stereoselectivities.

Nickel-catalyzed allyl ether cleavage   A convenient method for the cleavage of allyl ethers was developed by Ogasawara [70]. Treatment of various allyl ethers with ^{i}Bu$_2$AlH or NaBH$_4$ in the presence of 1 mol% NiCl$_2$(dppp) [dppp = bis(diphenylphosphino)propane] in an aprotic solvent led to the clean deallylation and formation of the corresponding alcohols in high yields after aqueous workup. No deallylation occurred in the absence of the nickel catalyst. The reaction can be applied to aromatic, benzylic, primary, secondary, and tertiary alcohols. Ester and non-allylic ether functionalities are stable (see Table 2-4). The reaction proceeds presumably through a nickel-catalyzed hydroalumination of the allylic C=C bond yielding **36**, followed by β-alkoxy elimination and release of propene. The oxygen is probably responsible for directing the regioselectivity of the addition. This method might be a valuable extension of the methods for the selective removal of the allyl protective group for alcohols in organic synthesis (Scheme 2-16) [71].

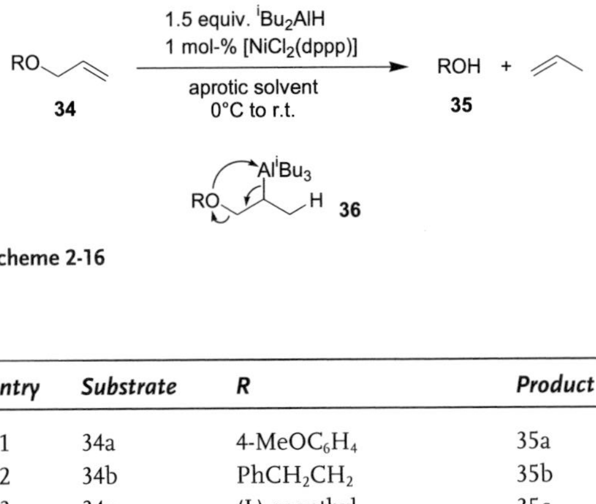

Scheme 2-16

Entry	Substrate	R	Product	Yield/%
1	34a	4-MeOC$_6$H$_4$	35a	90
2	34b	PhCH$_2$CH$_2$	35b	85
3	34c	(L)-menthyl	35c	97
4	34d	1-adamantyl	35d	80
5	34e	^{i}BuMe$_2$SiOCH$_2$)$_4$O	35e	92
6	34f	4-MeOC$_6$H$_4$O(CH$_2$)$_4$O	35f	95
7	34g	BnO(CH$_2$)$_4$O	35g	95
8	34h	MOMO(CH$_2$)$_4$O	35h	90
9	34i	THPO(CH$_2$)$_4$O	35i	89
10	34j	AcO(CH$_2$)$_4$O	35j	73
11	34k	PivO(CH$_2$)$_4$O	35k	85
12	34l	BnO(CH$_2$)$_4$O	35l	80

Tab. 2-4 Nickel catalyzed cleavage of allylic ethers

2.4.1.2 Enantioselective Hydroalumination of Alkenes

In addition to the enhanced rate of hydroalumination reactions in the presence of metal catalysts, tuning of the metal catalyst by the choice of appropriate ligands offers the possibility to influence the regio- and stereochemical outcome of the overall reaction. In particular, the use of chiral ligands has the potential to control the absolute stereochemistry of newly formed stereogenic centers. While asymmetric versions of other hydrometallation reactions, in particular hydroboration and hydrosilylation, are already well established in organic synthesis, the scope and synthetic utility of enantioselective hydroalumination reactions are only just emerging [72].

Early attempts by Pino and Giacomelli to resolve racemic 3,7-dimethyl-1-octene (37) by treatment with 0.3 equiv. of triisobutylaluminum in the presence of bis[(S)-sec-butylsalicylideneimine]nickel {Ni[(S)-sec-busal]$_2$} and subsequent hydrolysis gave (S)-2,6-dimethyloctane (38) with an enantiomerical excess of 1.2% along with the unreacted starting material (S)-37 with 1.8% ee, as judged by optical rotation (Schemes 2-17 and 2-18) [28].

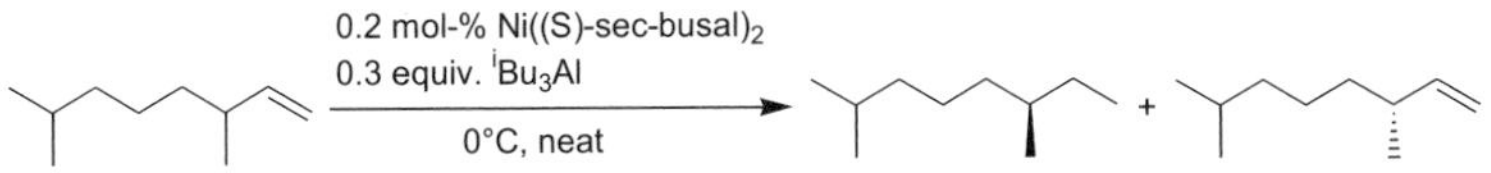

0.2 mol-% Ni((S)-sec-busal)$_2$
0.3 equiv. ^{i}Bu$_3$Al
0°C, neat

(R/S) 37   →   (S) 38   +   (S) 37
                 1.2 % ee   1.8 % ee

Scheme 2-17

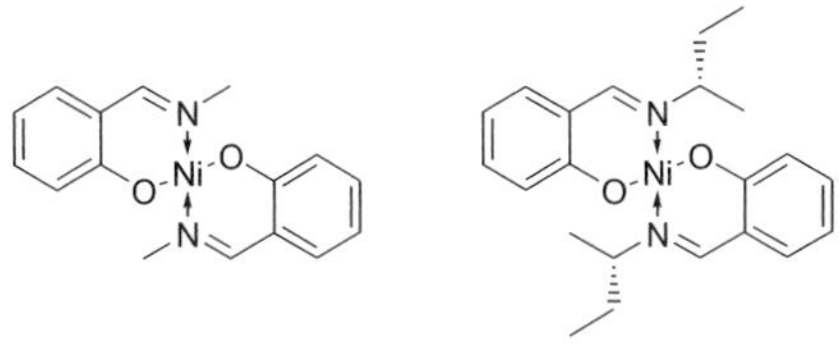

Ni(mesal)$_2$ Ni((S)-sec-busal)$_2$

Scheme 2-18

In spite of the modest asymmetric induction it was concluded that at least one of the chiral ligands is coordinated to the nickel in the catalytically active species. An alternative interpretation was given by Wilke and coworkers [29]. They could show that (methylsalicylidene)dimethylaluminum forms a stable adduct with nickel(0) complexes. It was concluded that the asymmetric induction in Pino's experiment might be attributed to a complex in which the chiral ligand is complexed to the Lewis acidic aluminum.

This interpretation was supported by further investigations by Giacomelli and coworkers [73]. Racemic 4-phenyl-1-hexene was kinetically resolved by isomerization of the double bond using a catalyst system consisting of Al^{i}Bu$_3$, (R)-N,N-dimethyl-1-phenylethylamine and Ni(mesal)$_2$. Very poor enantioselectivities (ee < 0.3%) were observed for both the isomerization product and the unreacted alkene. The authors note that it is essential to first react the alane with the chiral amine. No

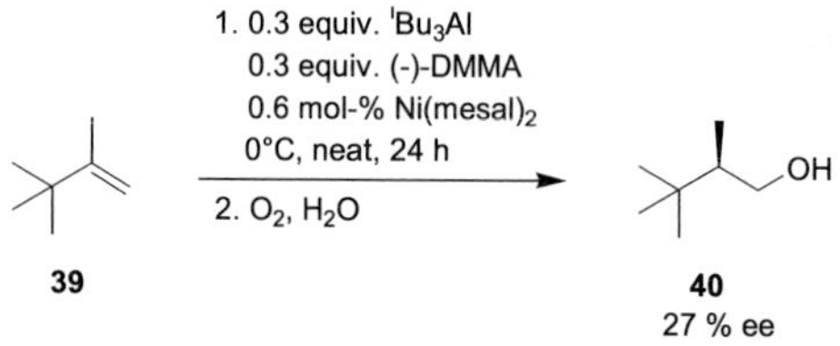

1. 0.3 equiv. ^{i}Bu$_3$Al
   0.3 equiv. (-)-DMMA
   0.6 mol-% Ni(mesal)$_2$
   0°C, neat, 24 h
   ───────────────→
2. O$_2$, H$_2$O

39

40
27 % ee

Scheme 2-19

induction was observed if Ni(mesal)$_2$ was treated with the amine prior to the addition of ^{i}Bu$_3$Al. In addition, (DIOP)NiCl$_2$ failed to give any asymmetric induction.

The asymmetric nickel-catalyzed hydroalumination of prochiral terminal alkenes using adducts of ^{i}Bu$_3$Al and chiral amines was reported in 1981 [74]. Among the different amines investigated, (–)-*N,N*-dimethylmenthylamine (DMMA) gave the best enantioselectivities. For example, reaction of 2,3,3-trimethyl-1-butene (**39**) at room temperature with 0.33 equiv. of the DMMA/iBu3Al adduct in the presence of 0.6 mol% of Ni(mesal)$_2$ gave, after oxidation of the intermediate organoaluminum compounds, 2,3,3-trimethyl-1-butanol **40** in 76% yield and 27% *ee* (Scheme 2-19).

An enantioselective version of the aluminum hydride-induced reductive ring opening reaction (see Scheme 2-15) was developed by Lautens and Rovis. Oxabicyclic [2.2.1] substrates are readily desymmetrized using an Ni(COD)$_2$/BINAP catalyst system to give cyclohexenols in high yields and enantiomeric excess [75]. The reactions were initially carried out at room temperature using 14 mol% Ni(COD)$_2$ and 21 mol% BINAP. Slow addition of the ^{i}Bu$_2$AlH was found to be essential for obtaining high *ee*s. Thus, cyclohexenol **42** was obtained from oxabicyclic diether **41** in 95% yield and 97% *ee* when the alane was added over a period of 1 h. In contrast, fast addition (< 10 s) led to a dramatic decrease of the enantioselectivity (56% *ee*). While an exhaustive survey of ligands has not been carried out, BINAP remains the best ligand found to date for this reaction. As little as 2 mol% of the catalyst can be used, but it is important to maintain a 1:1.5 ratio of the metal:ligand in order to achieve *ee*s >95%. As the amount of catalyst is decreased it usually helps to slow the rate of addition of ^{i}Bu$_2$AlH (Scheme 2-20).

A broad range of functional groups are tolerated under the reaction conditions including cyclopropyl carbinyl ethers, benzyl, and silyl ethers (see Table 2-5) [76]. However, substituents at the bridgehead position of the bicyclic substrate led to a decrease in the enantioselectivity (compare entry 2). The structurally related oxabenzonorbornenes, which contain a strained [2.2.1] oxabicyclic ring system fused to an aromatic ring, can also undergo enantioselective reductive ring opening [77]. Because of the lability of the resulting dihydronaphthols toward oxidation and elimination lead-

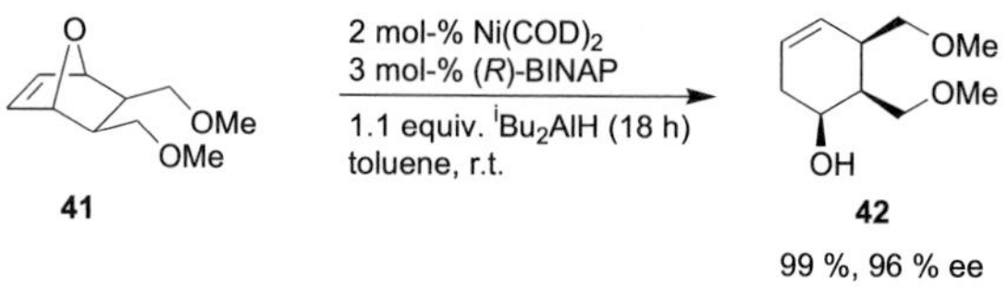

2 mol-% Ni(COD)$_2$
3 mol-% (*R*)-BINAP
─────────────────→
1.1 equiv. ^{i}Bu$_2$AlH (18 h)
toluene, r.t.

41

42
99 %, 96 % ee

Scheme 2-20

Tab. 2-5 Enantioselective reductive ring opening of oxabicyclic subrates

Entry	Substrate	Product	Yield/%	EE
	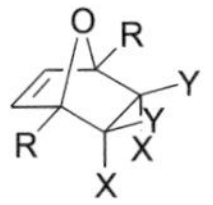	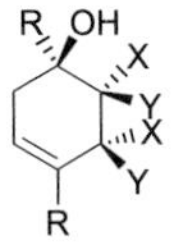		
1	43 R=H, X=H, Y=CH₂OBn	44	96	67
2	45 R=Me, X=H, Y=CH₂OMe	46	81	84
3	47 R=H, X=CH₂OMe, Y=–CH₂–	48	88	91
4	49 R=H, X=CH₂OMe, Y=H	50	50	86
5	51 R=H, X=–OCMe₂O, Y=H	52	50	78
	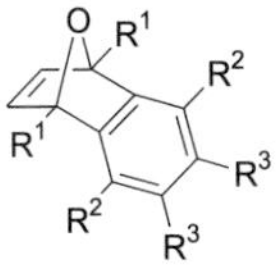	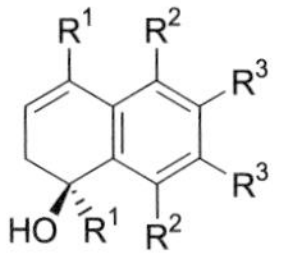		
6	53 R¹=H, R²=H, R³=H	54	88	98
7	55 R¹=Me, R²=H, R³=H	56	66	73
8	57 R¹=H, R²=Me, R³=H	58	73	88
9	59 R¹=H, R²=H, R³=OCH₂O	60	58	94
10	61 R¹=H, R²=H, R³=F	62	84	96
		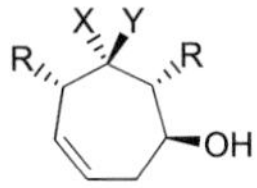		
11	63 R=Me, X=OMe, Y=H	64	83-95	97
12	65 R=Me, X=OTIPS, Y=H	66	87	>95
13	67 R=H, X=OH, Y=H	68	99	99.5
14	69 R=H, X=OMe, Y=H	70	67	95
15	71 R=H, X=H, Y=OBn	72	88	95

ing to naphthols and naphthalenes, respectively, milder reaction conditions have to be applied. THF, which decreases the Lewis acidity of the ^{i}Bu₂AlH, was a far better solvent, allowing the synthesis of dihydronaphthols in high yields and enantioselectivities up to 98% (entries 6–10). Again, substitution at the bridgehead position was found to significantly decrease the enantioselectivity (entry 7). In contrast, electronic effects seem to have only marginal effects (entries 9 and 10). In these substrates as little as 2 mol% of Ni(COD)₂ has been used and the *ee* of **54** was still >90%.

Oxabicyclo [3.2.1]octenes give a poor yield of the ring-opened product in low *ee* if reacted at room temperature with ^{i}Bu₂AlH in the presence of the Ni(COD)₂/BINAP catalyst in either toluene or THF solution (Scheme 2-21). However, if the reaction is carried out at 60°C, high yields and enantioselectivities are obtained [78]. Various substitution patterns and protective groups are tolerated (entries 11–15).

Although some preliminary studies have been carried out [79], uncertainties still exist about the mechanism of this transformation. The hydroalumination step was

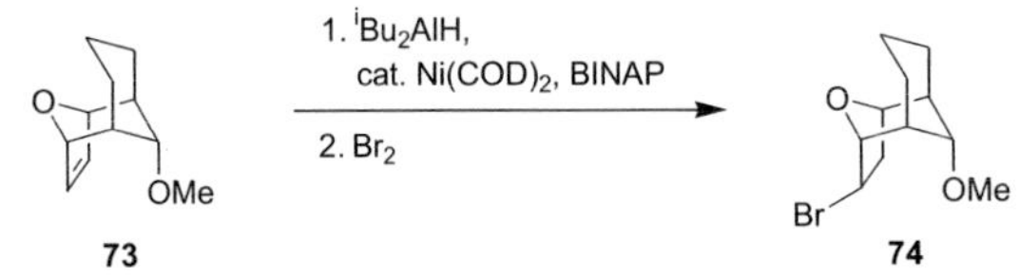

Scheme 2-21

examined for benzonorbornene and compound **73**. In both cases the intermediate alane was trapped using Br_2 and the products were shown to be racemic. The hydroalumination of the C=C bond thus occurs with little or no enantioselectivity. It is therefore the subsequent ring opening step *via* β-elimination of the oxygen bridge which proceeds with a high degree of enantioselectivity. The BINAP ligand was clearly shown to be bound to nickel by ^{31}P NMR spectroscopy which implies that this reaction step is assisted by the transition metal. To obtain high selectivities it appears to be essential to establish conditions where the non-selective hydrometallation step is reversible.

The asymmetric reductive ring opening of oxabenzonorbornene **53** was applied as a key step in the total synthesis of serotonin re-uptake inhibitor sertraline [77, 80].

2.4.2
Alkynes

Alkynes are much more reactive toward hydroalumination than alkenes. Hence, they readily react with both dialkylaluminum hydrides and $LiAlH_4$ under mild conditions in the absence of a catalyst [1]. However, it is not always possible to avoid side reactions and subsequent transformation of the vinylalanes formed in this transformation [81, 82]. In addition, *cis-trans*-isomerization of the metallated C=C bond can take place, thereby reducing the stereoselectivity of the overall reaction [83].

In general, transition metal-catalyzed hydroaluminations of alkynes occur in a *syn* fashion, i.e., both aluminum and hydride are added to the same face of the π-bond. Isomerization of the initially formed vinylalane is usually not observed under the mild reaction conditions used for these transformations.

The regiochemistry of Al–H addition to unsymmetrically substituted alkynes can be significantly altered by the presence of a catalyst. This was first shown by Eisch and Foxton in the nickel-catalyzed hydroalumination of several disubstituted acetylenes [26, 32]. For example, the product of the uncatalyzed reaction of 1-phenyl-propyne (**75**) with $^iBu_2AlH$ was exclusively *cis*-β-methylstyrene (**76**). Quenching the intermediate organoaluminum compounds with D_2O revealed a regioselectivity of 82:18. In the nickel-catalyzed reaction, *cis*-β-methylstyrene was also the major product (66%), but it was accompanied by 22% of *n*-propylbenzene (**78**) and 6% of (E,E)-2,3-dimethyl-1,4-diphenyl-1,3-butadiene (**77**). The selectivity of Al–H addition was again studied by deuterolytic workup; a ratio of **76a:76b** = 56:44 was found in this case. Hydroalumination of other unsymmetrical alkynes also showed a decrease in the regioselectivity in the presence of a nickel catalyst (Scheme 2-22).

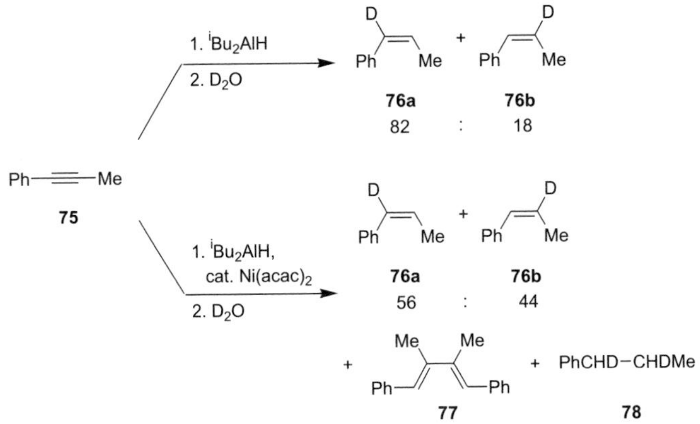

Scheme 2-22

A Cp$_2$ZrCl$_2$-catalyzed addition of ^{i}Bu$_2$AlH to terminal alkynes has been applied in the synthesis of (E)-vinyl phosphonates [84]. 1-Hexyne and 1-octyne were hydroaluminated at 0°C and the resulting vinylalanes transformed into the respective aluminate complexes by treatment with methyllithium. Subsequent addition of oxazaphospholidinone **79**, derived from (–)-ephedrine, lead to the homochiral vinyl phosphonates in yields of ca. 75% (Scheme 2-23).

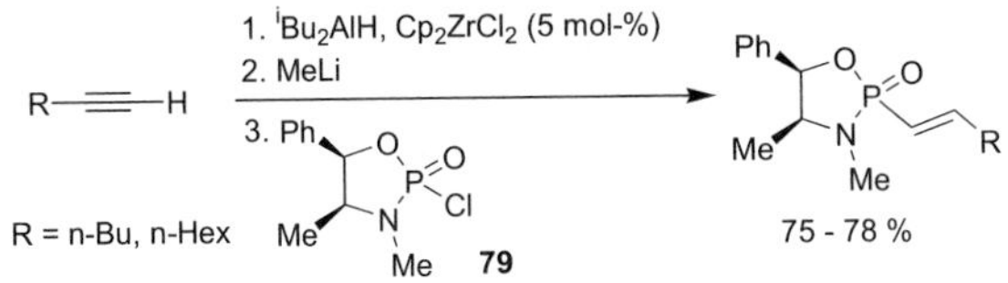

Scheme 2-23

Bis(diisopropylamino)alane is also able to add to alkynes in the presence of Cp$_2$TiCl$_2$ [85, 86]. The reactions are carried out at 0°C to room temperature in benzene in the presence of 5 mol% of the titanium catalyst. Both terminal and internal alkynes react essentially quantitatively. Upon hydrolysis, the corresponding Z-alkenes are obtained in good yields, accompanied with small amounts (<4%) of the E-isomer. The reaction of terminal alkynes is hampered by the formation of a significant amount of the overreduced alkane (ca. 45–50%). Workup of the reaction mixtures with D$_2$O gave a high degree of deuterium incorporation in the resulting alkenes, indicating that the formation of vinyl aluminum compounds occurs in high yield. Differences in the regioselectivity of Al–H addition were found compared to the uncatalyzed hydroalumination. For example, the catalyzed reaction of the aminoalane with 1-phenylpropyne (**80**) occurs with a 90:10 selectivity towards the vinylmetal intermediate, which bears the aluminum at the carbon distal to the phenyl group (Scheme 2-24). The non-catalyzed hydroalumination leads to the formation of the other regioisomer as the major product (**81a** : **81b** = 80 : 20) [87]. A similar reversal in the regioselectivity is observed in the case of 1-trimethylsilyloctyne **82**.

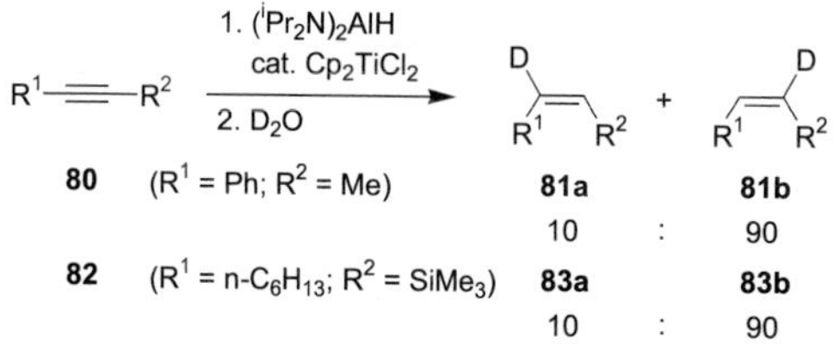

Scheme 2-24

^{i}Bu$_2$AlCl has been used as a hydrometallating reagent for internal alkynes in the presence of Cp$_2$TiCl$_2$ as a catalyst [88]. While diphenylacetylene gave, after hydrolysis, *cis*-1,2-diphenylethene in 92% yield, the hydroalumination of 4-octene and 5-decene was accompanied by significant diene formation. Hydroalumination of unsymmetrically substituted alkynes was highly regioselective in the case of 1-phenylbutyne (84) as judged by subsequent coupling of the vinylalkylchloroalanes with allyl chloride (Scheme 2-25).

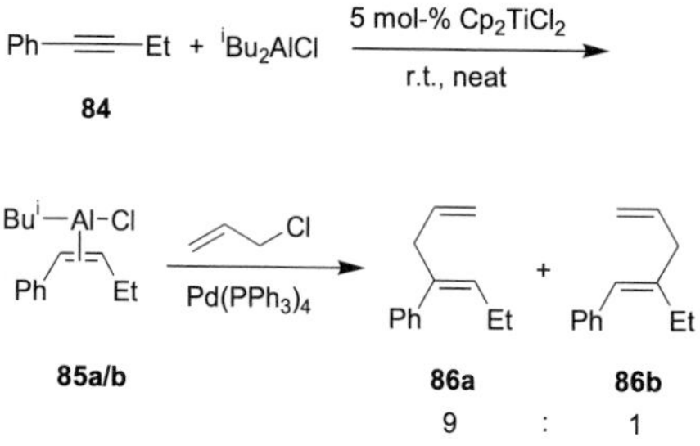

Scheme 2-25

The hydroalumination of internal alkynes by complex aluminum hydrides such as LiAlH$_4$, NaAlH$_4$, LiAlMe$_3$H, NaAlMe$_3$H, NaAl(OCH$_2$CH$_2$OCH$_3$)$_2$H$_2$ (vitride), LiAlH$_2$(NR$_2$)$_2$ and NaAlH$_2$(NR$_2$)$_2$ catalyzed by Cp$_2$TiCl$_2$ proceeds smoothly at room temperature in THF solution [57, 89]. Bis(alkyl)substituted acetylenes are reduced to give, after aqueous workup, *cis*-alkenes in almost quantitative yields. Quenching with D$_2$O leads to >98% deuterium incorporation, which shows again the almost quantitative formation of vinylaluminum intermediates. The regioselectivity of Al–H addition to unsymmetrically substituted alkynes is low. Terminal alkynes such as 1-octyne and phenylethyne as well as 1-phenylpropyne give only poor to moderate yields of the respective *cis*-alkenes. In the case of diphenylacetylene the formation of a significant amount of the *trans*-alkene can be observed [89].

It should be noted that the selective reduction of phenylacetylene and diphenylacetylene to either the *cis*-alkene or the alkane was achieved using LiAlH$_4$ in the presence of FeCl$_2$ or NiCl$_2$ as a catalyst [90, 91]. However, deuterolytic workup of the reaction mixtures gave deuterium incorporations <26%, indicating that these reagent systems are not well suited for the synthesis of vinyl- or alkylaluminum compounds from alkynes.

2.5
Conclusions

Catalytic hydroalumination of alkenes and alkynes has the potential to be a powerful tool for the functionalization of carbon-carbon multiple bonds in organic synthesis. Among the different catalyst systems developed to date, titanium, zirconium and nickel catalysts have found the widest application. However, most studies were carried out on relatively simple organic substrates. This might be attributed to the low compatibility of the aluminum reagents with numerous functional groups. A promising approach to solve this problem might be the *in situ* generation of the reactive metal hydride species from trialkylalanes, but further investigations are clearly necessary to establish the catalytic hydroalumination as a widely applicable synthetic transformation.

To date, synthetically useful enantioselective hydroalumination is limited to the asymmetric reductive ring-opening reaction of bicyclic ethers. In spite of the fact that further studies are necessary to get a detailed understanding of the reaction mechanism, this reaction provides a new route to various cycloalkenol derivatives, which are useful intermediates in the preparation of biologically active compounds.

References

1 J. J. Eisch in *Comprehensive Organic Synthesis*, B. M. Trost, I. Fleming (eds.), Pergamon Press, Oxford, **1991**, Vol. 8, p. 733

2 J. J. Eisch in *Comprehensive Organometallic Chemistry*, G. Wilkinson, F. G. A. Stone, E. W. Abel (eds.), Pergamon Press, Oxford, **1982**, Vol. 1, p. 555

3 H. Lehmkuhl, K. Ziegler, H. G. Gellert in *Methoden der Org. Chem. (Houben-Weyl*, E. Müller (ed.), **1970**, *XIII/4*, 23

4 U. M. Dzhemilev, O. S. Vostrikova, G. A. Tolstikov, *Russ. Chem. Rev.* **1990**, *59*, 1157

5 U. M. Dzhemilev, O. S. Vostrikova, G. A. Tolstikov, *J. Organomet. Chem.* **1986**, *304*, 17

6 J. R. Zietz, G. C. Robinson, K. L. Lindsay in *Comprehensive Organometallic Chemistry*, G. Wilkinson, F. G. A. Stone, E. W. Abel (eds.), Pergamon Press, Oxford, **1982**, Vol. 7, p. 384

7 for recent examples see:
a) J. Tanaka, S. Kanemasa, *Bull. Chem. Soc. Jpn.* **1990**, *63*, 51; G. Zweifel, W. Leong, *J. Am. Chem. Soc.* **1987**, *109*, 6409; M. Al-Hassan, *Synth. Commun.* **1989**, *19*, 1677
b) J. M. Chong, M. A. Heuft, *Tetrahedron* **1999**, *55*, 14243; S. Ma, F. Liu, E.-I. Negishi, *Tetrahedron Lett.* **1997**, *38*, 3829; E.-I. Negishi, S. Ma, J. Amanfu, C. Copéret, J. A. Miller, J. M. Tour, *J. Am. Chem. Soc.* **1996**, *118*, 5919; E. V. Gorobets, O. V. Shitikova, S. I. Lomakina, G. A. Tolstikov, A. V. Kuchin, *Russ. Chem. Bull.* **1993**, *42*,

1573; A. P. Khrimyan, O. A. Garibyan, G. A. Panosyan, N. S. Mailyan, F. S. Kinoyan, G. M. Makaryan, S. O. Badanyan, *Russ. J. Org. Chem.* **1993**, *29*, 1961; J. A. Miller, *J. Org. Chem.* **1989**, *54*, 998; A. Alexakis, J. M. Duffault, *Tetrahedron Lett.* **1988**, *29*, 6243; M. Al-Hassan, *Synth. Commun.* **1987**, *17*, 1787
c) E. V. Gorobetz, A. N. Kasatkin, A. V. Kutchin, G. A. Tolstikov, *Russ. Chem. Bull.* **1994**, *43*, 466; V. N. Odinokov, G. Y. Ishmuratov, R. Y. Kharisov, E. P. Serebryakov, G. A. Tolstikov, *Russ. Chem. Bull.* **1993**, *42*, 98; V. N. Odinokov, G. Y. Ishmuratov, R. Y. Kharisov, E. P. Serebryakov, G. A. Tolstikov, *Russ. Chem. Bull.* **1993**, *42*, 100
d) E. B. Baker, H. H. Sister, *J. Am. Chem. Soc.* **1953**, *75*, 5193
e) R. L. Danheiser, H. Sard, *J. Org. Chem.* **1980**, *45*, 4810
f) H. Matsutani, H. Poras, T. Kusumoto, T. Hiyama, *Synlett* **1998**, 1353; E.-I. Negishi, *Tetrahedron Lett.* **1988**, *29*, 3903
g) A. Alexakis, D. Jachiet, *Tetrahedron* **1989**, *45*, 6197
h) R. D. A. Hudson, S. A. Osborne, G. R. Stephenson, *Tetrahedron* **1997**, *53*, 4095; P. V. Ramachandran, M. V. R. Reddy, M. T. Rudd, *J. Chem. Soc., Chem. Commun.* **1999**, 1979; L.-H. Xu, E. P. Kündig, *Helv. Chim. Acta* **1994**, *77*, 1480
i) T. Ishikawa, A. Ogawa, T. Hirao, *J. Am. Chem. Soc.* **1998**, *120*, 5124; B. H. Lipshutz, G. Bulow, R. F. Lowe, K. L.

STEVENS, *Tetrahedron* **1996**, *52*, 7265; M. J. DABDOUB, T. M. CASSOL, S. L. BARBOSA, *Tetrahedron Lett.* **1996**, *37*, 831; M. J. DABDOUB, T. M. CASSOL, *Tetrahedron* **1995**, *51*, 12971; M. HAVRANEK, D. DVORÁK, *Synthesis* **1998**, 1264; M. AL-HASSAN, J. *Organomet. Chem.* **1990**, *386*, 395; M. AL-HASSAN, J. *Organomet. Chem.* **1987**, *321*, 119; B. L. GROH, A. F. KREAGER, J. B. SCHNEIDER, *Synth. Commun.* **1991**, *21*, 2065

8 E. C. ASHBY, J. J. LIN, J. *Org. Chem.* **1978**, *43*, 2567

9 J. J. EISCH, S. R. SEXSMITH, K. C. FICHTER, J. *Organomet. Chem.* **1990**, *382*, 273

10 G. FOLCHER, J. F. LE MARÉCHAL, H. MARQUET-ELLIS, J. *Chem. Soc., Chem. Commun.* **1982**, 323

11 J. F. LE MARÉCHAL, M. EPHERITIKHINE, G. FOLCHER, J. *Organomet. Chem.* **1986**, *309*, C1

12 F. ASINGER, B. FELL, R. JANSSEN, *Chem. Ber.* **1964**, *97*, 2515

13 F. SATO, S. SATO, M. SATO, J. *Organomet. Chem.* **1976**, *122*, C25

14 F. SATO, S. SATO, M. SATO, J. *Organomet. Chem.* **1977**, *131*, C26

15 F. SATO, S. SATO, H. KODAMA, M. SATO, J. *Organomet. Chem.* **1977**, *142*, 71

16 K. ISAGAWA, K. TATSUMI, Y. OTSUJI, *Chem. Lett.* **1976**, 1145

17 K. ISAGAWA, K. TATSUMI, H. KOSUGI, Y. OTSUJI, *Chem. Lett.* **1977**, 1017

18 E. C. ASHBY, S. A. NODING, *Tetrahedron Lett.* **1977**, 4579

19 E. C. ASHBY, S. A. NODING, J. *Org. Chem.* **1979**, *44*, 4355

20 R. A. KOVAR, E. C. ASHBY, *Inorg. Chem.* **1971**, *10*, 893

21 E.-I. NEGISHI, T. YOSHIDA, *Tetrahedron Lett.* **1980**, *21*, 1501

22 J. SCHWARTZ, J. *Organomet. Chem. Library* **1976**, *1*, 461

23 J. J. BARBER, C. WILLIS, G. M. WHITESIDES, J. *Org. Chem.* **1979**, *44*, 3603

24 A. G. IBRAGIMOV, I. V. ZAGREBEL'NAYA, K. G. SATENOV, L. M. KHALILOV, U. M. DZHEMILEV, *Russ. Chem. Bull.* **1998**, *47*, 691

25 K. ZIEGLER, H.-G. GELLERT, E. HOLZKAMP, G. WILKE, *Brennst.-Chem.* **1954**, *35*, 321

26 J. J. EISCH, M. W. FOXTON, J. *Organomet. Chem.* **1968**, *12*, P33.

27 K. ZIEGLER, H.-G. GELLERT, E. HOLZKAMP, G. WILKE, E. W. DUCK, W.-R. KROLL, *Justus Liebigs Ann. Chem.* **1960**, *629*, 172.

28 L. LARDICCI, G. P. GIACORNELLI, P. SALVADORI, P. PINO, J. *Am. Chem. Soc.* **1971**, *93*, 5794

29 K. FISCHER, K. JONAS, P. MISBACH, R. STABBA, G. WILKE, *Angew. Chem. Int. Ed.* **1973**, *12*, 943

30 K. FISCHER, K. JONAS, A. MOLLBACH, G. WILKE, *Z. Naturforsch.* **1984**, *39b*, 1011

31 K.-R. PÖRSCHKE, W. KLEIMANN, Y.-H. TSAY, C. KRÜGER, G. WILKE, *Chem. Ber.* **1990**, *123*, 1267

32 J. J. EISCH, S. R. SEXSMITH, K. C. FICHTER, J. *Organomet. Chem.* **1990**, *382*, 273

33 J. J. EISCH, K. C. FICHTER, J. *Am. Chem. Soc.* **1974**, *96*, 6815

34 J. J. EISCH, X. MA, M. SINGH, G. WILKE, J. *Organomet. Chem.* **1997**, *527*, 301

35 E. C. ASHBY, J. J. LIN, *Tetrahedron Lett.* **1977**, 4481

36 E. C. ASHBY, J. J. LIN, J. *Org. Chem.* **1978**, *43*, 2567

37 E. C. ASHBY, S. A. NODING, J. *Org. Chem.* **1979**, *44*, 4364

38 S. GAGNEUR, H. MAKABE, E.-I. NEGISHI, *Tetrahedron Lett.* **2001**, *42*, 785

39 K. MARUOKA, H. SANO, K. SHINODA, S. NAKAI, H. YAMAMOTO, J. *Am. Chem. Soc.* **1986**, *108*, 6063

40 K. MARUOKA, H. SANO, K. SHINODA, H. YAMAMOTO, *Chem. Lett.* **1987**, 73

41 K. MARUOKA, K. SHINODA, H. YAMAMOTO, *Synth. Commun.* **1988**, *18*, 1029

42 S. NAGAHARA, K. MARUOKA, Y. DOI, H. YAMAMOTO, *Chem. Lett.* **1990**, 1595

43 S. NAGAHARA, K. MARUOKA, H. YAMAMOTO, *Bull. Chem. Soc. Jpn.* **1993**, *66*, 3783

44 J. TSUJI, Y. YAMAGAWA, T. MANDAI, *Tetrahedron Lett.* **1978**, 833

45 J. TSUJI, T. MANDAI, *Chem. Lett.* **1977**, 975

46 F. SATO, Y. TOMURA, H. ISHIKAWA, M. SATO, *Chem. Lett.* **1980**, 99

47 K. ZIEGLER, *Angew. Chem.* **1956**, *68*, 721

48 F. SATO, Y. MORI, M. SATO, *Chem. Lett.* **1978**, 833

49 F. SATO, Y. MORI, M. SATO, *Chem. Lett.* **1979**, 1405

50 F. SATO, H. KODAMA, M. SATO, *J. Organomet. Chem.* **1978**, *157*, C30

51 F. SATO, H. KODAMA, M. SATO, *Chem. Lett.* **1978**, 789

52 K. ISAGAWA, M. OHIGE, K. TATSUMI, Y. OTSUJI, *Chem. Lett.* **1978**, 1155

53 F. SATO, H. KODAMA, Y. TOMURO, M. SATO, *Chem. Lett.* **1979**, 623

54 F. SATO, T. OIKAWA, M. SATO, *Chem. Lett.* **1979**, 167

55 G. BOIREAU, A. KORENOVA, A. DEBERLY, D. ABENHAIM, *Tetrahedron Lett.* **1985**, *26*, 4181

56 H. MATSUTANI, H. PORAS, T. KUSUMOTO, T. HIYAMA, *Synlett* **1998**, 1353

57 E. C. ASHBY, S. A. NODING, *J. Org. Chem.* **1980**, *45*, 1035

58 U. M. DZHEMILEV, O. S. VOSTRIKOVA, A. G. IBRAGIMOV, G. A. TOLSTIKOV, *Izv. Akad. Nauk SSSR, Ser. Khim.* **1980**, 2134

59 G. A. TOLSTIKOV, U. M. DZHEMILEV, O. S. VOSTRIKOVA, *Bull. Acad. Sci. USSR, Div. Chem. Sci.* **1982**, *31*, 596

60 U. M. DZHEMILEV, A. G. IBRAGIMOV, *J. Organomet. Chem.* **1994**, *466*, 1

61 U. M. DZHEMILEV, *Tetrahedron* **1995**, *51*, 4333

62 D. Y. KONDAKOV, E.-I. NEGISHI, *J. Am. Chem. Soc.* **1996**, *118*, 1577

63 U. M. DZHEMILEV, A. G. IBRAGIMOV, O. S. VOSTRIKOVA, G. A. TOLSTIKOV, L. M. ZELENOVA, *Bull. Acad. Sci. USSR, Div. Chem. Sci.* **1981**, *30*, 281

64 U. M. DZHEMILEV, A. G. IBRAGIMOV, O. S. VOSTRIKOVA, E. V. VASILÍEVA, G. A. TOLSTIKOV, *Bull. Acad. Sci. USSR, Div. Chem. Sci.* **1987**, *36*, 1004

65 A. G. IBRAGIMOV, D. L. MINSKER, A. A. BERG, O. V. SHITIKOVA, S. I. LOMAKINA, U. M. DZHEMILEV, *Bull. Russ. Acad. Sci., Div. Chem. Sci.* **1992**, *41*, 2217

66 P. CHIU, M. LAUTENS, *Top. Curr. Chem.* **1997**, *190*, 3 and references cited therein

67 M. LAUTENS, P. CHIU, J. T. COLLUCCI, *Angew. Chem. Int. Ed. Engl.* **1993**, *32*, 281

68 M. LAUTENS, P. CHIU, S. MA, T. ROVIS, *J. Am. Chem. Soc.* **1995**, *117*, 532

69 M. LAUTENS, S. MA, P. CHIU, *J. Am. Chem. Soc.* **1997**, *119*, 6478

70 T. TANIGUCHI, K. OGASAWARA, *Angew. Chem. Int. Ed. Engl.* **1998**, *37*, 1136

71 F. GRUIBE, *Tetrahedron* **1997**, *53*, 13509

72 M. LAUTENS, T. ROVIS in *Comprehensive Asymmetric Catalysis*, E. N. JACOBSEN, A. PFALTZ, H. YAMAMOTO (eds.), Springer Verlag, **1999**, Vol. 1, p. 337

73 G. GIACOMELLI, L. LARDICCI, R. MENICAGLI, L. BERTERO, *J. Chem. Soc., Chem. Commun.* **1979**, 633

74 G. GIACOMELLI, L. BERTERO, L. LARDICCI, *Tetrahedron Lett.* **1981**, *22*, 883

75 M. LAUTENS, P. CHIU, S. MA, T. ROVIS, *J. Am. Chem. Soc.* **1995**, *117*, 532

76 M. LAUTENS, T. ROVIS, *Tetrahedron* **1998**, *54*, 1107

77 M. LAUTENS, T. ROVIS, *J. Org. Chem.* **1997**, *62*, 5246

78 M. LAUTENS, T. ROVIS, *J. Am. Chem. Soc.* **1997**, *119*, 11090

79 T. ROVIS, PhD Thesis, University of Toronto, **1998**, p. 185-232

80 M. LAUTENS, T. ROVIS, *Tetrahedron* **1999**, *55*, 8967

81 G. WILKE, H. MÜLLER, *Justus Liebigs Ann. Chem.* **1960**, *629*, 222

82 J. J. EISCH, W. C. KASKA, *J. Am. Chem. Soc.* **1966**, *88*, 2213

83 J. J. EISCH, W. C. KASKA, *J. Am. Chem. Soc.* **1963**, *85*, 2165

84 T. TAAPKEN, S. BLECHERT, *Tetrahedron Lett.* **1995**, *36*, 6659

85 E. C. ASHBY, S. A. NODING, *Tetrahedron Lett.* **1977**, 4579

86 E. C. ASHBY, S. A. NODING, *J. Organomet. Chem.* **1979**, *177*, 117

87 J. J. EISCH, W. C. KASKA, *J. Organomet. Chem.* **1964**, *2*, 184

88 U. M. DZHEMILEV, A. G. IBRAGIMOV, I. R. RAMAZANOV, R. M. SULTANOV, L. M. KHALILOV, R. R. MUSLUKHOV, *Russ. Chem. Bull.* **1996**, *45*, 2610

89 H. S. LEE, *Bull. Korean Chem. Soc.* **1987**, *8*, 484

90 E. C. ASHBY, J. J. LIN, *Tetrahedron Lett.* **1977**, 4481

91 E. C. ASHBY, J. J. LIN, *J. Org. Chem.* **1978**, *43*, 2567

3
Asymmetric Hydrosilylation

Jun Tang and Tamio Hayashi

3.1
Introduction

Hydrosilylation is the reaction where hydrosilanes add to unsaturated substrates [1, 2]. From the viewpoint of asymmetric heterofunctionalization, it generally refers to the addition of silicon-hydrogen bond across carbon-carbon multiple bond under catalysis with transition metal complexes forming both valid hydrogen-carbon and silicon-carbon bonds. On this subject, there have been several pertinent reviews which have made a comprehensive survey [3, 4, 5]. Like other asymmetric methodologies, asymmetric hydrosilylation can be realized by incorporation of a chiral ligand with the metal catalyst. The optically active hydrosilylation products are very powerful intermediates in organic synthesis *via* some efficient transformations [6, 7]. Especially, based on the diastereoselective reactions of chiral allyl- and allenylsilanes with C=O and C=N bonds [8, 9], more than one chiral centers can be constructed in sophisticated molecules. The challenge of asymmetric hydrosilylation, in many cases, concerns the design of efficient catalytic systems for the achievement of not only the enantioselectivity but also the regioselectivity. Much effort has been dedicated to this purpose, and some efficient approaches have been established including the Palladium/MOP system for hydrosilylation of olefins and the Rh/BINAP system for intramolecular hydrosilylation. Asymmetric hydrosilylation has been extended to cyclization/hydrosilylation of 1,6-dienes giving a five-membered cyclic silylated product and the reaction of butadiynes giving optically active allenylsilanes, which demonstrates the utility of asymmetric hydrosilylation for the synthesis of key building blocks not available using other methodologies. This review is an attempt to present all of these fruitful successes and new tendencies in asymmetric hydrosilylation for heterofunctionalization.

3.2
Mechanism of Transition Metal-Catalyzed Hydrosilylation

The early research of catalytic hydrosilylation was concentrated on platinum-catalyzed reaction with a wide variety of hydrosilanes and alkenes [1, 10]. A transition metal complex, MLn (L = ligand), especially an electron-rich complex of a late transition metal such as Co(I), Rh(I), Ni(0), Pd(0), or Pt(0) as a precatalyst, activates both hydrosilanes, HSiR₃, and a variety of substrates, typically alkenes. A catalytic cycle is considered to involve successive two steps as depicted in Scheme 3-1. The hydrosilylation of alkenes catalyzed by $H_2PtCl_6 \cdot 6H_2O/i\text{-PrOH}$ (called the Speier catalyst) generally proceeds by the Chalk-Harrod mechanism (Scheme 3-1, cycle A) [11, 12]. Reversible oxidative addition of a hydrosilane to the metal-alkene complex gives a hydrido-silyl complex **I**. The complex **I** undergoes rapid reversible migratory insertion of the alkene into the M–H bond (hydrometallation) to give the alkyl-silyl species **II**. Irreversible reductive elimination of the alkyl and silyl ligands from **II** forms the hydrosilylation product. The plausibility of each step has been well established by many model reactions [13].

Although the Chalk-Harrod mechanism (an H-D exchange between deuterosilanes and alkenes as well as the observed regioselectivity always associated with the catalytic hydrosilylation) can account for an alkene isomerization, an alternative mechanism has been proposed which involves preferentially an alkene insertion into the M–Si bond (silylmetallation) in order to account for the formation of a vinylsilane observed in the hydrosilylation of ethylene with triethylsilane catalyzed by Fe(CO)₅. The reaction with a high molar ratio of alkene to silane gives triethylvinylsilane almost quantitatively. In this case, silylmetallation was suggested to give β-silylalkyl-hydrido intermediate **III**, followed by reductive elimination to complete the hydrosilylation [14, 15, 16] or subsequent β-hydride elimination from the insertion product **III** to generate the vinysilane (Scheme 3-1, cycle B). Recently it has become evident that the silylmetallation is a fairly common process with a number of

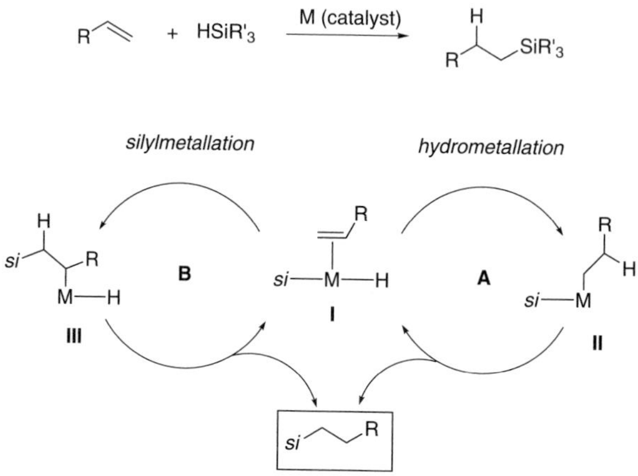

Scheme 3-1

other catalyst systems of rhodium, cobalt, and palladium catalyst. Some of them will be discussed in the following parts. It is worth mentioning that hydrosilylation exhibits a wide spectrum of reactivities in the oxidative addition step depending on the substituents on the silicon atom and the nature of the metal catalyst. Thus, Pt complexes tolerate any hydrosilanes, such as $HSiCl_nMe_{3-n}$ (n = 1–3), $HSi(OR)_3$, or H_{n-} SiR_{4-n} (n = 1–3; R = alkyl or Ph) in the hydrosilylation, while, Pd complexes are applicable mostly to $HSiCl_nR_{3-n}$ (n = 2, 3) and Rh complexes to preferably $HSiR_3$ [2].

3.3
Asymmetric Hydrosilylation of Olefins

3.3.1
Hydrosilylation of 1,1-Disubstituted and Monosubstituted Olefins

Catalytic asymmetric hydrosilylation has been developed alongside the development of chiral phosphine ligands. At the initial stage, phosphine ligands where the chiral center is on the phosphorus atom were used. In the first report, a platinum complex coordinated with (R)-benzylmethylphenylphosphine (1), cis-PtCl$_2$(C$_2$H$_4$)(1), was used for the reaction of 2-phenylpropene (3) with methyldichlorosilane at 40°C to give (R)-1-(methyldichlorosilyl)-3-phenylpropane (4) of 5% ee [17, 18] (Scheme 3-2). With a platinum catalyst of (R)-methylphenylpropylphosphine (2) the enantioselectivity was lower (1% ee). Use of trans-NiCl$_2$(1)$_2$ catalyst bearing the phosphorus chiral ligand 1 for the hydrosilylation of 3 improved the enantioselectivity up to 18% ee [19, 20]. Cationic rhodium complexes coordinated with (R)-benzylmethylphenylphosphine (1) or (–)-DIOP (5) as a ligand catalyzed the hydrosilylation of 2-phenyl-

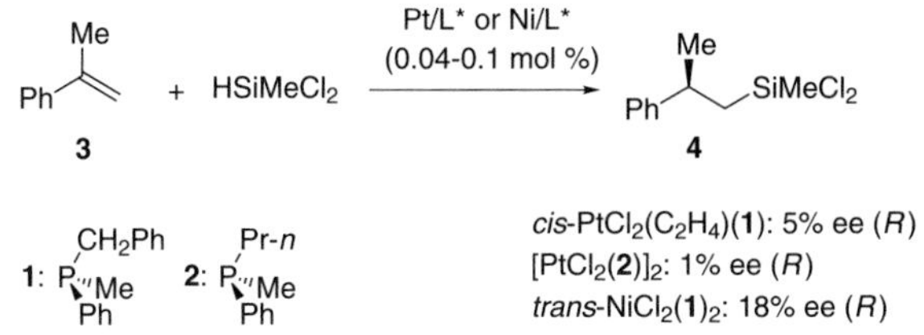

Scheme 3-2

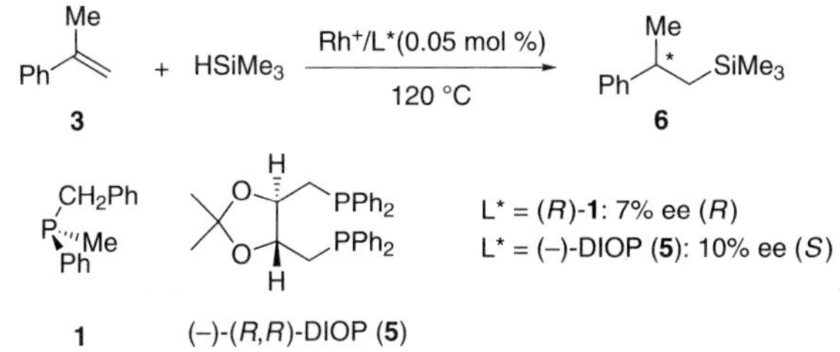

Scheme 3-3

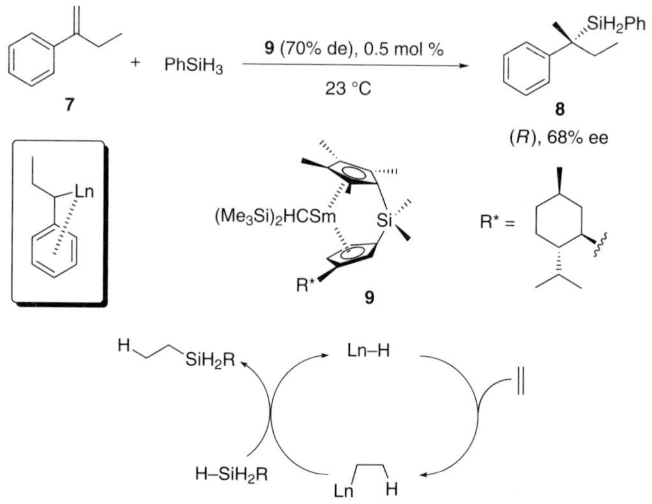

Scheme 3-4

propene (**3**) with trimethylsilane to give 1-(trimethylsilyl)-3-phenylpropane (**6**) of 7% and 10% *ee*, respectively (Scheme 3-3) [20].

Organolanthanide and Group 3 metallocene complexes, which were originally developed as olefin polymerization catalysts, show distinctive characteristics for olefin hydrosilylation [21, 22]. Organolanthanide-catalyzed hydrosilylation of α-substituted styrenes gave the reverse regioselectivity because of an interaction of the electrophilic lanthanide center with the arene π system (Scheme 3-4). It was demonstrated that a chiral organosamarium precatalyst coordinated with a menthylcyclopentadienyl ligand **9** (70% enantiopure) effects the hydrosilylation of 2-phenyl-1-butene (**7**) with high enantioselectivity [22]. The mechanism of the lanthanide-catalyzed hydrosilylation is different from that discussed above. Based on kinetic and mechanistic studies, a catalytic cycle involving an olefin insertion into metal-hydride bond followed by metathesis between metal-carbon and hydrogen-silicon bonds is proposed, which is consistent with most experimental results [22].

In general, hydrosilylation has been studied most extensively with platinum complexes. However, palladium-catalyzed hydrosilylation has been neglected because it is difficult to avoid reduction of palladium complex to inactive metal in this process. Recently a palladium complex coordinated with an axially chiral monodentate phosphine ligand, MeO-MOP (**10a**) or its analogs [23], has been demonstrated to exhibit extremely high activity for the asymmetric hydrosilylation of alkyl-substituted terminal olefins along with an unprecedented regioselectivity (Scheme 3-5) [24, 25]. Simple terminal olefins **11** were transformed efficiently into the corresponding optically active 2-alkanols **14** with enantio-selectivities ranging between 94% and 97% *ee* by the catalytic hydrosilylation-oxidation procedure. For example, the reaction of 1-octene (**11b**) with trichlorosilane in the presence of 0.1 mol% of palladium catalyst generated from [PdCl(π-C$_3$H$_5$)]$_2$ and (*S*)-MeO-MOP (**10a**) at 40°C for 24 h gave 2-octylsilane (**12b**) and 1-octylsilane (**13b**) in a ratio of 93 to 7. The branch isomer was

	12/13	% ee of 14
a: R = *n*-C_4H_9:	89/11	94 (*R*)
b: R = *n*-C_6H_{13}:	93/7	95 (*R*)
c: R = *n*-$C_{10}H_{21}$:	94/6	95 (*R*)
d: R = CH_2CH_2Ph:	81/19	97 (*S*)
e: R = *cyclo*-C_6H_{11}:	66/43	96 (*R*)

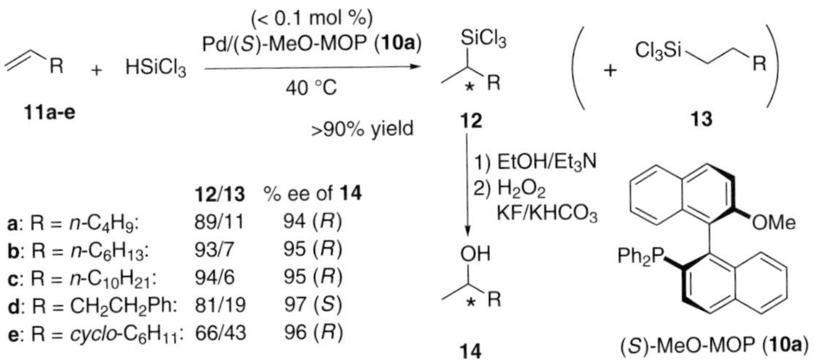

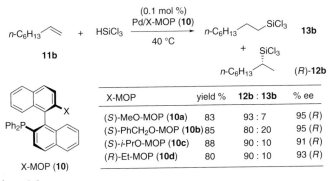

Scheme 3-5

oxidized into (*R*)-2-octanol (**14a**) of 95% *ee*. It is noteworthy that the reaction of simple terminal alkenes with MeO-MOP ligand proceeds with high branch-selectivity. The predominant formation of 2-silylalkanes from purely aliphatic 1-alkenes in hydrosilylation has never been observed with any transition-metal catalysts. Asymmetric hydrosilylation of 4-pentenyl benzoate and 1,5-heptadiene gave the corresponding 2-alkanols of 90% *ee* and 87% *ee*, respectively. The ester carbonyl and the internal double bond remained intact [25]. The high selectivity was also observed with MOP ligands **10b**, **10c**, and **10d**, which have other substituents than methoxy at the 2′ position [25] (Scheme 3-6). Thus, the hydrosilylation of 1-octene (**11b**) with MOP ligands substituted with benzyloxy or isopropoxy gave over 91% enantioselectivity and over 80% branch selectivity, suggesting that the steric bulkiness of the 2′-substituents has little influence on this asymmetric hydrosilylation. The presence of an alkoxy group at the 2′ position is not essential for the high selectivity because replacement of the alkoxy group by an alkyl group did not affect the selectivity.

3.3.2
Hydrosilylation of Styrene and its Derivatives

Palladium-catalyzed hydrosilylation of styrene derivatives usually proceeds with high regioselectivity to produce benzylic silanes, 1-aryl-1-silylethanes, because of the

X-MOP	yield %	**12b** : **13b**	% ee
(*S*)-MeO-MOP (**10a**)	83	93 : 7	95 (*R*)
(*S*)-$PhCH_2O$-MOP (**10b**)	85	80 : 20	95 (*R*)
(*S*)-*i*-PrO-MOP (**10c**)	88	90 : 10	91 (*R*)
(*R*)-Et-MOP (**10d**)	80	90 : 10	93 (*R*)

Scheme 3-6

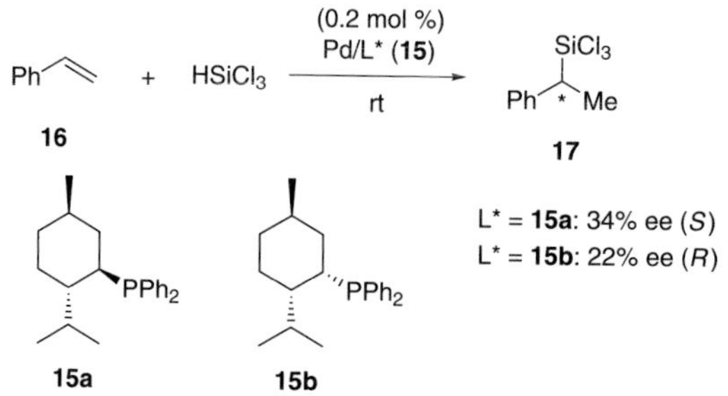

Scheme 3-7

contribution of π-benzylic palladium intermediates [1, 5]. It is known that bisphosphine-palladium complexes are catalytically much less active than monophosphine-palladium complexes, and hence asymmetric synthesis has been attempted by the use of chiral monodentate phosphine ligands. In the first report, menthyldiphenylphosphine (15a) and neomenthyldiphenylphosphine (15b) were used for the palladium-catalyzed reaction of styrene (16) with trichlorosilane. They gave 1-(trichlorosilyl)-1-phenylethane (17) of 34% and 22% *ee*, respectively (Scheme 3-7) [26, 27]. Use of ferrocenylmonophosphine (*R*)-(*S*)-PPFA (18a) for the same reaction improved the enantioselectivity [28, 29, 30]. Here the hydrosilylation product was oxidized into (*S*)-1-phenylethanol (19) of 52% *ee* (Scheme 3-8). Among the ferrocenylphosphine ligands, a ligand 18b, which has a phenyl substituent at the 2 position, gave the best result (73% yield, 70% *ee*) in the hydrosilylation of styrene [31]. The ferrocenylmonophosphine (18c) supported on Merrifield polystyrene has also been used for the hydrosilylation of styrene, though the enantioselectivity was lower (15% *ee*) [32]. Several chiral (β-*N*-sulfonylaminoalkyl)phosphines (20) were prepared from (*S*)-valinol and used for the asymmetric hydrosilylation of styrene and cyclopentadiene [33]. For styrene, phosphine 20a, which contains a methylsulfonyl group, was most effec-

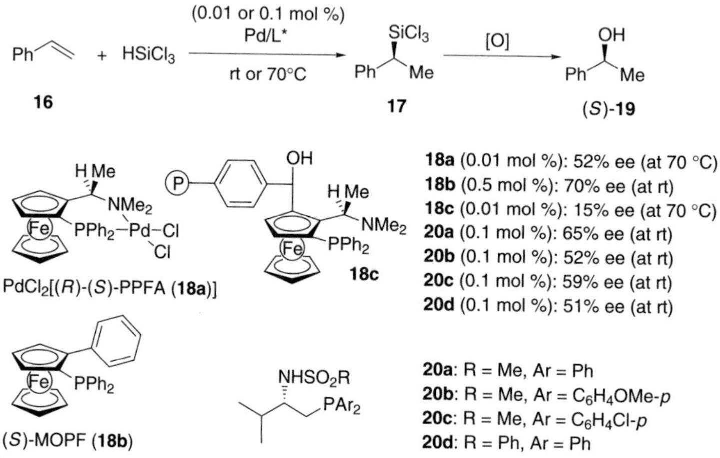

Scheme 3-8

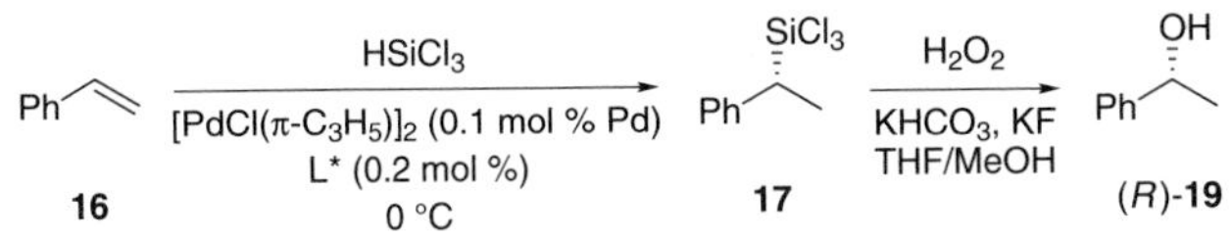

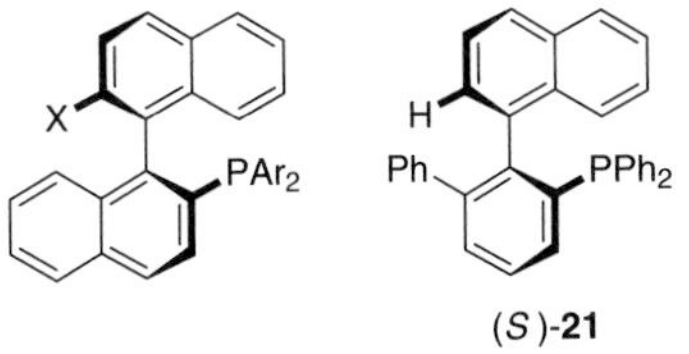

L*	% ee of **19**
(*R*)-MeO-MOP (**10a**)	14% ee
(*S*)-Et-MOP (**10d**)	18% ee
(*R*)-CN-MOP (**10e**)	26% ee
(*S*)-H-MOP (**10f**)	94% ee
(*S*)-**21**	91% ee

X = OMe, Ar = phenyl: (*R*)-MeO-MOP (**10a**)
X = Et, Ar = phenyl: (*S*)-Et-MOP (**10d**)
X = CN, Ar = phenyl: (*R*)-CN-MOP (**10e**)
X = H, Ar = phenyl: (*S*)-H-MOP (**10f**)

Scheme 3-9

tive, giving (*S*)-1-phenylethanol (**19**) of 65% *ee*. Other amidophosphines **20b-c** are also fairly effective for this asymmetric hydrosilylation (Scheme 3-8).

Axially chiral monophosphine ligand, MeO-MOP (**10a**), was not as effective for styrene derivatives as for simple terminal olefins [34]. The palladium-catalyzed hydrosilylation of styrene (**16**) with trichlorosilane in the presence of (*R*)-MeO-MOP (**10a**) ligand under standard conditions (without solvent) followed by oxidation gave (*R*)-1-phenylethanol (**19**) of only 14% *ee* (Scheme 3-9). The use of benzene as solvent for the hydrosilylation improved the enantioselectivity to 71%. For substituted styrenes such as *o*-chlorostyrene or β-methylstyrene, the enantioselectivity was around 80% with MeO-MOP ligand. The substituents at the 2′ position in MOP ligands strongly affected the enantioselectivity [35]. Ligand H-MOP (**10f**), which has the same 1,1′-binaphthyl skeleton as MeO-MOP but lacks the methoxy group, is particularly effective for the palladium-catalyzed hydrosilylation of styrene, giving (*R*)-**19** of 94% *ee*. On the other hand, the enantiomeric purities of alcohol **19** obtained with Et-MOP (**10d**) and CN-MOP (**10e**) were much lower, 18% *ee* (*R*) and 26% *ee* (*R*), respectively. The monophosphine (*S*)-**21**, which was prepared through the catalytic

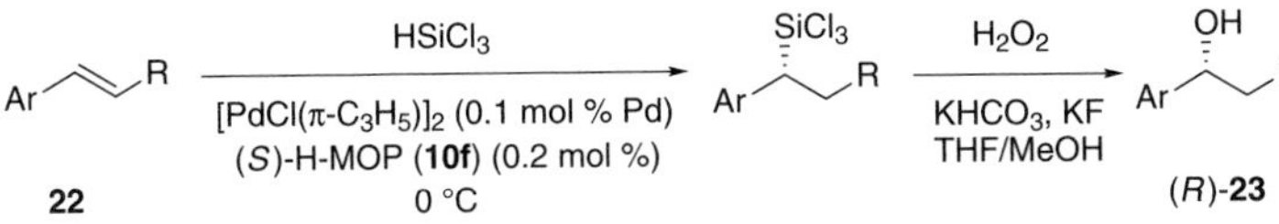

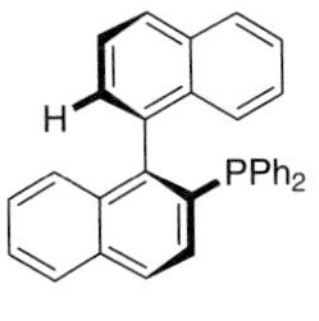

(*S*)-H-MOP (**10f**)

Ar and R in **22**	% ee of **23**
Ar = 4-$CF_3C_6H_4$, R = H	96% ee
Ar = 4-ClC_6H_4, R = H	94% ee
Ar = 4-MeC_6H_4, R = H	89% ee
Ar = 4-$MeOC_6H_4$, R = H	61% ee
Ar = Ph, R = Me	89% ee (at 20 °C)
Ar = Ph, R = Bu-*n*	92% ee (at 20 °C)

Scheme 3-10

asymmetric cross-coupling [36], was as effective as (*S*)-H-MOP (**10f**) for the hydrosi-lylation of styrene giving (*R*)-**19** of 91% *ee*. These results suggest that the small size of the hydrogen at the 2' position in H-MOP (**10f**) is important for high enantiose-lectivity. Asymmetric hydrosilylation of styrenes **22** substituted on the phenyl ring or β-position catalyzed by palladium/H-MOP (**10f**) also proceeded with high enan-tioselectivity giving the corresponding optically active benzylic alcohols **23** of high enantiomeric purity (Scheme 3-10) [35].

In all of these cases, palladium-catalyzed hydrosilylation proceeds *via* hydropalla-dation followed by reductive elimination of alkyl- and silyl group from the palladi-um. In the reaction of o-allylstyrene (**24**) with trichlorosilane, which gives hydrosily-lation products on the styrene double bond **25** and cyclized product **26**, the hy-dropalladation process is supported by the absence of side products which would re-sult from the intermediate of the silylpalladation process (Scheme 3-11) [37].

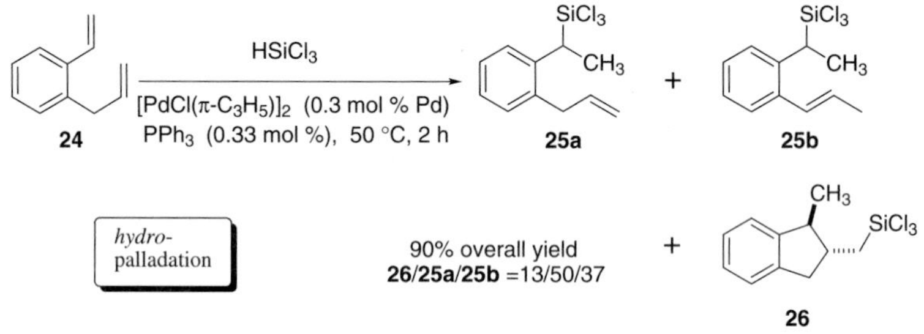

Scheme 3-11

3.3.3
Hydrosilylation of Cyclic Olefins

Asymmetric synthesis through a selective monofunctionalization of enantiotopic positions is one of the most attractive strategies for one-step construction of multi-ple chiral carbon centers [38, 39]. Asymmetric hydrosilylation of norbornene (**27**) was first attempted by the use of palladium catalyst coordinated with ferrocenyl-monophosphine, (*R*)-(*S*)-PPFA (**18a**) [30]. The hydrosilylation of **27** with trichlorosi-lane gave (1*R*,2*R*,4*S*)-*exo*-2-(trichlorosilyl)norbornane (**28**) of about 50% *ee* (Scheme 3-12). Treatment of **28** with potassium fluoride followed by oxidation of the resulting pentafluorosilicate with MCPBA or NBS gave *exo*-2-norbornanol (**29**) or *endo*-2-bromonorbornane (**30**), respectively. A fine tuning of chiral ferrocenylphosphine ligand by introduction of 3,5-bis(trifluoromethyl)phenyl group and a pyrazole **18d** gave the corresponding product of up to 99% *ee* [40]. The palladium–MeO-MOP (**10a**) complex shows high enantioselectivity and catalytic activity in most cases [41]. The hydrosilylation of norbornene (**27**) with trichlorosilane took place at 0°C in the presence of 0.01 mol% of the MOP/palladium catalyst to give a quantitative yield of *exo*-2-(trichlorosilyl)norbornane (**28**) as a single product (Scheme 3-13). Direct oxi-dation of **28** with hydrogen peroxide in the presence of a large excess of potassium

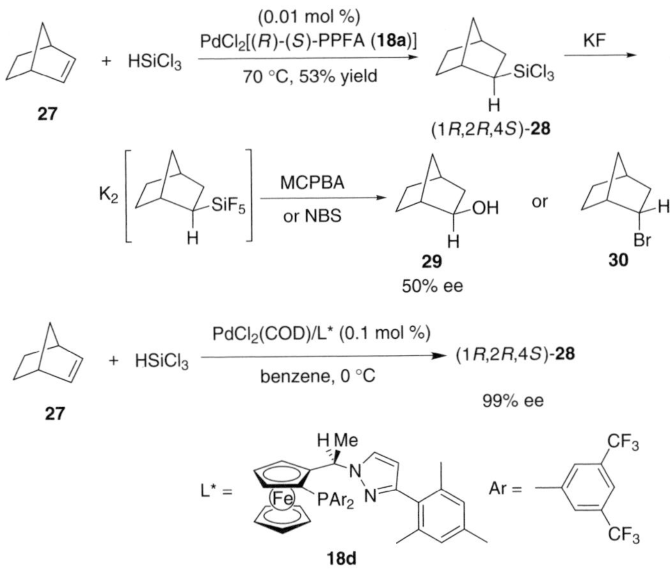

Scheme 3-12

fluoride and potassium bicarbonate gave (1*S*,2*S*,4*R*)-*exo*-2-norbornanol (**29**) of 93% *ee* in high yield. The hydrosilylation carried out at –20°C raised the enantiomeric excess to 96% *ee*. Bicyclo[2.2.2]octene, a diester of norbornenedicarboxylic acid, and 2,5-dihydrofuran derivatives were also successfully subjected to the asymmetric hydrosilylation-oxidation under similar reaction conditions to give the corresponding optically active alcohols, the enantioselectivity being in excess of 92% [42].

It is remarkable that the monofunctionalization of norbornadiene (**31**), giving *exo*-5-trichlorosilyl-2-norbornene (**32a**), is effected by the palladium-MOP catalyst with high chemo- and enantioselectivity [41] (Scheme 3-14). Thus, the reaction of **31** with

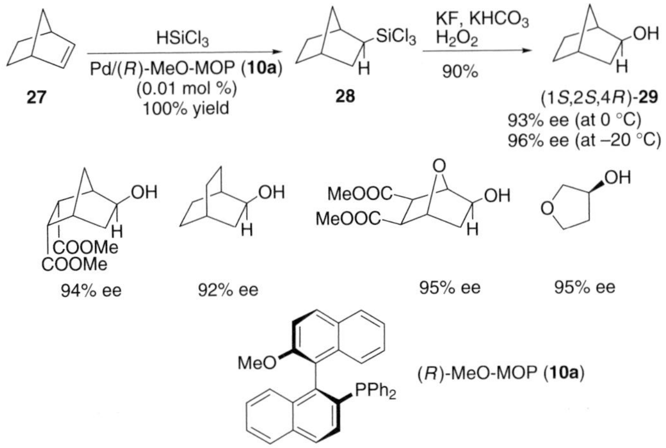

Scheme 3-13

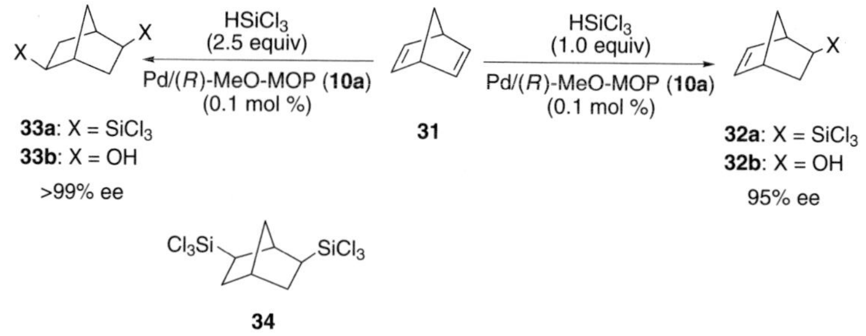

33a: X = SiCl₃
33b: X = OH

>99% ee

31

32a: X = SiCl₃
32b: X = OH

95% ee

34

Scheme 3-14

1.0 equiv. of trichlorosilane and the palladium/MeO-MOP catalyst followed by hydrogen peroxide oxidation gave (1*R*,4*R*,5*S*)-*exo*-5-hydroxy-2-norbornene (**32b**) with 95% *ee*. Reacting **31** with 2.5 equiv. of trichlorosilane induced enantioselective hydrosilylation in both double bonds, thus giving a 78% yield of chiral disilylnorbornane **33a** and the meso isomer **34** in a ratio of 18 : 1. The oxidation of **33a** gave the diol (1*R*,2*S*,4*R*,5*S*)-**33b** with more than 99% *ee*, the high enantiomeric purity being due to the expected double stereoselection.

Very recently, the yttrium hydride {[2,2'-bis(tert-butyldimethylsilylamido)-6,6'-dimethylbiphenyl]YH(THF)}₂ (**36**), conveniently generated in situ from [2,2'-bis(tert-butyldimethylsilylamido)-6,6'-dimethylbiphenyl]YMe(THF)₂ (**35**) demonstrated its high catalytic activity in olefin hydrosilylation. This system represents the first use of a d⁰ metal complex with non-Cp ligands for the catalytic hydrosilylation of olefins. Hydrosilylation of norbornene with PhSiH₃ gave the corresponding product (**37**) of 90% *ee* (Scheme 3-15) [43].

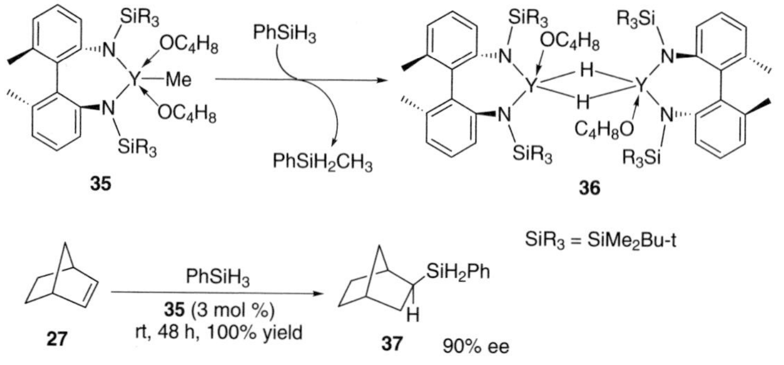

35

36

SiR₃ = SiMe₂Bu-t

27

PhSiH₃

35 (3 mol %)
rt, 48 h, 100% yield

37   90% ee

Scheme 3-15

3.4
Asymmetric Hydrosilylation of Dienes

3.4.1
Hydrosilylation of 1,3-Dienes

Palladium-catalyzed hydrosilylation of 1,3-dienes is one of the important synthetic methods for allylic silanes, and considerable attention has been paid to their asymmetric synthesis by this catalytic method [44]. The optically active allylic silanes have been used as chiral allylating reagents in the S_E' reactions with electrophiles, typically aldehydes [45, 46]. In the presence of Pd catalysts, the reaction with hydrosilanes containing electron-withdrawing heteroatoms or substituents on the silicon usually proceeds in a 1,4-fashion giving allylic silanes [47, 48]. Asymmetric hydrosilylation of cyclopentadiene (38), forming optically active 3-silylcyclopentene (39), has been most extensively studied (Scheme 3-16). In the first report, hydrosilylation of cyclopentadiene (38) with methyldichlorosilane in the presence of 0.01 mol% of palladium–(R)-(S)-PPFA (18a) as a catalyst gave allylsilane (S)-39a of 24% ee [49]. The use of ferrocenylphosphines 18d, e containing perfluoroalkyl groups on the side chain for the reaction of 38 with trichlorosilane increased the enantioselectivity (up to 60% ee see Table 3-1) [50]. Some of (β-N-sulfonylaminoalkyl)phosphines (20) [33] and phosphetane ligand 40 are also useful for the asymmetric hydrosilylation of 38, which gave 39b of 71% ee [51, 52]. The highest enantioselectivity so far reported for cyclopentadiene is 80% ee, which was obtained with MOP-phen ligand 41 [53].

For the asymmetric hydrosilylation of 1,3-cyclohexadiene (42) (Scheme 3-17), the enantioselectivity is higher in the reaction with phenyldifluorosilane than that with trichlorosilane or methyldichlorosilane. The reaction of 42 with phenyldifluorosilane in the presence of a palladium catalyst coordinated with ferrocenylphosphine

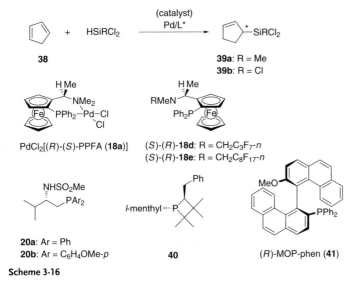

Scheme 3-16

Tab. 3-1 Asymmetric Hydrosilylation of Cyclopentadiene (38)

Ligand L*	Catalyst (mol %)	HSiX₃	Temp (°C)	Time (h)	Yield (%)	% EE Of 39	Ref.
(R)-(S)-PPFA (18a)	0.01	HSiMeCl₂	30	20	87	25 (S)	[49]
(S)-(R)-18d	0.02	HSiCl₃	25	90	73	57 (R)	[50]
(S)-(R)-18d	0.02	HSiCl₃	0	20	7	60 (R)	[50]
(S)-(R)-18e	0.02	HSiCl₃	25	90	41	55 (R)	[50]
(S)-20a	0.1	HSiCl₃	0	–	82	61 (S)	[33]
(S)-20a	0.1	HSiCl₃	–20	–	35	71 (S)	[33]
(S)-20b	0.1	HSiCl₃	0	–	74	62 (S)	[33]
40	0.03	HSiCl₃	70	30	26	44 (S)	[51, 52]
40	0.03	HSiCl₃	25-30	2	70	54 (S)	[51, 52]
(R)-MOP-phen (41)	0.1	HSiCl₃	20	120	99	80 (R)	[53]
(R)-MOP-phen (41)	0.1	HSiCl₃	40	45	85	72 (R)	[53]
(R)-MeO-MOP (10a)	0.1	HSiCl₃	20	14	100	39 (R)	[53]

18f gave allylsilane (S)-**43c** of 77% *ee* (Table 3-1) [54, 55]. In the asymmetric hydrosilylation of 1,3-cyclohexadiene, axially chiral ligands only exhibited the same level of asymmetric induction under optimized conditions as ferrocenyl ligands. The use of phenyldifluorosilane in place of trichlorosilane did not improve the enantioselectivity, but the reaction with deuterium-labeled silane (DSiF₂Ph) gave significant insight into the mechanism of palladium-catalyzed hydrosilylation of 1,3-dienes. The reaction of 1,3-cyclohexadiene with DSiF₂Ph gave *cis*-3-(phenyldifluorosilyl)-6-deuterio-cyclohexene as a single isomer without any diastereo- or regioisomers, demonstrating that 1,4-*cis*-addition is an exclusive pathway. The π-allylpalladium intermediate, which is formed by the addition of PdD(Si)L* species to the diene, has the silyl group located at the *trans* position to the π-allyl carbon next to the deuterated carbon, rapidly undergoes reductive elimination to form compound (**44**) before *trans-cis* isomerization of intermediate can occur.

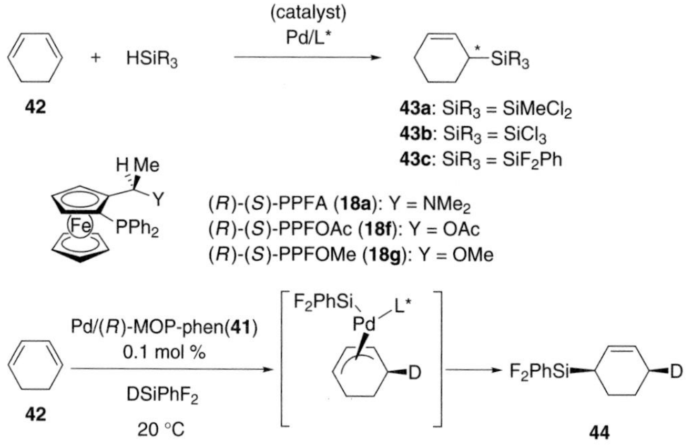

Scheme 3-17

Tab. 3-2 Asymmetric Hydrosilylation of 1,3-Cyclohexadiene (**42**)

Ligand L*	Catalyst (mol %)	HSiX₃	Temp (°C)	Time (h)	Yield (%)	% EE Of 43	Ref.
(R)-(S)-PPFA (**18a**)	0.01	HSiMeCl₂	30	20	95	2 (S)	[49]
(R)-MOP-phen (**41**)	0.1	HSiCl₃	20	150	99	51 (R)	[53]
(R)-(S)-PPFOAc (**18f**)	1	HSiPhF₂	rt	20	58	77 (S)	[54, 55]
(R)-(S)-PPFOMe (**18g**)	1	HSiPhF₂	rt	20	50	54 (S)	[54, 55]

Linear 1,3-dienes have also been subjected to palladium-catalyzed asymmetric hydrosilylation. Reaction of 1-phenyl-1,3-butadiene (**45**) with HSiCl₃ catalyzed by palladium–(R)-(S)-PPFA (**18a**) gave a mixture of regioisomeric allylsilanes **46a** and **47a** in a ratio of 94 : 6, the major isomer **46a** and the minor isomer **47a** being 66% *ee* (S) and 30% *ee* (R), respectively (Scheme 3-18) [56]. π-Allylpalladium intermediate **48** was proposed for this hydrosilylation. The use of phenyldifluorosilane in place of trichlorosilane slightly improved the enantioselectivity [54, 57]. Hydrosilylation of alkyl-substituted 1,3-dienes **49** and **50** in the presence of a ferrocenylmonophosphine-palladium catalyst also proceeded with high regioselectivity to give the corre-

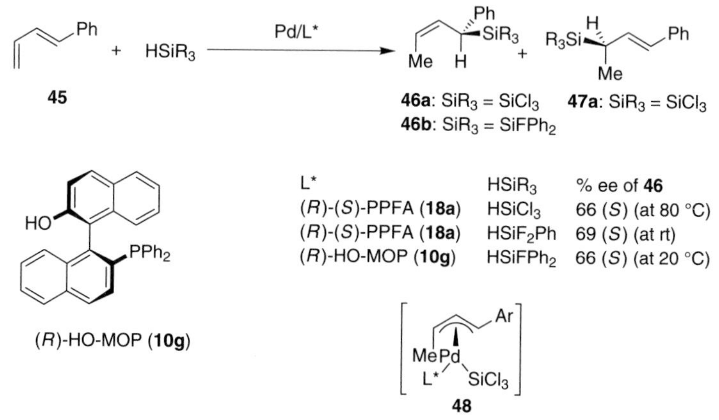

L*	HSiR₃	% ee of **46**
(R)-(S)-PPFA (**18a**)	HSiCl₃	66 (S) (at 80 °C)
(R)-(S)-PPFA (**18a**)	HSiF₂Ph	69 (S) (at rt)
(R)-HO-MOP (**10g**)	HSiFPh₂	66 (S) (at 20 °C)

Scheme 3-18

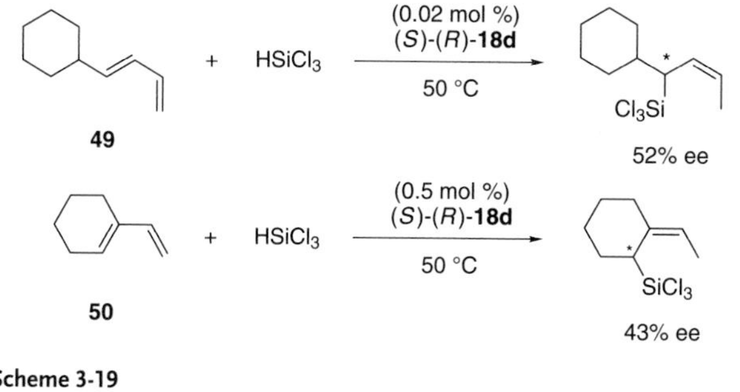

Scheme 3-19

sponding 1,4-addition products with moderate enantioselectivity (Scheme 3-19) [50, 58].

Asymmetric hydrosilylation can be extended to 1,3-diynes for the synthesis of optically active allenes, which are of great importance in organic synthesis, and few synthetic methods are known for their asymmetric synthesis with chiral catalysts. Catalytic asymmetric hydrosilylation of butadiynes provides a possible way to optically allenes, though the selectivity and scope of this reaction are relatively low. A chiral rhodium complex coordinated with (2S,4S)-PPM turned out to be the best catalyst for the asymmetric hydrosilylation of butadiyne to give an allene of 22% *ee* (Scheme 3-20) [59].

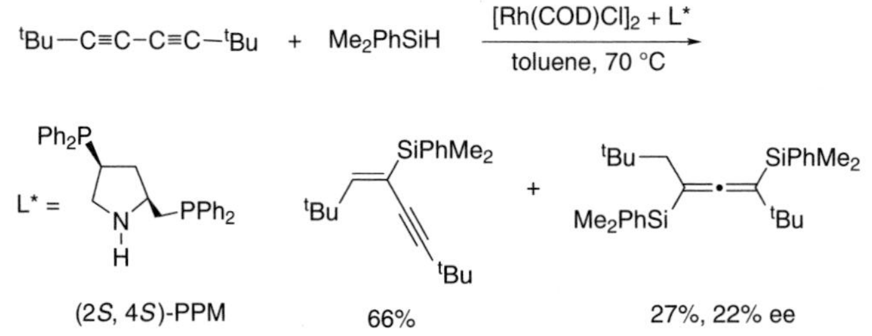

^{t}Bu$-$C$\equiv$C$-$C$\equiv$C$-$^{t}Bu + Me$_2$PhSiH $\quad\xrightarrow[\text{toluene, 70 °C}]{\text{[Rh(COD)Cl]}_2 + \text{L}^*}$

L* = (drawn structure with Ph$_2$P, N-H, PPh$_2$)

(2S, 4S)-PPM

(allene product, ^{t}Bu, SiPhMe$_2$, ^{t}Bu) 66%

+

(^{t}Bu, SiPhMe$_2$, Me$_2$PhSi, ^{t}Bu) 27%, 22% ee

Scheme 3-20

3.4.2
Cyclization/Hydrosilylation of 1,6-Dienes

Hydrosilylation of dienes accompanied by cyclization is emerging as a potential route to the synthesis of functionalized carbocycles. However, the utility of cyclization/hydrosilylation has been limited because of the absence of an asymmetric protocol. One example of asymmetric cyclization/hydrosilylation has been reported very recently using a chiral pyridine-oxazoline ligand instead of 1,10-phenanthroline of the cationic palladium complex (**53**) [60]. As shown in Scheme 3-21, the pyridine-oxazoline ligand is more effective than the bisoxazoline ligand in this asymmetric cyclization/hydrosilylation of a 1,6-diene.

A plausible mechanism proposed for this reaction involves migratory insertion of an olefin into the Pd–Si bond of a palladium-silyl intermediate **I** followed by migratory insertion of the pendant olefin into the resulting Pd–C bond of **II** forming palladium-alkyl intermediate **III**. Reaction of **III** with hydrosilane releases the carbocycle to regenerate the palladium-silyl complex **I** (Scheme 3-21) [61].

3.5
Asymmetric Intramolecular Hydrosilylation

Intramolecular asymmetric hydrosilylation-oxidation of (alkenyloxy)hydrosilanes provides an efficient method for the preparation of optically active polyols from allylic alcohols. Cyclization of silyl ethers **54** of a *meso*-type allylic alcohol in the pres-

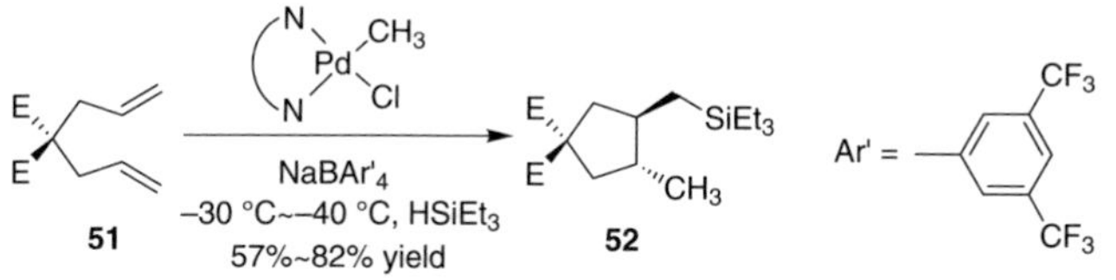

Scheme 3-21

N N	R	de of **52** (%)	% ee of **52**
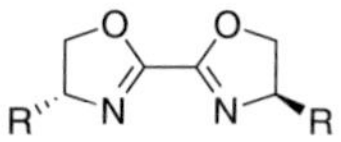 (oxazoline)	Bn	>95	72
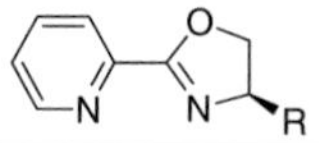 (pyridine-oxazoline)	i-Pr	>95	87
	t-Bu	91	91

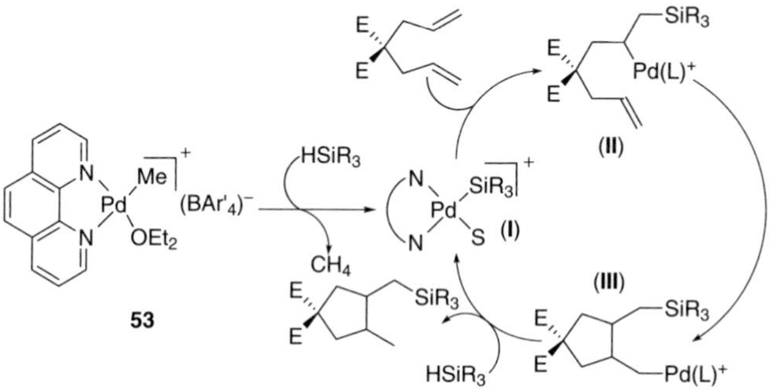

ence of rhodium-DIOP (**5**) as a catalyst proceeded with high diastereoselectivity and high enantioposition-selectivity. Oxidation of the carbon-silicon bond in the resulting sila-oxa-cyclopentane derivatives **55** gave syn-2,4-dimethyl-4-pentene-1,3-diol (**56**) of high enantiomeric excess (Scheme 3-22) [62]. The enantioselectivity was dependent on the alkyl groups on the silicon, sterically hindered 3,5-dimethylphenyl group giving the highest selectivity (93% ee).

Asymmetric cyclization was also successful in the rhodium-catalyzed hydrosilylation of silyl ethers **57** derived from allyl alcohols. High enantioselectivity (up to 97%

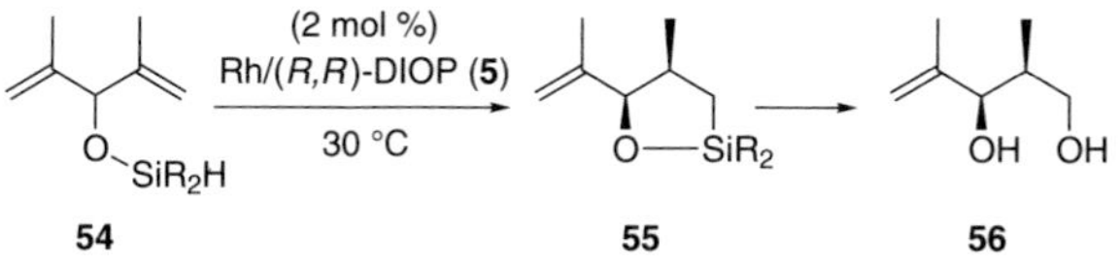

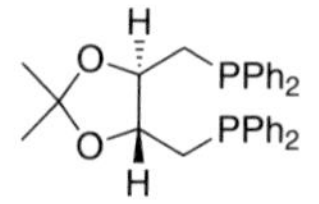

R	syn/anti	% ee
Me	95/5	80
Ph	99/1	83
3,5-Me$_2$C$_6$H$_3$	>99/1	93

(R,R)-DIOP (**5**)

Scheme 3-22

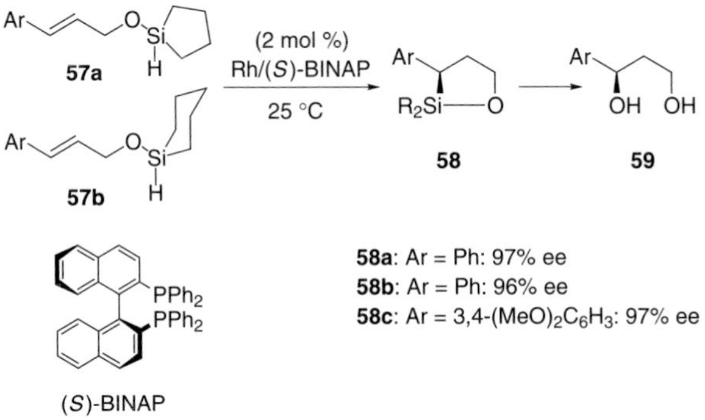

Scheme 3-23

58a: Ar = Ph: 97% ee
58b: Ar = Ph: 96% ee
58c: Ar = 3,4-(MeO)$_2$C$_6$H$_3$: 97% ee

(S)-BINAP

ee) was observed in the reaction of silyl ethers containing a bulky group on the silicon atom in the presence of a rhodium-BINAP catalyst (Scheme 3-23) [63]. The cyclization products **58** were readily converted into 1,3-diols **59** by the oxidation. During studies on this asymmetric hydrosilylation, the silyl-rhodation pathway in the catalytic cycle was demonstrated by a deuterium scrambling method [64].

Axially chiral spirosilane **61** was efficiently prepared by double intramolecular hydrosilylation of bis(alkenyl)dihydrosilane **60**. By use of SILOP ligand, a C_2 symmetric spirosilane which is almost enantiomerically pure was obtained with high diastereoselectivity (Scheme 3-24) [65]. SILOP ligand is much more stereoselective for this asymmetric hydrosilylation than DIOP (**5**) though they have similar structure.

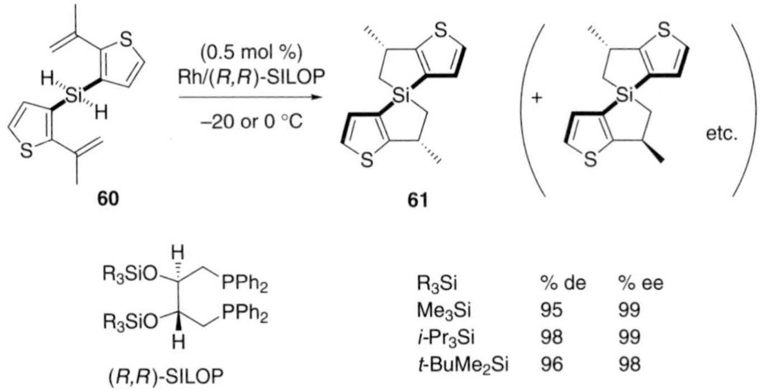

R$_3$Si	% de	% ee
Me$_3$Si	95	99
i-Pr$_3$Si	98	99
t-BuMe$_2$Si	96	98

(R,R)-SILOP

Scheme 3-24

References

1 I. OJIMA in *The Chemistry of Organic Silicon Compounds*, S. PATAI and Z. RAPPOPORT (eds.), John Wiley, Chichester, **1989**, pp 1479-1526.

2 B. MARCINIEC (ed.), *Comprehensive Handbook on Hydrosilylation*, Pergamon, Oxford, **1992**.

3 T. HAYASHI in *Comprehensive Asymmetric Catalysis (I)* E. N. JACOBSON, A. PFALTZ and H. YAMAMOTO (eds.), Springer-Verlag, Berlin Heidelberg, **1999**, Vol. 1, pp 319-333.

4 K. YAMAMOTO, T. HAYASHI in *Transition Metals for Organic Synthesis (II)* M. BELLER and C. BOLM (eds.), Wiley-VCH, Weinheim, **1998**, Vol. 2, pp 120-131.

5 T. HIYAMA, T. KUSUMOTO in *Comprehensive Organic Synthesis* B. M. TROST and I. FLEMING (eds.), Pergamon, Oxford, **1991**, Vol. 8, pp 763-792.

6 K. TAMAO in *Advances in silicon chemistry*, G. L. LARSON (ed.), JAI Press, London, **1996**, pp 1- 62.

7 K. TAMAO in *Organosilicon and Bioorganosilicon Chemistry*, H. SAKURAI, E. HORWOOD (eds.), Chichester, **1985**, pp 231-242.

8 C. E. MASSE, J. S. PANE, *Chem. Rev.* **1995**, *95*, 1293-1316.

9 a) S. KOBAYASHI, K. NISHINO, *J. Org. Chem.* **1994**, *59*, 6620-6628. b) S. KOBAYASHI, R. HIRABAYASHI, *J. Am. Chem. Soc.* **1999**, *121*, 6942-6943.

10 J. L. SPEIER, J. A. WEBSTER, G. H. BARNES, *J. Am. Chem. Soc.* **1957**, *79*, 974-979.

11 A. J. CHALK, J. F. HARROD, *J. Am. Chem. Soc.* **1965**, *87*, 16-21.

12 T. D. TILLEY in *The Chemistry of Organic Silicon Compounds*, S. PATAI and Z. RAPPOPORT (eds.), John Wiley, Chichester, **1989**.

13 F. OZAWA, T. HIKIDA, T. HAYASHI, *J. Am. Chem. Soc.* **1994**, *116*, 2844-2849.

14 B. BOSNICH, *Acc. Chem. Res.* **1998**, *31*, 667-674.

15 S. DUCKETT, R. N. PERUTZ, *Organometallics* **1992**, *11*, 90-98.

16 M. BROOKHART, B. E. GRANT, *J. Am. Chem. Soc.* **1993**, *115*, 2151-2156.

17 K. YAMAMOTO, T. HAYASHI, M. KUMADA, *J. Am. Chem. Soc.* **1971**, *93*, 5301-5302.

18 K. YAMAMOTO, T. HAYASHI, M. ZEMBAYASHI, M. KUMADA, *J. Organomet. Chem.* **1976**, *118*, 161-181.

19 K. YAMAMOTO, Y. URAMOTO, M. KUMADA, *J. Organomet. Chem.* **1971**, *31*, C9-C10.

20 K. YAMAMOTO, T. HAYASHI, Y. URAMOTO, R. ITO, M. KUMADA, *J. Organomet. Chem.* **1976**, *118*, 331-348.

21 G. A. MOLANDER, E. D. DOWDY, B. C. NOLL, *Organometallics* **1998**, *17*, 3754-3758.

22 P-F, FU, L. BRARD, Y. LI, T. J. MARKS, *J. Am. Chem. Soc.* **1995**, *117*, 7157-7168.

23 T. HAYASHI, *Acta Chem. Scand.* **1996**, *50*, 259-266.

24 Y. UOZUMI, T. HAYASHI, *J. Am. Chem. Soc.* **1991**, *113*, 9887-9888.

25 Y. UOZUMI, K. KITAYAMA, T. HAYASHI, K. YANAGI, E. FUKUYO, *Bull. Chem. Soc. Jpn.* **1995**, *68*, 713-722.

26 Y. KISO, K. YAMAMOTO, K. TAMAO, M. KUMADA, *J. Am. Chem. Soc.* **1972**, *94*, 4373-4374.

27 K. Yamamoto, Y. Kiso, R. Ito, K. Tamao, M. Kumada, *J. Organomet. Chem.* **1981**, *210*, 9-17.

28 T. Hayashi in *Ferrocenes*, A. Togni and T. Hayashi (eds.), VCH, Weinheim, **1995**.

29 T. Hayashi, *Pure Appl. Chem.* **1988**, *60*, 7-12.

30 T. Hayashi, K. Tamao, Y. Katsuro, I. Nakae, M. Kumada, *Tetrahedron Lett.* **1980**, *21*, 1871-1874.

31 H. L. Pedersen, M. Johannsen, *Chem. Commun.* **1999**, 2517-2518.

32 W. R. Cullen, N. F. Han, *J. Organomet. Chem.* **1987**, *333*, 269-280.

33 T. Okada, T. Morimoto, K. Achiwa, *Chem. Lett.* **1990**, 999-1002.

34 Y. Uozumi, K. Kitayama, T. Hayashi, *Tetrahedron: Asymmetry* **1993**, *4*, 2419-2422.

35 K. Kitayama, Y. Uozumi, T. Hayashi, *J. Chem. Soc., Chem. Commun.* **1995**, 1533-1534.

36 T. Hayashi, S. Niizuma, T. Kamikawa, N. Suzuki, Y. Uozumi, *J. Am. Chem. Soc.* **1995**, *117*, 9101-9102.

37 Y. Uozumi, H. Tsuji, T. Hayashi, *J. Org. Chem.* **1998**, *63*, 6137-6140.

38 J. D. Morrison (ed). *Asymmetric Synthesis*, Vol. 1–5. London: Academic Press, 1983 – 1985.

39 M. Nógrádi, *Stereoselective Synthesis*, New York: Weinheim, **1987**.

40 G. Pioda, A. Togni, *Tetrahedron: Asymmetry* **1998**, *9*, 3903-3910.

41 Y. Uozumi, S.-Y. Lee, T. Hayashi, *Tetrahedron Lett.* **1992**, *33*, 7185-7188.

42 Y. Uozumi, T. Hayashi, *Tetrahedron Lett.* **1993**, *34*, 2335-2338.

43 T. I. Gountchev, T. D. Tilley, *Organometallics* **1999**, *18*, 5661-5667.

44 I. Ojima, *Catalytic Asymmetric Synthesis*, VCH, New York, **1993**.

45 I. Fleming, J. Dunogues, R. Smithers, *Organic Reactions*, **1989**, Vol. 37.

46 A. Hosomi, H. Sakurai, *J. Org. Synth. Chem. Jpn.* **1985**, *43*, 406-418.

47 J. Tsuji, M. Hara, K. Ohno, *Tetrahedron* **1974**, *30*, 2143-2146.

48 I. Ojima, M. Kumagai, *J. Organomet. Chem.* **1978**, *157*, 359-372.

49 T. Hayashi, K. Kabeta, T. Yamamoto, K. Tamao, M. Kumada, *Tetrahedron Lett.* **1983**, *24*, 5661-5664.

50 T. Hayashi, Y. Matsumoto, I. Morikawa, Y. Ito, *Tetrahedron: Asymmetry* **1990**, *1*, 151-154.

51 A. Marinetti, *Tetrahedron Lett.* **1994**, *35*, 5861-5864.

52 A. Marinetti, L. Ricard, *Organometallics* **1994**, *13*, 3956-3962.

53 K. Kitayama, H. Tsuji, Y. Uozumi, T. Hayashi, *Tetrahedron Lett.* **1996**, *37*, 4169-4172.

54 H. Ohmura, H. Matsuhashi, M. Tanaka, M. Kuroboshi, T. Hiyama, Y. Hatanaka, K-I. Goda, *J. Organomet. Chem.* **1995**, *499*, 167-171.

55 T. Hiyama, H. Matsuhashi, A. Fujita, M. Tanaka, K. Hirabayashi, M. Shimizu, A. Mori, *Organometallics* **1996**, *15*, 5762-5765.

56 T. Hayashi, K. Kabeta, *Tetrahedron Lett.* **1985**, *26*, 3023-3026.

57 Y. Hatanaka, K-I. Goda, F. Yamashita, T. Hiyama, *Tetrahedron Lett.* **1994**, *35*, 7981-7982.

58 T. Hayashi, S. Hengrasmee, Y. Matsumoto, *Chem. Lett.* **1990**, 1377-1380.

59 A. Tillack, D. Michalik, C. Koy, M. Michalik, *Tetrahedron Lett.* **1999**, *40*, 6567-6568.

60 N. S. Perch, R. A. Widenhoefer, *J. Am. Chem. Soc.* **1999**, *121*, 6960-6961.

61 R. A. Widenhoefer, M. A. DeCarli, *J. Am. Chem. Soc.* **1998**, *120*, 3805-3806.

62 K. Tamao, T. Tohma, N. Inui, O. Nakayama, Y. Ito, *Tetrahedron Lett.* **1990**, *31*, 7333-7336.

63 S. E. Bergens, P. Noheda, J. Whelan, B. Bosnich, *J. Am. Chem. Soc.* **1992**, *114*, 2121-2128.

64 S. H. Bergens, P. Noheda, J. Whelan, B. Bosnich, *J. Am. Chem. Soc.* **1992**, *114*, 2128-2135.

65 K. Tamao, K. Nakamura, H. Ishii, S. Yamaguchi, M. Shiro, *J. Am. Chem. Soc.* **1996**, *118*, 12469-12470.

4
Catalytic Hydroamination of Unsaturated Carbon-Carbon Bonds

J. J. Brunet and D. Neibecker

4.1
Introduction

Ammonia, whose name seems to come from that of the Egyptian god Amon, was obtained for the first time in 1612 by Kunchel by heating ammonium chloride from animal excrements with calcium hydroxide [1]. This gas, described as stifling, has been studied by Scheele and Priestley, and its formula (NH_3) was established by Berthollet in 1875 [1, 2].

It was only around 1850 that the first amines were discovered by Wurtz [2], who considered them as alkylated (or arylated) derivatives of NH_3. Nowadays, it is well known that the amine function is widespread among biologically important compounds, but mostly it is present in polyfunctional molecules such as amino acids, alkaloids, etc. Simple amines are very rare in nature, with the exception of triethylamine and the trimethylammonium ion which come from the putrefaction of proteins.

Amines are important industrial chemicals which are involved in everyday life [3, 4]. Apart from the usual classification into primary, secondary, and tertiary amines, the distinction is often made between 'light' amines (less than six-carbon substituents) and 'fatty' amines. Light amines are intermediates for the synthesis of drugs, herbicides, cosmetics, etc. [3]. They also find use as vulcanization accelerators and extraction agents. Fatty amines are involved in the synthesis of corrosion inhibitors and cationic surfactants, which are used in ore flotation processes and are good fabric softeners and antistatic agents [4–6].

Since the discovery of NH_3, chemists have developed synthetic methods for the preparation of amines, and many methods are now available [7]. However, most of these reactions are unsuitable for industrial purposes, mainly because they involve too expensive starting materials or reagents. Ammonia thus appeared as the best reagent for the introduction of the amine function into an organic molecule. From this point of view, an important landmark was the design, by Haber and Bosch in the early 20th century, of a process for the synthesis of NH_3 directly from nitrogen and hydrogen in the presence of catalysts [2].

Ammonia has always been the starting material for the synthesis of aliphatic amines. Thus, processes have been developed for the condensation of NH_3 with alkyl halides (Hoffman reaction) or with alcohols in the presence of various catalysts. The latter reaction, first discovered by Sabatier in 1909 [8, 9] is nowadays the main method of industrial production of light amines (e.g. methylamines: 600 000 t/yr) [5].

With the exception of methanol and ethanol, most alcohols are produced from olefins, either through hydration [10] or *via* a hydroformylation-hydrogenation sequence (Scheme 4-1) [11].

It is thus obvious that the direct transformation of a simple alkene into an amine would be a more economic process, since it would suppress at least one step without formation of co-products (atom efficiency) [Scheme 4-1, paths (c) and (d)].

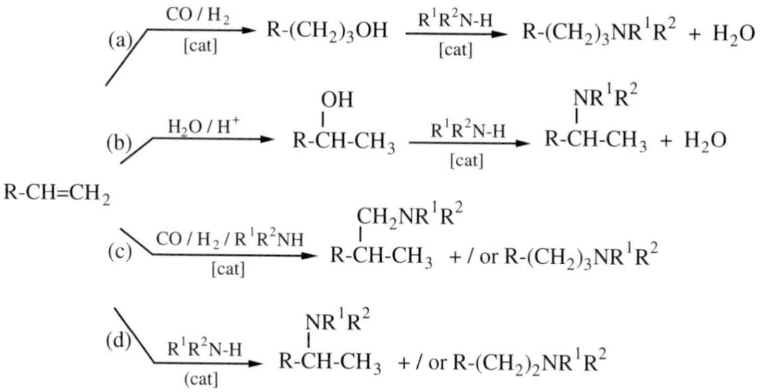

Scheme 4-1 From alkenes to amines

Hydroaminomethylation of alkenes [path (c)] will not be considered [12]. This review deals exclusively with the hydroamination reaction [path (d)], i.e. the direct addition of the N–H bond of NH_3 or amines across unsaturated carbon-carbon bonds. It is devoted to the state of the art for the *catalytic* hydroamination of alkenes and styrenes but also of alkynes, 1,3-dienes and allenes, with no mention of activated substrates (such as Michael acceptors) for which the hydroamination occurs without catalysts. Similarly, the reaction of the N–H bond of amine derivatives such as carboxamides, tosylamides, ureas, etc. will not be considered.

NH_3 and amines are moderate bases ($pK_b = 5$–6) and weak acids ($pK_a = 30$–35). The N–H bond enthalpy is 107 kcal/mol for NH_3, 88–100 kcal/mol for primary amines and 87–91 kcal/mol for secondary amines [13].

From a thermodynamic point of view, the addition of NH_3 and amines to olefins is feasible. For example, the free enthalpy for the addition of NH_3 to ethylene is $\Delta G°$ ≈ -4 kcal/mol [14]. Calculations showed that the enthalpies for the hydroamination of higher alkenes are in the range -7 to -16 kcal/mol and that the exothermicities of both hydration and hydroamination of alkenes are closely similar [15]. Such N–H additions, however, are characterized by a high activation barrier which prevents the reaction from occurring under normal conditions. Furthermore, the reaction entropy is highly negative, so that increasing the temperature tends to shift the equilibrium to the starting materials [16, 17]. Therefore, catalysts must be used to decrease the activation barrier.

Both heterogeneous and homogeneous catalysts have been found which allow the hydroamination reaction to occur. For heterogeneously catalyzed reactions, it is very difficult to determine which type of activation is involved. In contrast, for homogeneously catalyzed hydroaminations, it is often possible to determine which of the reactants has been activated (the unsaturated hydrocarbon or the amine) and to propose reaction mechanisms (catalytic cycles).

Since the review by Lattes et al. on the 'amination of alkenes' in 1983 [18] and our first review on the 'catalytic amination of monoolefins' in 1989 [19], several review articles have appeared in the literature [13, 14, 17, 20–23]. The present review corresponds to an analysis of literature data up to the end of 1999.

4.2
Hydroamination of Alkenes

4.2.1
Heterogeneous Catalysis

4.2.1.1 Catalysis by Transition Metals
The first example of a heterogeneously catalyzed hydroamination of an alkene appeared in a 1929 patent in which it is claimed that NH_3 reacts with ethylene (450°C, 20 bar) over a reduced ammonium molybdate to give $EtNH_2$ [24]. An intriguing reaction was also reported by Bersworth, who reacted oleic acid with NH3 in the presence of catalysts like palladium or platinum black or copper chromite to give the hydroamination product in quantitative yields [25]. However, this result could not be reproduced [26].

Teter et al. filed a series of patents aimed at the 'production of organic compounds containing nitrogen' or the 'production of nitriles and amines from ammonia and olefins' by passing mixtures of olefin and NH3 over transition metals, mainly cobalt deposited on various supports at 250–370°C and 100–200 bar [27–43]. With cobalt on asbestos, a mixture of amine, nitrile, olefin hydrogenation product, polymers, and cracking products is obtained (Eq. 4.1) [31].

At higher temperatures, propene and NH_3 react over basic catalysts to afford a mixture of nitriles (Eq. 4.2) [42].

$$\text{(propene)} + NH_3 \xrightarrow[\substack{100 \text{ bar, } 370°C}]{\substack{25\%Co/MgO \\ 10\text{-}25\% \text{ celite } 337}} MeCN + EtCN + n\text{-PrCN} + i\text{-PrCN} + \ldots \qquad (4.2)$$

up to 36% yield 25% 44% 5% 18% 8%

It was thought that propionitrile came from dehydrogenation of the anti-Markovnikov hydroamination product, n-PrNH$_2$. Propionitrile can break down to ethylene and HCN, the former reacting with NH_3 to generate acetonitrile via ethylamine, the latter adding to propene to form the butyronitriles [26, 37].

PhNH$_2$ reacts with ethylene in the presence of alkali metals, e.g., sodium deposited on alumina, to afford the hydroamination product in good yield but with a low turnover frequency (TOF = mol of product synthesized per mol of catalyst in 1 h) (Eq. 4.3) [44].

$$= + PhNH_2 \xrightarrow[\substack{80 \text{ bar, } 260°C}]{\substack{10\% \text{ Na/Al}_2O_3}} PhNH \text{(alkyl)} + \text{higher alkylates} \qquad (4.3)$$

conversion = 74% 94% 6%
TOF = 1h^{-1}

Other catalysts are 20% Na/C, 5% Li/Al$_2$O$_3$ or 10% Na/8% MoO$_3$-Al$_2$O$_3$ or 10% NaH/Al$_2$O$_3$ [44]. Similarly, gaseous mixtures of olefins and NH_3 have been claimed to give hydroaminated products over a ternary K/graphite/Al$_2$O$_3$ catalyst [45].

Last, McClain disclosed the gas phase hydroamination of ethylene and propene with NH_3 over palladium on alumina (Eq. 4.4) [46].

$$R\text{(alkene)} + NH_3 \xrightarrow[\substack{1\text{-}7 \text{ bar, } 120°C}]{\substack{2\% \text{ Pd/Al}_2O_3}} R\text{-}NH_2 + R\text{-}NH_2 \qquad (4.4)$$

% ?

4.2.1.2 Acid Catalysis

The first example of acid catalysis appeared in a 1934 patent in which it is claimed that 'surface catalysts, particularly hydrosilicates of large surface area', known at that time under the trade name Tonsil, Franconit, Granisol, etc. lead to 'a smooth addition of the olefine to the molecule of the primary aromatic amine'. Aniline and cyclohexene were reacted over Tonsil at 230–240°C to give, *inter alia*, the hydroamination product, N-cyclohexylaniline [47].

The hydroamination of alkenes has been performed in the presence of heterogeneous acidic catalysts such as zeolites, amorphous aluminosilicates, phosphates, mesoporous oxides, pillared interlayered clays (PILCs), amorphous oxides, acid-treated sheet silicates or NafioN-H resins. They can be used either under batch conditions or in continuous operation at high temperature (above 200°C) under high pressure (above 100 bar).

The advent of zeolites gave a new impetus to the research on the hydroamination of alkenes. Zeolites, i.e. either naturally occurring crystalline aluminosilicates or synthetic zeolites, are suitable catalysts since they are usually heat-stable, have a large number of catalytically active centers in combination with a large surface area, and can be extremely shape selective, thus permitting highly selective syntheses in the field of intermediates and fine chemicals [48, 49]. The zeolites used differ in type and in the aftertreatments, which modify nearly at will the acid strength and the number of acid centers (Brønsted and Lewis acid centers), the importance of which was recognized early [50]. Conversion of olefins to the corresponding amines is limited by thermodynamic equilibria. Amination is favored by low temperatures, high pressures and a high amine/olefin ratio. However, high reaction temperatures are required to activate simple olefins [51]. The type of zeolite used depends on the olefin to be aminated. Usually an excess of NH_3 or amine (e.g., NH_3/olefin > 1.3) is used to avoid formation of olefin oligomers or polymers and of coke, leading to catalyst deactivation. In all cases, as a result of acid catalysis, the Markovnikov hydroamination products are obtained.

Large-pore zeolites such as Y zeolites are efficient for the hydroamination of several olefins. For example, propene reacts with NH_3 over SK-500 (a pelleted lanthanum-exchanged zeolite) or La-Y or H-Y zeolites with 6–15% conversion to give i-PrNH$_2$ with high selectivity (95–100%) (Eq. 4.5) [50].

$$\text{(alkene)} + NH_3 \xrightarrow[\text{340-370°C, 50 bar}]{\text{La-Y}} \text{(product)}-NH_2 \qquad (4.5)$$

conversion = 10-15%   100%

Small to medium pore size zeolites, such as H-clinoptilolite, H-offretite or H-erionite, are efficient for the hydroamination of ethylene [51–54]. Ethylene and NH3 react at 360°C and 50 bar over H-clinoptilolite to give EtNH$_2$ only (11.4% conversion). There is a clear shape selectivity since propene and 1-butene as well as higher amines give rise to extremely low conversions [52]. In contrast to H-clinoptilolite or H-erionite, H-offretite is effective for proprene hydroamination with NH_3 (7.2% conversion, 90% i-PrNH$_2$ + 8% i-Pr$_2$NH) [55]. Small pore size H-erionite is the best catalyst in terms of lifetime, conversion and selectivity for the synthesis of ethylamine [56]. The efficiency of H-clinoptilolite can be improved by acid or base plus acid treatment of natural clinoptilolite (18% conversion, EtNH$_2$/Et$_2$NH>20) [57].

With perhaps the exception of H-erionite, the above-mentioned alumina-containing zeolites suffer from the disadvantage of rapid deactivation. Zeolites of the pentasil type, namely aluminosilicates of high silica content (SiO_2/Al_2O_3>90), iron silicates (SiO_2/Fe_2O_3>10) [58], borosilicates, and borogermanates of high silica content [59] are characterized by longer lifetimes, similar conversion, and as high selectivity (>95%) as the aluminosilicates of the Y type. The best results were obtained in the synthesis of t-BuNH$_2$ with a borosilicate (94%SiO$_2$/3%B$_2$O$_3$) molded with high-surface-area, high-silica-content aluminosilicates (Eq. 4.6) [59].

$$\diagdown\!\!\!\!= \;+\; NH_3 \;\xrightarrow[\text{300°C, 300 bar}]{\text{borosilicate}}\; \diagup\!\!\!\!\diagdown\!\!-NH_2 \tag{4.6}$$

conversion = 17.3%   97.3%

Many studies have been devoted to the hydroamination of isobutene with NH_3 since BASF started the production of t-$BuNH_2$ in Antwerp in 1986 (6000 t/yr) [60, 61]. These studies were aimed mainly at improving conversion, selectivity, catalyst lifetime and space time yields, using less expensive catalysts than zeolites, decreasing the NH_3/isobutene ratio to nearly 1/1, and recycling of the NH_3/isobutene mixtures.

Particularly effective catalysts are H-Y or rare earth-exchanged Y zeolites [51, 62], boro- or ferrosilicates of the pentasil type impregnated with 2% of chromium [63], dealuminated Y zeolites [64], the H form of gallium silicates of the pentasil type [65], H-ZSM-5 of high silica content (H-MFI-81>H-MFI-51>H-MFI41>H-MFI-25) [66–69], oxide mixtures consisting of a combination made up of at least two kinds of oxides (except a combination of silica and alumina) [70], crystalline silicoaluminophosphates or metal-containing crystalline aluminophosphates [71], acid-modified smectite clays such as Montmorillonite clays treated with mineral acids or modified with Lewis acids, fluorophosphoric acid or heteropolyacids (tungsto- and molybdophosphoric acids, tungsto- and molybdosilicic acids) on an inert support (titania) [72], H-Y and H-ZSM-5 zeolites, if necessary dealuminated [73], highly acidic high-silica-containing ($SiO_2/Al_2O_3 > 60$) high-surface-area (> 300 m^2/g) H-β zeolites (CP-704-3 or C-861-β) if necessary modified by impregnation with transition metals or F- [74], dealuminated H-Y zeolites (LZY-82) [75], mesoporous or microporous X-ray amorphous, high-silica-containing high-surface-area (>500 m^2/g) catalysts of the composition $aMO_2.bQ_2O_3.cP_2O_5$, where M stands for Si, Ti, Ge and Q for Al, B, Cr, Fe, Ga, the most active being aluminosilicates ($SiO_2/Al_2O_3>30$) or borosilicates ($SiO_2/B_2O_3>50$) [76], H-β zeolites [77, 78], acid-washed pillared clays with TiO_2, ZrO_2, or Al_2O_3 pillars [79], boron β zeolites [80], the H-form of zeolite SSZ-337 [81], hexagonal faujasites (H-EMT), or their mixed crystals with cubic faujasites (EMT-FAU) [82], mesoporous oxides called Si-MPO, Si-B-MPO, Si-Al-MPO, Cs/Si-MPO, or MCM-41 [83], high-surface-area high-silica-containing oxides of Group 4 or 6 or a mixture thereof on a carrier, especially WO_3/SiO_2 or WO_3+ZrO_2/SiO_2 [84], crystalline oxides based on aluminum phosphates and having faujasite structures such as H-SAPO-37 [85], zeolites with NES structure (two-dimensional channel system) like H-NU-87 [86], H-MCM-49 and H-MCM-56 [87], zeolites having a multidimensional channel system like H-NU-85 [88], boron-MCM-22 zeolites (H-ERB-1) [89], zeolites of the type SSZ-26 (aluminosilicate) or SSZ-33 or CIT-1 (borosilicate) [90] and zeolites having the specific structure PSH-3, MCM-22 or SSZ-25 [91].

H-Nafion resins [92] or ammonium halides in the presence of a catalyst promoter on an inert support (e.g. $NH_4I+CrCl_3$ on silica or NH_4I/C) appear less promising catalysts [93].

4.2.2
Homogeneous Catalysis

4.2.2.1 Activation of the Alkene

The first record of amine addition to an alkene in homogeneous medium is Hickinbottom's publication in 1932. $PhNH_2$ or p-toluidine were reacted with cyclohexene
in the presence of the hydrochloride of the amine to give *inter alia* the N-cyclohexy-
larylamines [94]. Low yields but high selectivities to the anti-Markovnikov monohydroamination products were obtained in the presence of light (λ = 160–220 nm) or
in the presence of light (λ > 220 nm) and a photocatalyst such as NH_4Cl, NH_4I,
NH_4I + NBu_4I, Mo_6Cl_{12} or $Fe(CO)_5$+$P(OEt)_3$ [95, 96].

Ammonium salts have also been used as catalysts in the synthesis of t-$BuNH_2$
from NH_3 and isobutene in water (Eq. 4.7). The co-produced t-BuOH can be recycled [97].

$$\text{(structure)} + NH_3 \xrightarrow[\substack{H_2O, 250°C \\ \sim 100 \text{ bar}}]{NH_4Fe(SO_4)_2 \ (5\%)} \text{(structure)}-NH_2 + \text{(structure)}-OH \qquad (4.7)$$

5 equiv.

conversion = 40%   50%, TOF = 1.25 h^{-1} 50%

Hegedus et al. have thoroughly studied the homogeneous hydroamination of
olefins in the presence of transition metal complexes. However, most of these reactions are either promoted or assisted, i.e. are stoichiometric reactions of an amine
with a coordinated alkene [98–101] or, if catalytic, give rise to the oxidative hydroamination products, as for example in the cyclization of o-allylanilines to 2-alkylindoles
[102, 103], i.e. are relevant to Wacker-type chemistry [104].

The first transition metal-catalyzed hydroamination of an olefin was reported in
1971 by Coulson who used rhodium(I), rhodium(III) or iridium(III) catalysts (Eq.
4.8) [105, 106].

$$= + \text{(structure)}NH \xrightarrow[\text{THF, 180°C}]{RhCl_3 \cdot 3H_2O \ (1\%)} \text{(structure)}N- \qquad (4.8)$$

1.5 equiv.

65%, TOF = 22 h^{-1}

This reaction is restricted to ethylene and to secondary amines of high basicity (nucleophilicity) and low steric bulk (Me_2NH, pyrrolidine, piperidine). No high molecular
weight products are formed. However, the same catalysts [107, 108] as well as $PdCl_2$ [108]
also exhibit some activity for the hydroamination of ethylene with $PhNH_2$ (Eq. 4.9).

$$= + PhNH_2 \xrightarrow[\text{200°C}]{RhCl_3 \cdot 3H_2O \ (0.25\%)} \text{(structure)} + \text{(structure)} \qquad (4.9)$$

100 bar

TOF = 0.5 h^{-1} 7.5%   2.5%

The reaction can be made selective for the synthesis of N-ethylaniline (150°C, 10 bar) but at the expense of catalytic activity [107, 108].

Study of the mechanism of the rhodium-catalyzed hydroamination of ethylene with secondary amines indicated that the piperidine complex *trans*-RhCl(C$_2$H$_4$)(piperidine)$_2$ can serve as a catalyst precursor [109, 110].

It was elegantly shown later that the hydroamination of ethylene with piperidine or Et$_2$NH can be greatly improved using cationic rhodium complexes at room temperature and atmospheric pressure to afford a high yield of hydroaminated products (Eq. 4.10) [111]. However, possible deactivation of the catalyst can be questioned [17].

$$\overset{\text{1 bar}}{=\!=} \ + \ R_2NH \ \xrightarrow[\text{THF, r.t.}]{[Rh(Me_2CO)(C_2H_4)(PPh_3)_2]PF_6 \ (1\%)} \ R_2N\diagup \qquad (4.10)$$

R$_2$NH = Et$_2$NH, piperidine   >99%, TOF = 50 h^{-1}

Mechanistic studies, i.e. model studies of the elementary steps of the catalytic cycle, are currently under way [112].

Ruthenium and iron compounds have been claimed to catalyze the hydroamination of olefins with NH$_3$, primary and secondary amines (120–190°C, 10–20 bar) [113, 114]. Ethylene is the most reactive olefin either with ruthenium (Eq. 4.11) or with iron catalysts (Eq. 4.12).

$$\overset{\text{20 bar}}{=\!=} \ + \ Me_2NH \ \xrightarrow[\text{dioxane, 150-190°C}]{Ru(NH_3)_4(OH)Cl.2H_2O \ (1.1\%)} \ Me_2N\diagup \ + \ MeN\diagdown \qquad (4.11)$$

conversion = 21%, TOF = 3 h^{-1} 6%   94%

$$\overset{\text{30 bar}}{=\!=} \ + \ Et_2NH \ \xrightarrow[\text{150-170°C}]{Fe(CO)_5 \ (4\%)/P(OPh)_3 \ (4\%)} \ Et_2N\diagup \ + \ EtNH_2 \qquad (4.12)$$

conversion = 48%, TOF = 2 h^{-1} 89%   11%

As shown in Eqs. (4.11) and (4.12), some dealkylation of the starting amine and redistribution of alkyl groups occur. Hydroamination of ethylene with PhNH$_2$ affords a low-yield mixture of N-ethylaniline, N,N-diethylaniline, 2-methylquinoline, 2-(1-butenyl)aniline, and N-ethyltoluidine [113, 114].

4.2.2.2 Activation of the Amine

4.2.2.2.1 Activation by Bases

The first examples were reported in a patent by Gresham et al. in 1950 [115]. NH$_3$ reacts with propene in the presence of catalytic amounts of sodium in benzene at 950–1070 bar and 267-278°C to give mainly i-PrNH$_2$ (Eq. 4.13). Under similar conditions (650–700 bar, 265°C), isobutene yields t-BuNH$_2$.

$$\text{} + NH_3 \xrightarrow[\text{650-700 bar, 270°C}]{\text{Na (10\%), benzene}} \text{} -NH_2 + \text{} -NH- \text{} \qquad (4.13)$$

$$\text{TOF} = 0.7 \text{ h}^{-1} \qquad 23\% \qquad\qquad 4.5\%$$

Further examples were reported by Whitman et al. for the hydroamination of ethylene [116, 117]. The critical factors for these reactions are the pressure, which must be higher than 500 bar (max yields at 800–1200 bar), and the temperature, which must be higher than 150°C (best yields at 170–300°C). NH_3 reacts with α-olefins according to Markovnikov's rule. Ethylene is the more reactive α-olefin. It can be hydroaminated with NH_3 using sodium, but also lithium, potassium, sodium hydride, and lithium hydride. Examples of ethylene hydroamination with primary amines under high pressure were also reported. In most cases, large quantities of high-boiling materials containing nitrogen were also formed.

A significant improvement was obtained by using alkali or alkaline-earth metal amides as catalysts [118]. These include inorganic amides MNH_2 or $M'(NH_2)_2$ (M = Na, K, Li, Rb, Cs, M(= Mg, Ca, Sr, Ba), but preferably organic amides NHR or MNR_2. The latter may be prepared by a variety of reactions, including the reaction of the metal with the amine at elevated temperatures [118, 119]. An elegant method consists in the reaction of the metal with the amine in the presence of 1,3-butadiene, whereby the metal amide is produced and the diene is hydrogenated to butane. In most cases, the catalyst used is the sodium amide derived from the amine to be alkylated. It must be prepared before the alkylation step.

Using the above *preformed* catalysts, ethylene can be hydroaminated by primary and secondary amines under much lower pressures (3–55 atm) than those required for the reactions catalyzed by alkali metals (800–1200 atm). The example of N-ethylation of piperidine has been described in full details in Organic Syntheses (Eq. 4.14) [120].

$$\text{} + \text{}NH \xrightarrow[\text{100°C}]{C_5H_{10}NNa\ (4\%),\ pyridine} \text{}N- \qquad (4.14)$$

$$28 \text{ bar} \qquad\qquad 77\text{-}83\%, \text{ TOF} = 8.3 \text{ h}^{-1}$$

However, the use of higher olefins does not appear promising [119].

Lehmkuhl et al. demonstrated the beneficial effect of TMEDA (N,N,N',N'-tetramethylethylenediamine) on the addition of n-$BuNH_2$ to ethylene catalyzed by n-BuNHLi (from n-$BuNH_2$ and EtLi) [121]. This is also true for secondary amines. The efficiency of this system is exemplified by the hydroamination of ethylene with Et_2NH (Eq. 4.15).

$$\text{} + Et_2NH \xrightarrow[\text{140°C}]{Et_2NLi\ (6\%)\ /\ TMEDA} Et_3N \qquad (4.15)$$

$$70 \text{ bar} \qquad\qquad 82.5\%, \text{ TOF} = 1\text{h}^{-1}$$

Using higher concentrations of the above Et_2NLi-TMEDA system allows operation under even milder conditions (6–10 bar, 70–90°C) [122]. However, the importance of TMEDA has been questioned [123, 124]. Within the range of conditions studied, an empirical rate law for the formation of Et_3N has been formulated as $v = k[Et_2NH]^0[C_2H_4]^{-1}[Et_2NLi]^0$ [122]. The suggested catalytic cycle is represented in Scheme 4-2. The rate-determining step is considered to be the reversible formation of the $(TMEDA)_nLi^+[C_2H_4NEt_2]^-$ species.

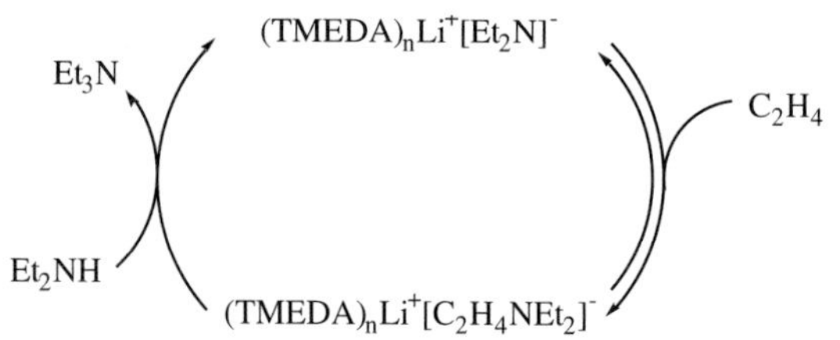

Scheme 4-2 General scheme for the base-catalyzed hydroamination of ethylene with Et_2NH

In a related study, Lehmkuhl et al. found that TMEDA may be a source of lithium dimethylamide by reaction with alkyllithiums at high temperatures [121]. Thus, heating TMEDA in the presence of EtLi under propene pressure at 150–200°C produces a mixture of dimethylisopropylamine and dimethylvinylamine (Eq. 4.16).

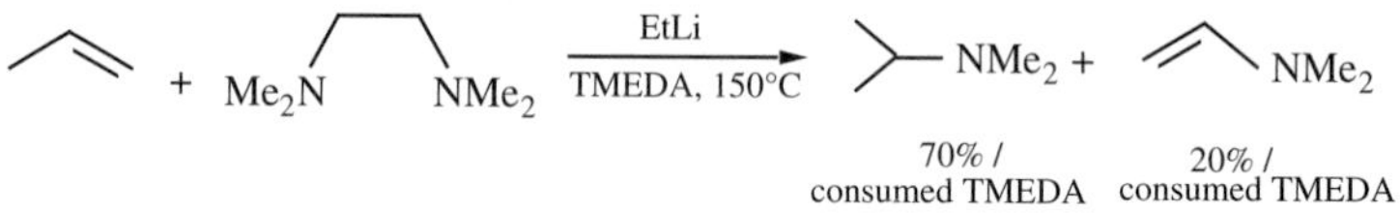

The reaction of Et_2NH with ethylene can also be catalyzed by NEt_2 [122].

$CsNH_2$ and $RbNH_2$ are active catalysts for the hydroamination of ethylene with NH_3 (100°C, 120 bar, TOF $\approx$ 1–3 h^{-1}) [122, 125]. Under the same conditions, $NaNH_2$ and KNH_2 are not catalysts. The higher activity of cesium and rubidium amides has been explained by their higher solubility in liquid NH_3. However, a KNH_2-$2NaNH_2$ eutectic mixture (m.p. 92°C) exhibits a reasonable activity.

4.2.2.2.2 Activation by Early Transition Metals
Attempts to hydroaminate ethylene, allylbenzene, and norbornene with $ArNH_2$ in the presence of zirconium bisamides $Cp_2Zr(NHAr)_2$ (Ar = 2,6-Me_2C_6H3, o-MeC_6H_4) at temperatures up to 160°C have been unsuccessful [126].

4.2.2.2.3 Activation by Lanthanides
Although efficient for the intramolecular hydroamination/cyclization (abbreviated IH below) of aminoalkenes (see below), organolanthanides exhibit a much lower catalytic activity for the intermolecular hydroamination of alkenes, as exemplified by the reaction of n-$PrNH_2$ with 1-pentene catalyzed by a neodymium complex (Eq. 4.17) [127].

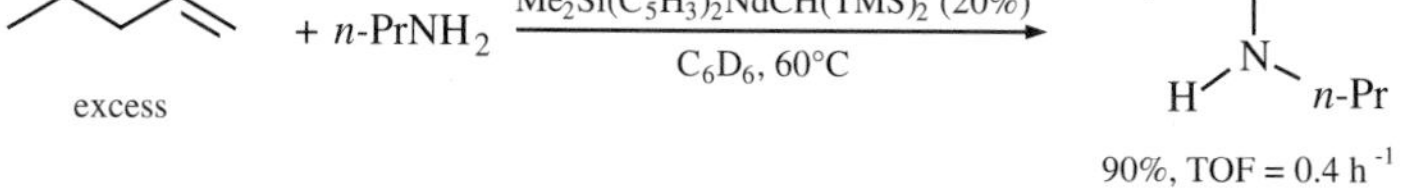

Nevertheless, hydroamination of ethylene, propene, 1-butene, and the like with NH_3, primary and secondary amines using $Cp^*_2Sm(thf)_2$ (0.5-1%) has been claimed in a patent [128].

The efficiency of organolanthanides (e.g. $Cp^*_2LaH)_2$, $Cp^* = C_5Me_5$) as catalyst precursors for the IH of aminoolefins was reported by Marks et al. in 1989 [129]. One of the noteworthy features of these reactions includes complete conversion of the substrate to afford heterocycles with > 99% regioselectivity *via* 5- or 6-*Exo-Trig* processes according to Baldwin's nomenclature (Eq. 4.18) [130]. All reactions catalyzed by organolanthanides must be conducted under *rigorous exclusion of air and moisture* (glove boxes) since these complexes are extremely oxophilic [129].

$$H_2N{\underset{n}{\diagdown\diagup\diagdown\diagup}}\quad\xrightarrow[\text{toluene, 60°C}]{Cp^*_2LaCH(TMS)_2\ (3\%)}\quad \overset{\overset{H}{N}}{\bigcirc}\Big)_n \qquad (4.18)$$

$$n = 1,\ TOF = 140\ h^{-1}$$
$$n = 2,\ TOF = 5\ h^{-1}$$

The efficiency of the more readily accessible Cp^*_2Sm and $Cp^*_2Sm(thf)_2$ complexes was also recognized [128, 129]. Further studies indicate that active catalyst precursors include organolanthanide complexes of the general formula Cp^*_2Ln-R (Ln = La, Nd, Sm, Yh, Lu) with R = H, η_3- C_3H_5, $CH(TMS)_2$, $N(TMS)_2$ [131]. Generation of the catalytically active species is believed to occur *via* protonolysis of the Ln–R bond by the amine (Eq. 4.19).

$$Cp^*Ln-R + H_2N-(CH_2)_n-CH{=}CH_2 \longrightarrow Cp^*_2Ln-NH-(CH_2)_n-CH{=}CH_2 + RH \quad (4.19)$$

In the case of the Cp^*_2Sm precursor, similar amidosamarium(III) complexes are believed to be generated *via* initial allylic C–H attack [128, 129].

Organolanthanides are catalyst precursors for the formation of five-, six-, and even seven-membered heterocycles. These IHs are also effective with secondary and aromatic amines (e.g. Eq. 4.20).

$$\underset{NH_2}{\bigcirc\!\!\!\diagup\diagdown}\quad\xrightarrow[\text{toluene, 80°C}]{Cp^*_2LaCH(TMS)_2\ (3\%)}\quad \underset{\overset{|}{H}}{\bigcirc\!\!\!\bigcirc\!\!N}\diagdown \qquad (4.20)$$

$$TOF = 12\ h^{-1}$$

TOF varies from 0.3 h^{-1} (seven-membered ring) to 140 h^{-1} (five-membered ring). As in the case of the cyclization of $H_2NCH_2C(Me)_2CH_2CH{=}CH_2$, the TOF decreases with decreasing ionic radius of the lanthanide. Thorough kinetic studies ($v = k[\text{catalyst}]^1[\text{substrate}]^0$) and deuterium labeling experiments have been performed. These reactions ($\Delta H^{\ddagger} = 12.7$ kcal/mol and $\Delta S^{\ddagger} = -27$ eu) are zero order in the substrate, suggesting that the turnover-limiting step is the intramolecular olefin insertion into the Ln–N bond [131]. A four-centered transition state has been suggested, by analogy with the chemistry developed for azametallacyclobutenes in the

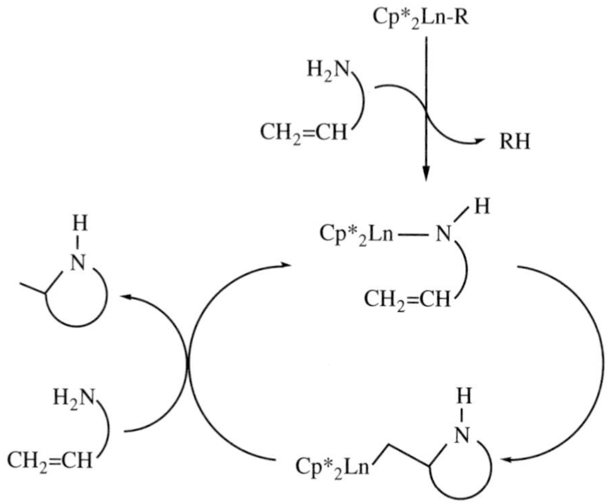

Scheme 4-3 Catalytic cycle for the organolanthanide-catalyzed IH of aminoalkenes

zirconium series (see below). The proposed catalytic cycle is represented in Scheme 4-3.

The most active catalyst precursors are the monocyclopentadienyl complexes [Me$_2$Si(C$_5$Me$_4$)(t-BuN)]LnN(TMS)$_2$ [132]. Comparative reactivities for the IH of H$_2$NCH$_2$C(Me)$_2$CH$_2$CH=CH$_2$ to give 2,4,4-trimethylpyrrolidine are given in Table 4-1.

Tab. 4-1 Compared reaction rates for the IH of H$_2$NCH$_2$C(Me)$_2$CH$_2$CH=CH$_2$

Catalyst precursor	Reaction temp. (°C)	TOF (h^{-1})	Ref.
[Me$_2$Si(C$_5$Me$_4$)(t-BuN)]SmN(TMS)$_2$	25	181	[132]
[Me$_2$Si(C$_5$Me$_4$)(t-BuN)]NdN(TMS)$_2$	25	200	[132]
[Me$_2$Si(C$_5$Me$_4$)(t-BuN)]YbN(TMS)$_2$	25	10	[132]
Me$_2$Si(C$_5$Me$_4$)$_2$LuCH(TMS)$_2$	25	90	[132]
(C$_5$Me$_5$)$_2$LaCH(TMS)$_2$	25	95	[132]
(C$_5$Me$_5$)$_2$SmCH(TMS)$_2$	60	48	[132]
(C$_5$Me$_5$)$_2$LuCH(TMS)$_2$	80	<1	[132]
(EBI)YbN(TMS)$_2$	25	0.7	[133]

IH of hindered aminoalkenes has been developed as a route to heterocycles containing quaternary centers α to the nitrogen atom (Eq. 4.21) [134]. For these reactions, the easy-to-prepare catalyst precursors [(Me$_3$SiCp)$_2$LnMe]$_2$ prove very efficient.

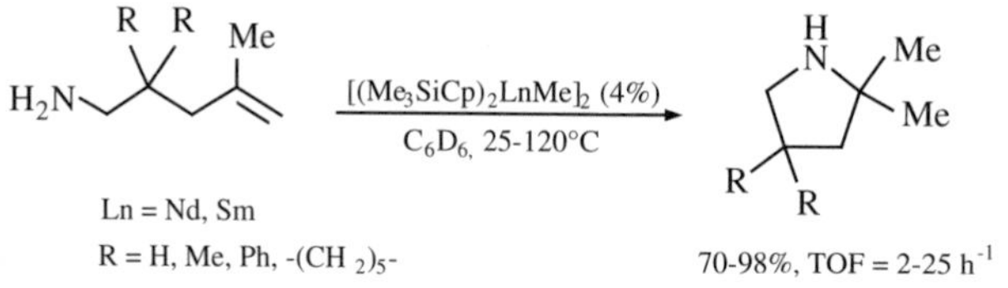

$$\text{(4.21)}$$

Ln = Nd, Sm
R = H, Me, Ph, -(CH$_2$)$_5$-

70-98%, TOF = 2-25 h^{-1}

The IH of exocyclic alkenes allows the construction of bicyclic amines bearing one methyl group at the ring junction. Although long reaction times (2–7 days) are necessary (Eq. 4.22), amines are obtained in good yield. In contrast, endocyclic aminoalkenes are resistant to cyclization [134].

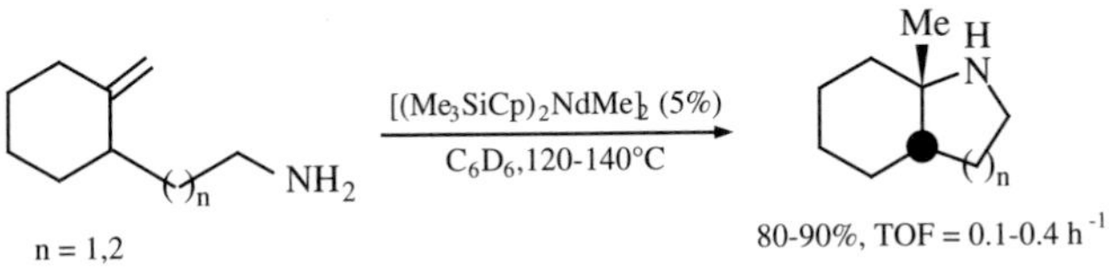

$$(4.22)$$

$$n = 1,2$$

$$80\text{-}90\%, TOF = 0.1\text{-}0.4\ h^{-1}$$

The above catalyst precursors have been used with success for the IH of several other hindered aminoalkenes, including 5-methylenecyclooctylamine (Eq. 4.23) [134].

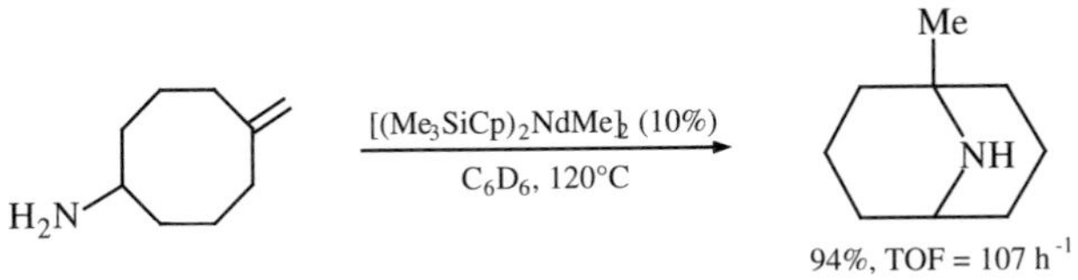

$$(4.23)$$

$$94\%, TOF = 107\ h^{-1}$$

Chiral organolanthanides $Me_2Si(C_5Me_4)(C_5H_3R^*)LnE(SiMe_3)_2$ (E = N or CH), R* = (–)-menthyl or (+)-neomenthyl (Fig. 4-1) efficiently catalyze the enantioselective IH of aminoalkenes to chiral pyrrolidines and piperidines with *ee* of up to 74% (Eq. 4.24) [135, 136].

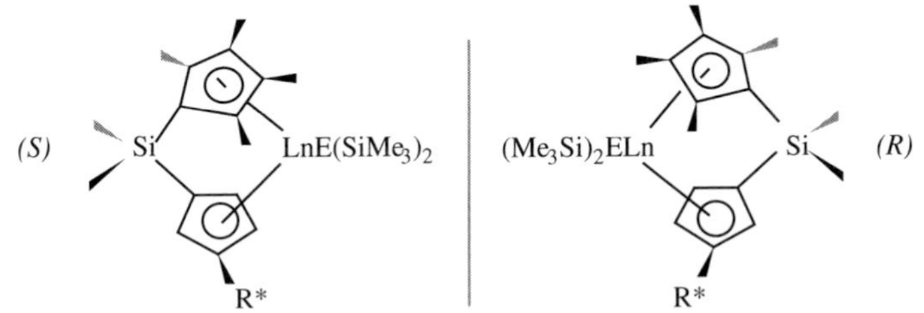

Figure 4-1 Chiral organolanthanides.

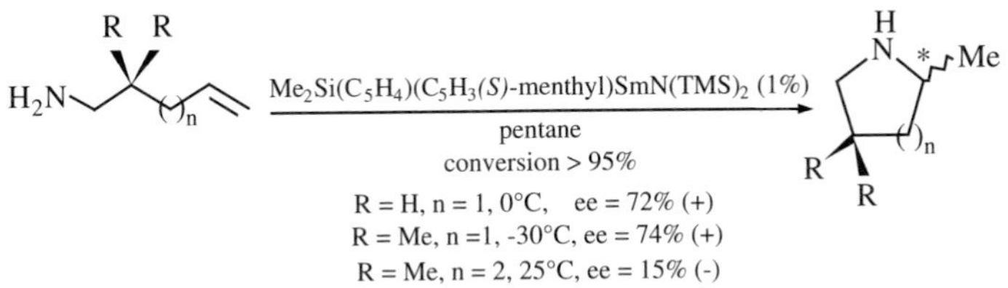

$$(4.24)$$

R = H, n = 1, 0°C, ee = 72% (+)
R = Me, n =1, -30°C, ee = 74% (+)
R = Me, n = 2, 25°C, ee = 15% (-)

With $(Me_2Si)(C_5Me_4)(C_5H_3R^*)LnE(TMS)_2$ as catalyst precursors, TOF are, under identical conditions, ca. 10 times those of the $Cp^*_2Ln–R$ catalysts presented above. In the case of 2-amino-5-hexene, *trans*-2,5-dimethylpyrrolidine is obtained with > 95% diastereoselectivity [136].

An interesting extension of the above reactions in the achiral series is the facile, regioselective, one-pot bicyclization of aminodialkenes leading to a variety of poly-cyclic heteroatom-containing skeletons (Eq. 4.25) [137].

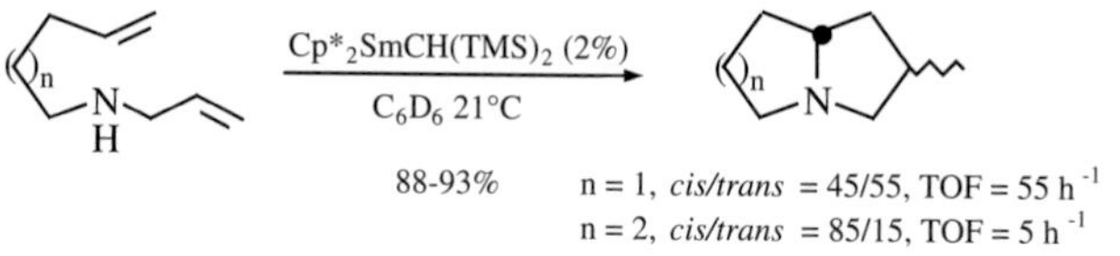

$$\text{(4.25)}$$

$$\text{Cp*}_2\text{SmCH(TMS)}_2 \ (2\%)$$
$$\text{C}_6\text{D}_6 \ 21°\text{C}$$

88-93% $n = 1$, *cis/trans* $= 45/55$, TOF $= 55 \ h^{-1}$
$n = 2$, *cis/trans* $= 85/15$, TOF $= 5 \ h^{-1}$

The above catalyst and $\text{Me}_2\text{Si(C}_5\text{Me}_4)_2\text{NdCH(SiMe}_3)_2$ are also efficient for the in-tramolecular hydroamination/bicyclization of aminoalkenynes, as exemplified in Eq. (4.26) [137, 138].

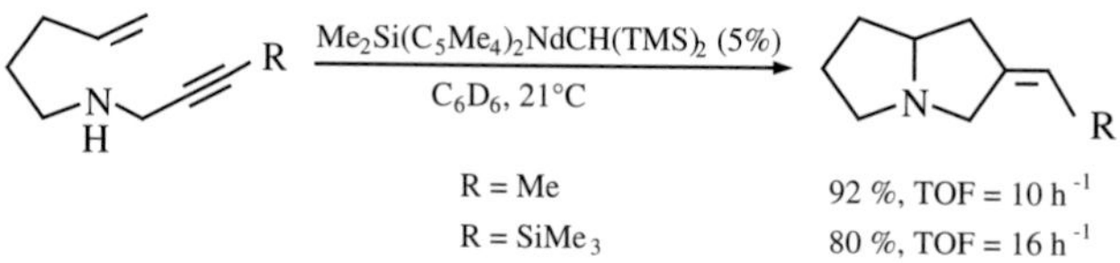

$$\text{(4.26)}$$

$$\text{Me}_2\text{Si(C}_5\text{Me}_4)_2\text{NdCH(TMS)}_2 \ (5\%)$$
$$\text{C}_6\text{D}_6, \ 21°\text{C}$$

$R = \text{Me}$ 92 %, TOF $= 10 \ h^{-1}$
$R = \text{SiMe}_3$ 80 %, TOF $= 16 \ h^{-1}$

Ansa-metallocenes with a bridged tethered donor functionality, $\text{R(Me)Si(C}_5\text{Me}_4)_2\text{Ln CH(SiMe3)2}$ (Ln = Y, Sm, R = $(\text{CH}_2)n\text{OMe})$, have been synthe-sized [139]. Use of these new catalyst precursors results in a significantly enhanced activity (2- to 5-fold) while the diastereoselectivity (e.g., for the IH of 2-amino-5-hex-ene) is essentially unchanged.

4.2.2.2.4 Activation by Late Transition Metals

Although the oxidative addition of the N–H bond of NH_3 and amines to transition metal complexes had been known for some time [140], it was only in the late 1980s that Milstein et al. succeeded in designing a homogeneously catalyzed hydroamina-tion reaction involving such an activation process (Eq. 4.27) [141].

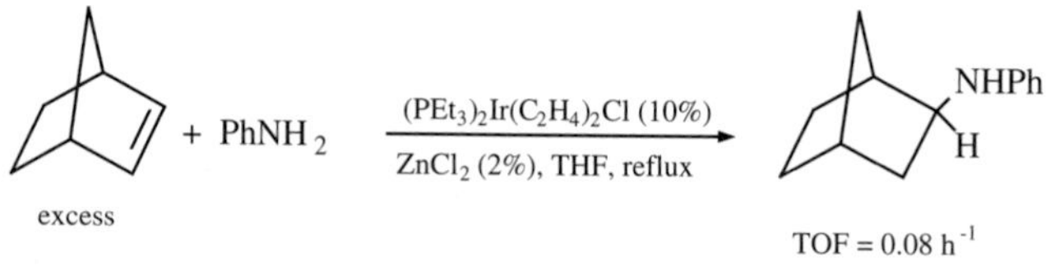

$$\text{(4.27)}$$

$$+ \ \text{PhNH}_2$$

$$\frac{(\text{PEt}_3)_2\text{Ir(C}_2\text{H}_4)_2\text{Cl} \ (10\%)}{\text{ZnCl}_2 \ (2\%), \ \text{THF, reflux}}$$

NHPh
H

excess

TOF $= 0.08 \ h^{-1}$

The reaction mechanism clearly involves the oxidative addition of aniline to an unsaturated Ir(I) complex (Scheme 4-4). Interestingly, the azametallacyclobutane in-termediate could be characterized by single-crystal X-ray diffraction [141]. This re-sult confirms that insertion of an olefin into the M–H bond is less favorable than in-sertion into the M–N bond [142].

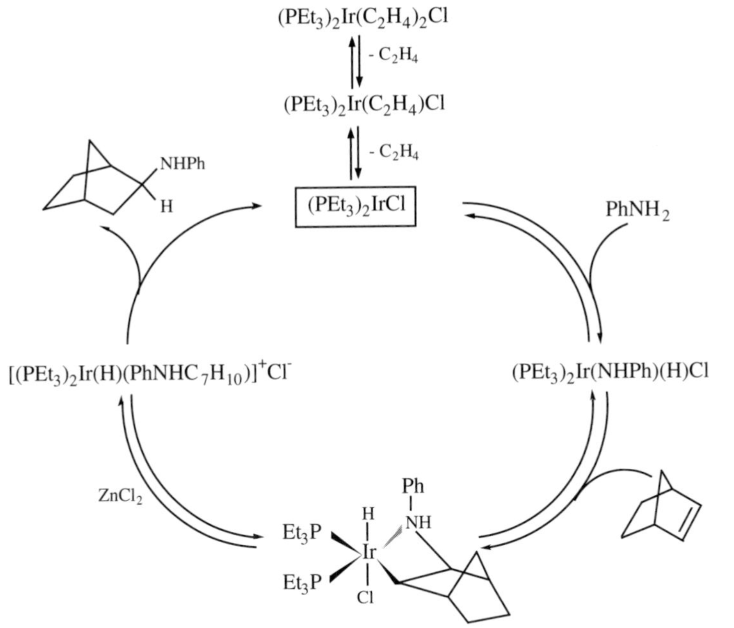

Scheme 4-4 Hydroamination of norbornene *via* N–H activation of $PhNH_2$

Although a maximum of 6 turnovers in 3 days (TOF ≈ 0.08 h^{-1}) were reached before loss of activity, this is the first successful demonstration of hydroamination of an alkene via a transition metal-catalyzed N–H activation process.

An interesting breakthrough was reported later by Togni et al., who tested binuclear chloro-bridged iridium complexes [IrCl(diphosphine)]$_2$ containing chiral chelating diphosphines such as Josiphos-type ligands, BINAP, and Biphemp [143, 144]. They pointed out a remarkable 'naked' fluoride ion effect on both the activity and the enantioselectivity of the asymmetric hydroamination of norbornene with $PhNH_2$ (Eq. 4.28). The fluoride ion is introduced as 1,1,1,3,3,3-hexakis(dimethylamino)diphosphazenium fluoride. Enantiomeric excesses up to 95% and TOFs up to 3.4 h^{-1} could be reached.

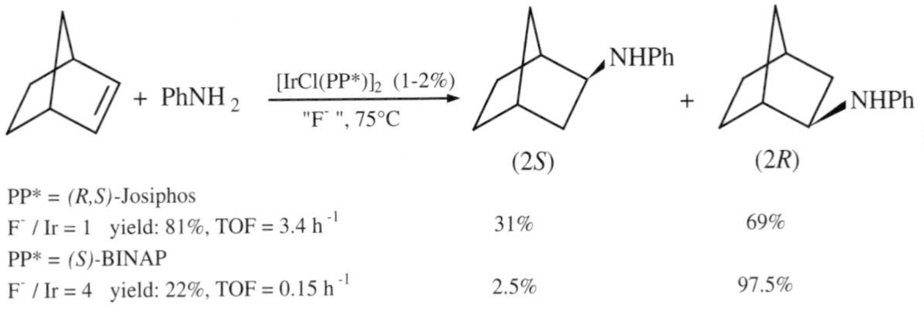

PP* = *(R,S)*-Josiphos
F$^-$ / Ir = 1   yield: 81%, TOF = 3.4 h^{-1} 31%   69%
PP* = *(S)*-BINAP
F$^-$ / Ir = 4   yield: 22%, TOF = 0.15 h^{-1} 2.5%   97.5%

The analogous rhodium complexes exhibit very low activity, even in the presence of fluoride ions [143]. In contrast, (PR$_3$)$_3$RhCl (R = Ph, Et) and the corresponding chloro-bridged dinuclear Rh(I) complexes exhibit catalytic activity when used in the presence of PhNHLi (5 equiv./Rh) [145, 146]. Under these conditions, an anionic bis(anilido)Rh(I) complex, $[(PR_3)_2Rh(NHPh)_2]^-Li^+$ is generated, which seems to be the resting state of the active catalyst [147, 148]. This catalytic system has been used for the condensation of $PhNH_2$ with norbornene to give a mixture of the hydroamination product and an unexpected hydroarylation product in a 30/70 ratio (Eq. 4.29).

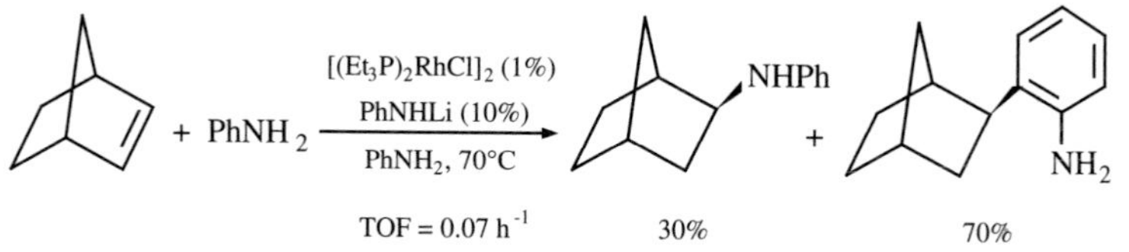

$$\text{(4.29)}$$

Although the reaction rate is low, it is interesting to note that no loss of activity has been observed during 20 days reaction at 70°C [146].

4.3
Hydroamination of Styrenes

4.3.1
Heterogeneous Catalysis

Gaseous mixtures of styrene and NH_3 have been claimed to give hydroaminated products over a ternary $K/graphite/Al_2O_3$ catalyst [45].

4.3.2
Homogeneous Catalysis

4.3.2.1 Activation of the Styrenes
To the best of our knowledge, no example has been reported.

4.3.2.2 Activation of the Amine

4.3.2.2.1 Activation by Bases
The first example of hydroamination of styrene in the presence of an alkali metal appeared in a patent in 1948, albeit with a low catalytic activity (Eq. 4.30) [149]. The anti-Markovnikov addition product was obtained.

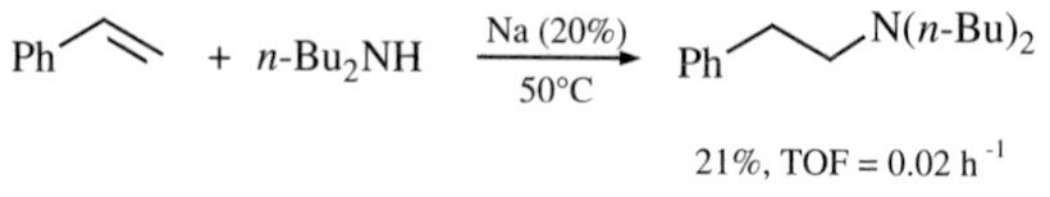

$$\text{(4.30)}$$

Similarly, NH_3, primary and secondary amines [150, 151], $PhNH_2$ [151], and aziridines [150, 152, 153] add to styrene in the presence of sodium (e.g., Eq. 4.31).

Typical conditions are 150–180°C using 1-2% sodium metal [151], but in some cases (e.g., the butylamines, aziridine), lower temperatures appear to be convenient [154]. The TOF ranges from 1 to 10 h^{-1}.

$$Ph\diagup\diagdown + PhNH_2 \xrightarrow[\text{reflux}]{\text{Na (1-3\%)}} Ph\diagup\diagdown\diagup^{NHPh} \qquad (4.31)$$

$$70\%, \text{TOF} = 9\,h^{-1}$$

Lithium alkylamides (*in situ* generated from the amine and either *n*-BuLi or *sec*-BuLi) generally give higher yields under milder conditions. Thus, *n*-BuLi (5%) catalyzes the addition of primary and secondary amines to styrene to afford (-phenethylamines in moderate to good yields (e.g., Eq. 4.32) [155]. NH_3 and $PhNH_2$, however, do not add to styrene under these conditions. α-Methylstyrene and 1,1-diphenylethylene can also be hydroaminated.

$$Ph\diagup\diagdown + \underset{NH}{\bigcirc} \xrightarrow[\text{THF, 50°C}]{\textit{n}\text{-BuLi (5\%)}} Ph\diagup\diagdown\diagup^{N}\bigcirc \qquad (4.32)$$

$$88\%, \text{TOF} = 4\,h^{-1}$$

Kinetic studies of the addition of Et_2NH to styrene catalyzed by Et_2NLi indicated the rate law $v = k[\text{styrene}][Et_2NLi]^x$, the reaction order x depending on the initial $[Et_2NH]_0/[Et_2NLi]_0$ ratio. For a ratio of 3, $x = 1.3$ and for a ratio of 10, $x = 1$ [156].

The *n*-BuLi-catalyzed hydroamination of styrene has been extended to 4-arylpiperazines [157, 158]. For example, *N*-(4-fluorophenyl)piperazine reacts with styrene to afford the hydroamination product in 99% yield (Eq. 4.33).

$$Ph\diagup\diagdown + p\text{-}FC_6H_4-N\bigcirc NH \xrightarrow[\text{THF, 90°C}]{\textit{n}\text{-BuLi (2.5\%)}} Ph\diagup\diagdown\diagup^{N}\bigcirc N-p\text{-}FC_6H_4 \qquad (4.33)$$

$$99\%, \text{TOF} = 40\,h^{-1}$$

As previously mentioned [155], $PhNH_2$ does not react with styrene under the above conditions. However, Beller et al. discovered that the hydroamination of styrene could be achieved in excellent yield by using either a *n*-BuLi-K_2CO_3 mixture or, better, *t*-BuOK as catalysts [159]. Using *t*-BuOK (10%) in THF at 120°C (pressure tube), styrene is hydroaminated with aniline (5 equiv.) to give the anti-Markovnikov product in 96% yield (Eq. 4.34), $R^1 = R^2 = R^3 = R^4 = H$, TOF = 0.5 h^{-1}]. The scope of this new base-catalyzed hydroamination has been demonstrated by the condensation of substituted anilines with different vinylaromatics (Eq. 4.34), giving easy access to interesting pharmaceuticals or intermediates [159].

$$\text{(structure with } R^2, R^3, R^4 \text{) } + \text{(aniline with } R^1, NH_2) \xrightarrow[\text{THF, 120°C}]{\textit{t}\text{-BuOK}} \text{(product with } R^2, R^3, R^4, H, N, R^1) \qquad (4.34)$$

Although the above system does not work for the IH of, e.g., 2-isopropenylaniline, a somewhat unexpected extension has been pointed out [159]. Reaction of anilines with 2-chlorostyrene in the presence of *t*-BuOK (1.5 equiv.) affords *N*-arylindolines in a one-pot reaction (53–58% yield) (Eq. 4.35) [159].

$$
\begin{array}{c}
\underset{\text{Cl}}{\overset{\text{CH=CH}_2}{\bigcirc}} + \text{Ar-NH}_2 \xrightarrow[\text{toluene, 135°C}]{t\text{-BuOK}} \bigcirc\!\!-\!\!\bigcirc\text{N}-\text{Ar}
\end{array} \tag{4.35}
$$

Ar = C$_6$H$_5$, 4F-C$_6$H$_4$, 2MeO-C$_6$H$_4$

The reaction mechanism is believed to involve a domino hydroamination-intramolecular arynic condensation. Indeed, the same results are obtained from either 2- and 3-chlorostyrenes [159].

Cesium hydroxide, CsOH.H$_2$O (20%) also catalyzes the addition of PhNH$_2$ and PhNHMe to styrene (NMP, 120°C), although in moderate yields (42–69%, TOF = 0.2–0.3 h^{-1}) [160].

4.3.2.2.2 Activation by Early Transition Metal
To the best of our knowledge, no example has been reported.

4.3.2.2.3 Activation by Lanthanides
The only reported example of such hydroaminations concerns the IH of α-(aminoalkyl)styrenes (5-*Exo-Trig* process) (Eq. 4.36) [134].

$$
\text{H}_2\text{N}\underset{R}{\overset{Ph}{\diagdown}} \xrightarrow[\text{C}_6\text{D}_6,\ 120°C]{[(\text{Me}_3\text{SiCp})_2\text{NdMe}]_2\ (5\%)} \underset{\underset{H}{N}}{\overset{R}{\diagdown}}\overset{Ph}{\underset{Me}{\diagup}} \tag{4.36}
$$

R = H, 90%, TOF = 0.12 h^{-1}
R = Me, 100%, TOF = 0.4 h^{-1}

4.3.2.2.4 Activation by Late Transition Metals
The anionic rhodium complexes [(PR$_3$)$_2$Rh(NHPh)$_2$]$^-$Li$^+$ (see above) have been shown to catalyze (TON = 21) the condensation of aniline with styrene to give the expected amine and the corresponding imine (oxidative amination) (Eq. 4.37, TOF ~ 0.07 h^{-1}) [161].

$$
\text{Ph}\diagdown + \text{PhNH}_2 \xrightarrow[\text{PhNH}_2,\ 70°C]{\substack{[(\text{PEt}_3)_2\text{RhCl}]_2\ (1\%) \\ \text{PhNHLi}\ (10\%)}} \underset{30\%}{\text{Ph}\overset{\text{NHPh}}{\diagup}} + \underset{65\%}{\text{Ph}\overset{\text{NPh}}{\diagup}} \tag{4.37}
$$

Both products result from a Markovnikov-type addition. The same regioselectivity was obtained using 1-hexene as olefin [161].

More recently, Beller et al. reported the first example of an oxidative amination of styrenes by secondary amines to give enamines corresponding to an anti-

Markovnikov addition [162]. The catalyst precursor is a cationic complex, $[Rh(cod)_2]^+BF_4^-$, used in the presence of 2 equiv. of PPh_3. An equivalent amount of the hydrogenated vinylarene is formed (Eq. 4.38) [163–165].

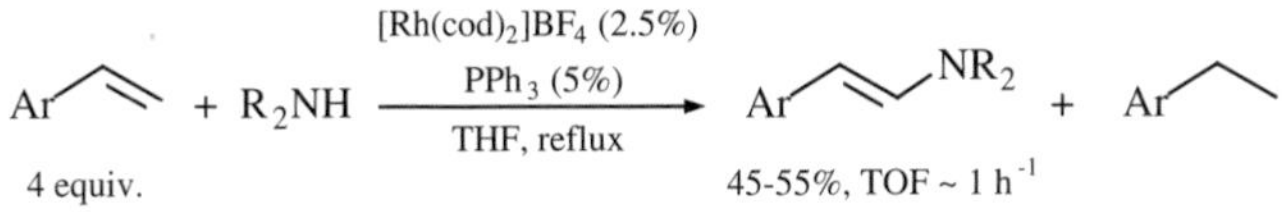

$$Ar\diagup\diagdown + R_2NH \xrightarrow[\text{THF, reflux}]{\substack{[Rh(cod)_2]BF_4\ (2.5\%)\\PPh_3\ (5\%)}} Ar\diagup\diagdown\diagup NR_2 + Ar\diagup\diagdown \qquad (4.38)$$

4 equiv.   45-55%, TOF ~ 1 h^{-1}

With some secondary amines, especially morpholine, the reaction leads to a mixture of the oxidative amination product and of the hydroamination product, both corresponding to an anti-Markovnikov addition (Eq. 4.39) [166].

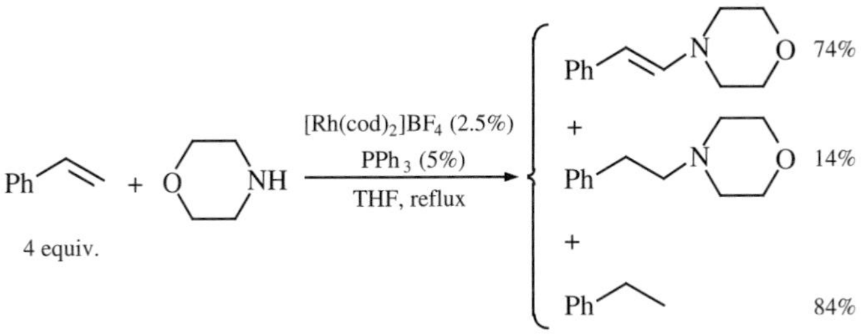

$$Ph\diagup\diagdown + O\diagdown NH \xrightarrow[\text{THF, reflux}]{\substack{[Rh(cod)_2]BF_4\ (2.5\%)\\PPh_3\ (5\%)}} \left\{ \begin{array}{l} Ph\diagup\diagdown\diagup N\diagdown O \quad 74\% \\ + \\ Ph\diagup\diagdown\diagup N\diagdown O \quad 14\% \\ + \\ Ph\diagup\diagdown \qquad\qquad 84\% \end{array} \right. \qquad (4.39)$$

4 equiv.

Although N-(2-phenylethyl)morpholine is formed in only 14% yield (TOF = 0.3 h^{-1}), this is the first example of a *transition metal-catalyzed* anti-Markovnikov hydroamination of a non-activated olefin. Concerning the reaction mechanism, labeling experiments led the authors to favor activation of the N–H bond over olefin activation [166].

Although the hydroamination of Michael systems is beyond the scope of this review, it is interesting to note the high yield (98%, TOF = 2 h^{-1}) obtained using the above cationic rhodium complexes for the hydroamination of 2-vinylpyridine with morpholine. Indeed, without catalyst, the hydroamination yield is only 5% [167].

4.4
Hydroamination of 1,3-Dienes

4.4.1
Heterogeneous Catalysis

Examples of heterogeneously catalyzed hydroamination of 1,3-dienes are scarce. Primary amines ($CyNH_2$, n-$BuNH_2$) or secondary amines (Et_2NH, morpholine, piperidine) react with 1,3-butadiene or isoprene in the presence of catalytic systems comprising graphite and alkali metals (KC_8, KC_{24}, graphite Na, LiC_{32}) to produce 1,4-hydroamination products (Eq. 4.40) [168].

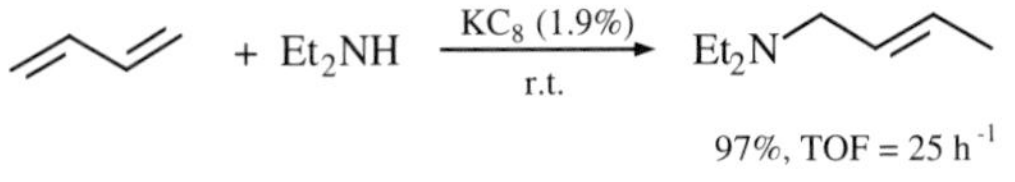

$$\diagup\diagdown\diagup + Et_2NH \xrightarrow[\text{r.t.}]{KC_8\ (1.9\%)} Et_2N\diagup\diagdown\diagup \qquad (4.40)$$

97%, TOF = 25 h^{-1}

A gaseous mixture of butadiene and NH_3 (5/95) has been claimed to react over K/Al_2O_3/graphite at 40°C to give a butenylamine as the major product [45].

Addition of primary and secondary amines to 1,3-butadiene and isoprene at 0 to 180°C over solid bases such as MgO, CaO, SrO, La_2O_3, ThO2, and ZrO_2 has also been studied. CaO exhibits the highest activity, while ZrO_2 is inactive. Me_2NH is the most reactive amine, giving primarily the 1,4-addition product which undergoes isomerization to the enamine *N,N*-dimethyl-1-butenylamine. It has been proposed that addition of amines to 1,3-dienes on basic catalysts proceeds *via* aminoallyl carbanion intermediates which result from addition of amide ions to the dienes [169, 170].

4.4.2
Homogeneous Catalysis

4.4.2.1   Activation of the Diene
The hydroamination of 1,3-dienes has been attempted with catalyst precursors of nearly all transition metals. The most active are those based on Group 9 and 10 transition metals.

1,3-Butadiene and isoprene give rise to mixtures of what are usually called telomers, namely 1:1 telomers between the amine and the 1,3-diene (true hydroamination products), 1:2 telomers and even higher homologs together with oligomers of the diene as exemplified in Eq. (4.41).

$$\nearrow\!\!\!\!\diagup + HNR_2 \longrightarrow H\text{-}[C_4H_6]\text{-}NR_2 + H\text{-}[C_4H_6]_2\text{-}NR_2 + H\text{-}[C_4H_6]_3\text{-}NR_2 + ... \quad (4.41)$$

The ratio of formation of 1:1 *vs.* 1:2 telomers does not appear to be solely a function of the metal but is also determined by the nature of the added ligands and of the co-catalysts (additives). Higher dienes like 1,3-pentadiene, 2,3-dimethylbutadiene, etc. afford 1:1 telomers only.

In fact, catalytic systems which effect solely the hydroamination of 1,3-butadiene and isoprene are rare and usually specific to the diene and to the amine. Thus morpholine adds to 1,3-butadiene in the presence of $RhCl_3.3H_2O$ to give a mixture of 1,2-(Markovnikov) and 1,4-hydroamination products in good overall yield (Eq. 4.42) [171, 172].

$$\nearrow\!\!\!\!\diagup + O\!\!\diagdown\!\!NH \xrightarrow[\text{EtOH, 75°C}]{RhCl_3.3H_2O\ (0.8\%)} \quad (4.42)$$

3 equiv.
60%, TOF = 5 h^{-1} 25%, TOF = 2 h^{-1}

Although less active, the cobalt system $CoCl_2/PPh_3/NaBH_4$ affords a 90/10 mixture of 1,2- and 1,4-addition products in 98% yield at room temperature [172]. Palladium catalysts are more efficient (Eq. 4.43) [173].

$$\nearrow\!\!\!\!\diagup + O\!\!\diagdown\!\!NH \xrightarrow[\substack{PhOH\ (2.5\%) \\ 140\text{-}150°C}]{Pd(dppe)_2\ (0.25\%)} \quad (4.43)$$

1.5 equiv.
65%, TOF = 260 h^{-1} 5%, TOF = 20 h^{-1}

A mixture of $PdBr_2/PhONa/PhOH$ can also be used to generate the required Pd(0) catalyst. The presence of phenol greatly improves catalyst activity and selectivity, favoring the formation of the 1,4-addition product. The reaction could be extended to isoprene, 1,3-pentadiene and 2,4-hexadiene to afford mixtures of 1,2- and 1,4-addition products, albeit in lower yields. In all these reactions, some high-boiling products are also formed [173]. Similar results have been obtained in the hydroamination of isoprene with Et_2NH catalyzed by $Pd(acac)_2/dppe/BF_3 \cdot OEt_2$ [174].

Reduced nickel salts in the presence of a large amount of phosphine catalyze the hydroamination of 1,3-butadiene with Et_2NH (Eq. 4.44) [175, 176].

$$\text{[CH}_3\text{(CH}_2\text{)}_{10}\text{CO}_2\text{]}_2\text{Ni (0.44\%)}$$

$$\bigwedge\!\!\!\diagup + \text{Et}_2\text{NH} \xrightarrow[\substack{\text{NaBH}_4\ (5.4\%) \\ \text{PhP(O}i\text{-Pr)}_2\ (3.5\%) \\ 90°C}]{} \text{Et}_2\text{N}\diagdown\!\!\diagup\!\!\diagdown\!\!\diagup + \text{Et}_2\text{N}[\text{C}_4\text{H}_6]_2\text{H} \quad (4.44)$$

$$\text{45\%, TOF} = 25\ \text{h}^{-1} \qquad\qquad 0.8\%$$

Similarly, a mixture of $Ni(cod)_2/P(n\text{-}Bu)_3/CF_3CO_2H$ catalyzes the reaction of 1,3-butadiene with morpholine to give a mixture of 1,2- and 1,4-addition products (90%, TOF = 13 h^{-1}) [177].

It was also shown that a cationic allylnickel complex $[(\eta_3\text{-crotyl})Ni\{P(OEt)_3\}_2]PF_6$ catalyzes this reaction without added acid and that *cis*- and *trans*-1,3-pentadienes react with morpholine to give a mixture of 1,2- and 1,4-addition products [178].

$NiCl_2/PhP(Oi\text{-}Pr)_2$ or $Ni(acac)_2/NaBH_4/PhP(Oi\text{-}Pr)_2$ combinations catalyze the reaction of 1,3-butadiene with morpholine or $n\text{-}BuNH_2$ at 100°C in an autoclave to a mixture of 1,2- and 1,4-addition products. However, by passing 1,3-butadiene gas into a solution of $NiBr_2/PhP(Oi\text{-}Pr)_2$ in morpholine, 1-(N-morpholino)-2-butene could be obtained highly selectively (Eq. 4.45) [179]. In the early stages of the reaction, a mixture of 3-(N-morpholino)-1-butene and 1-(N-morpholino)-2-butene is formed, which fully isomerizes into 1-(N-morpholino)-2-butene [179].

$$\bigwedge\!\!\!\diagup + \text{O}\bigcirc\text{NH} \xrightarrow[\substack{\text{NiBr}_2\ (2\%) \\ \text{PhP(O}i\text{-Pr)}_2\ (2.2\%) \\ \text{r.t.}}]{} \text{O}\bigcirc\text{N}\diagdown\!\!\diagup\!\!\diagdown \quad (4.45)$$

$$\text{98\%, TOF} = 74\ \text{h}^{-1}$$

Dzhemilev and Tolstikov et al. have studied the influence of phosphines and additives in the hydroamination of 1,3-butadiene with morpholine catalyzed by the $Ni(acac)_2/phosphine/AlEt_3/$ additive (1/3/3/10) system [180]. With CF_3CO_2H as additive, the reaction is highly selective (>93%) for the formation of 1-(N-morpholino)-2-butene (>80% yield) by using $P(n\text{-}Bu)_3$ or $P(OEt)_3$ as phosphine. With the same system and PCl_3 as phosphine, a mixture of 3-(N-morpholino)-1-butene (70%) and 1-(N-morpholino)-2-butene (30%) is formed (TOF = 21 h^{-1}).

Several additives were tested in the case of the $Ni(acac)_2/P(n\text{-}Bu)_3/AlEt_3/$additive (1/3/3/10) system. $BF_3 \cdot OEt_2$ or $TiCl_4$ or morpholinium chloride induces the fully selective formation of 3-(N-morpholino)-1-butene without formation of 1:2 telomers or oligomers (Eq. 4.46) [180].

$$\text{(4.46)}$$

Reaction conditions: Ni(acac)$_2$(1%), AlEt$_3$ (3%), P(n-Bu)$_3$ (3%), [OC$_4$H$_9$NH]$^+$Cl$^-$ (10%), 80°C; 75%, TOF = 25 h^{-1}

The same catalytic system has been tested for the hydroamination of 1,3-butadiene with cyclic amines from the three-membered ring aziridine to the seven-membered ring perhydroazepine. Although arizidine does not lead to a hydroamination reaction, all other cyclic amines give rise to a mixture of 1:1 telomers in fair to excellent yields (e.g., Eq. 4.47) [181].

$$\text{(4.47)}$$

Reaction conditions: Ni(acac)$_2$(1%), AlEt$_3$ (3%), P(n-Bu)$_3$ (3%), CF$_3$CO$_2$H (10%), 100°C; 60% + 40%; 90%, TOF = 18 h^{-1}

The palladium salt/dppe combination catalyzes the hydroamination of 1,3-butadiene with NH$_3$ to give a mixture of 1:1, bis-(1:1) and tris-(1:1) hydroamination products (Eq. 4.48) [182].

$$\text{(4.48)}$$

Reaction conditions: Pd(OAc)$_2$(1.4%), dppe (1.8%), EtOH, 145°C; NH$_3$ 6 equiv.; conversion = 90%, TOF = 64 h^{-1}; 49% (62/38); 27% (50/50); 8% + oligomers (16%)

Attempts to prepare optically active amines by replacing dppe with optically active diphosphines such as DIOP or DIPAMP in the Pd/dppe system [183] or by replacing PhP(Oi-Pr)$_2$ with dimenthyl phenylphosphinite in the Ni/PhP(Oi-Pr)$_2$ system [184] have met with little success.

Platinum complexes Pt(PPh$_3$)$_4$ and PtCl$_2$(PPh$_3$)$_2$, either alone or in combination with CF$_3$CO$_2$H, catalyze the selective monohydroamination of 1,3-butadiene with BnNH$_2$ or PhNH$_2$ (Eq. 4.49) in moderate yields [185]. Pd(OAc)$_2$/CF$_3$CO$_2$H is also an efficient but less active catalytic system for this reaction [185].

$$\text{(4.49)}$$

Reaction conditions: Pt(PPh$_3$)$_4$ (0.5%), CF$_3$CO$_2$H (4%), 150°C; PhNH$_2$ 2 equiv.; 34%, TOF = 3.7 h^{-1}; 9%, TOF = 1 h^{-1}

In the presence of triethylammonium iodide, Pd(OAc)$_2$/PPh$_3$ catalyzes the selective 1,4- hydroamination of 1,3-butadiene with Et$_2$NH to afford the same product as in Eq. (4.44) (45%, TOF = 4 h^{-1}) [186].

Isoprene or 2,3-dimethyl-1,3-butadiene or 1,3-cyclohexadiene (with Et$_2$NH), 2,3-dimethyl-1,3-butadiene (with *n*-BuNH$_2$ or piperidine) and 1,3-hexadiene or 2,4-hexadiene (with PhNH$_2$) similarly give 1:1 telomers in fair to good yields [186].

More substituted 1,3-dienes or trienes containing a 1,3-diene unit react with amines to give 1,2-addition products in the presence of nickel (Eq. 4.50) or palladium (Eq. 4.51) catalysts [187, 188].

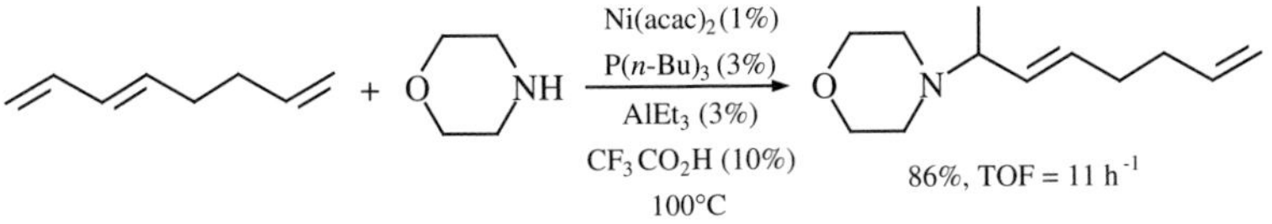

$$\begin{array}{c} \text{Ni(acac)}_2\,(1\%) \\ \text{P(}n\text{-Bu)}_3\,(3\%) \\ \hline \text{AlEt}_3\,(3\%) \\ \text{CF}_3\text{CO}_2\text{H}\,(10\%) \\ 100^\circ\text{C} \end{array} \qquad 86\%,\ \text{TOF} = 11\ \text{h}^{-1} \tag{4.50}$$

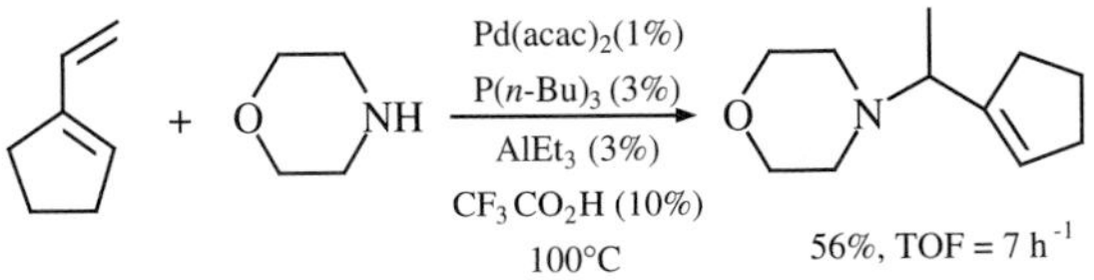

$$\begin{array}{c} \text{Pd(acac)}_2\,(1\%) \\ \text{P(}n\text{-Bu)}_3\,(3\%) \\ \hline \text{AlEt}_3\,(3\%) \\ \text{CF}_3\text{CO}_2\text{H}\,(10\%) \\ 100^\circ\text{C} \end{array} \qquad 56\%,\ \text{TOF} = 7\ \text{h}^{-1} \tag{4.51}$$

The formation of telomers rests on η^3-allylic intermediates, and the ratio of formation of 1:1 *vs.* 1:2 telomers stems from the reaction of the amine on a C$_4$-allylic complex *vs.* a C$_8$-allylic complex, an excess of phosphine and the presence of an acidic (Brønsted or Lewis) co-catalyst favoring the η^3-crotyl complex (Scheme 4-5) [178, 180, 189–196].

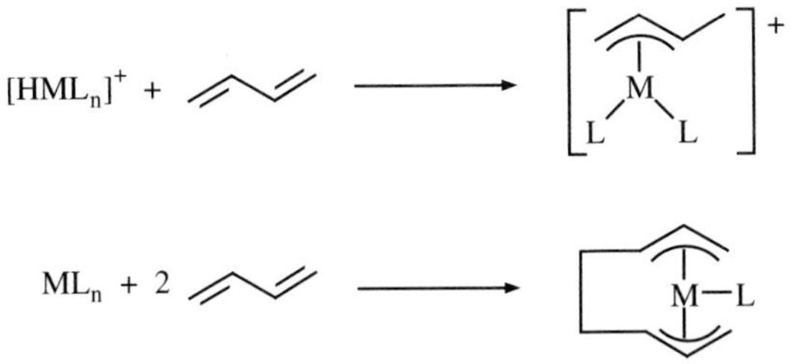

Scheme 4-5 Intermediates in the hydroamination *vs.* telomerization of 1,3-dienes

4.4.2.2 Activation of the Amine

4.4.2.2.1 Activation by Bases

The first examples of hydroamination of 1,3-dienes catalyzed by alkali metals appeared as early as in 1928 for the production of 'pest destroying agents' [197]. For example, reacting NH3 with 1,3-butadiene in the presence of sodium for more than 10 days yields 45% tri(butenyl)amine and 55% of 'high boiling bases rich in carbon' (Eq. 4.52).

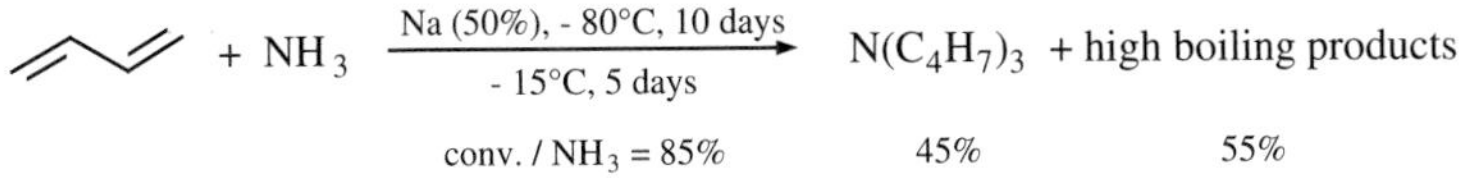

$$\begin{array}{c} \text{Na (50\%), - 80}^\circ\text{C, 10 days} \\ \hline \text{- 15}^\circ\text{C, 5 days} \\ \text{conv. / NH}_3 = 85\% \end{array}$$

$$+\ \text{NH}_3 \longrightarrow \text{N(C}_4\text{H}_7)_3\ +\ \text{high boiling products} \tag{4.52}$$

$$45\% \qquad\qquad 55\%$$

Several other examples (50–60% yields) were provided, including hydroamination with primary amines, piperidine, anilines and hexamethylenediamine [197–199]. The special case of aziridine was reported to afford a mixture of isomers 'in a good yield' [200].

Similar reactions are performed more efficiently by preparing separately the desired amide, which is then used in catalytic amounts (Eq. 4.53) [201, 202].

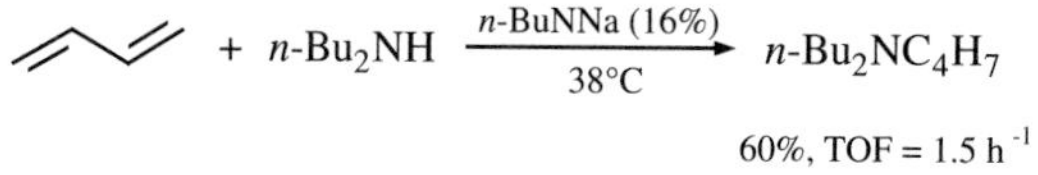

$$\text{CH}_2=\text{CH-CH}=\text{CH}_2 + n\text{-Bu}_2\text{NH} \xrightarrow[\text{38°C}]{n\text{-BuNNa (16\%)}} n\text{-Bu}_2\text{NC}_4\text{H}_7 \qquad (4.53)$$

$$60\%, \text{TOF} = 1.5 \text{ h}^{-1}$$

In all the above reactions, secondary amines give good yields of butenyl compounds, whereas with primary amines products of high molecular weight, i.e. containing more than 2 butadiene units per mol of amine, are also formed. The latter have been shown to be aldimine derivatives containing three C_4 units [203].

Further studies showed that the alkylamide catalysts can also be generated *in situ* from the amine and *n*-BuLi or *sec*-BuLi [155]. The resulting product mixture (> 99% 1,4-adduct) depends on the nature of the amine and the nature of the solvent (Scheme 4-6).

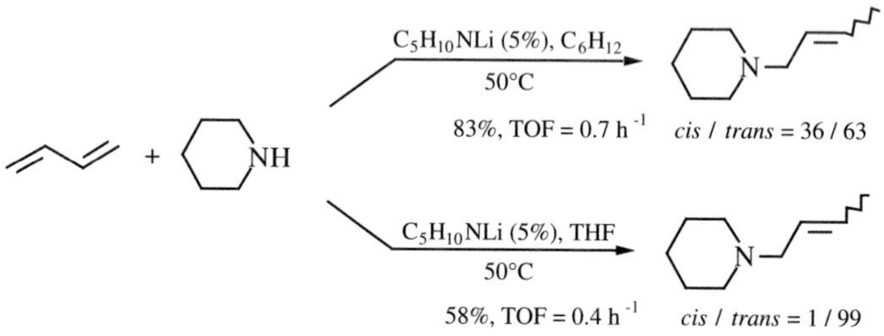

Scheme 4-6   Influence of the solvent on the condensation of C5H10NH with 1,3-butadiene

In fact, the stereochemistry of the hydroamination seems to depend strongly on the experimental conditions. For example, for the condensation of Et$_2$NH with 1,3-butadiene, either *cis*-1-diethylamino-2-butene (*n*-BuLi, C$_6$H$_6$-Et$_2$O) [204, 205], or *trans*-1-diethylamino-2-butene (*sec*-BuLi, THF) [155] can be obtained in ca. 98% yield stereoselectivity. In some cases, telomerization products are also formed [205].

Studies of the reaction of Et$_2$NLi with 1,3-butadiene gave evidence that the reaction occurs only in the presence of free Et$_2$NH (i.e. Et$_2$NLi alone does not add to 1,3-butadiene), the best initial ratio being Et$_2$NH/Et$_2$NLi = 3 [204, 206]. From kinetic and spectroscopic (IR) studies, it was concluded that an Et$_2$NLi-Et$_2$NH (1/2) complex plays an important role in the reaction.

Sodium alkylamides can also be generated *in situ* by reacting the amine with sodium naphthalene in THF [207, 208]. This procedure has been used for the condensation of secondary amines with different 1,3-dienes to give high yields of monoalkylated products without formation of 1:2 telomers. In the case of primary amines, mixtures of mono- and dialkylated products are obtained together with unidentified products. The case of isoprene is representative (Eq. 4.54). The regioselectivity depends upon the isoprene/amine ratio [208].

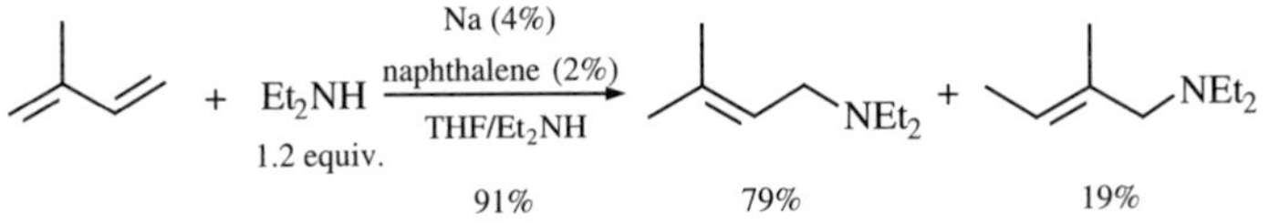

$$
\text{(4.54)}
$$

The above system has been used for the reaction of Et$_2$NH with myrcene to give a mixture of hydroamination products (53% yield) containing 80% of N,N-diethylgeranylamine [208], a key intermediate for the synthesis of industrially important monoterpenes [208, 209-211], including (–)-menthol (Tagasako process) [212].

The synthesis of N,N-diethylgeranylamine with a still higher selectivity (92–95%) has been fully described in Organic Syntheses by using lithium cut or n-BuLi (Eq. 4.55) [213, 214].

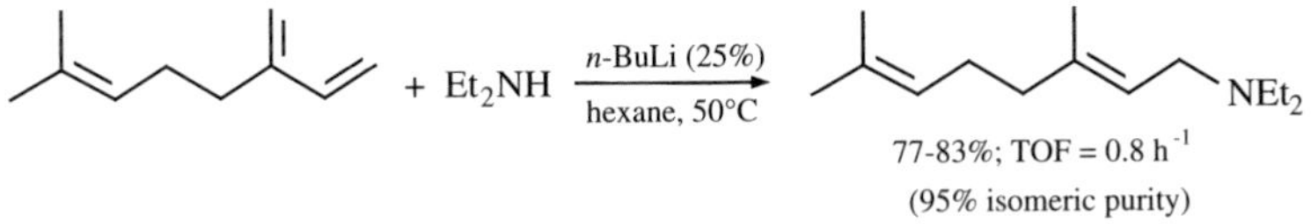

$$
\text{(4.55)}
$$

The isomeric N,N-diethylnerylamine could be obtained in 99% isomeric purity (77–86% yield) by telomerization of isoprene with diethylamine in the presence of n-BuLi [215].

4.5
Hydroamination of Alkynes

4.5.1
Heterogeneous Catalysis

Since 1877 and until the mid of the 20th century, numerous patents and a few published articles have dealt with the condensation of acetylene with NH$_3$ with the aim of manufacturing 'valuable nitrogenous products' [216]. With the exception of the first attempts [216, 217], the reaction has been carried out with catalysts at temperatures from 200 to 600°C. Usually di- and triethylamine, acetonitrile, pyridine, picolines, higher nitrogen heterocyles and higher hydrocarbons, sometimes hydrogen, coke, soot, and tars are formed in varying amounts.

The condensation of NH$_3$ with acetylene can give rise to vinylamine, which is probably an intermediate (together with ethylideneimine) in the formation of acetonitrile and other condensation products (Scheme 4-7).

Acetonitrile can be obtained in 50–90% yield on passing mixtures of NH$_3$ and acetylene at 300–500°C over mixtures of oxides or nitrates of thorium and zinc on

$$HC \equiv CH + NH_3 \longrightarrow CH_2 = CHNH_2 \longrightarrow CH_3CH = NH \longrightarrow CH_3CN + H_2$$

Scheme 4-7 Possible intermediates in the reaction of acetylene and NH_3 leading to acetonitrile

silica [218-220], zinc and chromium oxides on alumina [221], zinc oxide on kaolin [222, 223], and ZnCl2-based molten salts [224].

Nearly quantitative yields of acetonitrile can be obtained by passing mixtures of NH3 and acetylene over zircon at 400–500°C [225], over Cr_2O_3 on γ-alumina at 360°C [226] or by passing mixtures of NH_3, acetylene and hydrogen at 400–420°C over a mixture of zinc and thorium oxides on silica [227] or at 300–450°C over zinc oxide or zinc sulfate or zinc chloride on silica [228, 229]. In such reactions, the role of traces of water has often been questioned. However, acetonitrile could be obtained under rigorously anhydrous conditions, thus demonstrating the direct amination of acetylene with NH_3. It was also reported that ethylideneimine can be obtained in up to 26% yield [225]. However, in the light of more recent work [230, 231] the product was most probably 2,4,6-trimethyl-1,3,5-hexahydrotriazine.

Propionitrile can be obtained as the major product (70%) on passing mixtures of propyne and NH_3 over $ZnCl_2$/silica gel at 360–380°C [229].

Nicodemus obtained mixtures of ethylamines on passing mixtures of NH3 and acetylene at 280–290°C over zinc chloride on pumice [232, 233] or at 350°C over a mixture of zinc nitrate, tin chloride, silica and charcoal on diatomaceous earth (Eq. 4.56) [234].

$$HC \equiv CH + NH_3 \xrightarrow[\text{on diatomaceous earth}]{\text{Zn(NO}_3)_2\text{-SnCl}_2\text{-SiO}_2\text{-C}} EtNH_2 + \underbrace{Et_2NH + Et_3N} + CH_3CN + \text{pyridinic bases}$$

$$ 350°C 10\% 14\% 39\% (4.56)$$

Mixtures of Et_2NH and Et_3N and pyridinic bases can be obtained from the reaction of acetylene and $EtNH_2$ under similar conditions [235]. Ethylideneimine was proposed as an intermediate in the formation of ethylamines and acetonitrile.

It was later shown that aziridine reacts over mixtures of zinc and chromium oxides on alumina at 400°C to give the same products as those obtained from mixtures of NH_3 and acetylene [221]. Aziridine, which would form by addition of NH_3 to acetylene followed by IH (Scheme 4-8), was thus postulated to be an intermediate in the formation of acetonitrile (by dehydrogenation), monoethylamine (by hydrogenation) and all other heterocyclic bases (by ammonolysis and subsequent reactions) [221].

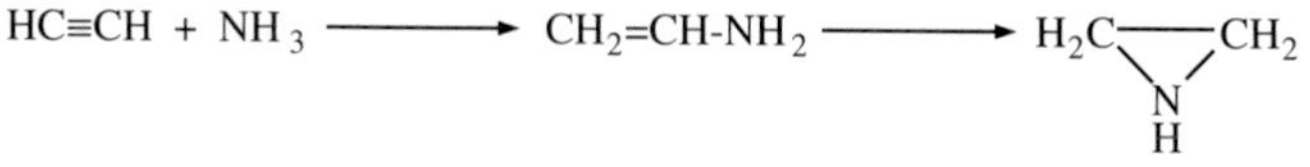

$$HC \equiv CH + NH_3 \longrightarrow CH_2 = CH\text{-}NH_2 \longrightarrow \underset{\underset{H}{N}}{H_2C \overset{\diagdown}{\underset{\diagup}{\qquad}} CH_2}$$

Scheme 4-8 Further intermediates in the reaction of acetylene and NH_3

The reaction of amines with acetylene or higher alkynes has seldom been described. n-BuNH$_2$ reacts with acetylene at 160°C in the presence of $CdSO_4$ or CuCl or $ZnCl_2$ to give n-butylpyrroline (Eq. 4.57) [236].

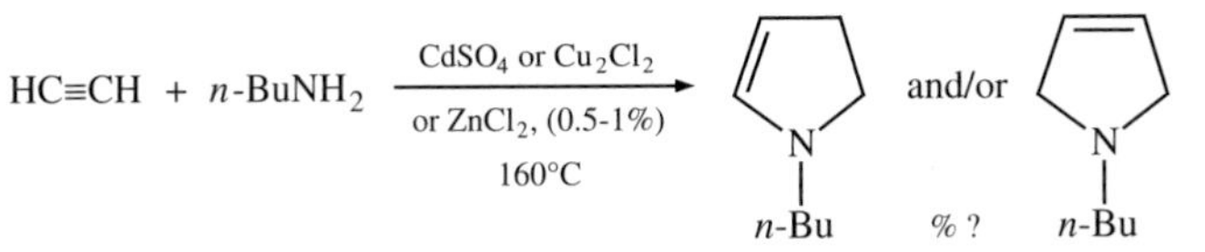

$$HC\equiv CH \ + \ n\text{-BuNH}_2 \quad \xrightarrow[\substack{\text{or ZnCl}_2, (0.5\text{-}1\%) \\ 160°C}]{\text{CdSO}_4 \text{ or Cu}_2\text{Cl}_2} \quad \underset{n\text{-Bu}}{\boxed{}} \quad \text{and/or} \quad \underset{n\text{-Bu}}{\boxed{}} \qquad (4.57)$$

%?

Last, N-vinylpyrrole can be obtained in 'good yield' from the reaction of pyrrole and acetylene over soda lime at 265°C [237].

Only one publication describes the use of a zeolite as catalyst in the hydroamination of alkynes. MeNH$_2$ reacts with 1-propyne over a Zn(II) ion-exchanged Y zeolite (ZnY) at room temperature and atmospheric pressure to give a mixture of *N-iso*-propylidenemethylamine and *N-n*-propylidenemethylamine (Eq. 4.58) [238].

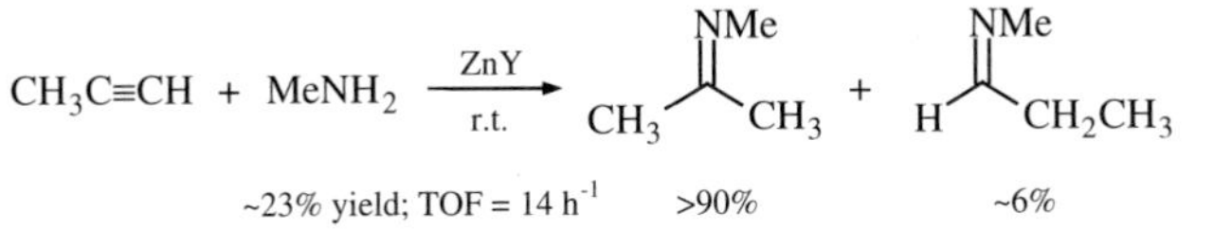

$$CH_3C\equiv CH \ + \ MeNH_2 \quad \xrightarrow[\text{r.t.}]{\text{ZnY}} \quad \underset{CH_3}{\overset{NMe}{\underset{}{\|}}}CH_3 \quad + \quad \underset{H}{\overset{NMe}{\underset{}{\|}}}CH_2CH_3 \qquad (4.58)$$

~23% yield; TOF = 14 h^{-1} >90%   ~6%

Only ZnY and cadmium(II) ion-exchanged Y zeolite (CdY) are active catalysts. Two enamines resulting from Markovnikov and anti-Markovnikov additions are the initial products of the reaction. Extension was made to the hydroamination of 1-propyne (with EtNH$_2$, *i*-PrNH$_2$, or PhNH$_2$) of 2-butyne (with MeNH$_2$ or *i*-PrNH$_2$) and of 1-hexyne (with MeNH$_2$), but with lower activity and selectivity. The hydroamination fails in the case of 1-propyne with Me$_2$NH, *t*-BuNH$_2$, or NH$_3$ and of acetylene with MeNH$_2$ [238].

4.5.2
Homogeneous Catalysis

4.5.2.1   Activation of the Alkyne

It was probably Kozlov et al., studying the condensation of acetylene with aniline, who observed the first homogeneously catalyzed hydroamination of an alkyne [239]. They give experimental evidence that N-phenylethylideneamine is an intermediate, which evolves to an aldol-type condensation product (Scheme 4-9), known as an Ekstein-Eibner base [240–242].

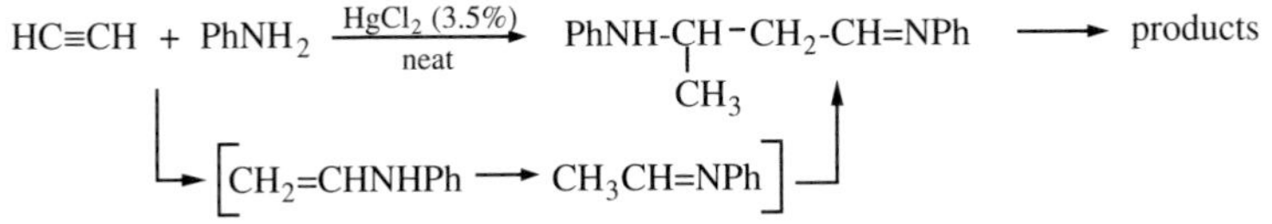

$$HC\equiv CH \ + \ PhNH_2 \quad \xrightarrow[\text{neat}]{\text{HgCl}_2 \ (3.5\%)} \quad PhNH\text{-}\underset{CH_3}{\underset{|}{CH}}\text{-}CH_2\text{-}CH=NPh \quad \longrightarrow \quad \text{products}$$

$$\xrightarrow{\hspace{1cm}} \big[CH_2=CHNHPh \longrightarrow CH_3CH=NPh\big]$$

Scheme 4-9   Hydroamination products as intermediates in the reaction of acetylene and PhNH$_2$

Et$_2$NH reacts with acetylene in the presence of copper(I) chloride or acetylide to give 3-diethylamino-1-butyne (Scheme 4-10) [243–247].

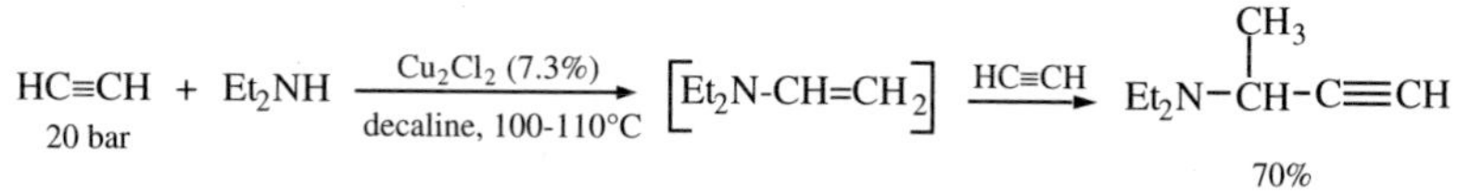

$$HC{\equiv}CH \ + \ Et_2NH \ \xrightarrow[\text{decaline, 100-110°C}]{Cu_2Cl_2 \ (7.3\%)} \ \left[Et_2N-CH{=}CH_2 \right] \ \xrightarrow{HC{\equiv}CH} \ Et_2N-\overset{\overset{\textstyle CH_3}{|}}{CH}-C{\equiv}CH$$

20 bar   70%

Scheme 4-10 Hydroamination product as intermediate in the reaction of acetylene and Et$_2$NH

Similar reactions with primary and secondary amines result in the formation of 3-alkylamino- or 3-dialkylamino-1-butyne in 30–80% yield (TON = 3–9) [243–247]. In one example, the TOF could be estimated as 0.2 h^{-1} [246]. Enamines were proposed as reaction intermediates. It was later shown that enamines effectively react with terminal alkynes, including acetylene, to afford the expected aminoalkynes without catalyst or, more rapidly, sometimes exothermically, in the presence of Cu$_2$Cl$_2$ [248]. Aromatic amines do not react under the same conditions. However, in the presence of organic acids, e.g. acetic acid, 3-arylamino-1-butynes can be isolated in moderate yields (Eq. 4.59) [246, 247, 249].

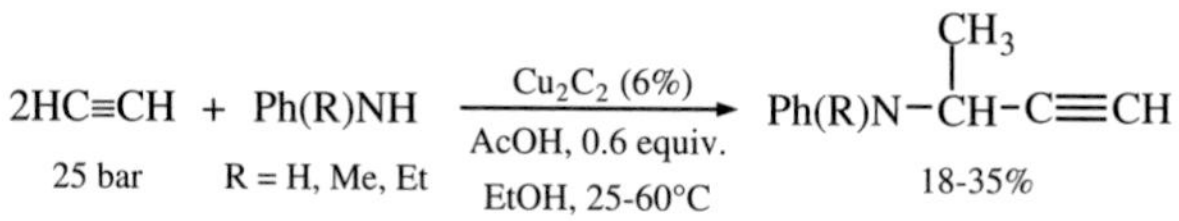

$$2HC{\equiv}CH \ + \ Ph(R)NH \ \xrightarrow[\substack{AcOH, 0.6 \ equiv. \\ EtOH, 25\text{-}60°C}]{Cu_2C_2 \ (6\%)} \ Ph(R)N-\overset{\overset{\textstyle CH_3}{|}}{CH}-C{\equiv}CH \qquad (4.59)$$

25 bar   R = H, Me, Et   18-35%

The reaction of amines and acetylene takes a different course in the presence of Zn(OAc)$_2$/Cd(OAc)$_2$. Thus, for example, EtNH$_2$ gives N-ethylethylideneamine in moderate yield (Eq. 4.60) [250], not ethylvinylamine as Reppe et al. had reported earlier [251].

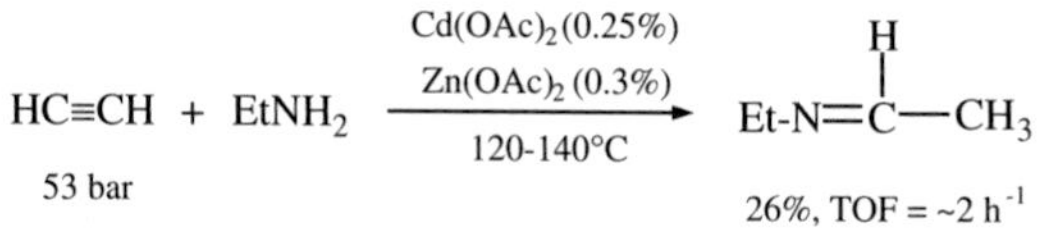

$$HC{\equiv}CH \ + \ EtNH_2 \ \xrightarrow[\substack{Zn(OAc)_2 \ (0.3\%) \\ 120\text{-}140°C}]{Cd(OAc)_2 \ (0.25\%)} \ Et-N{=}\overset{\overset{\textstyle H}{|}}{C}-CH_3 \qquad (4.60)$$

53 bar   26%, TOF = ~2 h^{-1}

However, like N-phenylethylideneamine [239], N-alkylethylideneamines are not stable and undergo either aldol-type condensations to give Eckstein-Eibner bases [240–242] or condensation-elimination to give alkenylideneamines [252–254] as shown (Eq. 4.61) [250] or even higher vinylogs [250].

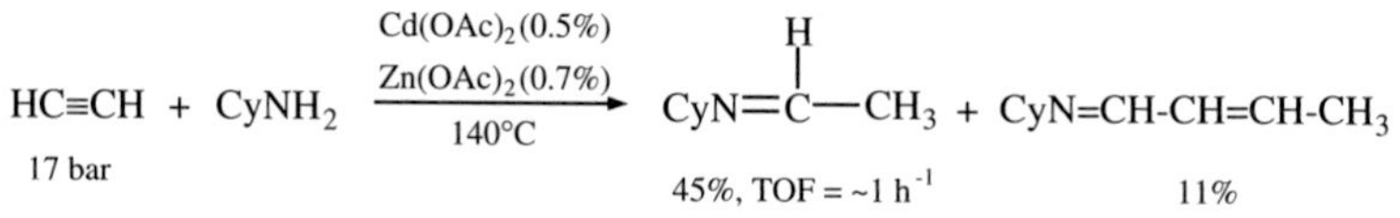

$$HC{\equiv}CH \ + \ CyNH_2 \ \xrightarrow[\substack{Zn(OAc)_2 \ (0.7\%) \\ 140°C}]{Cd(OAc)_2 \ (0.5\%)} \ CyN{=}\overset{\overset{\textstyle H}{|}}{C}-CH_3 \ + \ CyN{=}CH\text{-}CH{=}CH\text{-}CH_3 \quad (4.61)$$

17 bar   45%, TOF = ~1 h^{-1} 11%

Imidazole reacts with acetylene in the presence of RuCl$_3$ to afford N-vinylimidazole in fair yield (Eq. 4.62) [255].

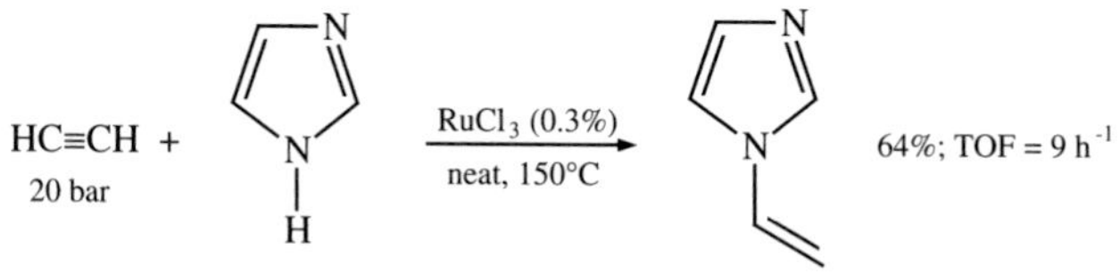

$$HC{\equiv}CH \ + \ \text{(imidazole)} \ \xrightarrow[\text{neat, 150°C}]{RuCl_3 \ (0.3\%)} \ \text{(N-vinylimidazole)} \qquad 64\%; \ TOF = 9 \ h^{-1} \qquad (4.62)$$

20 bar

With terminal alkynes, the direction of the reaction depends on the nature of the substituent, the type of amine and the catalyst. Thus, dialkylamines can react with propyne to give 4-dialkylamino-4-methyl-2-pentynes traced from the *fully* regioselective formation of 2-dialkylaminopropene, i.e. the Markovnikov hydroamination product (Scheme 4-11) [256].

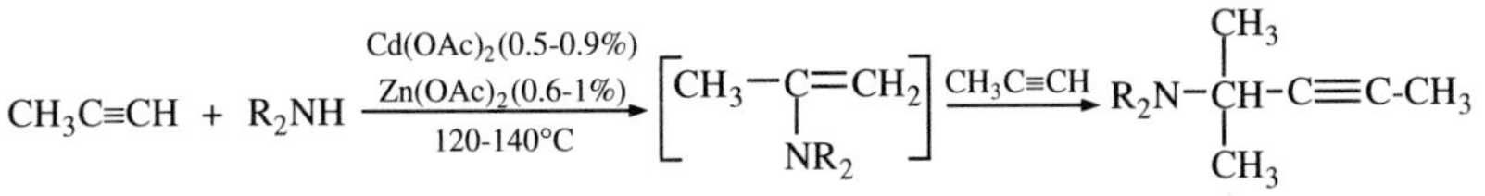
Cd(OAc)2(0.5-0.9%)
Zn(OAc)2(0.6-1%)
120-140°C
CH3C≡CH + R2NH
CH3-C=CH2
NR2
CH3C≡CH
R2N-CH-C≡C-CH3
CH3
CH3

$R = CH_3$, 71%, TOF = 7 h^{-1}
$R = C_2H_5$, 80%, TOF = 4.5 h^{-1}

Scheme 4-11 Hydroamination product as intermediates in the reaction of propyne and R_2NH

In contrast, the reaction of morpholine with phenylacetylene is less regioselective. Two products are formed, arising from two vinylamine intermediates (Scheme 4-12) [257].

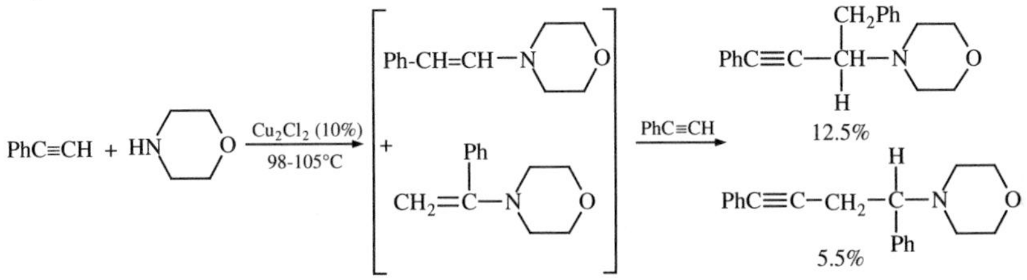
PhC≡CH + HN O
Cu2Cl2 (10%)
98-105°C
Ph-CH=CH-N O
+
Ph
CH2=C-N O
PhC≡CH
CH2Ph
PhC≡C-CH-N O
H
12.5%
H
PhC≡C-CH2-C-N O
Ph
5.5%

Scheme 4-12 Hydroamination products as intermediates in the reaction of phenylacetylene and morpholine

$PhNH_2$ reacts with alkynes in the presence of HgO-BF_3 to give *N*-phenylalkylideneamines arising from the isomerization of the initially formed vinylamines (Eq. 4.63) [258].

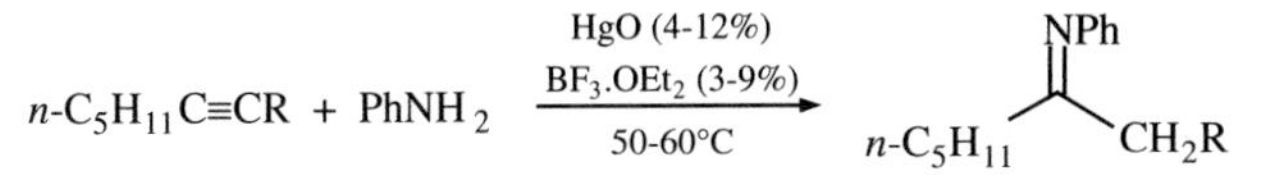
n-C5H11C≡CR + PhNH2
HgO (4-12%)
BF3.OEt2 (3-9%)
50-60°C
NPh
n-C5H11
CH2R

(4.63)

R = H, 26-34%, TOF = 2-8.5 h^{-1}
R = Et, 19%, TOF = 1.5-4.5 h^{-1}

PhNHEt and 1-heptyne afford a *N*-vinylamine as a stable Markovnikov hydroamination product (Eq. 4.64) [258].

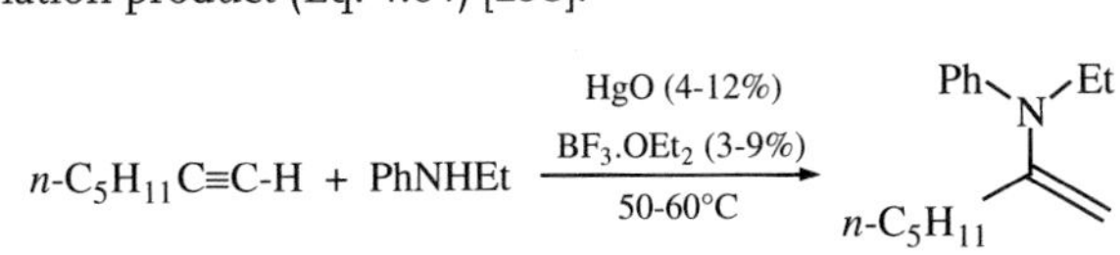
n-C5H11C≡C-H + PhNHEt
HgO (4-12%)
BF3.OEt2 (3-9%)
50-60°C
Ph
N
Et
n-C5H11

(4.64)

Barluenga et al. have extensively studied the hydroamination of alkynes catalyzed by mercury compounds, especially mercury(II) chloride. Terminal alkynes and

PhNH$_2$ react in the presence of HgCl$_2$ (5%) to produce the same anils as those obtained with the HgO/BF$_3$.OEt$_2$ system (Eq. 4.63), but at room temperature and with
a higher TOF (12 h^{-1}) [259]. Phenylacetylene and secondary aromatic amines afford
enamines like those of Eq. (4.64), but aliphatic terminal alkynes and secondary aromatic amines give rise to a mixture of enamines (Eq. 4.65) [259].

$$RCH_2C\equiv C\text{-}H + PhNHEt \xrightarrow[60°C]{HgCl_2\ (5\%)}$$

5 equiv.

R = *n*-C$_5$H$_{11}$

>95   <5

84%, TOF = 17 h^{-1}

(4.65)

Under the same conditions, the hydroamination of acetylene with primary or secondary aromatic amines brings about the formation of dimerization-cyclization
products since the generated imines or enamines, respectively, are not stable.

The hydroamination of alkynes with primary and secondary aliphatic amines ne-
cessitates the use of higher amounts of catalyst (17%) and higher temperatures, and
TOFs are low (<1 h^{-1}) [260]. With aliphatic and aromatic terminal alkynes and a 5-
fold excess of primary aliphatic amines, the products are the corresponding imines
(40–78% yield, TOF up to 0.3 h^{-1}). In contrast to the Cu$_2$Cl$_2$-catalyzed reaction of
phenylacetylene and secondary aliphatic amines (Scheme 4-12), the HgCl$_2$-catalyzed
reaction is fully regioselective for the Markovnikov hydroamination products which
do not evolve under the reaction conditions (Eq. 4.66) [260].

$$PhC\equiv C\text{-}H + HN\underset{\text{5 equiv.}}{\bigcirc}O \xrightarrow[\text{dioxane, 100°C}]{HgCl_2\ (17\%)}$$

50%
TOF = 0.2 h^{-1}

(4.66)

As in the case of secondary aromatic amines, aliphatic terminal alkynes and secondary aliphatic amines give rise to mixtures of enamines like those of Eq. (4.65)
[260].

Mechanistic studies indicated the possibility of alkynylmercury chlorides as intermediates. They would react with amines to give 2-aminovinylmercury chlorides
which could be protonated to give enamines (or imines in the case of primary
amines) (Scheme 4-13) [260].

The hydroamination of phenylacetylene with primary or secondary aromatic
amines is also catalyzed by Tl(OAc)$_3$ to give imines or enamines, respectively, in low
to good yields (10–89%) with TOF up to 6 h^{-1} [261].

Substituted 3-alken-1-ynes can be hydroaminated with primary or secondary
aliphatic or aromatic amines at the alkynyl sites or at the alkynyl *and* at the alkenyl
sites in the presence of Hg(II) salts. However, the reaction is essentially stoichiometric in nature, even if the mercury compound can be recycled without apparent
loss of activity [262–264].

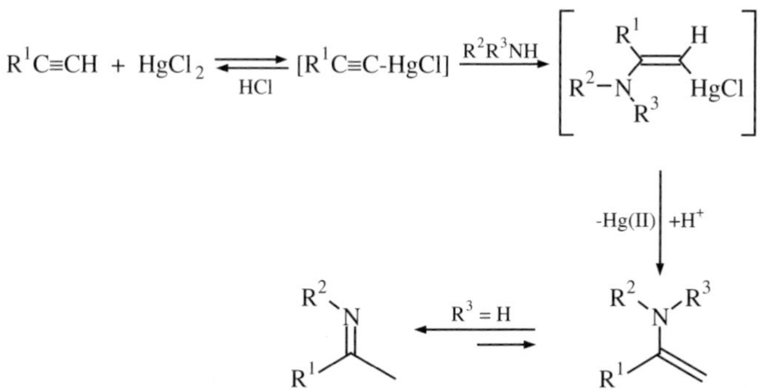

$$R^1C{\equiv}CH + HgCl_2 \underset{HCl}{\rightleftharpoons} [R^1C{\equiv}C\text{-}HgCl] \xrightarrow{R^2R^3NH} \left[\begin{array}{c} R^1 \qquad H \\ \diagdown \quad / \\ \diagup \quad \diagdown \\ R^2\text{-}N \qquad HgCl \\ | \\ R^3 \end{array} \right]$$

$$\downarrow -Hg(II) \; +H^+$$

$$\underset{R^1}{\overset{R^2}{\diagdown}}\!\!N \underset{}{\Vert} \quad \xleftarrow{\quad R^3 = H \quad} \quad \underset{R^1}{\overset{R^2 \diagdown \quad N \diagup R^3}{\diagdown}}$$

Scheme 4-13 Mechanism of the HgCl$_2$-catalyzed hydroamination of terminal alkynes

Hexafluoro-2-butyne, vinylacetylene, and 1,3-butadiyne can be hydroaminated without catalyst [265–267]. However, the reaction of 1,3-butadiyne with Et$_2$NH is accelerated and gives a better yield in the presence of silver or silver diacetylide (29% *vs.* 19% without catalyst) (Eq. 4.67) [268].

$$HC{\equiv}C\text{-}C{\equiv}CH + Et_2NH \xrightarrow[C_6H_6, \, 45°C]{Ag \, (0.3\%)} HC{\equiv}C\text{-}CH{=}CH\text{-}NEt_2 \qquad (4.67)$$
$$29\%, \text{TOF} \sim 50 \text{ h}^{-1}$$

1,3-butadiyne, mono- and disubstituted 1,3-butadiynes react with NH$_3$ and primary alkyl- and arylamines in the presence of Cu$_2$Cl$_2$, affording a wide variety of 1,2-, 2,5- and 1,2,5-substituted pyrroles in good yields (e.g., Eq. 4.68) [269–271]. This reaction has been applied to the synthesis of polypyrroles and of the porphyrin ring [272, 273].

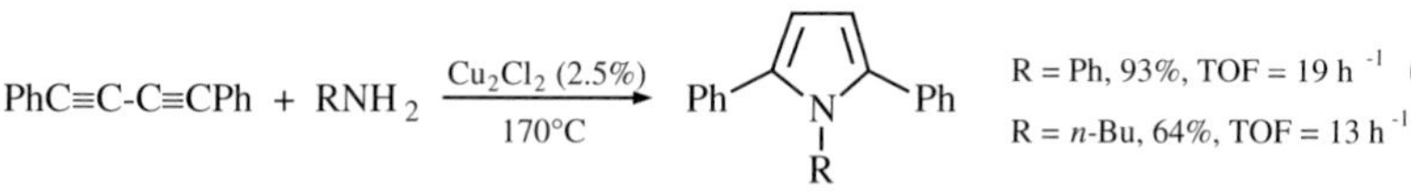

$$PhC{\equiv}C\text{-}C{\equiv}CPh + RNH_2 \xrightarrow[170°C]{Cu_2Cl_2 \, (2.5\%)} Ph\!\!\diagup\!\!\underset{R}{\overset{\diagdown}{N}}\!\!\diagdown\!\!Ph \qquad \begin{array}{l} R = Ph, 93\%, \text{TOF} = 19 \text{ h}^{-1} \quad (4.68) \\ R = n\text{-Bu}, 64\%, \text{TOF} = 13 \text{ h}^{-1} \end{array}$$

IH of alkynylamines has been performed with a variety of catalytic systems based on palladium [274–281], cobalt, rhodium, iridium, ruthenium, platinum, copper, silver, zinc, cadmium, mercury [279–281], nickel [279–282], gold [279–281, 283], and molybdenum [284] derivatives.

In the case of 3-alkynylamines, IH proceeds exclusively in a 5-*Endo-Dig* process to give substituted 1-pyrrolines. The best catalysts are palladium complexes (Eq. 4.69); the reaction fails for terminal alkyne owing to the formation of a stable palladium acetylide [278].

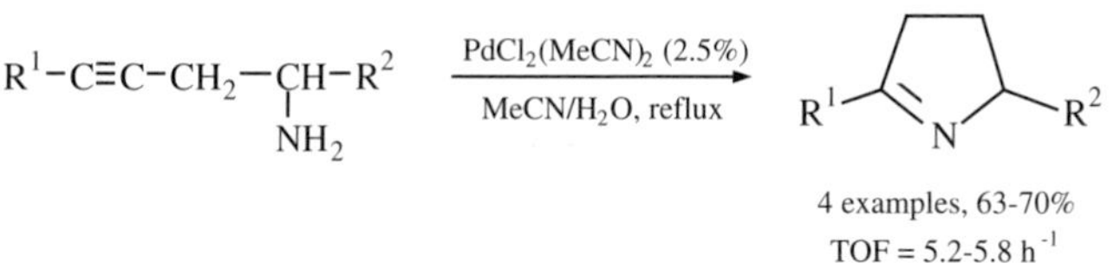

$$R^1\text{-}C{\equiv}C\text{-}CH_2\text{-}\underset{NH_2}{\overset{|}{C}H}\text{-}R^2 \xrightarrow[MeCN/H_2O, \, reflux]{PdCl_2(MeCN)_2 \, (2.5\%)} R^1\!\!\diagup\!\!\underset{N}{\overset{\diagdown}{}}\!\!\diagdown\!\!R^2 \qquad (4.69)$$
$$4 \text{ examples, } 63\text{-}70\%$$
$$\text{TOF} = 5.2\text{-}5.8 \text{ h}^{-1}$$

$Ni(CO)_2(PPh_3)_2$ (under CO pressure) is also an efficient catalyst, but $[Rh(CO)_2Cl]_2$ is not [282].

For 4-alkynylamines, IH in the presence of $PdCl_2(MeCN)_2$ proceeds by two competitive processes; a major 5-*Exo-Dig* cyclization and a minor 6-*Endo-Dig* one (Eq. 4.70) [278].

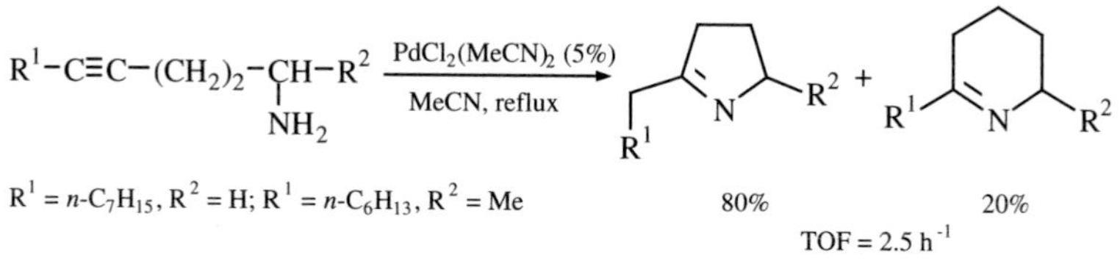

$$R^1-C\equiv C-(CH_2)_2-\underset{NH_2}{CH}-R^2 \xrightarrow[\text{MeCN, reflux}]{PdCl_2(MeCN)_2\ (5\%)}$$

$R^1 = n\text{-}C_7H_{15}, R^2 = H; R^1 = n\text{-}C_6H_{13}, R^2 = Me$ 80%   20%   (4.70)

TOF = 2.5 h^{-1}

With $[Pd(MeCN)_4](BF_4)_2$ [281] or $Ni(CO)_2(PPh_3)_2$ under CO pressure [282], only substituted pyrrolines are formed (Pd: 48-70%, TOF = 6-8 h^{-1}; Ni: 40%, TOF = 0.5 h^{-1}).

IH of 5-alkynylamines proceeds in a 6-*Exo-Dig* manner exclusively, whatever the catalyst [278–281, 283]. In the case of 6-amino-1-hexyne, the best catalysts by far are $[Cu(MeCN)_4]PF_6$, $Zn(O_3SCF_3)_2$, $Cd(NO_3)_2.4H_2O$, or $Hg(NO_3)_2.H_2O$ (Eq. 4.71) [280].

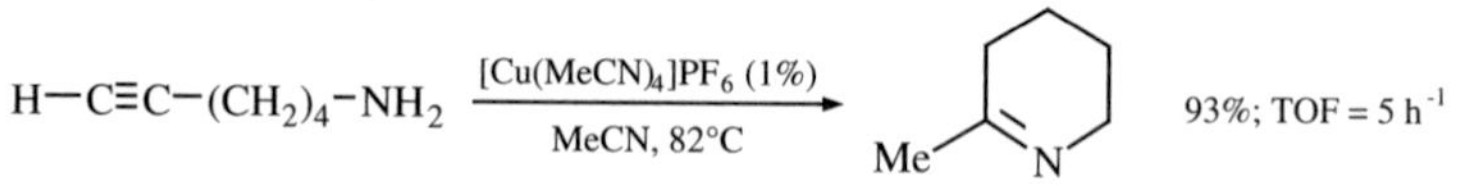

$$H-C\equiv C-(CH_2)_4-NH_2 \xrightarrow[\text{MeCN, 82°C}]{[Cu(MeCN)_4]PF_6\ (1\%)}$$

93%; TOF = 5 h^{-1} (4.71)

For substituted 5-alkynylamines, gold derivatives are the most efficient catalysts (Eq. 4.72) [283].

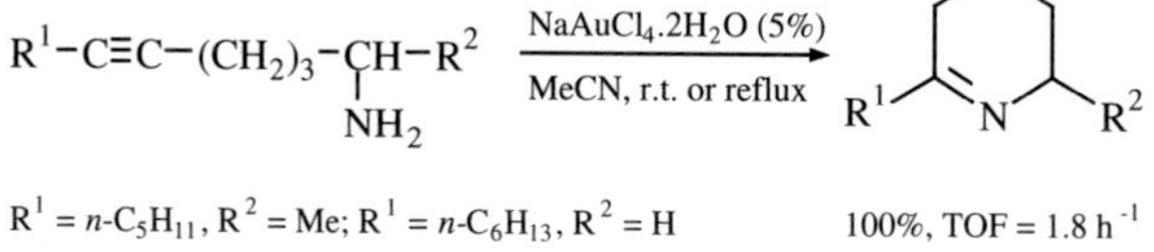

$$R^1-C\equiv C-(CH_2)_3-\underset{NH_2}{CH}-R^2 \xrightarrow[\text{MeCN, r.t. or reflux}]{NaAuCl_4.2H_2O\ (5\%)}$$

(4.72)

$R^1 = n\text{-}C_5H_{11}, R^2 = Me; R^1 = n\text{-}C_6H_{13}, R^2 = H$ 100%, TOF = 1.8 h^{-1}

These IHs have been applied to the synthesis of several pyrrolines and 2,3,4,5-tetrahydropyridines known as venom constituents from ant species [278].

IH of 2-alkynylanilines has been also studied in the presence of molybdenum [284], palladium [276, 277, 281], and gold catalysts [276]. It provides indoles in low to good yields as a result of 5-*Endo-Dig* cyclizations (Eq. 4.73) [277].

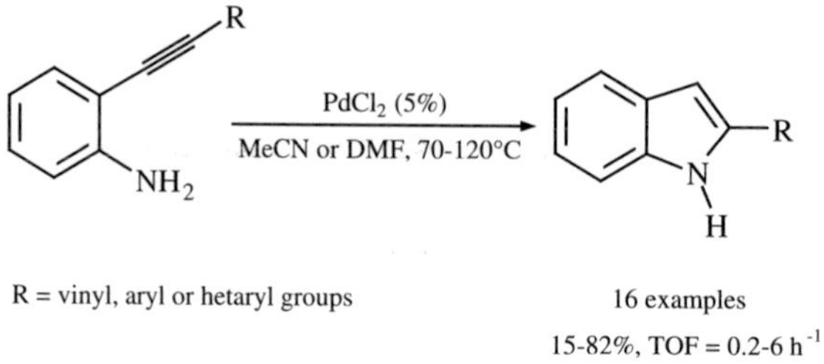

$$\xrightarrow[\text{MeCN or DMF, 70-120°C}]{PdCl_2\ (5\%)}$$

(4.73)

R = vinyl, aryl or hetaryl groups

16 examples
15-82%, TOF = 0.2-6 h^{-1}

The mechanism of these IHs has been studied in the case of 6-amino-1-hexyne. The oxidation state of the metal is of the utmost importance since no conversion is observed with complexes in a different oxidation state. Catalytically active species are electron-rich d^8 or d^{10} complexes. A general catalytic cycle has been proposed on the basis of deuterium-labeling experiments (Scheme 4-14) [280]. It is believed to occur for all the catalysts used.

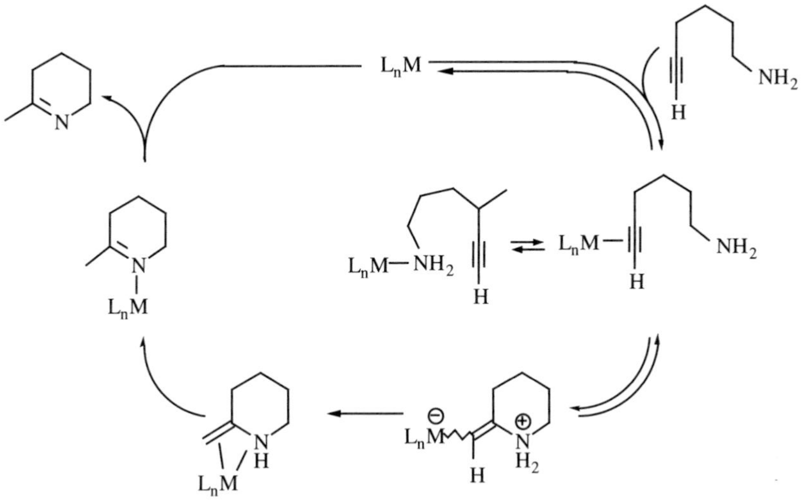

Scheme 4-14 Catalytic cycle for the IH of aminoalkynes *via* activation of the alkyne

4.5.2.2 Activation of the Amine

4.5.2.2.1 Activation by Bases
The first example of base-catalyzed hydroamination of alkynes is due to Reppe et al. in 1935 (Eq. 4.74) [285].

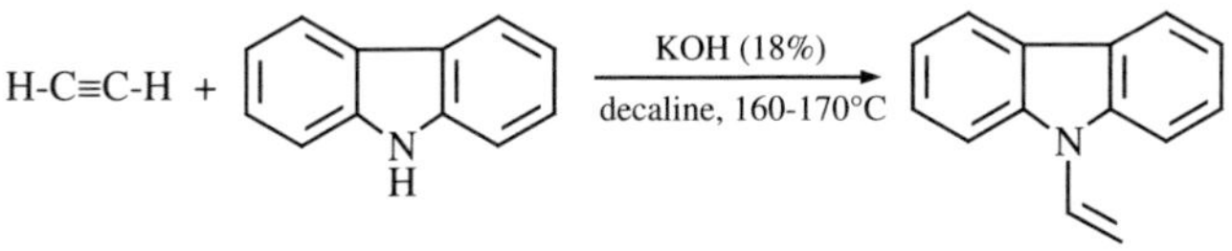

$$(4.74)$$

Similarly, pyrrole, indole, and tetrahydrocarbazole [285] as well as diarylamines give the corresponding *N*-vinyl compounds [286]. Several improvements of yields and reaction rates were observed by conducting the reactions in the presence of additives [287, 288]. The vinylation of imidazole and benzimidazole was reported to be catalyzed by KOH in the presence of zinc or cadmium salts [289]. The above reactions were reviewed in 1965 [290].

Highly basic systems such as KOH/DMSO (superbasic Trofimov conditions [291]) have been used for the *N*-vinylation of 3-vinylpyrrole (Eq. 4.75) [292].

$$H-C\equiv C-H + \underset{\underset{H}{|}}{\overset{}{\text{(pyrrole)}}} \xrightarrow[\text{DMSO, 120°C}]{\text{KOH (30\%)}} \text{(N-vinyl pyrrole)} \quad 55\%; \text{TOF} = 1\ h^{-1} \qquad (4.75)$$

10 bar

CsOH.H$_2$O has also been shown to be a catalyst for the hydroamination of phenylacetylene with anilines (Eq. 4.76) and heterocyclic secondary amines [160].

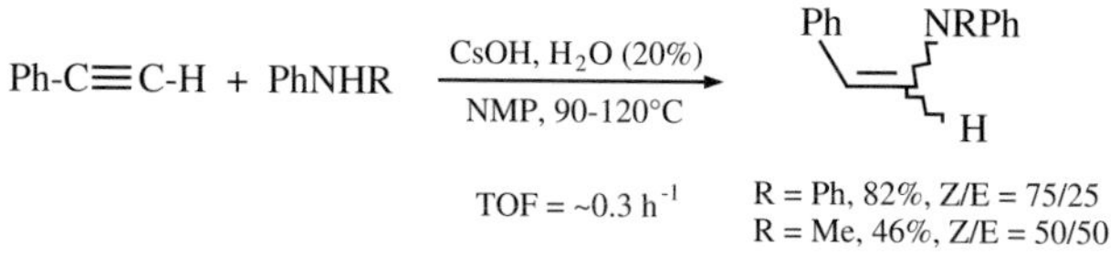

$$Ph-C\equiv C-H + PhNHR \xrightarrow[\text{NMP, 90-120°C}]{\text{CsOH, H}_2\text{O (20\%)}} \quad \underset{H}{\overset{Ph\quad NRPh}{\diagup}} \qquad (4.76)$$

TOF = ~0.3 h^{-1} R = Ph, 82%, Z/E = 75/25
R = Me, 46%, Z/E = 50/50

4.5.2.2.2 Activation by Early Transition Metals

In 1992, Bergman et al. reported that zirconium bisamides Cp$_2$Zr(NHR)$_2$ catalyze the intermolecular hydroamination of alkynes with *sterically hindered* primary amines to give enamines or their tautomeric imines (e.g., Eq. 4.77) [126].

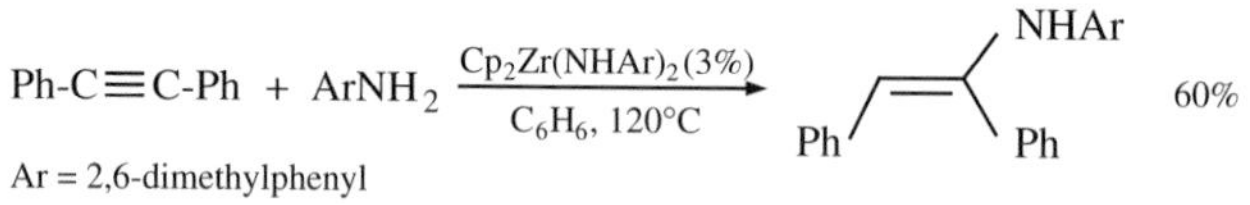

$$Ph-C\equiv C-Ph + ArNH_2 \xrightarrow[\text{C}_6\text{H}_6,\ 120°C]{\text{Cp}_2\text{Zr(NHAr)}_2\ (3\%)} \quad \underset{Ph}{\overset{NHAr}{\diagup}}_{Ph} \quad 60\% \qquad (4.77)$$

Ar = 2,6-dimethylphenyl

The reaction rate is low (TOF = 0.04-0.2 h^{-1}), but, interestingly, the catalyst seems indefinitely stable under the reaction conditions. Kinetic data (v =

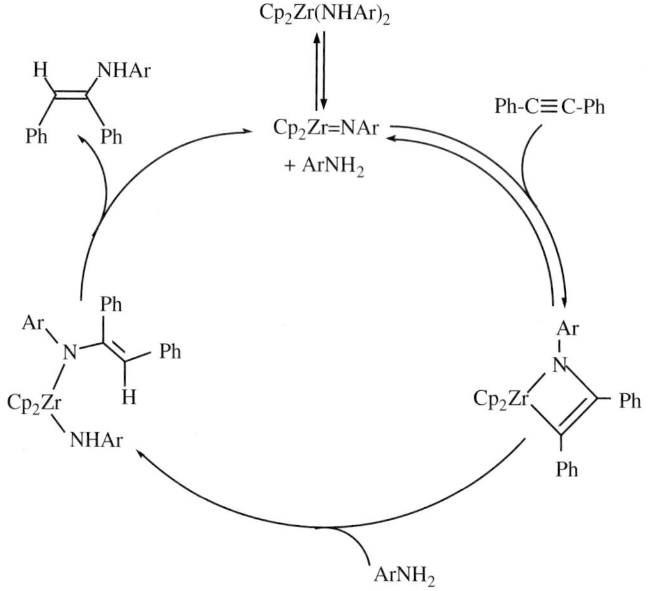

Scheme 4-15 Catalytic cycle for the zirconium(bisamide)-catalyzed hydroamination of alkynes

$k[\text{cat.}]^1[\text{alkyne}]^0[\text{ArNH}_2]^{-1})$ are consistent with a reversible rate-determining α-elimination of amine to generate a transient imido complex $Cp_2Zr=NAr$. A [2+2] cycloaddition of the latter with the alkyne then generates an azametallacyclobutene [293], which is cleaved at the Zr–C bond by protonolysis by the amine. Regeneration of the imido complex then occurs by α-elimination of the enamine (Scheme 4-15).

The stoichiometric hydroamination of unsymmetrically disubstituted alkynes is highly regioselective, generating the azametallacycle with the larger alkyne substituent α to the metal center [294, 295]. In others words, the enamine or imine formed results from an anti-Markovnikov addition. Unfortunately, this reaction could not be applied to less sterically hindered amines.

More recently, Doye et al. used Cp_2TiMe_2 as catalyst precursor [296]. In the presence of a primary amine RNH_2, Cp_2TiMe_2 loses methane to generate the titanium bisamide or titanium imido complexes $[Cp_2Ti(NHR)_2$ or $Cp_2Ti=NR]$, which are active for the hydroamination of alkynes (R = aryl, *t*-Bu, Cy). Although the potentially hydrolyzable imines formed could be isolated, more representative yields are given by hydrolysis to the corresponding ketones or reduction to the amine (e.g., Scheme 4-16, TOF ≈ 0.5 h^{-1}) [296].

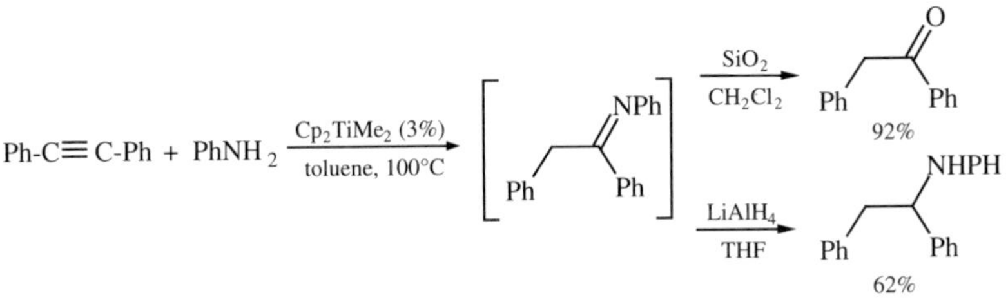

Scheme 4-16   Hydroamination of diphenylacetylene with PhNH$_2$

For unsymmetrically disubstituted alkynes, the regioselectivity is 100% anti-Markovnikov, phenylpropyne being the most reactive alkyne (Eq. 4.78) [296].

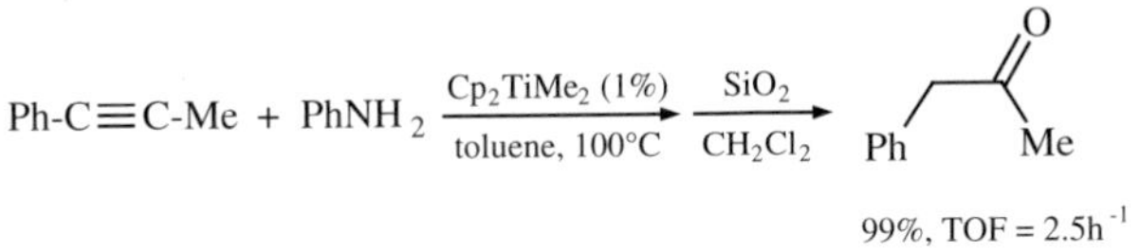

$$\text{Ph-C}\equiv\text{C-Me} + \text{PhNH}_2 \xrightarrow[\text{toluene, 100°C}]{\underset{\text{CH}_2\text{Cl}_2}{Cp_2TiMe_2\ (1\%)\quad SiO_2}} \qquad \qquad \qquad \qquad (4.78)$$

99%, TOF = 2.5h^{-1}

The proposed catalytic cycle is based on that reported by Bergman et al. with zirconium-imido complexes (see above, Scheme 4-15).

IH of aminoalkynes can be performed under mild conditions using $CpTiCl_3$ or $CpTi(Me)_2Cl$ as catalyst precursors to give dihydropyrrole and tetrahydropyridine derivatives (regioselective *Exo-Dig* processes) in high yield with TOF near 10 h^{-1} (e.g., Eq. 4.79) [297].

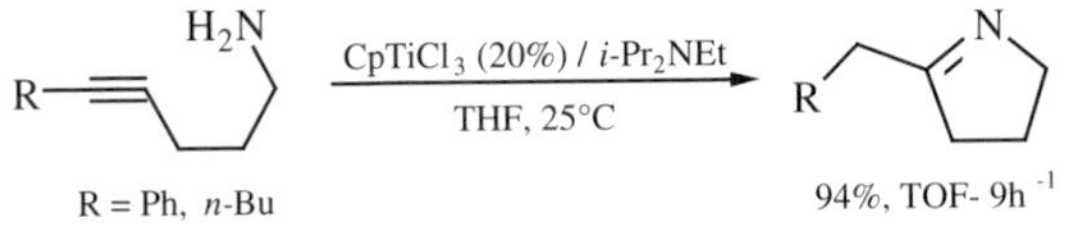

$$(4.79)$$

R = Ph, *n*-Bu   94%, TOF- 9h^{-1}

The reaction mechanism is believed to involve *in situ* generation of an imidochlorotitanium complex (CpTi(Cl)=NR), which then undergoes a [2+2] cycloaddition.

This catalytic process has been used for the key step of a convergent synthesis of (±)-monomorine [298] and for a stereoselective total synthesis of (+)-preussin [299].

(Aminotroponiminato)yttrium amides also catalyze the regioselective IH of primary aminoalkynes of Eq. (4.79), but the catalytic activity is lower (TOF < 1 h^{-1}) [300].

4.5.2.2.3 Activation by Lanthanides and Actinides

Organolanthanides are effective catalyst precursors for the regioselective hydroamination of internal alkynes with primary amines (Eq. 4.80) [127].

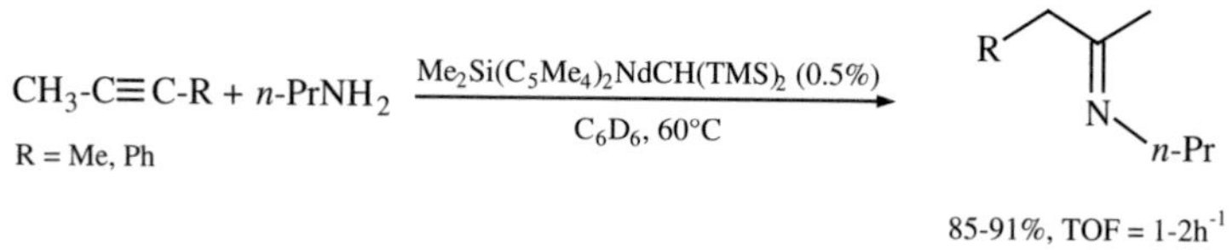

$$CH_3\text{-}C\equiv C\text{-}R + n\text{-}PrNH_2 \xrightarrow[C_6D_6, 60°C]{Me_2Si(C_5Me_4)_2NdCH(TMS)_2 \ (0.5\%)}$$

R = Me, Ph

(4.80)

85-91%, TOF = 1-2h^{-1}

In the case of silyl-substituted alkynes (R = Me$_3$Si), the initially formed imine undergoes a subsequent 1,3-sigmatropic silyl shift yielding the corresponding enamine (Eq. 4.81).

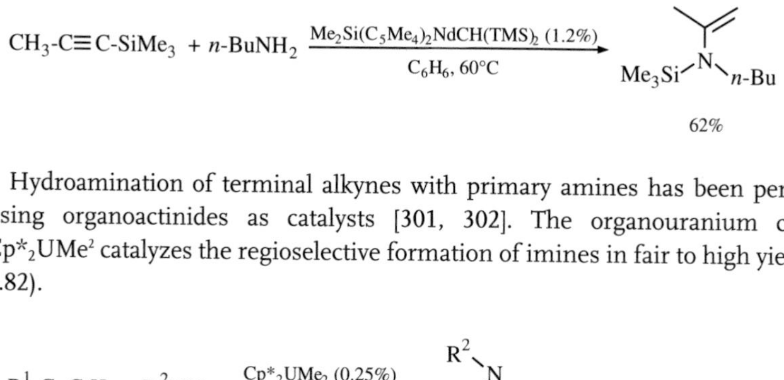

$$CH_3\text{-}C\equiv C\text{-}SiMe_3 + n\text{-}BuNH_2 \xrightarrow[C_6H_6, 60°C]{Me_2Si(C_5Me_4)_2NdCH(TMS)_2 \ (1.2\%)}$$

(4.81)

62%

Hydroamination of terminal alkynes with primary amines has been performed using organoactinides as catalysts [301, 302]. The organouranium complex Cp*$_2$UMe^2 catalyzes the regioselective formation of imines in fair to high yields (Eq. 4.82).

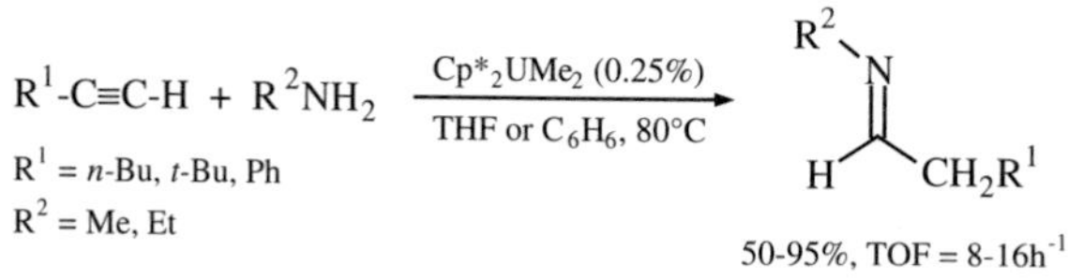

$$R^1\text{-}C\equiv C\text{-}H + R^2NH_2 \xrightarrow[THF \ or \ C_6H_6, 80°C]{Cp*_2UMe_2 \ (0.25\%)}$$

R^1 = n-Bu, t-Bu, Ph
R^2 = Me, Et

(4.82)

50-95%, TOF = 8-16h^{-1}

The Cp*$_2$ThMe$_2$ analog catalyzes the hydroamination of acetylene with EtNH$_2$ (Eq. 4.83) [301].

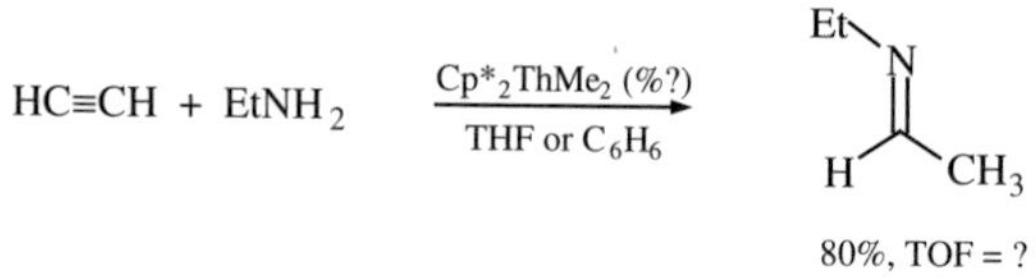

$$HC\equiv CH + EtNH_2 \xrightarrow[THF \ or \ C_6H_6]{Cp*_2ThMe_2 \ (\%?)}$$

(4.83)

80%, TOF = ?

However, with 1-hexyne or phenylacetylene, the thorium catalyst induces a dramatic inversion in regioselectivity giving imines with various amounts of dimerized alkyne (e.g., Eq. 4.84) [301].

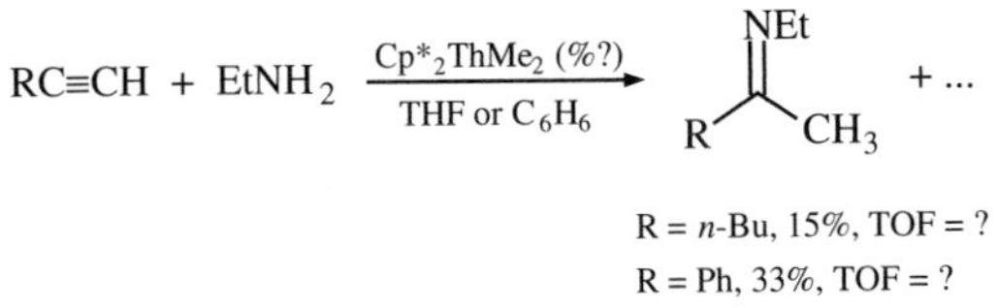

$$RC\equiv CH + EtNH_2 \xrightarrow[\text{THF or } C_6H_6]{Cp*_2ThMe_2 (\%?)} \underset{R}{\overset{NEt}{\|}}_{CH_3} + ... \tag{4.84}$$

R = *n*-Bu, 15%, TOF = ?
R = Ph, 33%, TOF = ?

Coupled intermolecular hydroaminations and intramolecular cyclizations give pyrroles resulting from isomerization of the initially formed exomethylene dihydropyrrole derivatives (Eq. 4.85) [137].

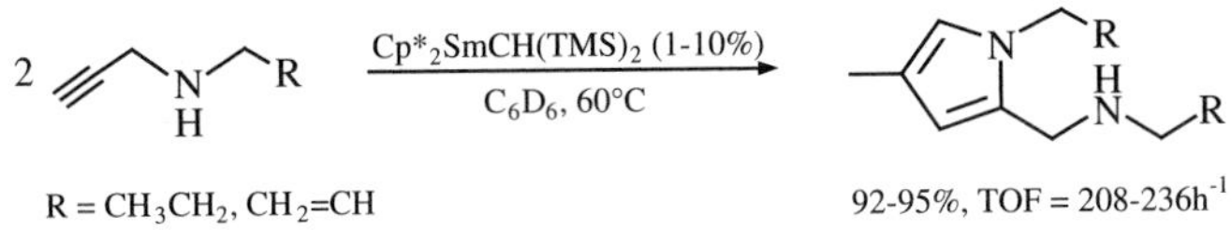

$$2 \underset{H}{\overset{}{\diagup}}_{N}^{R} \xrightarrow[C_6D_6, 60°C]{Cp*_2SmCH(TMS)_2 (1-10\%)} \tag{4.85}$$

R = CH$_3$CH$_2$, CH$_2$=CH   92-95%, TOF = 208-236h^{-1}

Organolanthanides are efficient catalyst precursors for the regioselective (>95%) IH of primary aminoalkynes, forming five-, six-, and seven-membered cyclic imines via *Exo-Dig* processes (Eq. 4.86) [303, 304].

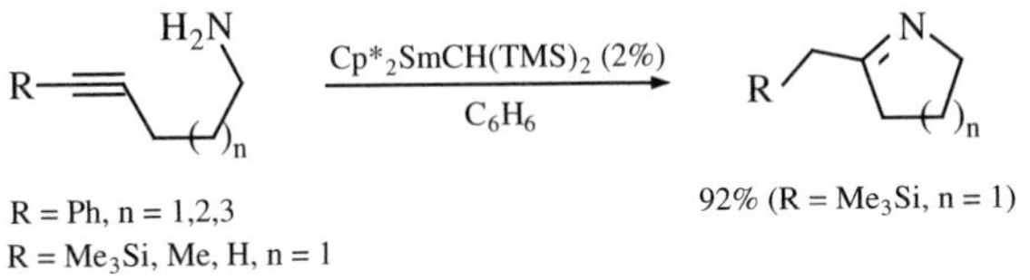

$$R-\!\!\!\equiv\!\!\!\underset{H_2N}{\overset{}{\diagdown}}_{n} \xrightarrow[C_6H_6]{Cp*_2SmCH(TMS)_2 (2\%)} R \xrightarrow{} \underset{N}{\overset{}{\diagdown}}_{n} \tag{4.86}$$

R = Ph, n = 1,2,3   92% (R = Me$_3$Si, n = 1)
R = Me$_3$Si, Me, H, n = 1

These reactions (Eq. 4.86) are 10–100 times more rapid than the corresponding IH of primary aminoalkenes with the same catalyst. TOFs up to 7600 h^{-1} have been observed (Cp*$_2$SmCH(TMS)$_2$, R = Me$_3$Si, n = 1) [303].

The initial reaction has been extended to the IH of secondary aminoalkynes to give cyclic enamines with an exocyclic C=C bond [304].

As in the case of aminodialkenes (see above), hydroamination/bicyclizations of aminoalkenynes allow the regiospecific synthesis of pyrrolizidine skeletons (Eq. 4.87) [138, 303].

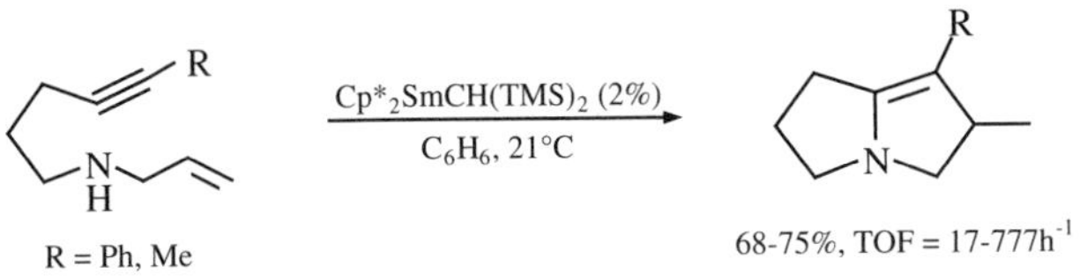

$$\underset{H}{\overset{}{\diagdown}}_{N}\!\!-\!\!\overset{R}{\diagup} \xrightarrow[C_6H_6, 21°C]{Cp*_2SmCH(TMS)_2 (2\%)} \overset{R}{\underset{N}{\diagup}} \tag{4.87}$$

R = Ph, Me   68-75%, TOF = 17-777h^{-1}

4.5.2.2.4 Activation by Late Transition Metals.

The intermolecular hydroamination of alkynes catalyzed by late transition metals was reported for the first time in 1999. Ruthenium carbonyl catalyzes the Markovnikov hydroamination of terminal alkynes with PhNHMe to give enamines (Eq. 4.88) [305].

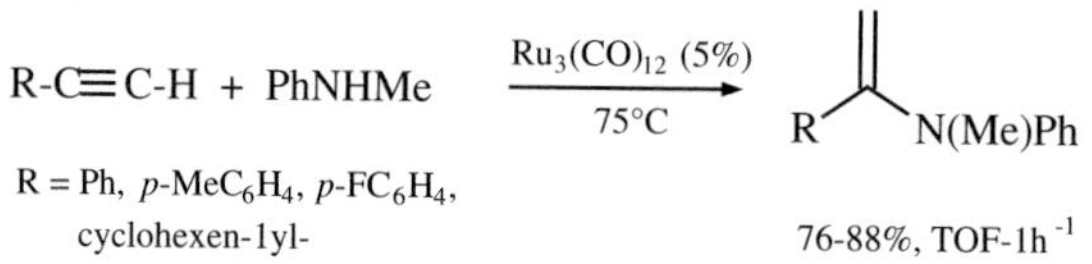

R-C≡C-H + PhNHMe $\xrightarrow[\text{75°C}]{\text{Ru}_3(\text{CO})_{12}\ (5\%)}$ (488)

R = Ph, p-MeC$_6$H$_4$, p-FC$_6$H$_4$,
    cyclohexen-1yl-

$$R \overset{\text{(structure)}}{\underset{}{\bigwedge}} N(Me)Ph$$

76-88%, TOF-1h^{-1}

Since activation of the N–H bond of PhNH$_2$ by Ru$_3$(CO)$_{12}$ has been reported to take place under similar conditions [306], it has been proposed that the reaction mechanism involves (i) generation of an anilido ruthenium hydride, (ii) coordination of the alkyne, (iii) intramolecular nucleophilic attack of the nitrogen lone pair on the coordinated triple bond, and (iv) reductive elimination of the enamine with regeneration of the active Ru(0) center [305].

The above ruthenium carbonyl gives low yield (<5%) for the hydroamination of phenylacetylene with PhNH$_2$ [307]. An important breakthrough was obtained by using the same catalyst in the presence of additives, especially strong acids (HPF$_6$, HBF$_4$) or their ammonium salts, to give the aromatic ketimines (Eq. 4.89) [307].

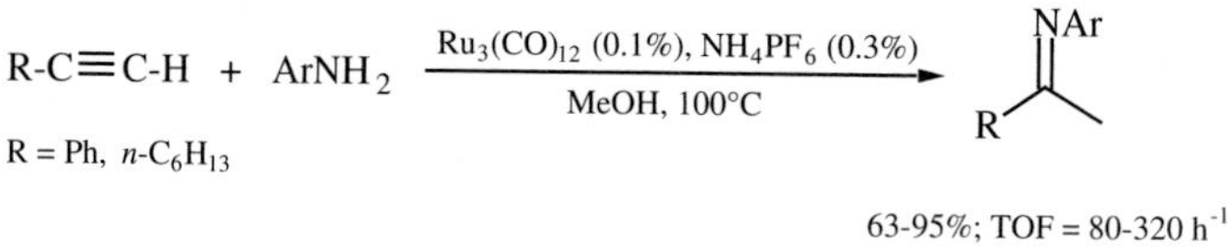

R-C≡C-H + ArNH$_2$ $\xrightarrow[\text{MeOH, 100°C}]{\text{Ru}_3(\text{CO})_{12}\ (0.1\%),\ \text{NH}_4\text{PF}_6\ (0.3\%)}$ (4.89)

R = Ph, n-C$_6$H$_{13}$

$$R \overset{\text{NAr}}{\underset{}{\bigwedge}}$$

63-95%; TOF = 80-320 h^{-1}

The reason for the remarkable enhancement effect of additives remains to be clarified. However, it is clear that both protons and the associated anions play a role [307].

A very interesting feature of the above system is that the reactions can be conducted under air, without solvent. Preparative scale (40 g) hydroamination of phenylacetylene with PhNH$_2$ yielded the acetophenone N-phenylimine in 92% isolated yield in 3 h (TOF = 100 h^{-1}).

4.6
Hydroamination of Allenes

4.6.1
Heterogeneous Catalysis

To the best of our knowledge, no example has been reported.

4.6.2
Homogeneous Catalysis

4.6.2.1 Activation of the Allene
The hydroamination of allene with morpholine or allylamines has been attempted with palladium-based catalysts. Usually, a mixture of 1:1 telomers (hydroamination products) and 1:2 telomers is obtained, the latter being the major [308, 309] or only

products [310]. However, by using palladium complexes/Brønsted acid combinations, it is possible to direct the reaction so that the hydroamination product becomes the major or sole product. Thus, allenes react with secondary amines to give a mixture of allylic amines (hydroamination product) and 1:2 telomers in the presence of a Pd(dba)$_2$/PPh$_3$/Et$_3$NHI catalytic system (dba = dibenzylideneacetone) (Eq. 4.90) [311].

$$
\text{R}\diagdown\!\!=\!\!=\!\!\diagup + \text{R}^1\text{R}^2\text{NH} \xrightarrow[\substack{\text{PPh}_3\,(10\%)\\ \text{Et}_3\text{NHI}\,(20\%)\\ \text{THF, 65°C}}]{\text{Pd(dba)}_2(5\%)} \text{R}\diagup\!\!\diagdown\!\!\diagup\text{NR}^1\text{R}^2 + \underset{\underset{\text{R}}{\big|}}{\overset{\text{R}}{\diagup\!\!\diagdown}}\text{NR}^1\text{R}^2 \tag{4.90}
$$

R = n-C$_7$H$_{15}$, Ph

R^{1}R^{2}NH = Et$_2$NH, pyrrolidine, piperidine

53-89% 0-19%

TOF = 0.5-0.8 h^{-1}

These reactions have been improved (yields, reaction rates, and selectivity) to give only allylic amines by using 1,1'-bis(diphenylphosphino)ferrocene and acetic acid instead of PPh$_3$ and Et$_3$NHI, respectively (Eq. 4.91) [312].

$$
\text{R}\diagdown\!\!=\!\!=\!\!\diagup + \text{R}^1\text{R}^2\text{NH} \xrightarrow[\substack{\text{dppf}\,(10\%)\\ \text{AcOH}\,(20\%)\\ \text{THF, 80°C}}]{\text{Pd}_2\text{(dba)}_3.\text{CHCl}_3\,(5\%)} \text{R}\diagup\!\!\diagdown\!\!\diagup\text{NR}^1\text{R}^2 \tag{4.91}
$$

R = Ph, p-MeC$_6$H$_4$, p-CF$_3$OC$_6$H$_4$

R^1 = R^2 = Bn, Ph; R^1 = Ph, R^2 = 2-naphtyl

62-99%

TOF = 1.5-2.5 h^{-1}

An analogous catalytic system has been applied to the IH of γ- and δ-allenic amines which cyclize smoothly in the 5-*Exo-Trig* or 6-*Exo-Trig* mode, giving vinylpyrrolidines and vinylpiperidines, respectively (Eq. 4.92) [313].

$$
\text{Ph}\diagdown\!\!\underset{\text{NH}_2}{\overset{(\text{)}_n}{\diagup}}\!\!=\!\!=\!\!\diagup \xrightarrow[\substack{\text{dppf}\,(10\%)\\ \text{AcOH}\,(15\%)\\ \text{THF, reflux}}]{[(\eta^3\text{-C}_3\text{H}_5)\text{PdCl}]_2\,(5\%)} \text{Ph}\diagdown\!\!\underset{\underset{\text{H}}{\overset{|}{\text{N}}}}{\overset{(\text{)}_n}{\diagup}}\!\!\diagup\!\!\diagup \tag{4.92}
$$

n = 1, 41%, TOF = 2.7 h^{-1}, *cis/trans* = 62/38 or 50/50?

n = 2, 52%, TOF = 2.5 h^{-1}, *cis/trans* = >95/<5

IH of allenic amines also occurs in the presence of silver salts. IH of α-allenic amines proceeds in good yields in the presence of AgBF$_4$ and provides a useful method for 3-pyrrolines synthesis *via* Endo-Trig processes (Eq. 4.93) [314].

$$
\text{RCH=C=CHCH}_2\text{NHPh} \xrightarrow[\text{CHCl}_3,\,\text{r.t.}]{\text{AgBF}_4\,(5\%)} \underset{\underset{\text{Ph}}{\overset{|}{\text{N}}}}{\diagup\!\!\diagdown}\!\!\text{R} \tag{4.93}
$$

R = H, 90%, TOF = 3.6 h^{-1}

R = n-C$_3$H$_7$, 85%, TOF = 3.4 h^{-1}

Although less efficient (TOF = 0.04 h^{-1}), similar IH of of γ- and δ-allenic amines in the presence of AgNO$_3$ give 2-alkenylpyrrolidines and 2-alkenylpiperidines, respectively (5 [or 6]-*Exo-Trig* processes) [315]. These reactions have been applied to the synthesis of ($\pm$)-pinidine [316] and -($-$)(R) coniine [317].

The [(η^3-C$_3$H$_5$)PdCl]$_2$/dppf/AcOH catalytic system has been used for the bis(hydroamination) of 3-alken-1-ynes to alkenic 1,4-diamines (Eq. 4.94), a reaction which seems to be mechanistically related to the hydroamination of allenes since an α-allenic amine CH$_2$=C=CH(R^1)CH$_2$NR$_2$ is believed to be an intermediate [318].

$$[(\eta^3\text{-C}_3\text{H}_5)\text{PdCl}]_2 \ (5\%)$$

R = Bn, allyl

R^1 = Me, *n*-hexyl, SiMe$_3$

$$\text{(4.94)}$$

+ HNR$_2$, 4 equiv. → dppf (10%), AcOH (1 equiv.), THF, 80°C → R$_2$N...R^1...NR$_2$

30-70%, TOF = 0.3-0.8 h^{-1}

A Pd(0)/benzoic acid system has been found to catalyze the hydroamination of certain arylalkynes with secondary amines, a reaction which is also mechanistically related to the hydroamination of allenes, affording high yields of allylic amines (Eq. 4.95) [319].

$$\text{Ph-C}\equiv\text{C-CH}_3 + \text{Bn}_2\text{NH} \xrightarrow[\substack{\text{PhCO}_2\text{H (10\%)} \\ \text{dioxane, 100°C}}]{\text{Pd(PPh}_3)_4 \ (5\%)} \text{Ph}\diagup\diagdown\text{NBn}_2 \qquad \text{(4.95)}$$

98%, TOF = 1.5 h^{-1}

Morpholine also gives the allylic amine in high yield. The reaction is thought to involve a known hydridopalladium-catalyzed isomerization of alkynes to allenes followed by reaction of the latter with the hydridopalladium complex to give 1-phenyl-substituted η^3-allylpalladium complexes. These complexes react with amines affording the allylic amines. Primary amines give the diallylic amines. An intramolecular version has been developed for the synthesis of 2-(2-phenyl)-pyrrolidines and -piperidines [319].

4.6.2.2 Activation of the Amine

4.6.2.2.1 Activation by Bases
To the best of our knowledge, no example has been reported.

4.6.2.2.2 Activation by Early Transition Metals
Although zirconium bisamides Cp$_2$Zr(NHAr)$_2$ do not catalyze the hydroamination of alkenes (see above), they are catalyst precursors for the hydroamination of the more reactive double bond of allenes to give the anti-Markovnikov addition product (Eq. 4.96) [126].

$$\text{CH}_2\text{=C=CH}_2 + \text{ArNH}_2 \xrightarrow[\text{90°C}]{\text{Cp}_2\text{Zr(NHAr)}_2 \ (3\%)} \begin{array}{c} \text{Me} \qquad \text{Ar} \\ \diagdown\text{C=N}\diagup \\ \text{Me}\diagup \end{array} \qquad \text{(4.96)}$$

Ar = 2,6-dimethylphenyl

83%, TOF = 0.15h^{-1}

4.6.2.2.3 Activation by Organolanthanides and Actinides

IH of aminoallenes appeared only recently [320, 321]. For monosubstituted allenes, the reaction generally gives a mixture of two regioisomers (Eq. 4.97).

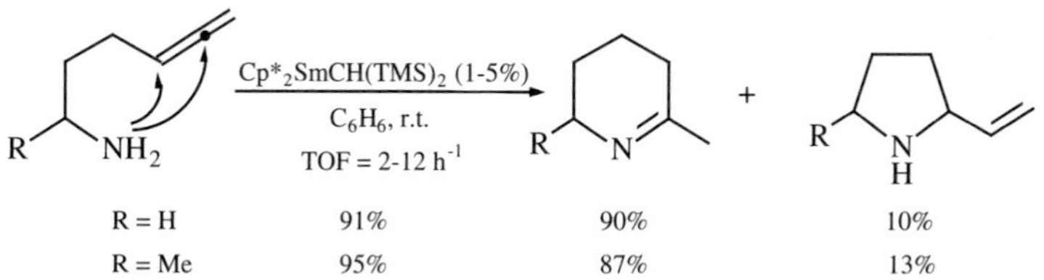

$$\text{(4.97)}$$

R = H	91%	90%	10%
R = Me	95%	87%	13%

In contrast, the reaction of 1,3-disubstituted aminoallenes proceeds exclusively through an exocyclic pathway (Eq. 4.98).

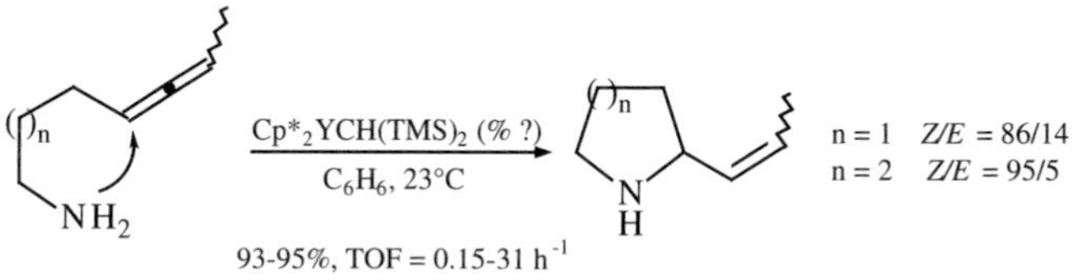

n = 1   Z/E = 86/14
n = 2   Z/E = 95/5

$$\text{(4.98)}$$

The ring-size dependence of the cyclization rate (TOF) is 5 > 6, as already observed for IH of aminoalkenes and aminoalkynes.

Interestingly, in the case of enantiomerically pure internal allenes, a diastereoselective IH is observed (e.g., Scheme 4-17), as shown by the catalytic hydrogenation of the Z/E mixture to give the *trans*-2,5-disubstituted pyrrolidine as a single compound.

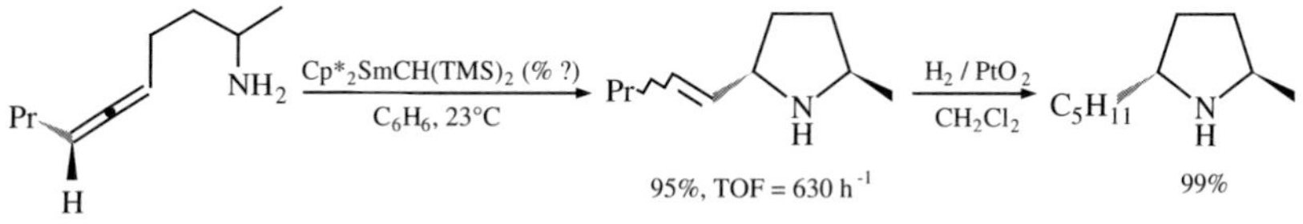

Scheme 4-17   Diastereoselective IH of internal aminoallenes

These stereoselective IHs of internal aminoallenes have been used for the key step of the enantioselective total synthesis of (+)-pyrrolidine 197B and (+)-xenovenine [322].

With regard to TOF, the IH of aminoallenes is significantly more rapid than that of the corresponding aminoalkenes, but slower (ca. 5–20 times) than that of the corresponding aminoalkynes.

4.7
Summary and Conclusions

The design of a general and efficient process for the hydroamination of alkenes would be a very important (economic) breakthrough for the production of amines. A lot of work, both from academical and industrial research groups, has been devoted to this reaction, but, with the exception of the BASF *tert*-butylamine process, this goal has not yet been reached, even if the knowledge of the transition metal-catalyzed hydroamination has been greatly improved. The main problem still remains catalyst efficiency, a prerequisite for industrial applications (TOF > 500 h^{-1}) [23].

Interestingly, significant progress has been made for the hydroamination of more reactive substrates such as styrenes, alkynes, dienes, and allenes. Specifically, highly selective catalysts have been discovered for the synthesis of fine chemicals (pharmaceuticals, natural products, chemical intermediates). In this area however, the problem of catalyst stability can also be questioned in several cases.

In 1993, ten challenges faced the catalysis research community. One of these was the 'anti-Markovnikov addition of water or ammonia to olefins to directly synthesize primary alcohols or amines' [323]. Despite some progress, the direct addition of N–H bonds across unsaturated C–C bonds, an apparently simple reaction, still remains a challenging fundamental and economic task for the coming century.

Acknowledgements   The authors wish to thank the Centre National de la Recherche Scientifique (CNRS) for financial support, and Dr. Vladimir Arion and Dr. R. Mathieu from the Laboratoire de Chimie de Coordination for the translation of publications written in Russian and for the reading of the manuscript, respectively.

References

1 V. GRIGNARD in *Traité de Chimie Organique*, Masson & Cie, Paris **1935**, Vol. XII.

2 *Encyclopedia Universalis*, France SA **1980**.

3 A. E. SCHWEIZER, R. L. FOWLKES, J. H. MCCLAIN, T. E. WHYTE JR, (eds.) *Kirk-Othmer Encyclopedia of Chemistry and Technology*, Wiley, New York **1978**, Vol. 2, 272-283.

4 H. B. BATHINA, R. A. RECK (eds.) *Kirk-Othmer Concise Encyclopedia of Chemistry and Technology*, Wiley, New York **1985**, 83.

5 K. WEISSERMEL, H.-J. HARPE, *Industrial Organic Chemistry*, 2nd ed., VCH, Weinheim **1993**.

6 P. J. CHENIER (ed.) *Survey of Industrial Chemistry*, 2nd ed., VCH, Weinheim **1992**.

7 J. MARCH, *Advanced Organic Chemistry*, 4th ed.; J. Wiley & Sons, New York **1992**.

8 P. SABATIER, A. MAILHE, *C. R. Acad. Sci.* **1909**, *148*, 898.

9 A. BAIKER, J. KIJENSKI, *Catal. Rev.-Sci. Eng.* **1985**, *27*, 653-697.

10 P. D. SHERMAN JR, P. R. KAVASMANECK (eds.) *Kirk-Othmer Encyclopedia of Chemistry and Technology* Wiley, New-York **1980**, Vol. 2, 338.

11 J. FALBE, *New Syntheses with Carbon Monoxide* Springer Verlag, Berlin **1980**.

12 see for example : B. ZIMMERMANN, J. HERTWIG, M. BELLER, *Angew. Chem. Int. Ed. Engl.* **1999**, *38*, 2372-2375 and references therein.

13 D. M. ROUNDHILL, *Chem. Rev.* **1992**, *92*, 1-27.

14 T. E. MÜLLER, M. BELLER, *Chem. Rev.* **1998**, *98*, 675-703.

15 H. F. KOCH, L. A. GIRARD, D. M. ROUNDHILL, *Polyhedron* **1999**, *18*, 2275-2279.

16 D. STEINBORN, R. TAUBE, *Z. Chem.* **1986**, 349-359.

17 R. TAUBE in *Applied Homogeneous Catalysis with Organometallic Complexes*, B. CORNILS, W. HERRMANN (eds.) VCH, Weinheim **1996**, *Vol. 2*, 507-521.

18 M. B. GASC, A. LATTES, J.-J. PERIÉ, Tetrahedron Report 144, *Tetrahedron* **1983**, *39*, 703-731.

19 J.-J. BRUNET, D. NEIBECKER, F. NIEDERCORN, *J. Mol. Catal.* **1989**, *49*, 235-259.

20 D. M. ROUNDHILL, *Catal. Today* **1997**, *37*, 155-165.

21 J.-J. BRUNET, *Gazz. Chim. Ital.* **1997**, *127*, 111-118.

22 E. HAAK, S. DOYE, *Chem. unserer Zeit* **1999**, *33*, 296-303.

23 T. E. MÜLLER, M. BELLER, *Transition Metals for Organic Synthesis*, M. BELLER, C. BOLM (eds.), Wiley-VCH, Weinheim **1998**, *Vol. 2*, 316-330.

24 I. G. Farbenindustrie A.G., DE 479,079, **1929**.

25 F. C. BERSWORTH, (Martin Dennis Company) US 2,294,442, **1942**.

26 D. D. DIXON, W. F. BURGOYNE, *Appl. Catal.* **1986**, *20*, 79-90.

27 J. W. TETER, (Sinclair Refining Company) US 2,381,470, **1945**.

28 J. W. TETER, (Sinclair Refining Company) US 2,381,471, **1945**.

29 J. W. TETER, (Sinclair Refining Company) US 2,381,472, **1945**.

30 J. W. Teter, (Sinclair Refining Company) US 2,381,473, **1945.**

31 F. A. Apgar, J. W. Teter, (Sinclair Refining Company) US 2,381,709, **1945.**

32 J. W. Teter, (Sinclair Refining Company) US 2,392,107, **1946.**

33 J. W. Teter, (Sinclair Refining Company) US 2,398,899, **1946.**

34 J. W. Teter, (Sinclair Refining Company) US 2,406,929, **1946.**

35 J. W. Teter, (Sinclair Refining Company) US 2,417,892 **1947.**

36 J. W. Teter, (Sinclair Refining Company) US 2,417,893 **1947.**

37 J. W. Teter, (Sinclair Refining Company) US 2,418,562 **1947.**

38 J. W. Teter, (Sinclair Refining Company) US 2,429,855 **1947.**

39 J. W. Teter, (Sinclair Refining Company) US 2,468,522 **1949.**

40 J. W. Teter, (Sinclair Refining Company) US 2,479,879, **1949.**

41 J. W. Teter, L. E. Olson, (Sinclair Refining Company) US 2,520,181, **1950.**

42 J. W. Teter, L. E. Olson, (Sinclair Refining Company) US 2,623,061, **1952.**

43 J. W. Teter, L. E. Olson, (Sinclair Refining Company) US 2,658,041, **1953.**

44 D. L. Esmay, P. Fotis, (Standard Oil Company) US 2,984,687, **1961.**

45 E. Tanaka, M. Ichikawa, (Kuraray Co. and Sagami Chemical Research Center) JP 7,452,790, **1974.**

46 D. M. McClain, (National Distillers and Chemical Corporation) US 3,412,158, **1968.**

47 I. G. Farbenindustrie A.G., GB 414,574, **1934.**

48 W. Hölderich, M. Hesse, F. Näumann, *Angew. Chem. Int. Ed. Engl.* **1988**, *27*, 226-246.

49 J. B. Nagy, P. Bodart, I. Hannus, I. Kiricsi, *Synthesis, Characterization and Use of Zeolitic Microporous Materials*, Z. Konya, V. Tubak (eds.), DecaGen Ltd, Szeged, Hungary, **1998.**

50 J. O. H. Peterson, H. S. Fales, (Air Products and Chemicals, Inc.) US 4,307,250, **1981**. H. S. Fales, J. O. H. Peterson, (Air Products and Chemicals, Inc.) EP 39,618, **1981**. J. O. H. Peterson, H. S. Fales, (Air Products and Chemicals, Inc.) US 4,375,002, **1983.**

51 M. Deeba, M. E. Ford, T. A. Johnson, *Catalysis of Organic Reactions*, D. W. Blackburn (ed.), Dekker, New York, **1990**, pp. 241-260.

52 M. Deeba, W. J. Ambs, (Air Products and Chemicals, Inc.) EP 77,016, **1982.**

53 M. Deeba, M. E. Ford, T. A. Johnson, *J. Chem. Soc., Chem. Commun.* **1987**, 562-563.

54 M. Deeba, M. E. Ford, T. A. Johnson, *Stud. Surf. Sci. Catal.* **1988**, *38*, 221-231.

55 W. J. Ambs, M. Deeba, J. F. Ferguson, (Air Products and Chemicals, Inc.) EP 101,921, **1984.**

56 M. Deeba, M. E. Ford, *Zeolites* **1990**, *10*, 794-797.

57 G. J. Hutchings, T. Themistocleous, R. G. Copperthwaite, (Zeofuels Research Ltd.) EP 510,825, **1992.**

58 V. Taglieber, W. Hölderich, R. Kummer, W. D. Mross, G. Saladin, (BASF A. G.) DE 3,326,579, **1985**; EP 132,736, **1985**; US 4,929,758, **1990.**

59 V. Taglieber, W. Hölderich, R. Kummer, W. D. Mross, G. Saladin, (BASF A. G.) DE 3,327,000, **1985**; EP 133,938, **1985**; US 4,929,759, **1990.**

60 A. Chauvel, B. Delmon, W. F. Hölderich, *Appl. Catal.* **1994**, *115*, 173-217.

61 BASF has production plants of 8000 and 6000 t/yr in Antwerp (Belgium) and Geismar (USA), put on stream at the end of 1993 and 1999, respectively. Personal communication from BASF, March 2000.

62 M. Deeba, M. E. Ford, *J. Org. Chem.* **1988**, *53*, 4594-4596.

63 W. Hölderich, V. Taglieber, H. H. Pohl, R. Kummer, K. G. Baur, (BASF A. G.) DE 3,634,247, **1987**; EP 263,462, **1988.**

64 M. Deeba, (Air Products and Chemicals, Inc.) EP 305,564, 1989.

65 M. Hesse, W. Steck, H. Lermer, M. Schwarzmann, R. Fischer, (BASF A. G.) DE 3,940,349, **1991**; M. Hesse, W. Steck, H. Lermer, R. Fischer, M. Schwarzmann, (BASF AG) EP 431,451, **1991.**

66 M. Tabata, N. Mizuno, M. Iwamoto, *Chem. Lett.* **1991**, 1027-1030.

67 N. Mizuno, M. Tabata, T. Uematsu, M. Iwamoto, *J. Catal.* **1994**, *146*, 249-256.

68 N. Mizuno, M. Tabata, T. Uematsu, M. Iwamoto, *Stud. Surf. Sci. Catal.* **1994**, *90*, 71-76.

69 M. Lequitte, F. Figueras, C. Moreau, S. Hub, *J. Catal.* **1996**, *163*, 255-261.

70 T. Kiyoura, (Mitsui Toatsu Chem. Inc.) JP 4,082,864, **1992**.

71 T. Kiyoura, (Mitsui Toatsu Chem. Inc.) JP 4,139,156, **1992**.

72 J. F. Knifton, N. J. Grice, (Texaco Chemical Company) EP 469,719, **1992**; US 5,107,027, **1992**; US 5,304,681, **1994**.

73 M. Berger, M. Nywlt, (Akzo, NV) DE 4,206,992, **1993**; WO 93/17995, **1993**; M. Berger, M. Nywlt, (Akzo Nobel, NV) US 5,648,546, **1997**.

74 J. F. Knifton, P.-S. E. Dai, B. L. Benac, (Texaco Chemical Company) CA 2,092,964, **1994**.

75 B. L. Benac, J. F. Knifton, P.-S. E. Dai, (Texaco Chemical Company) EP 587,424, **1994**.

76 U. Dingerdissen, G. Lauth, A. Henne, P. Stops, K. Eller, E. Gehrer, (BASF A. G.) De 4,431,093, **1996**; EP 699,653, **1996**.

77 U. Dingerdissen, R. Kummer, P. Stops, J. Herrmann, H.-J. Lützel, K. Eller, (BASF A. G.) DE 19,524,241, **1997**; EP 752,410, **1997**.

78 U. Dingerdissen, J. Herrmann, K. Eller, (BASF AG) DE 19,524,240, **1997**; EP 752,49, **1997**; US 5,763,668, **1998**.

79 U. Dingerdissen, K. Eller, (BASF AG) DE 19,524,242, **1997**; EP 752,411, **1997**.

80 U. Dingerdissen, R. Kummer, P. Stops, U. Müller, J. Herrmann, K. Eller, (BASF AG) DE 19,530,177, **1997**; WO 97/07088, **1997**; EP 844,991, **1998**.

81 K. Eller, R. Kummer, P. Stops, (BASF AG) DE 19,545,876, **1997**; EP 778,259, **1997**; US 5,739,405, **1998**.

82 K. Eller, R. Kummer, E. Gehrer (BASF AG) DE 19,602,709, **1997**; EP 786,449, **1997**; US 5,773,660, **1998**.

83 K. Eller, R. Kummer, U. Müller (BASF AG) DE 19,615,482, **1997**; EP 802,176, **1997**; US 5,780,680, **1998**.

84 K. Eller, R. Kummer, M. Hesse (BASF AG) EP 814,075, **1997**; DE 19,624,206, **1998**; US 5,780,681, **1998**.

85 K. Eller, R. Kummer, P. Stops (BASF AG) DE 19,601,409, **1997**; EP 785,185, **1997**; US 5,786,510, **1998**.

86 K. Eller, R. Kummer, M. Dembach (BASF AG) DE 19,630,670, **1998**; EP 822,179, **1998**; US 5,780,681, **1998**.

87 K. Eller, R. Kummer, (BASF AG) DE 19,649,944, **1998**; EP 846,875, **1998**; US 5,840,988, **1998**.

88 K. Eller, R. Kummer, M. Dembach, H.-J. Lützel, (BASF AG) DE 19,707,386, **1998**; EP 860,424, **1998**; US 5,874,621, **1999**.

89 K. Eller, R. Kummer, (BASF AG) DE 19,649,946, **1998**; EP 846,677, **1998**; US 5,877,352, **1999**.

90 K. Eller, R. Kummer, P. Stops (BASF AG) DE 19,545,875, **1997**; WO 97/21661, **1997**; EP 876,326, **1998**; US 5,886,226, **1999**.

91 K. Eller, R. Kummer, P. Stops (BASF AG) DE 19,526,502, **1997**; EP 754,676, **1997**; US 5,900,508, **1999**.

92 M. Deeba, (Air Products and Chemicals, Inc.) US 4,536,602, **1985**.

93 D. M. Gardner, P. J. Mc Elligott, R. T. Clark, EP 200,923, **1986**.

94 W. J. Hickinbottom, *J. Chem. Soc.* **1932**, 2646-2654.

95 D. M. Gardner, R. V. Gutowski, (Penwalt Corp.) US 4,459,191, **1984**.

96 D. M. Gardner, P. J. McElligott, (Penwalt Corp.) US 4,483,757, **1984**.

97 Y. Brigandat, J. Kervenal, (Atochem) EP 310,527, **1989**; US 4,937,383, **1990**.

98 E. W. Stern, *M. L. Spector, Proc. Chem. Soc.* **1961**, 370-370.

99 A. Panunzi, A. De Renzi, R. Palumbo, G. Paiaro, *J. Am. Chem. Soc.* **1969**, *91*, 3879-3883.

100 B. Akermark, J. E. Bäckvall, L. S. Hegedus, K. Zetterberg, K. Siirala-Hansen, K. Sjöberg, *J. Organomet. Chem.* **1974**, *72*, 127-138.

101 L. S. Hegedus, B. Akermark, K. Zetterberg, L. F. Olsson, *J. Am. Chem. Soc.* **1984**, *106*, 7122-7126.

102 L. S. Hegedus, G. F. Allen, J. J. Bozell, E. L. Waterman, *J. Am. Chem. Soc.* **1978**, *100*, 5800-5807.

103 L. S. Hegedus, *Angew. Chem. Int. Ed. Engl.* **1988**, *27*, 1113-1126.

104 R. C. Larock, T. R. Hightower, L. A. Hasvold, K. P. Peterson, *J. Org. Chem.* **1996**, *61*, 3584-3585 and references cited therein.

105 D. R. Coulson, *Tetrahedron Lett.* **1971**, 429-430.

106 D. R. Coulson, (E. I. du Pont de Nemours and Co.) US 3,758,586, **1973**.

107 S. E. Diamond, A. Szalkiewicz, F. Mares, *J. Am. Chem. Soc.* **1979**, *101*, 490-491.

108 S. E. Diamond, F. Mares, (Allied Chemical Corp.) US 4,215,218, **1980**.

109 D. Steinborn, R. Taube, *Z. Chem.* **1986**, *26*, 349-359.

110 D. Selent, D. Scharfenberg-Pfeiffer, G. Reck, R. Taube, *J. Organomet. Chem.* **1991**, *415*, 417-423.

111 E. Krukowka, R. Taube, D. Steinborn, (Technische Hochschule Leuna) DD 296,909, **1991**.

112 C. Hahn, M. Spiegler, E. Herdtweck, R. Taube, *Eur. J. Inorg. Chem.* **1999**, 435-440 and references therein.

113 D. M. Gardner, R. T. Clark, (Pennwalt Corp.) EP 39,061, **1981**.

114 D. M. Gardner, R. T. Clark, (Pennwalt Corp.) US 4,454,321, **1984**.

115 W. F. Gresham, R.E. Brooks, W.M. Bruner, (E. I. du Pont de Nemours and Co.) US 2,501,509, **1950**.

116 G. M. Whitman, (E. I. du Pont de Nemours and Co.) US 2,501,556, **1950**.

117 B. W. Howk, E. L. Little, S. L. Scott, G. M. Whitman, *J. Am. Chem. Soc.* **1954**, *76*, 1899-1902.

118 R. D. Closson, A. J. Kolka, (Ethyl Corp.) US 2,750,417, **1956**.

119 R. D. Closson, J. P. Napolitano, G. G. Ecke, A. J. Kolka, *J. Org. Chem.* **1957**, *22*, 646-649.

120 J. Wollensak, R. D. Closson, *Org. Synth.*, H.E. Baumgarten (ed.), J. Wiley, New York **1973**, Coll. Vol. 5, 575-577.

121 H. Lehmkuhl, D. Reinehr, *J. Organomet. Chem.* **1973**, *55*, 215-220.

122 G. Pez, J. E. Galles, *Pure Appl. Chem.* **1985**, *57*, 1917-1926.

123 D. Steinborn, B. Thies, I. Wagner, R. Taube, *Z. Chem.* **1989**, *29*, 333-335.

124 R. Taube *Applied Homogeneous Catalysis with Organometallic Compounds*, B. Cornils, W.A. Herrmann (eds.), VCH, Weinheim **1996**, Vol. 1, 507-520.

125 G. Pez, (Allied Chemical Corp.) US 4,302,603, **1981**.

126 P. J. Walsh, A. M. Baranger, R. G. Bergman, *J. Am. Chem. Soc.* **1992**, *114*, 1708-1719.

127 Y. Li, T. J. Marks, *Organometallics* **1996**, *15*, 3770-3772.

128 T. J. Marks, S. P. Nolan, R. Gagné, (Northwestern Univ.) US 5,110,948, **1992**.

129 M. R. Gagné, T. J. Marks, *J. Am. Chem. Soc.* **1989**, *11*, 4108-4109; M. R. Gagné, S. P. Nolan, T. J. Marks, *Organometallics* **1990**, 9, 1716-1718.

130 J. E. Baldwin, *J. Chem. Soc., Chem. Commun.* **1976**, 734-736; J. E. Baldwin, R. C. Thomas, L. I. Kruse, L., Silberman, *J. Org. Chem.* **1977**, *42*, 3846-3852.

131 M. R. Gagné, C. L. Stern, T. J. Marks, *J. Am. Chem. Soc.* **1992**, *114*, 275-294.

132 S. Tian, V. M. Arredondo, C. L. Stern, T. J. Marks, *Organometallics* **1999**, *18*, 2568-2570.

133 A. T. Gilbert, B. L. Davis, T. E. Emge, R. D. Broene, *Organometallics* **1999**, *18*, 2125-2132.

134 G. A. Molander, E. D. Dowdy, *J. Org. Chem.* **1998**, *63*, 8983-8988.

135 M. R. Gagné, L. Brard, V. P. Conticello, M. A. Giardello, C. Stern, T. J. Marks, *Organometallics* **1992**, *11*, 2003-2005.

136 M. A. Giardello, V. P. Conticello, L. Brard, M. R. Gagné, T. J. Marks, *J. Am. Chem. Soc.* **1994**, *116*, 10241-10254.

137 Y. Li, T. J. Marks, *J. Am. Chem. Soc.* **1998**, *120*, 1757-1771.

138 Y. Li, T. J. Marks, *J. Am. Chem. Soc.* **1996**, *118*, 707-708.

139 P. W. Roesky, C. L. Stern, T. J. Marks, *Organometallics* **1997**, *16*, 4705-4711.

140 see for example: M. D. Fryzuk, C. D. Montgomery, *Coord. Chem. Rev,* **1989**, *95*, 1-40.

141 A. L. Casalnuovo, J. C. Calabrese, D. Milstein, *J. Am. Chem. Soc.* **1988**, *110*, 6738-6744.

142 R. L. Cowan, W. C. Trogler, *Organometallics* **1987**, *6*, 2451-2453.

143 R. Dorta, P. Egli, F. Zürcher, A. Togni, *J. Am. Chem. Soc.* **1997**, *119*, 10857-10858.

144 R. DORTA, P. EGLI, N. H. BIELER, A. TOGNI, M. EYER, (Lonza AG) US 5,929,265, 1999.

145 J.-J. BRUNET, D. NEIBECKER, K. PHILIPPOT, J. Chem. Soc., Chem. Commun. 1992, 1215-1216.

146 J.-J. BRUNET, G. COMMENGES, D. NEIBECKER, K. PHILIPPOT, J. Organomet. Chem. 1994, 469, 221-228.

147 J.-J. BRUNET, G. COMMENGES, D. NEIBECKER, K. PHILIPPOT, L. ROSENBERG, Inorg. Chem. 1994, 33, 6373-6379.

148 J.-J. BRUNET, G. COMMENGES, D. NEIBECKER, L. ROSENBERG, J. Organomet. Chem. 1996, 522, 117-122.

149 J. D. DANFORTH, (Universal Oil Products Company) US 2,449,664, 1948.

150 H. BESTIAN, J. LIEBIGS. Ann. Chem. 1950, 566, 210-244.

151 R. WEGLER, G. PIEPER, Chem. Ber. 1950, 83, 1-6.

152 M. ERLENBACH, A. SEIGLITZ, GB 692,368, 1950.

153 R. K. RAZDAN, J. Chem. Soc., Chem. Commun. 1969, 770-771.

154 T. ASAHARA, M. SENO, S. TANAKA, N. DEN, Bull. Chem. Soc. Jpn. 1968, 42, 1996-2005.

155 R. J. SCHLOTT, J. C. FALK, K. W. NADURCY, J. Org. Chem. 1972, 37, 4243-4245.

156 T. NARITA, T. YAMAGUSHI, T. TSURUTA, Bull. Chem. Soc. Jpn. 1973, 46, 3825-3828.

157 M. BELLER, C. BREINDL, Tetrahedron 1998, 54, 6359-6368.

158 J. HERWIG, M. BELLER, C. BREINDL, (Hoechst Research and Technology Deutschland GmbH and Co., KG)WO 99/36,412, 1999.

159 M. BELLER, C. BREINDL, T. H. RIERMEIER, M. EICHBERGER, H. TRAUTHWEIN, Angew. Chem. Int. Ed. Engl. 1998, 37, 3389-3391.

160 D. TZALIS, C. KORADIN, P. KNOCHEL, Tetrahedron Lett. 1999, 40, 6193-6195.

161 J. J. BRUNET, D. NEIBECKER, K. PHILIPPOT, Tetrahedron Lett. 1993, 34, 3877-3880.

162 M. BELLER, M. EICHBERGER, H. TRAUTHWEIN, Chem. Ind. (Catalysis of Organic Reactions) 1998, 75, 319-332.

163 M. BELLER, M. EICHBERGER, H. TRAUTHWEIN, Angew. Chem. Int. Ed. Engl. 1997, 36, 2225-2227.

164 M. BELLER, M. EICHBERGER, J. HERWIG, H. TRAUTHWEIN, (Hoechst AG) DE 19722373, 1998.

165 M. BELLER, H. TRAUTHWEIN, M. EICHBERGER, C. BREINDL, T. E. MÜLLER, A. ZAPF, J. Organomet. Chem. 1998, 566, 277-285.

166 M. BELLER, H. TRAUTHWEIN, M. EICHBERGER, C. BREINDL, J. HERWIG, T. E. MÜLLER, O. R. THIEL, Chem. Eur. J. 1999, 5, 1306-1319.

167 M. BELLER, H. TRAUTHWEIN, M. EICHBERGER, C. BREINDL, T. E. MÜLLER, Eur. J. Inorg. Chem. 1999, 1121-1132.

168 M. ICHIKAWA, T. NOGUCHI, K. TAMARU, (Sagami Chemical Research Center) JP 7,392,913, 1973.

169 Y. KAKUNO, H. HATTORI, K. TANABE, Chem. Lett. 1982, 2015-2018.

170 Y. KAKUNO, H. HATTORI, J. Catal., 1984, 85, 509-518.

171 R. BAKER, D. E. HALLIDAY, Tetrahedron Lett. 1972, 2773-2776.

172 R. BAKER, A. ONIONS, R. J. POPPLESTONE, T. N. SMITH, J. Chem. Soc., Perkin Trans II 1975, 1133-1138.

173 K. TAKAHASHI, A. MIYAKE, G. HATA, Bull. Chem. Soc. Jpn. 1972, 45, 1183-1191.

174 W. KEIM, M. RÖPER, M. SCHIEREN, J. Mol. Catal. 1983, 20, 139-151.

175 D. ROSE, Tetrahedron Lett. 1972, 4197-4200.

176 D. ROSE, (Henkel and Cie, GmbH) DE 2,161,750, 1973.

177 J. KIJI, K. YAMAMOTO, E. SASAKAWA, J. FURUKAWA, J. Chem. Soc., Chem. Commun. 1973, 770.

178 J. KIJI, E. SASAKAWA, K. YAMAMOTO, J. FURUKAWA, J. Organomet. Chem. 1974, 77, 125-130.

179 R. BAKER, A. H. COOK, D. E. HALLIDAY, T. N. SMITH, J. Chem. Soc., Perkin Trans. II 1974, 1511-1517.

180 U. M. DZHEMILEV, A. Z. YAKUPOVA, G. A. TOLSTIKOV, Bull. Acad. Sci., Div. Chem. Sci. 1976, 25, 1691-1694.

181 U. M. DZHEMILEV, A. Z. YAKUPOVA, G. A. TOLSTIKOV, Bull. Acad. Sci., Div. Chem. Sci. 1978, 27, 923-927.

182 C. F. HOBBS, D. E. MCMACKINS, (Monsanto Company) US 4,120,901, 1978.

183 C. F. Hobbs, D. E. McMackins, (Monsanto Company) US 4,204,997, **1980**.

184 U. M. Dzhemilev, A. Z. Yakupova, G. A. Tolstikov, *Bull. Acad. Sci., Div. Chem. Sci.* **1975**, *24*, 2270-2270.

185 H. Watanabe, A. Nagai, M. Saito, H. Tanaka, Y. Nagai, *Reports Asahi Glass Foundation For Industrial Technology* **1981**, *38*, 111-121.

186 R. W. Armbruster, M. M. Morgan, J. L. Schmidt, C. M. Lau, R. M. Riley, D. L. Zabrowski, H. A. Dieck, *Organometallics* **1986**, *5*, 234-237.

187 U. M. Dzhemilev, A. Z. Yakupova, G. A. Tolstikov, *Bull. Acad. Sci., Div. Chem. Sci.* **1976**, *25*, 2190-2191.

188 U. M. Dzhemilev, A. Z. Yakupova, S. K. Minsker, G. A. Tolstikov, *J. Org. Chem. USSR*, **1979**, *15*, 1164-1169.

189 B. Åkermark, K. Zetterberg, *Tetrahedron Lett.* **1975**, 3733-3736.

190 B. Åkermark, G. Åkermark, C. Moberg, C. Björklund, K. Siirala-Hansén *J. Organomet. Chem.* **1979**, *164*, 97-105.

191 C. Moberg, *Tetrahedron Lett.* **1980**, 4539-4542.

192 J. E. Bäckvall, R. E. Nordberg, K. Zetterberg, B. Åkermark, *Organometallics* **1983**, *2*, 1625-1629.

193 T. Hayashi, M. Konishi, M. Kumada, *J. Chem. Soc., Chem. Commun.* **1984**, 107-108.

194 B. Åkermark, A. Vitagliano, *Organometallics* **1985**, *4*, 1275-1283

195 H. Kurozawa, *J. Organomet. Chem.* **1987**, *334*, 243-253.

196 T. Antonsson, A. Langlet, C. Moberg, *J. Organomet. Chem.* **1989**, *363*, 237-241.

197 O. Schmidt, F. A. Fries, L. Kollek, GB 313,934, **1928**; FR 662,431, **1929**; DE 528,466, **1931**.

198 J. E. Hyre, A. R. Bader, *J. Am. Chem. Soc.* **1958**, *80*, 437-439.

199 W. F. Gresham, R. E. Brooks, W. Bruner, US 2,501,509, **1950**.

200 M. Erlenbach, A. Sieglitz, GB 692,368, **1950**.

201 J. D. Danforth, (Universal Oil Products Company) CA 461,783, **1949**.

202 J. D. Danforth, (Universal Oil Products Company) FR 917,060, **1946**.

203 E. A. Zuech, R. Kleindschmidt, J. E. Mahan, *J. Org. Chem.* **1966**, *31*, 3713-3718.

204 N. Imai, T. Narita, T. Tsuruta, *Tetrahedron Lett.* **1971**, *38*, 3517-3520.

205 K. Takabe, T. Katagiri, J. Tanaka, *Tetrahedron Lett.* **1972**, *39*, 4009-1012.

206 T. Narita, N. Imai, T. Tsuruta, *Bull. Chem. Soc. Jpn.* **1973**, *46*, 1242-1246.

207 T. Fujita, K. Suga, S. Watanabe, *Chem. Ind.* **1973**, 231-232.

208 T. Fujita, K. Suga, S. Watanabe, *Aust. J. Chem.* **1974**, *27*, 531-535.

209 K. Takabe, T. Katagiri, J. Tanaka, *Tetrahedron Lett.* **1975**, 3005-3006.

210 K. Takabe, T. Yamada, T. Katagiri, *Chem. Lett.* **1982**, 1987-1988.

211 K. Tani, T. Yamagata, S. Otsuka, S. Akutagawa, H. Kumobayashi, T. Taketomi, H. Takaya, A. Miyashita, R. Noyori, *J. Chem. Soc., Chem. Commun* **1982**, 600-601.

212 I. Ojima *Catalytic Asymmetric Synthesis*, VCH, Weinheim, **1993**.

213 K. Takabe, T. Katagiri, J. Tanaka, *Bull. Chem. Soc. Jpn.* **1973**, *46*, 222-225.

214 K. Takabe, T. Katagiri, J. Tanaka, T. Fujita, S. Watanabe, K. Suga, *Org. Synth.*, J.P. Freeman (ed.), John Wiley & Sons, New York **1993**, *Coll. Vol. 8*, 188-190.

215 K. Takabe, T. Yamada, T. Katagiri, *Org. Synth.*, J.P. Freeman (ed.), John Wiley & Sons, New York, **1993**, *Coll. Vol. 8*, 190-192.

216 J. Dewar, *Jahresber. Chem.* **1877**, 445.

217 R. Meyer, H. Wesche, *Ber. Dtsch. Chem. Ges.* **1917**, *50*, 422-441.

218 L. Schlecht, H. Rötger, (I. G. Farbenindustrie AG) DE 477,049, **1929**.

219 I. G. Farbenindustrie AG, GB 295,276, **1929**.

220 I. G. Farbenindustrie AG, FR 658,614, **1929**.

221 F. Runge, H. Hummel, *Chem. Tech.* (Berlin) **1951**, *3*, 163-168.

222 T. Ishiguro, E. Kitamura, S. Kubota, N. Tabata, *J. Pharm. Soc. Jpn.* **1952**, *72*, 607-610.

223 Institute of Petroleum, Academica Sinica, *Hua Hsueh Tung Pao* **1958**, 659; *Chem. Abs.* **1962**, *56*, 4615.

224 R. S. Hanmer, S. Swann Jr., *Ind. Eng. Chem.* **1949**, *41*, 325-331.

225 J. Amiel, G. Nomine, *C. R. Acad. Sci.* **1947**, *224*, 483-484.

226 K.M. Akhmerov, A. B. Kuchkarov, D. Yusupov, A. Abdurakhmanov, *Tr. Tashk. Politekh. Inst.* **1973**, *91*, 50-55.

227 H. Rötger, (I. G. Farbenindustrie AG) DE 526,798, **1929**.

228 L. Schlecht, H. Rötger, (I. G. Farbenindustrie AG) DE 558,565, **1929**.

229 I. G. Farbenindustrie AG, FR 38,072, **1931**; addition to FR 658,614, **1929**.

230 A. T. Nielsen, R. L. Atkins, D. W. Moore, R. Scott, D. Mallory, J. M. Laberge, *J. Org. Chem.* **1973**, *38*, 3288-3295.

231 A. T. Nielsen, R. L. Atkins, J. DiPol, *J. Org. Chem.* **1974**, *39*, 1349-1355.

232 O. Nicodemus, (I. G. Farbenindustrie AG) DE 479,351, **1929**.

233 I. G. Farbenindustrie AG, GB 283,163, **1928**.

234 O. Nicodemus (I. G. Farbenindustrie AG) DE 547,518, **1932**.

235 O. Nicodemus, W. Schmidt, (I. G. Farbenindustrie AG) DE 516,765, **1931**.

236 I. G. Farbenindustrie AG, GB 451,794, **1936**.

237 W. Wolff, (I. G. Farbenindustrie AG) DE 651,734, **1937**.

238 R. S. Neale, L. Elek, R. E. Malz Jr., *J. Catal.* **1972**, *27*, 432-441.

239 N. Kozlov, D. Mitskevich, *J. Gen. Chem. USSR* **1937**, *7*, 1082-1085.

240 W. von Miller, J. Plöchl, *Ber. Dtsch. Chem. Ges.* **1892**, *25*, 2020-2071, in particular with F. Eckstein, 2029-2033.

241 W. von Miller, J. Plöchl, *Ber. Dtsch. Chem. Ges.* **1894**, *27*, 1296-1304, in particular with A. Eibner, 1299-1304.

242 A. Eibner, *J. Liebigs Ann. Chem.* **1901**, *318*, 58-89.

243 I. G. Farbenindustrie AG, GB 510,457, **1939**.

244 W. Reppe, H. Scholz, (General Aniline and Film Corporation) US 2,268,129, **1941**.

245 W. Reppe, H. Scholz, (I. G. Farbenindustrie AG) DE 730,850, **1942**.

246 C. Gardner, V. Kerrigan, J. D. Rose, B. C. L. Weedon, *J. Chem. Soc.* **1949**, 780-782.

247 W. Reppe and co-workers, *J. Liebigs Ann. Chem.* **1955**, *596*, 14-23.

248 K. C. Brannock, R. D. Burpitt, J. G. Thweatt, *J. Org. Chem.* **1963**, *28*, 1462-1464.

249 W. Reppe, O. Hecht, E. Gassenmeier, (General Aniline and Film Corporation) US 2,342,493, **1944**.

250 C. W. Kruse, R. F. Kleinschmidt, *J. Am. Chem. Soc.* **1961**, *83*, 213-216.

251 W. Reppe and co-workers, *J. Liebigs Ann. Chem.* **1956**, *601*, 81-138; more specifically 128-138.

252 R. Tiollais, H. Guillerm, *C. R. Acad. Sci.* **1953**, *236*, 1798.

253 R. W. Layer, *Chem. Rev.* **1963**, *63*, 489-510.

254 B. A. Shainyan, A. N. Mirskova, *Russ. Chem. Rev.* **1979**, *48*, 201-220.

255 M. Heider, J. Henckelmann, T. Ruehl, (B.A.S.F. A.G.) EP 646,271, **1995**; DE 4,333,237, **1995**.

256 C. W. Kruse, R. F. Kleinschmidt, *J. Am. Chem. Soc.* **1961**, *83*, 216-220.

257 J. D. Rose, R. A. Gale *J. Chem. Soc.* **1949**, 792-796.

258 J. A. Loritsch, R. R. Vogt, *J. Am. Chem. Soc.* **1939**, *61*, 1462-1463.

259 J. Barluenga, F. Aznar, R. Liz, R. Rodes, *J. Chem. Soc., Perkin Trans. I* **1980**, 2732-2737.

260 J. Barluenga, F. Aznar, R. Liz, R. Rodes, *J. Chem. Soc., Perkin Trans. I* **1983**, 1087-1091.

261 J. Barluenga, F. Aznar, *Synthesis* **1977**, 195-196.

262 J. Barluenga, F. Aznar, R. Liz, M.-P. Cabal, *J. Chem. Soc., Chem. Commun.* **1985**, 1375-1376.

263 J. Barluenga, F. Aznar, R. Liz, M.-P. Cabal, *Synthesis* **1986**, 960-962.

264 J. Barluenga, F. Aznar, C. Valdés, M.-P. Cabal, *J. Org. Chem.* **1991**, *56*, 6166-6171.

265 F. G. Pearson, B. Mawr (American Viscose Corp.) US 2,558,875, **1951**.

266 E. R. H. Jones, *Proc. Chem. Soc.* **1960**, 199-210.

267 M. F. Shostakovskii, I.A. Chekuleva, L. V. Kondrt'eva, B. V. Lopatin, *Izv. Akad. Nauk USSR, Otd. Khim. Nauk* **1962**, 2217 and references therein.

268 W. Franke, W. Thiele, (B.A.S.F.) DE 844,155, **1952**.

269 J. Reisch, K. E. Schulte, *Angew. Chem.* 1961, 73, 241-241.

270 K. E. Schulte, J. Reisch, H. Walker, *Chem. Ber.* 1965, 98, 98-103.

271 K. E. Schulte, J. Reisch, DE 1,189,080, 1965.

272 J. Reisch, *Dtsch. Apoth. Ztg.* 1963, 103, 1139-1144.

273 K. E. Schulte, J. Reisch, H. Walker, *Arch. Pharm. Ber. Dtsch. Pharm. Ges.* 1966, 299, 1-7.

274 K. Utimoto, H. Miwa, H. Nozaki, *Tetrahedron Lett.* 1981, 22, 4277-4278.

275 K. Utimoto, *Pure Appl.Chem.* 1983, 55, 1845-1852.

276 K. Iritani, S. Matsubara, K. Utimoto, *Tetrahedron Lett.* 1988, 29, 1799-1802.

277 A. Arcadi, S. Cacchi, F. Marinelli, *Tetrahedron Lett.* 1989, 30, 2581-2584.

278 Y. Fukuda, S. Matsubara, K. Utimoto, *J. Org. Chem.* 1991, 56, 5812-5816.

279 T. E. Müller, *Tetrahedron Lett.* 1998, 39, 5961-5962.

280 T. E. Müller, A.-K. Pleier, *J. Chem. Soc., Dalton Trans.* 1999, 583-587.

281 T. E. Müller, DE 19,816,479, 1999.

282 E. M. Campi, W. R. Jackson, *J. Organomet. Chem.* 1996, 523, 205-209.

283 Y. Fukuda, K. Utimoto, H. Nozaki, *Heterocycles* 1987, 25, 297-300.

284 F. E. McDonald, A. K. Chatterjee, *Tetrahedron Lett.* 1997, 38, 7687-7690.

285 W. Reppe, E. Keyssner, (I. G. Farbenindustrie AG) DE 618120, 1935.

286 W. Wolff, (I. G. Farbenindustrie AG) DE 636 213, 1936.

287 E. Keyssner, (I. G. Farbenindustrie AG) DE 642 939, 1937.

288 E. Keyssner, W. Wolff, (I. G. Farbenindustrie AG) DE 642 424, 1937.

289 W. Reppe, A. Hrubesch, O. Schlichting, (I. G. Farbenindustrie AG) DE 708 262, 1941.

290 I. A. Chekuleva, L. Kondrat'eva, *Russ. Chem. Rev.* 1965, 34, 669-680.

291 O. Tarasova, AG Mal'kina, A.I. Mikhaleva, L. Brandsma, B. A. Trofimov, *Synth. Commun.* 1994, 24, 2035-2037.

292 R. Settambolo M. Mariani, A. Caizzo, *J. Org. Chem.* 1998, 63, 10022-10026.

293 P. J. Walsh, F. J. Hollander, R. G. Bergman, *J. Am. Chem. Soc.* 1988, 110, 8729-8731.

294 A. M. Baranger, P. J. Walsh, R. G. Bergman, *J. Am. Chem. Soc.* 1993, 115, 2753-2763.

295 P. J. Walsh, F. J. Hollander, R. G. Bergman, *Organometallics* 1993, 12, 3705-3723.

296 E. Haak, I. Bytschkov, S. Doye, *Angew. Chem. Int. Ed. Engl.* 1999, 38, 3389-3391.

297 P. L. McGrane, M. Jensen, T. Livinghouse, *J. Am. Chem. Soc.* 1992, 114, 5459-5460.

298 P. L. McGrane, T. Livinghouse, *J. Org. Chem.* 1992, 57, 1323-1324.

299 P. L. McGrane, T. Livinghouse, *J. Am. Chem. Soc.* 1993, 115, 11485-11489.

300 M. Burgstein, H. Berberich, P. W. Roesky, *Organometallics* 1998, 17, 1452-1454.

301 A. Haskel, T. Straub, M. S. Eisen, *Organometallics* 1996, 15, 3773-3775.

302 M. S. Eisen, T. Straub, A. Haskel, *J. Alloys Comp.* 1998, 271-273, 116-122.

303 Y. Li, T. J. Marks, *Organometallics* 1994, 13, 439-440.

304 Y. Li, T. J. Marks, *J. Am. Chem. Soc.* 1996, 118, 9295-9306.

305 Y. Uchimaru, *Chem. Commun.* 1999, 1133-1134.

306 E. Sappa, L. Milone, *J. Organomet. Chem.* 1973, 61, 383-388

307 M. Tokunaga, M. Eckert, Y. Wakatsuki, *Angew. Chem. Int. Ed. Engl.* 1999, 38, 3222-3225.

308 B. P. Krymov, S. S. Zhukovskii, A. M. Taber, I. M. Zel'dis, I. V. Kalechits, V. E. Vasserberg, S. P. Chernykh, *Bull. Acad. Sci. USSR, Div. Chem. Sci.* 1982, 31, 1863-1866.

309 I. M. Zel'dis, S. S. Zhukovskii, A. M. Taber, I. V. Kalechits, V. E. Vasserberg, *Bull. Acad. Sci. USSR, Div. Chem. Sci.* 1983, 32, 1011-1013.

310 D. R. Coulson, *J. Org. Chem.* 1973, 38, 1483-1490.

311 L. Besson, J. Goré, B. Cazes, *Tetrahedron Lett.* 1995, 36, 3857-3860.

312 M. Al-Masum, M. Meguro, Y. Yamamoto, *Tetrahedron Lett.* 1997, 38, 6071-6074.

313 M. Meguro, Y. Yamamoto, *Tetrahedron Lett.* 1998, 39, 5421-5424.

314 A. Claesson, C. Sahlberg, K. Luthman, *Acta Chem. Scand.* **1979**, *B33*, 309-310.

315 S. Arseniyadis, J. Goré, *Tetrahedron Lett.* **1983**, *24*, 3997-4000.

316 S. Arseniyadis, J. Sartoretti, *Tetrahedron Lett.* **1985**, *26*, 729-732.

317 D. Mathbury, T. Gallagher, *J. Chem. Soc., Chem. Commun.* **1986**, 114-115.

318 U. Radhakrishan, M. Al-Masum, Y. Yamamoto, *Tetrahedron Lett.* **1998**, *39*, 1037-1040.

319 I. Kadota, A. Shibuya, L. M. Lutete, Y. Yamamoto, *J. Org. Chem.* **1999**, *64*, 4570-4571.

320 V. M. Arredondo, F. E. McDonald, T. J. Marks, *J. Am. Chem. Soc.* **1998**, *120*, 4871-4872.

321 V. M. Arredondo, F. E. McDonald, T. J. Marks, *Organometallics* **1999**, *17*, 1949-1960.

322 V. M. Arredondo, S. Tian, F. E. McDonald, T. J. Marks, *J. Am. Chem. Soc.* **1999**, *121*, 3633-3639.

323 A point of view of J. F. Roth from Air Products & Chemicals, *Chem. Eng. News* **1993**, *71*(22), 27.

5
Hydrophosphination and Related Reactions

Denyce K. Wicht and David S. Glueck

5.1
Introduction

Because organophosphorus compounds are important in the chemical industry and in biology, many methods have been developed for their synthesis [1]. This chapter reviews the formation of phosphorus-carbon (P–C) bonds by the metal-catalyzed addition of phosphorus-hydrogen (P–H) bonds to unsaturated substrates, such as alkenes, alkynes, aldehydes, and imines. Section 5.2 covers reactions of P(III) substrates (hydrophosphination), and Section 5.3 describes P(V) chemistry (hydrophosphorylation, hydrophosphinylation, hydrophosphonylation). Scheme 5-1 shows some examples of these catalytic reactions.

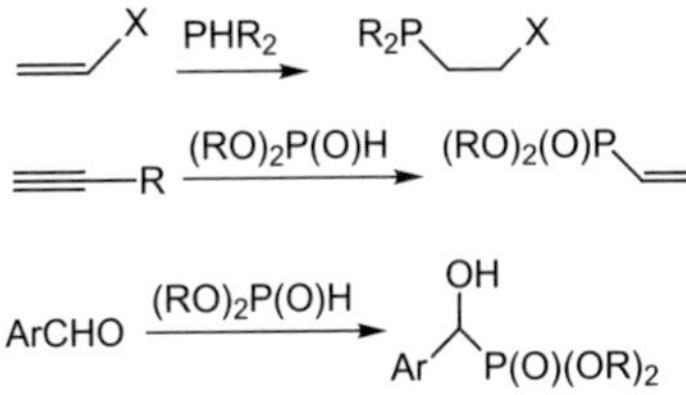

Scheme 5-1 Addition of P–H bonds to unsaturated substrates

The reaction of P–H bonds with unsaturated substrates often proceeds without a metal catalyst [2]. In addition, acid or base-catalyzed [3] as well as radical reactions [4] have been reported and extensively reviewed. Metal-catalyzed transformations like the ones described here, however, often offer improvements in rate, selectivity, and stereocontrol, as detailed below.

5.2
Metal-Catalyzed P(III)–H Additions: Hydrophosphination

Phosphine itself as well as primary and secondary phosphines undergo metal-catalyzed additions to olefins and formaldehyde. More than 40 years ago, some German patents described the addition of PH$_3$ to formaldehyde and acrylonitrile (Scheme 5-2) [5].

Although a wide variety of metals were claimed as active catalysts for formaldehyde hydrophosphination, platinum salts were preferred. Similarly, Group 10 metal salts were used to catalyze acrylonitrile hydrophosphination. Russian workers showed that Ni(II) or Co(II) salts in the presence of ammonia or amines would also catalyze the addition of phosphine to formaldehyde [6]. More recently, academic and industrial interest in these reactions was sparked by a series of papers by Pringle, who investigated late metal phosphine complexes as hydrophosphination catalysts. These and related studies are arranged below by substrate.

$$PH_3 + 3\ H_2C{=}O \xrightarrow{\text{catalyst}} P(CH_2OH)_3 \quad (1)$$

$$PH_3 + 3\ H_2C{=}CH(CN) \xrightarrow{\text{catalyst}} P(CH_2CH_2CN)_3 \quad (2)$$

(1) catalyst = Ni(II) or Co(II) + amine, or Pt(II) or Pt(IV) salts
(2) catalyst = NiCl$_2$ or PtCl$_4$

Scheme 5-2 Metal-catalyzed hydrophosphination of formaldehyde and acrylonitrile

5.2.1
Metal-Catalyzed P(III)–H Additions to Formaldehyde

Pringle found that zerovalent metal complexes were active catalysts (Scheme 5-3) for the addition of PH$_3$ to formaldehyde (see also Scheme 5-2, Eq. 1).

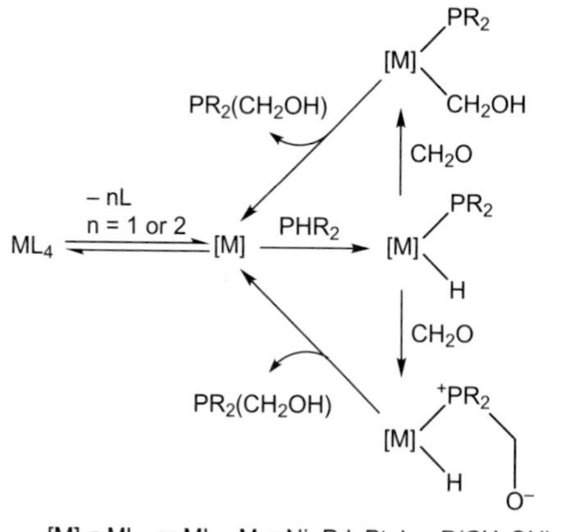

[M] = ML$_2$ or ML$_3$, M = Ni, Pd, Pt; L = P(CH$_2$OH)$_3$
R = H, CH$_2$OH, or CH$_2$OCH$_2$OH

Scheme 5-3 Proposed mechanism for metal-catalyzed hydrophosphination of formaldehyde

The precatalysts, Pt(IV) or Pt(II) salts, were found to be reduced in a series of steps by excess P(CH$_2$OH)$_3$ to give Pt[P(CH$_2$OH)$_3$]$_4$, which in aqueous solution exists in equilibrium with the five-coordinate cationic hydride [PtL$_4$H][OH] [L = P(CH$_2$OH)$_3$]. Since reaction mixtures are basic [rationalized by the formation of hemiacetals from P(CH$_2$OH)$_3$ and formaldehyde], the major Pt species present during catalysis is zerovalent PtL$_4$. Independently prepared PtL$_4$ and analogous PdL$_4$ and NiL$_4$ complexes catalyze the addition of PH$_3$ to formaldehyde. The platinum and palladium complexes were the most efficient catalysts, while the reaction catalyzed by the nickel complex was ten times slower. As an example of the catalytic activity in such systems, 0.1 mol% of K$_2$PtCl$_4$ catalyst was reported to give a quantitative yield of P(CH$_2$OH)$_3$ from PH$_3$ and aqueous formaldehyde at room temperature in about 2.5 h [7].

Pringle proposed a general mechanism for the reaction catalyzed by zerovalent Pt, Pd, or Ni (Scheme 5-3). First, P–H oxidative addition of PH$_3$ or hydroxymethyl-substituted derivatives gives a phosphido hydride complex. P–C bond formation was then suggested to occur in two possible pathways. In one, formaldehyde insertion into the M–H bond gives a hydroxymethyl complex, which undergoes P–C reductive elimination to give the product. Alternatively, nucleophilic attack of the phosphido group on formaldehyde gives a zwitterionic species, followed by proton transfer to form the O–H bond [7].

5.2.2
Metal-Catalyzed P(III)–H Additions to Acrylonitrile

Pringle showed that PtL$_3$ [L = P(CH$_2$CH$_2$CN)$_3$] catalyzes the addition of PH$_3$ to acrylonitrile to afford the tertiary phosphine P(CH$_2$CH$_2$CN)$_3$, along with the secondary and primary phosphines PH(CH$_2$CH$_2$CN)$_2$ and PH$_2$(CH$_2$CH$_2$CN) (Scheme 5-4, Eq. 1). A simpler process, the addition of PH(CH$_2$CH$_2$CN)$_2$ to acrylonitrile to give L, was found to be catalyzed by several late transition metal complexes ML$_3$ (M = Ni, Pd, Pt) or M(Cl)L$_3$ (M = Rh, Ir; Scheme 5-4, Eq. 2).

$$PH_3 + 3\ H_2C=CH(CN) \xrightarrow{\text{catalyst}} P(CH_2CH_2CN)_3 \quad (1)$$

catalyst = PtL$_3$ (L = P(CH$_2$CH$_2$CN)$_3$)

$$PH(CH_2CH_2CN)_2 + H_2C=CH(CN) \xrightarrow{\text{catalyst}} P(CH_2CH_2CN)_3 \quad (2)$$

catalyst = ML$_3$ (L = P(CH$_2$CH$_2$CN)$_3$, M = Ni, Pd, Pt)
catalyst = ML$_3$Cl (M = Rh, Ir)

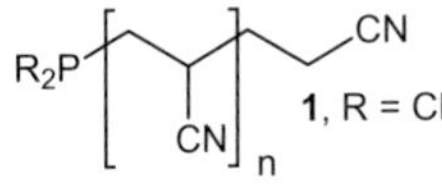

1, R = CH$_2$CH$_2$CN

Scheme 5-4 Platinum-catalyzed hydrophosphination of acrylonitrile using PH3 (Eq. 1) and PH (CH2CH2CN)2 (Eq. 2). The proposed structure of a telomeric by-product of this reaction (1) is also shown

This reaction was most efficiently catalyzed by platinum: 4 mol% of PtL$_3$ gave complete conversion of PH(CH$_2$CH$_2$CN)$_2$ to L after 6 h. The rate depended on solvent, increasing in the order acetonitrile < acetone ~ dimethylsulfoxide (DMSO). The solvent also affected the selectivity; in acetonitrile, less than 5% of a minor phosphine product, proposed to be the telomer 1 (Scheme 5-4), was produced. More of this by-product was formed in acetone (20%), while in DMSO it was the major product (60%) [8].

Kinetic studies of Pt catalysis of the reaction in Scheme 5-4, Eq. (2) showed that the rate was a complicated function of the concentration of the product $P(CH_2CH_2CN)_3$ (L). At low initial concentrations of this phosphine, increasing [L] reduced the rate, while at high initial concentrations, greater [L] increased the rate.

These observations were rationalized by invoking two parallel, [L]-dependent mechanisms for the catalysis, one *via* monomeric platinum species (Scheme 5-5)

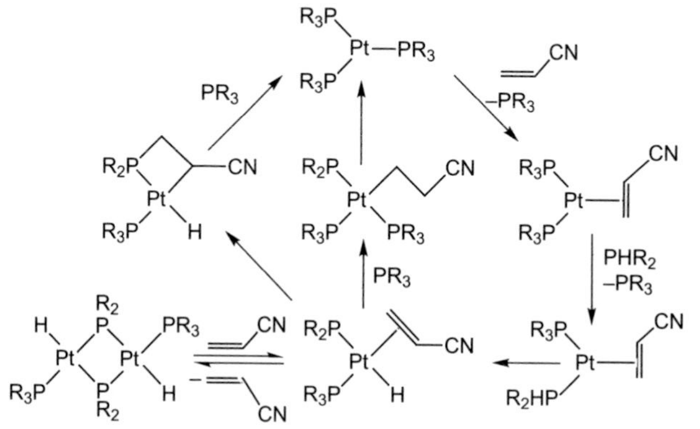

Scheme 5-5 Proposed mononuclear mechanism of Pt-catalyzed acrylonitrile hydrophosphination at high $[P(CH_2CH_2CN)_3]$

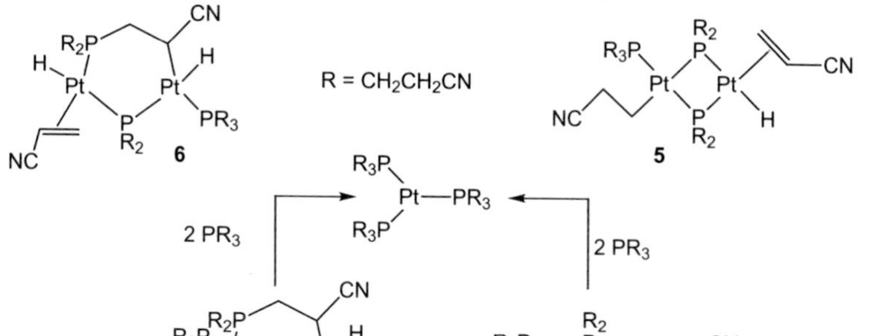

Scheme 5-6 Proposed dinuclear mechanism of Pt-catalyzed acrylonitrile hydrophosphination at low $[P(CH_2CH_2CN)_3]$. The main pathway illustrates pairwise P–C bond formation (intermediates **3** and **4**), but 'one-at-a-time' intermediates **5** and **6** were also proposed. Several mononuclear intermediates (see Scheme 5-5) have been removed for clarity

and the other involving the phosphido-bridged platinum dinuclear complex 2, which was prepared independently and shown to be a catalyst precursor (Scheme 5-6).

The mononuclear mechanism is similar to the one proposed for platinum-catalyzed hydrophosphination of formaldehyde (Scheme 5-3), but also includes a second P–C bond-forming pathway: nucleophilic (Michael) attack of the phosphido ligand on coordinated acrylonitrile. The binuclear mechanism is similar, but P–C bond formation is proposed to occur by cooperative action of two Pt centers, with complexes **4 – 6** as possible intermediates [8].

To avoid the complications posed by formation of dinuclear complexes, Glueck and coworkers studied the hydrophosphination of acrylonitrile using the Pt(0) catalysts Pt(diphos)(CH_2CHCN) (diphos = dppe or dcpe, Scheme 5-7).

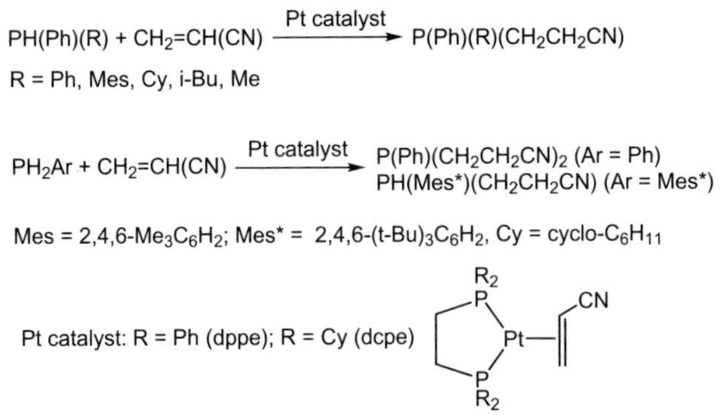
Pt catalyst
PH(Ph)(R) + CH2=CH(CN)
P(Ph)(R)(CH2CH2CN)
R = Ph, Mes, Cy, i-Bu, Me
PH2Ar + CH2=CH(CN)
P(Ph)(CH2CH2CN)2 (Ar = Ph)
PH(Mes*)(CH2CH2CN) (Ar = Mes*)
Mes = 2,4,6-Me3C6H2; Mes* = 2,4,6-(t-Bu)3C6H2, Cy = cyclo-C6H11
Pt catalyst: R = Ph (dppe); R = Cy (dcpe)
R2
P
CN
Pt
P
R2

Scheme 5-7 Pt(diphos)-catalyzed hydrophosphination of acrylonitrile with primary and secondary phosphines

In most cases, the dppe complex decomposed rapidly, but the more robust dcpe derivative was observed to be the catalyst resting state by NMR. Smaller phosphines reacted more quickly but less selectively in these systems, and selectivity for the hydrophosphination products ranged from 29 to >95%. It is likely that the phosphine byproducts included telomers such as **1** (Scheme 5-4), but they were not characterized. [9]

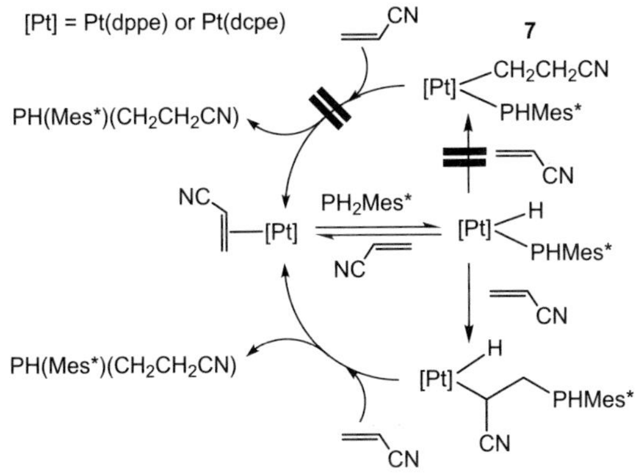
[Pt] = Pt(dppe) or Pt(dcpe)
CN
7
[Pt]
CH2CH2CN
PHMes*
PH(Mes*)(CH2CH2CN)
CN
NC
PH2Mes*
[Pt]
H
PHMes*
NC
CN
H
PH(Mes*)(CH2CH2CN)
PHMes*
CN
CN

Scheme 5-8 Proposed mechanism for Pt(diphos)-catalyzed hydrophosphination of acrylonitrile with the bulky primary phosphine PH2Mes* (Mes* = 2,4,6-(t-Bu)$_3C_6H_2$)

A detailed study of the addition of the very bulky primary phosphine Mes*PH2 {Mes* = [2,4,6-(*t*-Bu)$_3$C$_6$H$_2$]} to acrylonitrile concluded that P–C bond formation occurs not by P–C reductive elimination, but by insertion of acrylonitrile into a Pt–P bond (Scheme 5-8).

The P–H oxidative addition, acrylonitrile insertion, and C–H reductive elimination steps were observed directly with the dcpe catalyst, and the potential intermediates Pt(diphos)(PHMes*)(CH$_2$CH$_2$CN) (7, diphos = dppe, dcpe) were shown *not* to undergo P–C reductive elimination. The generality of this proposed mechanism for less bulky phosphine substrates, or for Pt catalysts supported by monodentate ligands, remains to be investigated [9].

Hydrophosphination catalyzed by a palladium(II) precursor has been reported.

The P–H bonds of a protected bis(phenylphosphino)pyrrolidine added to acrylonitrile in the presence of catalytic amounts of PdCl$_2$ and K$_2$CO$_3$ (Scheme 5-9). Without palladium, using KOH or K$_2$CO$_3$ as base gave only 30–40% yield. Similar catalytic chemistry was reported briefly for methyl acrylate [10].

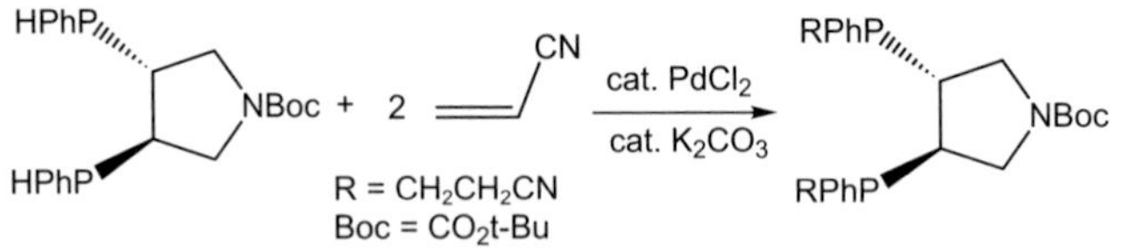

Scheme 5-9 Palladium-catalyzed hydrophosphination of acrylonitrile with a disecondary phosphine

5.2.3
Metal-Catalyzed P(III)–H Additions to Acrylate Esters

AIBN-promoted addition of PH$_3$ to ethyl acrylate (70 °C, 25 atm of PH$_3$) was reported to give a mixture of hydrophosphination products, including primary, secondary, and tertiary phosphines in ca. 1:1:1 ratio (Scheme 5-10, Eq. 1) [4c].

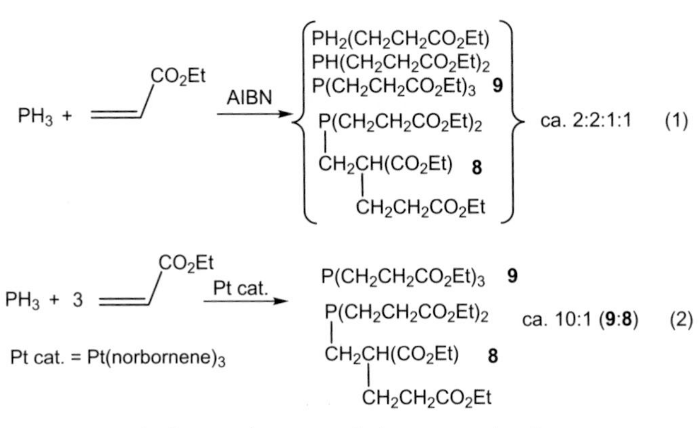

Scheme 5-10 Radical-initiated (Eq. 1) and platinum-catalyzed (Eq. 2) addition of PH$_3$ to ethyl acrylate

Notably, half of the tertiary product was the telomer **8**, which incorporates an additional equivalent of olefin. In contrast, the Pt(0) precatalyst Pt(norbornene)$_3$ (0.2 mol%) gave a 10:1 mixture of tertiary phosphine **9** and telomer **8** over 11 h at 55°C in toluene (Scheme 5-10, Eq. 2). The selectivity was higher (>95%) when only the final step [addition of PH(CH$_2$CH$_2$CO$_2$Et)$_2$ to ethyl acrylate] was monitored by NMR. In contrast, Pt[P(CH$_2$CH$_2$CF$_3$)$_3$]$_2$(norbornene) did *not* catalyze addition of PH$_3$ to CH$_2$=CHCF$_3$; thus, the olefin must be a Michael acceptor. [11]

The ethyl acrylate reaction mixtures contained several Pt(0) compounds (Scheme 5-11).

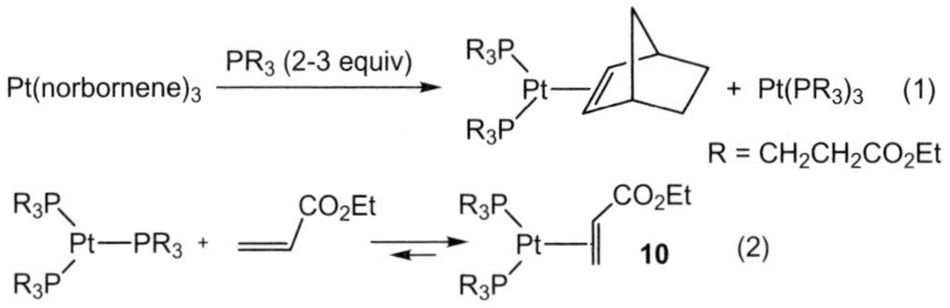

Scheme 5-11   Platinum complexes observed in hydrophosphination of ethyl acrylate using the precatalyst Pt(norbornene)$_3$

The acrylate complex **10** was suggested to be the major solution species during catalysis, since the equilibrium in Scheme 5-11, Eq. (2) lies to the right (K_{eq} > 100). Phosphine exchange at Pt was observed by NMR, but no evidence for four-coordinate PtL$_4$ was obtained. These observations help to explain why the excess of phosphine present (both products and starting materials) does not poison the catalyst. Pringle proposed a mechanism similar to that for formaldehyde and acrylonitrile hydrophosphination, involving P–H oxidative addition, insertion of olefin into the M–H bond, and P–C reductive elimination (as in Schemes 5-3 and 5-5) [11, 12].

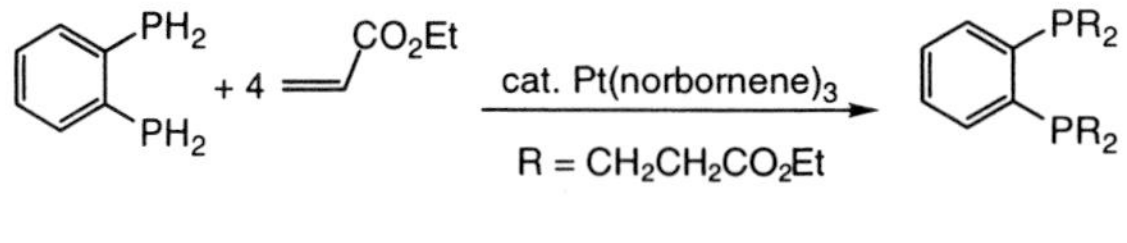

Even bidentate diphosphines do not act as catalyst poisons. For example, Pt(0) catalyzed smooth addition of ethyl acrylate to 1,2-diphosphinobenzene (Scheme 5-12, Eq. 1); no reaction took place in the absence of Pt(0), and radical initiators gave a mixture of products.

A ^{31}P NMR study of stoichiometric reactions using the di-primary phosphine H$_2$PCH$_2$CH$_2$CH$_2$PH$_2$ provided more information on the reaction mechanism (Scheme 5-12, Eq. 2). Norbornene was displaced from Pt(diphosphine)(norbornene) by ethyl acrylate. Reaction with the diphosphinopropane was very fast; this gave the hydrophosphination product, which, remarkably, did not bind Pt to give Pt(diphosphine)$_2$; instead, Pt(diphosphine)(norbornene) was observed [12].

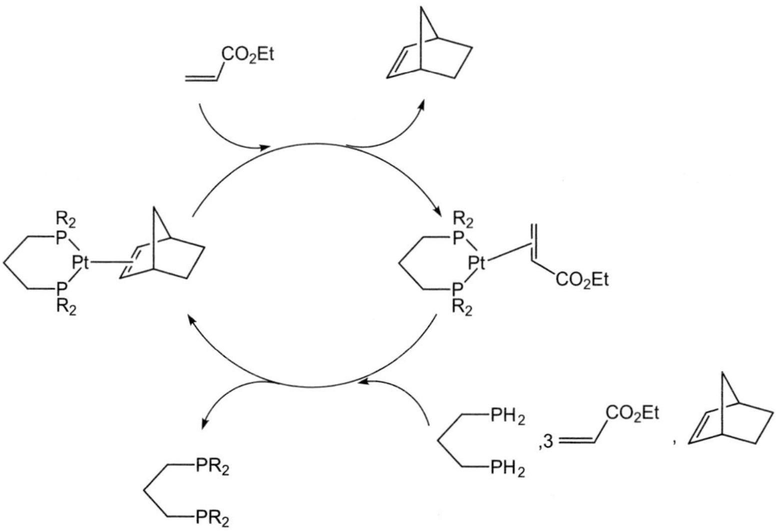

Scheme 5-12 Platinum-catalyzed hydrophosphination of ethyl acrylate as a route to bidentate diphosphines (Eq. 1). Stoichiometric reactions relevant to the proposed catalytic cycle (Eq. 2)

5.2.4
Platinum-Catalyzed Asymmetric Hydrophosphination

Recently, the chiral Pt(0) precatalyst Pt[(R, R)-Me-Duphos](*trans*-stilbene) (**11**) has been used to prepare enantiomerically enriched chiral phosphines *via* hydrophosphination of acrylonitrile, *t*-butyl acrylate and related substrates. This chemistry is summarized in Scheme 5-13.

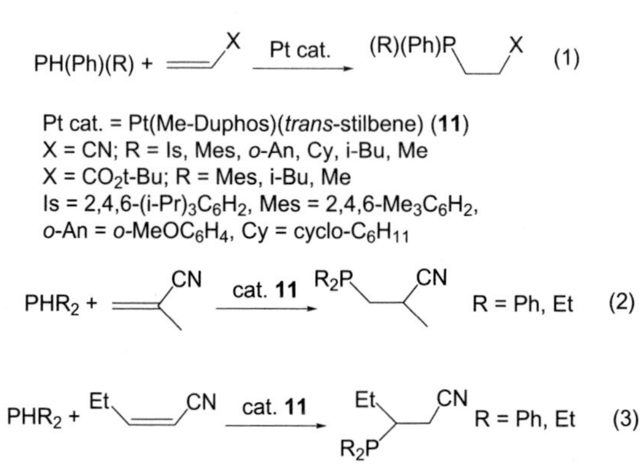

Scheme 5-13 Platinum(Me-Duphos)-catalyzed asymmetric hydrophosphination

As shown in Scheme 5-13, Eq. (1), addition of racemic secondary phosphines to acrylonitrile gave tertiary phosphines with control of phosphorus stereochemistry.

Similar catalytic reactions allowed stereocontrol at either of the olefin carbons (Scheme 5-13, Eqs. 2 and 3). As in related catalysis with achiral diphosphine ligands (Scheme 5-7), these reactions proceeded more quickly for smaller phosphine substrates. These processes are not yet synthetically useful, since the enantiomeric excesses (*ee*'s) were low (0–27%) and selectivity for the illustrated phosphine products ranged from 60 to 100%. However, this work demonstrated that asymmetric hydrophosphination can produce non-racemic chiral phosphines [13].

Consistent with the mechanism proposed earlier (Section 5.2.2, Scheme 5-8), in stoichiometric experiments, the intermediates phosphido hydride **12** and diastereomeric mixture of alkyl hydrides **13** could be isolated and observed by NMR at –20°C, respectively (Scheme 5-14).

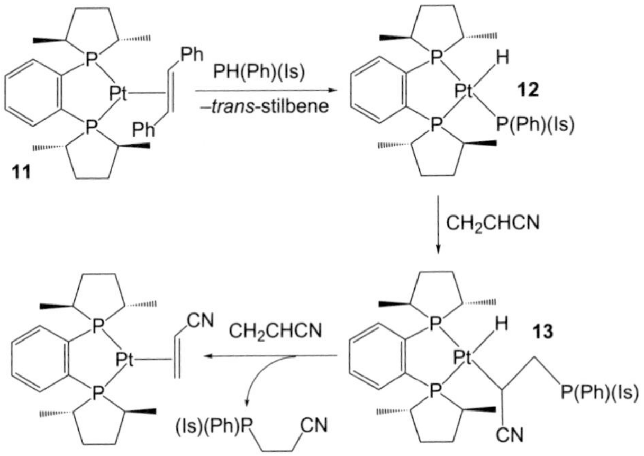

Scheme 5-14 Stoichiometric reactions of Pt(Me-Duphos) complexes relevant to the proposed catalytic cycle for asymmetric hydrophosphination

The reaction rate appears to be limited by the P–H oxidative addition because of tight binding of the olefin in the complexes Pt(Me-Duphos)(olefin). Increasing the reaction temperature to speed up this step, however, reduced the enantiomeric excess [13].

5.2.5
Organolanthanide-Catalyzed Hydrophosphination/Cyclization

Marks and Douglass recently reported organolanthanide-catalyzed hydrophosphination/cyclization of phosphino-alkenes and -alkynes (Scheme 5-15).

Unlike the late metal chemistry reviewed above, these reactions did not require Michael acceptor substrates, but the reactions were rather slow (turnover frequencies range from 2 to 13 h^{-1} at 22°C). For phosphino-alkenes (Scheme 5-15, Eqs. 1–3), a competing uncatalyzed reaction gave six-membered phosphorinane rings (Scheme 5-15, Eq. 6); this could be minimized by avoiding light and increased temperature. For phosphino-alkynes (Scheme 5-15, Eqs. 4 and 5), the products were unstable and could not be isolated [14].

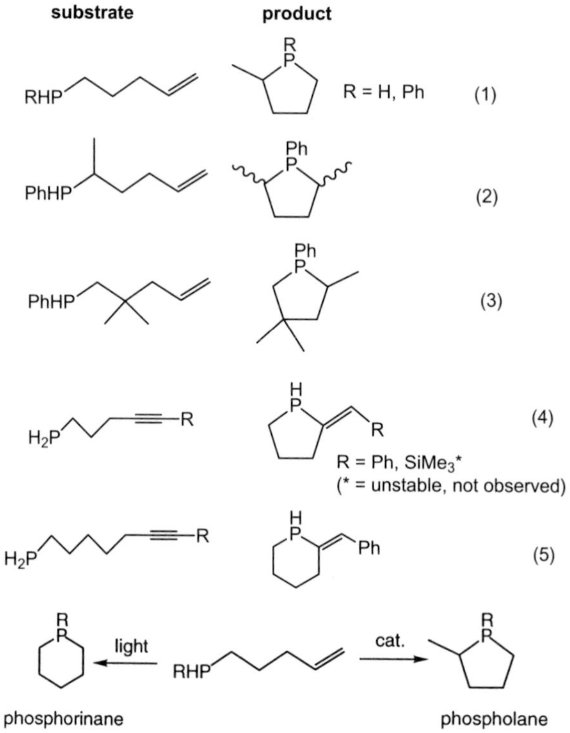

substrate   **product**

RHP~~~~ → R = H, Ph   (1)

PhHP~~~~ → Ph   (2)

PhHP~~~~ → Ph   (3)

H_2P~~~=-R → R   (4)
R = Ph, SiMe₃*
(* = unstable, not observed)

H_2P~~~=-R → Ph   (5)

phosphorinane ← light — RHP~~~~ — cat. → phosphorane

Scheme 5-15 Examples of organolanthanide-catalyzed hydrophosphination/cyclization. Eq. (6): Phosphinoalkenes can undergo organolanthanide-catalyzed hydrophosphination/cyclization to give phospholanes or an uncatalyzed reaction to give phosphorinanes

$Cp^*_2La[CH(SiMe_3)_2]$ was used as a precatalyst for most of the reactions; a proposed mechanism is shown in Scheme 5-16.

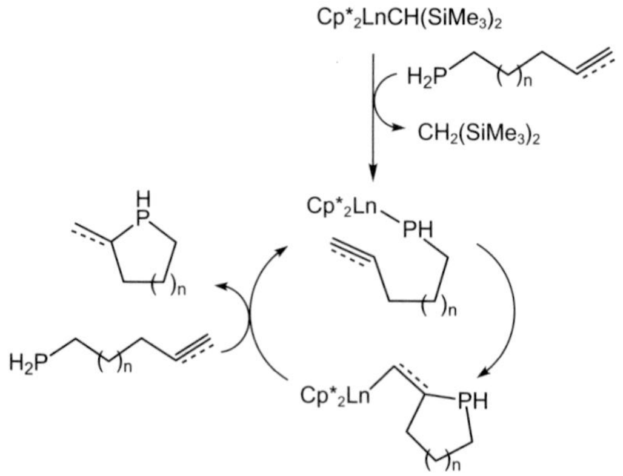

Scheme 5-16 Proposed mechanism of organolanthanide-catalyzed hydrophosphination/cyclization

σ-Bond metathesis of the Ln-alkyl with the phosphine gives a Ln-phosphido complex. (This initiation step was observed to be faster when the hydride derivative $[Cp^*_2LnH]_2$ was used.) Since the reactions were zero-order in substrate, the next step, insertion of the tethered alkene/alkyne into the Ln–P bond, was proposed to be rate-determining. Finally, phosphinolysis of the resulting Ln–C bond gives the product and regenerates the catalyst.

For the phosphino-alkyne $H_2P(CH_2)_3C\equiv CPh$, other organolanthanides Cp^*_2Ln [$CH(SiMe_3)_2$] were tested as catalysts; turnover frequencies decreased with the metal ion radius (rate order: La > Sm > Y). For this substrate, the constrained-geometry precatalyst $Me_2Si(Me_4C_5)(tBuN)Sm[N(SiMe_3)_2]$ gave faster rates than the metallocene derivative. The catalytic reactions were faster for primary phosphinoalkenes than for secondary ones, presumably because of steric effects, and alkynes reacted more quickly than alkenes. Mixtures of diastereomeric phosphine products were formed in these reactions, but stereocontrol may be possible using chiral organolanthanide catalysts [14].

5.2.6
Summary

Metal-catalyzed hydrophosphination has been explored with only a few metals and with a limited array of substrates. Although these reactions usually proceed more quickly and with improved selectivity than their uncatalyzed counterparts, their potential for organic synthesis has not yet been exploited fully because of some drawbacks to the known reactions. The selectivity of Pt-catalyzed reactions is not sufficiently high in many cases, and only activated substrates can be used. Lanthanide-catalyzed reactions have been reported only for intramolecular cases and also suffer from the formation of by-products. Recent studies of the mechanisms of these reactions may lead to improved selectivity and rate profiles. Further work on asymmetric hydrophosphination can be expected, since it is unlikely that good stereocontrol can be obtained in radical or acid/base-catalyzed processes.

5.3
Metal-Catalyzed P(V)–H Additions: Hydrophosphonylation

As in the P(III) chemistry above, both late metal (Pd) and lanthanide catalysts have been used for P(V)–H additions to alkynes, alkenes, aldehydes, and imines. In addition, titanium, aluminum, and zinc catalysts have been employed. Typical P(V) substrates include dialkyl phosphites $P(OR)_2(O)H$ and phosphine oxides $PR_2(O)H$.

5.3.1
Palladium-Catalyzed Hydrophosphorylation of Alkynes

Han and Tanaka reported that Pd(0) complexes catalyze the addition of dialkyl phosphites to terminal alkynes to give alkenylphosphonates (Scheme 5-17, hydrophosphorylation) [15].

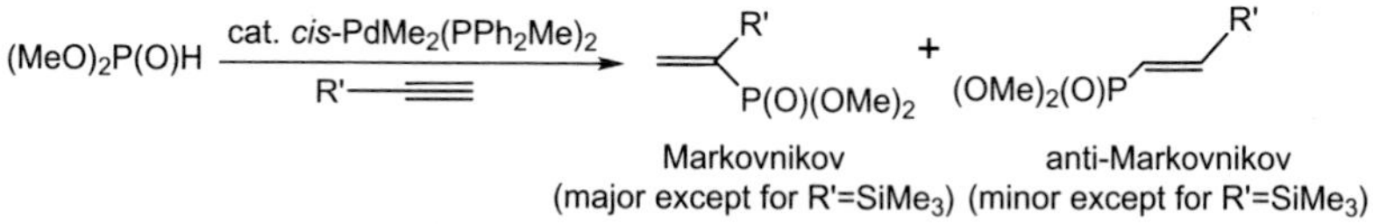

Scheme 5-17 Palladium-catalyzed hydrophosphorylation of alkynes

R' = n-C$_6$H$_{13}$, Ph, p-Tol, NC(CH$_2$)$_3$, HC≡C(CH$_2$)$_5$, 3-HC≡C-C$_6$H$_4$, 1-cyclohexenyl, SiMe$_3$

Palladium(0) or readily reduced palladium(II) complexes were the most efficient catalysts, giving higher yields than analogous Pt catalysts. The Markovnikov product was formed with high regioselectivity. In dialkynes, both C≡C bonds could be hydrophosphorylated, while the C=C double bond in a cyclohexenyl alkyne substituent did not react. With trimethylsilylacetylene, unusual anti-Markovnikov selectivity was observed.

Scheme 5-18 (Eq. 1) shows a proposed mechanism.

After formation of Pd(0) from the Pd(II) precursor, oxidative addition of the P–H bond could give a hydride complex. Insertion of the alkyne into either the Pd–P or Pd–H bond, followed by reductive elimination, gives the product. Consistent with this proposal, treatment of Pt(PEt$_3$)$_3$ with PH(O)(OEt)$_2$ gave the P–H oxidative addition product **14**, which reacted with phenylacetylene to give primarily (>99:1) the Markovnikov alkenylphosphonate (Scheme 5-18, Eq. 2).

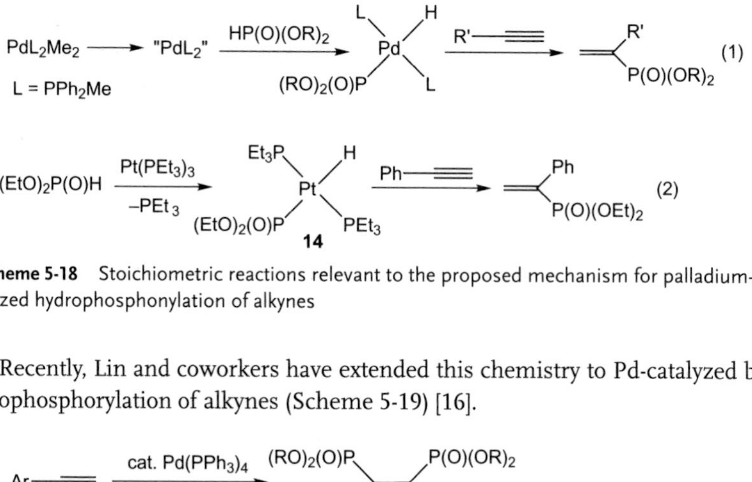

Scheme 5-18 Stoichiometric reactions relevant to the proposed mechanism for palladium-catalyzed hydrophosphonylation of alkynes

Recently, Lin and coworkers have extended this chemistry to Pd-catalyzed bis-hydrophosphorylation of alkynes (Scheme 5-19) [16].

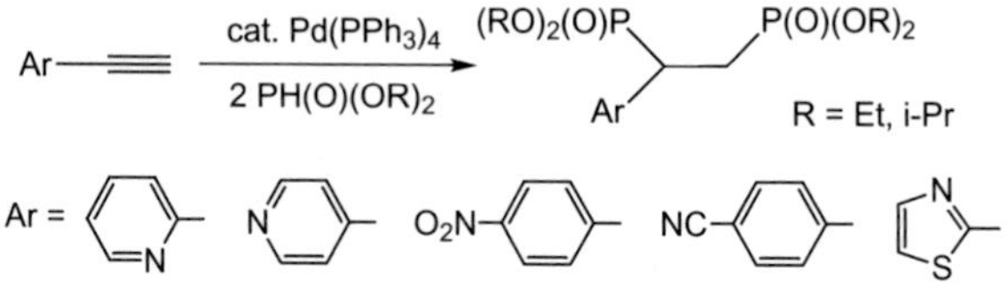

Scheme 5-19 Palladium-catalyzed bis-hydrophosphorylation of alkynes

Addition of dialkyl phosphites HP(O)(OR)$_2$ to terminal alkynes led to the formation of bis-phosphonates in 47-90% yield. These products result from addition of two P–H bonds across the C≡C triple bond. Interestingly, although diethyl phosphite gave good results, the isopropyl derivative gave lower yields, and only mono-hydrophosphorylation, yielding alkenylphosphonates, was observed for dimethyl phosphite. Electron-withdrawing aryl alkyne substituents (Scheme 5-19) were also required for the bis-hydrophosphorylation.

A mechanism similar to that of Tanaka (see Scheme 5-18) was proposed to explain these results (Scheme 5-20).

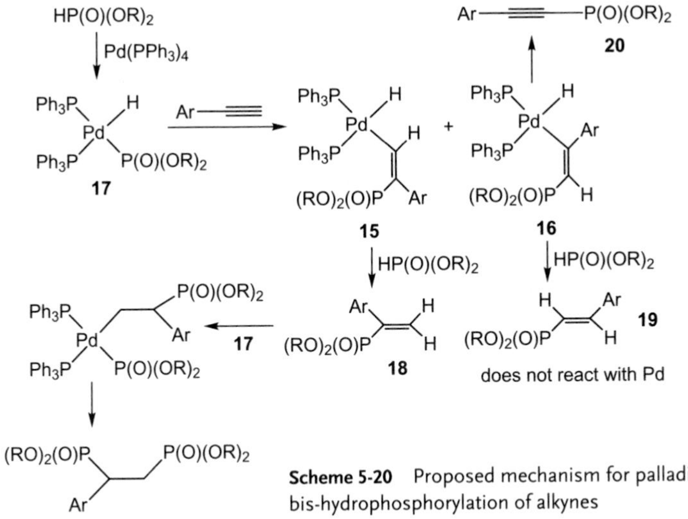

Scheme 5-20 Proposed mechanism for palladium-catalyzed bis-hydrophosphorylation of alkynes

P–H oxidative addition followed by alkyne insertion into a Pd–P bond gives the regio-isomeric alkenyl hydrides **15** and **16**. Protonolysis with dialkyl phosphite regenerates hydride **17** and gives alkenylphosphonate products **18** and **19**. Insertion of alkene **18** into the Pd–H bond of **17** followed by reductive elimination gives the bis-products, but alkene **19** does not react, presumably for steric reasons. β-Hydride elimination from **16** was invoked to explain formation of trace product **20**.

5.3.2
Palladium-Catalyzed Hydrophosphinylation of Alkynes

In closely related chemistry, Tanaka showed that Pd(PPh$_3$)$_4$ catalyzes addition of diphenylphosphine oxide to alkynes to give alkenyldiphenylphosphine oxides (hydrophosphinylation, Scheme 5-21).

The anti-Markovnikov product was formed with >95% regioselectivity at 35°C. The examples in Scheme 5-21, Eq. (1) show that cyano and hydroxyl functional groups are tolerated by the catalyst, and diphenylphosphine oxide can be added to both C≡C bonds in a di-alkyne. The reaction also worked for internal alkynes (Scheme 5-21, Eq. 2). Unusual Markovnikov selectivity was observed, however, for 1-ethynyl-cyclohexene (Scheme 5-21, Eq. 3) [17].

Stoichiometric reactions provided information on the mechanism of catalysis. Two equivalents of PH(O)Ph$_2$ reacted with M(PEt$_3$)$_3$ (M = Pd, Pt) to give *cis-*

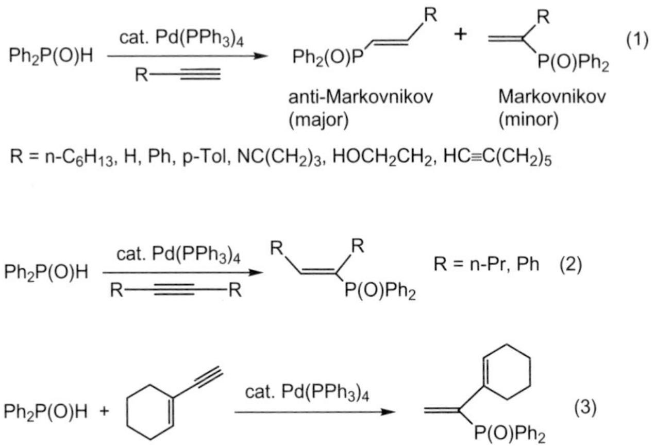

Scheme 5-21 Palladium-catalyzed hydrophosphinylation of alkynes

$M(H)[P(O)Ph_2][PPh_2(OH)](PEt_3)$, in which the O–H proton was involved in hydrogen bonding to both oxygen atoms (Scheme 5-22).

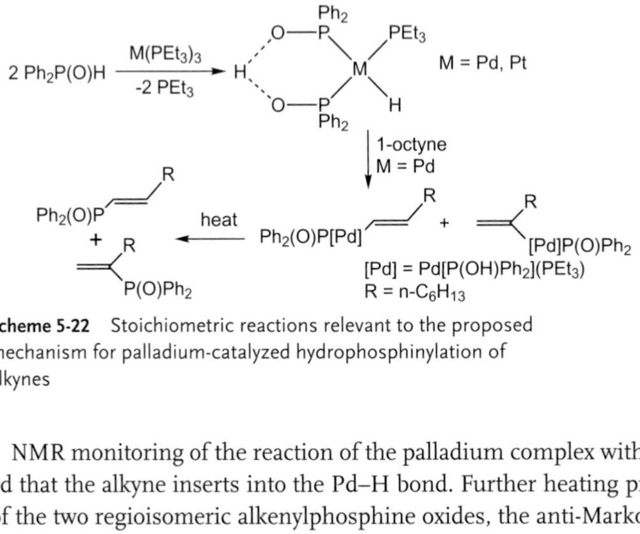

Scheme 5-22 Stoichiometric reactions relevant to the proposed mechanism for palladium-catalyzed hydrophosphinylation of alkynes

NMR monitoring of the reaction of the palladium complex with 1-octyne suggested that the alkyne inserts into the Pd–H bond. Further heating produced a mixture of the two regioisomeric alkenylphosphine oxides, the anti-Markovnikov adduct being the favored product (54:46, 65% yield).

The regioselectivity of these reactions was *reversed* (the Markovnikov product was now favored) simply by the addition of a catalytic amount of phosphinic acid (Scheme 5-23, Eq. (1); compare Scheme 5-21) [18].

Trimethylsilylacetylene is an exception, giving the anti-Markovnikov product. Internal alkynes also underwent the reaction, as observed without phosphinic acid (Scheme 5-23, Eq. 2).

A proposed mechanism which explains the role of phosphinic acid is shown in Scheme 5-24.

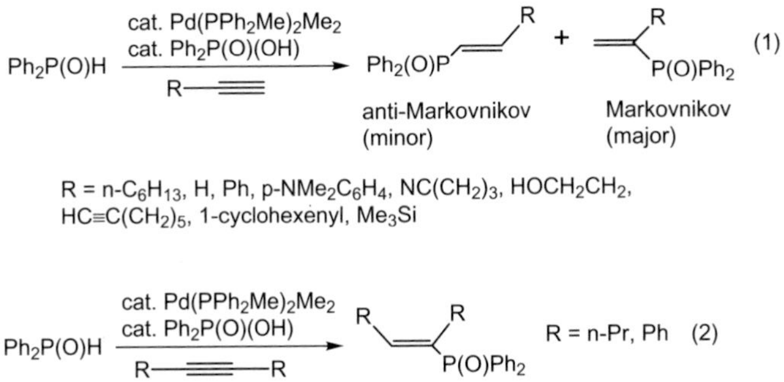

Scheme 5-23 The effect of added phosphinic acid on palladium-catalyzed hydrophosphinylation of alkynes

Protonolysis of the Pd–Me bonds by phosphinic acid and diphenylphosphine oxide gives complex **21**, which undergoes insertion of the alkyne into the Pd–P bond to give **22**. Protonolysis of the resulting Pd–C bond with $PPh_2(O)H$ regenerates **21** and forms the product. Consistent with this mechanism, treatment of Pd(dmpe) $(Me)_2$ (dmpe = $Me_2PCH_2CH_2PMe_2$) with phosphinic acid generated methane and $Pd(dmpe)(Me)[OP(O)Ph_2]$ (**23**, Scheme 5-24, Eq. 2). This complex reacted with diphenylphosphine oxide to give $Pd(dmpe)[OP(O)Ph_2][P(O)Ph_2]$ (**24**). While insertion of the alkyne into the Pd–P(O)Ph_2 bond was not observed directly, **24** catalyzed the addition of diphenylphosphine oxide to 1-octyne in the absence of additional phosphinic acid; the observed ratio of products corresponded to those observed in the original catalytic reaction [18].

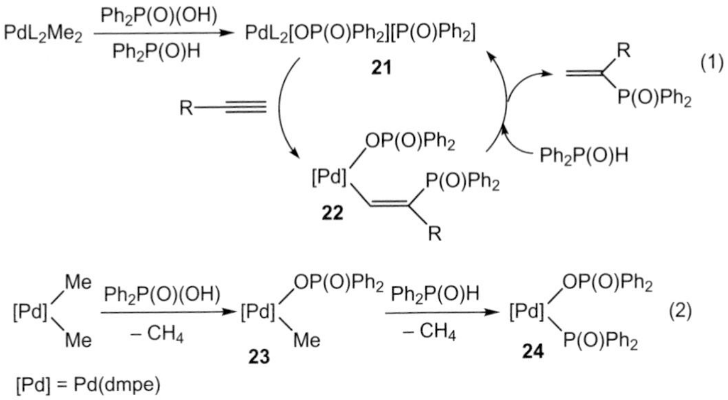

[Pd] = Pd(dmpe)

Scheme 5-24 Eq. (1): Proposed mechanism for phosphinic acid-modified palladium-catalyzed hydrophosphinylation of alkynes.   Eq. (2): Stoichiometric reactions relevant to the proposed mechanism

5.3.3
Asymmetric Hydrophosphonylation of Aldehydes and Imines

Enantioselective addition of P–H bonds in dialkyl phosphites to aldehydes and imines has been studied in detail. These reactions typically use early metal or lanthanide catalysts, whose activity is rationalized by their bifunctional character. The hard, oxophilic metal center metal binds the aldehyde or imine substrate, orienting it for nucleophilic attack which is enantioselective because of the chiral catalyst environment. The phosphite is activated for nucleophilic attack both by oxygen coordination to the metal and by deprotonation of the P–H group. Extensive development of these reactions has led to catalysts which deliver high yields and *ees*, and an industrial process practiced by Hokko for production of α-aminophosphonates [19].

5.3.3.1 Aldehydes: Initial Studies

Metal-catalyzed asymmetric addition of dialkyl phosphites to aldehydes (Pudovik reaction) has been extensively developed since the initial reports in 1993 by Shibuya. Scheme 5-25 illustrates the use of TiCl$_4$ to promote diastereoselective addition of diethyl phosphite to an α-amino aldehyde.

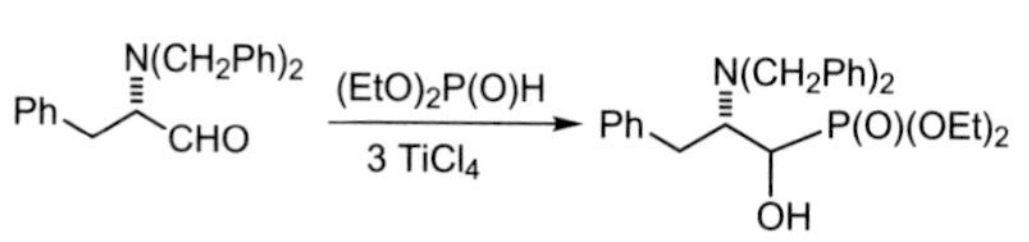

Scheme 5-25 Titanium-promoted hydrophosphonylation of an α-amino aldehyde

When 3 equiv. of TiCl$_4$ was used, a 93:7 ratio of diastereomers was formed, but with 1.2 equiv. of TiCl$_4$, no selectivity was observed and the yield was reduced [20].

The use of chiral titanium complexes as catalysts is also possible. As shown in Scheme 5-26, the Sharpless titanium catalyst gave *ees* up to 53% for the addition of diethyl phosphite to benzaldehyde.

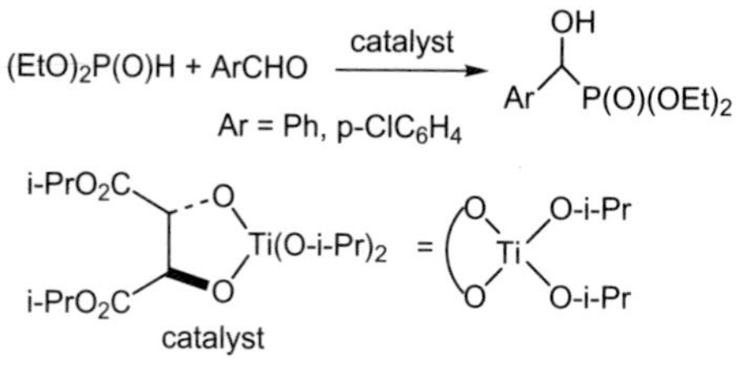

Scheme 5-26 Titanium alkoxide-catalyzed asymmetric hydrophosphonylation of arylaldehydes

The proposed mechanism (Scheme 5-27) suggests that the phosphite tautomerizes to **25** to allow Ti–O bond formation.

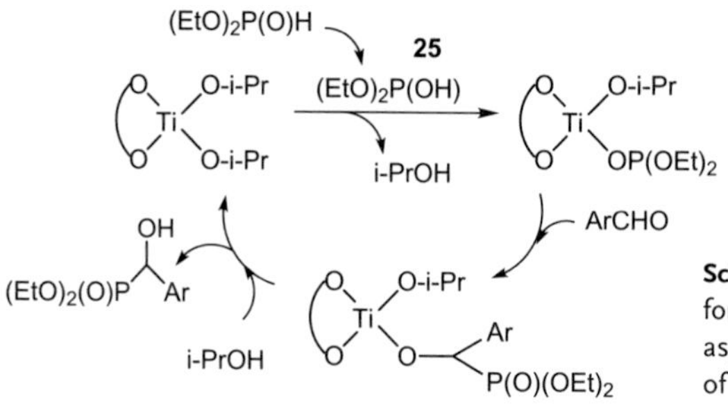

Scheme 5-27 Proposed mechanism for titanium alkoxide-catalyzed asymmetric hydrophosphonylation of arylaldehydes

Subsequent nucleophilic attack on the aldehyde then forms the P–C bond. Protonolysis with *i*-PrOH gives the product and regenerates the bis(isopropoxide) form of the catalyst; alternatively, reaction with **25** would reform the next intermediate [21].

Closely related transformations were also catalyzed by the *in situ* prepared lanthanum-lithium-BINOL (LLB) catalyst developed by Shibasaki (Scheme 5-28).

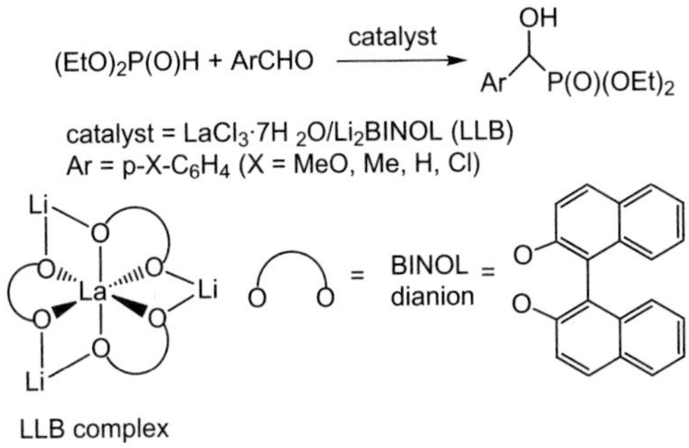

Scheme 5-28 Asymmetric hydrophosphonylation of arylaldehydes catalyzed by a heterobimetallic La/Li/BINOL catalyst (LLB)

The enantioselectivity depended on the benzaldehyde *para*-substituents, with the highest *ee* (82%) observed for *para*-anisaldehyde. The relatively large negative Hammett ρ value (–1.30) correlating the σ parameter and the *ee* suggested that aldehyde coordination was important in the enantiodetermining step [22]. Spilling independently applied the LLB catalyst to the addition of dimethyl phosphite to benzaldehyde and cinnamaldehyde (Scheme 5-29). For the latter, *ees* in the range 6–33% were observed; addition of dibenzyl phosphite to cinnamaldehyde gave 32% *ee* [23].

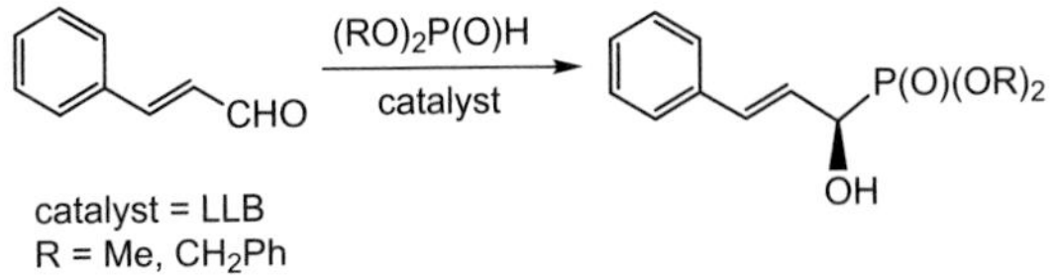

Scheme 5-29 LLB-catalyzed asymmetric hydrophosphonylation of cinnamaldehyde; LLB = La/Li/BINOL

Since these early reports, several groups have continued to improve these catalytic reactions. The most successful catalysts to date are rare earth/alkali metal/BINOL complexes like LLB, while titanium, aluminum, and zinc catalysts have also been described.

5.3.3.2 Titanium Catalysts

In 1997, Shibuya showed that in reactions involving the Sharpless catalyst (see Scheme 5-26 above), *ee* depended strongly on the solvent. The highest *ee* (53%) was obtained in ether, while no *ee* was observed in methylene chloride (Scheme 5-30) [24].

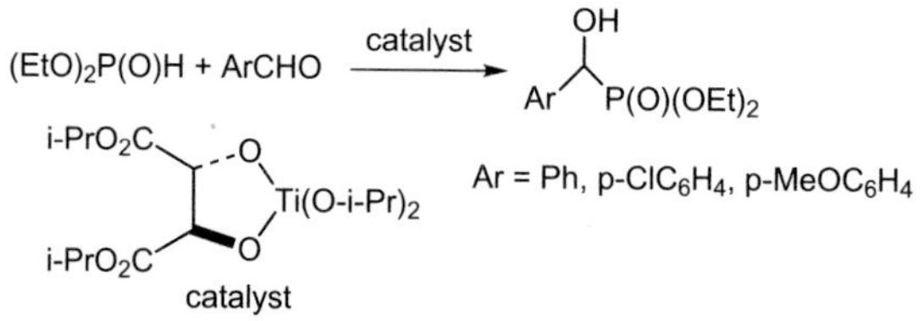

$$(EtO)_2P(O)H + ArCHO \xrightarrow{\text{catalyst}} \underset{Ar}{\overset{OH}{\diagup}} P(O)(OEt)_2$$

Ar = Ph, p-ClC$_6$H$_4$, p-MeOC$_6$H$_4$

i-PrO$_2$C, i-PrO$_2$C — Ti(O-i-Pr)$_2$

catalyst

Scheme 5-30 Titanium alkoxide-catalyzed asymmetric hydrophosphonylation of arylaldehydes

Spilling recently introduced various chiral diol ligands for Ti-catalyzed addition of dimethyl phosphite to cinnamaldehyde (Scheme 5-31).

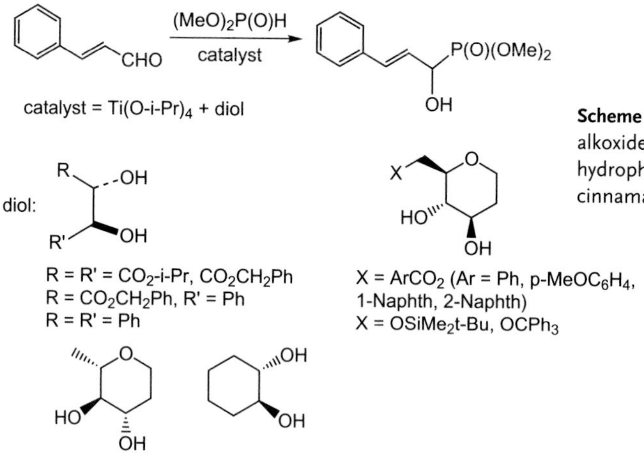

$$\underset{\text{CHO}}{PhCH=CH} \xrightarrow[\text{catalyst}]{(MeO)_2P(O)H} \underset{OH}{\diagup} P(O)(OMe)_2$$

catalyst = Ti(O-i-Pr)$_4$ + diol

Scheme 5-31 Titanium alkoxide-catalyzed asymmetric hydrophos-phonylation of cinnamaldehyde

diol:

R, R' —OH

R = R' = CO$_2$-i-Pr, CO$_2$CH$_2$Ph
R = CO$_2$CH$_2$Ph, R' = Ph
R = R' = Ph

X = ArCO$_2$ (Ar = Ph, p-MeOC$_6$H$_4$, 1-Naphth, 2-Naphth)
X = OSiMe$_2$t-Bu, OCPh$_3$

HO — OH

The best *ee*'s (67–70%) were obtained with cyclohexanediol, and this catalyst system was also studied with other aldehyde substrates (Scheme 5-32).

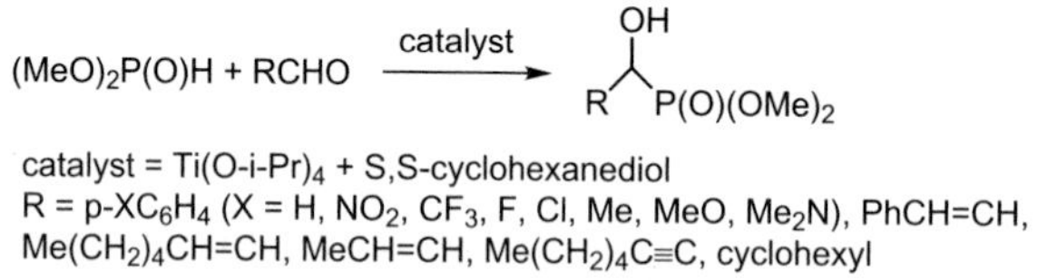

$$(MeO)_2P(O)H + RCHO \xrightarrow{\text{catalyst}} \underset{R}{\overset{OH}{\diagup}} P(O)(OMe)_2$$

catalyst = Ti(O-i-Pr)$_4$ + S,S-cyclohexanediol
R = p-XC$_6$H$_4$ (X = H, NO$_2$, CF$_3$, F, Cl, Me, MeO, Me$_2$N), PhCH=CH,
Me(CH$_2$)$_4$CH=CH, MeCH=CH, Me(CH$_2$)$_4$C≡C, cyclohexyl

Scheme 5-32 Titanium alkoxide-catalyzed asymmetric hydrophosphonylation of aldehydes

With para-substituted aromatic aldehydes, electronic factors had little effect on the *ee* (52–64%), and similar *ee*'s were obtained for the other substrates [25].

5.3.3.3 Heterobimetallic Binaphthoxide Catalysts

An improved preparation of Shibasaki's LLB catalyst allowed higher asymmetric induction in the chemistry shown in Scheme 5-28. The new recipe involved mixing LaCl$_3$•7H$_2$O (1 equiv.), BINOL-dilithium salt (2.7 equiv.) and NaOt-Bu (0.3 equiv.) in THF at 50°C. This catalyst allowed asymmetric hydrophosphonylation of aldehydes in high yields and up to 95% *ee* (Scheme 5-33, Eq. 1), and gave better results for aliphatic aldehydes than a related aluminum catalyst (ALB, see Scheme 5-37 below).

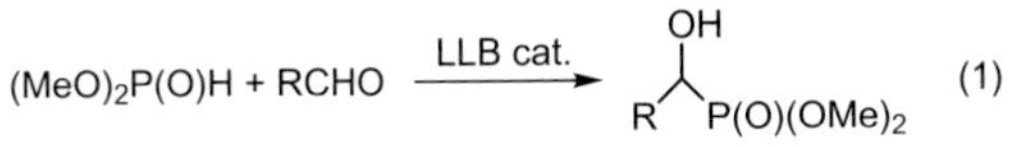
(MeO)2P(O)H + RCHO
LLB cat.
OH
R
P(O)(OMe)2
(1)

R = p-XC$_6$H$_4$ (X = H, NO$_2$, Cl, Me, MeO, Me$_2$N),
(E)-PhCH=CH, (E)-PhCH=C(Me), Me(CH$_2$)$_2$CH=CH, Me(CH$_2$)$_4$

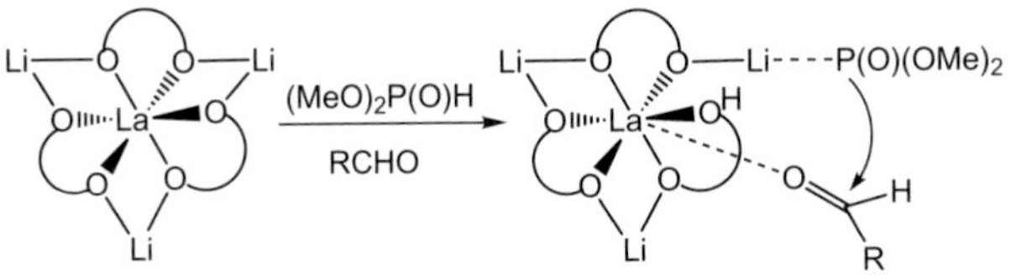
(MeO)2P(O)H
RCHO
Li---P(O)(OMe)2
La
Li
H
R

Scheme 5-33 LLB-catalyzed asymmetric hydrophosphonylation of aldehydes and a proposed mechanism; LLB = La/Li/BINOL

In the proposed mechanism (Scheme 5-33), both the phosphite and the aldehyde are coordinated to the catalyst before nucleophilic attack forms the P–C bond. Slow addition of the aldehyde was found to improve ees; it was suggested that this minimizes unselective attack of the activated phosphite on free aldehyde instead of the desired selective attack on complexed aldehyde [26]. In related chemistry with *para*-anisaldehyde, Shibuya found that *ees* also depended on the rare earth (La, Eu, Sm) in the heterobimetallic catalyst [24].

Heteroaromatic aldehydes undergo similar enantioselective hydrophosphonylation reactions (Scheme 5-34).

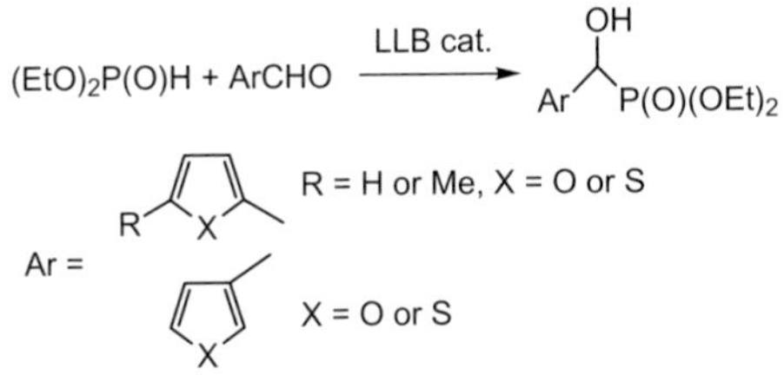
(EtO)2P(O)H + ArCHO
LLB cat.
OH
Ar
P(O)(OEt)2
Ar =
R = H or Me, X = O or S
X = O or S

Scheme 5-34 LLB-catalyzed asymmetric hydrophosphonylation of heteroaromatic aldehydes; LLB = La/Li/BINOL

Ees in these reactions increased with the electrophilic super-delocalizability [$S_r^{(E)}$] at the carbonyl oxygen, a measure of susceptibility to nucleophilic attack, consistent with the idea that Lewis acid coordination of the aldehyde is important for enantiocontrol [24].

Qian et al. explored LLB-catalyzed reactions using new BINOL ligands (Scheme 5-35).

Sterically bulky 3,3′ substituents reduced the enantioselectivity, while coordination between La and O atoms of *ortho*-substituents improved *ees*. 6,6′-diphenyl-BINOL gave the best results (69% *ee* for *para*-tolualdehyde); it was proposed that the phenyl substituents affected the Lewis acidity of the catalyst *via* electronic effects. With this catalyst, *ee* and yield depended strongly on solvent, THF being the most

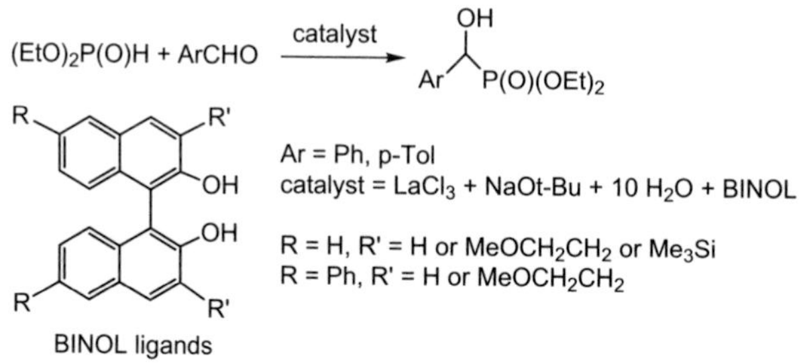

$$(EtO)_2P(O)H + ArCHO \xrightarrow{\text{catalyst}} \underset{Ar}{\overset{OH}{\bigwedge}} P(O)(OEt)_2$$

Ar = Ph, p-Tol
catalyst = LaCl$_3$ + NaOt-Bu + 10 H$_2$O + BINOL

R = H, R' = H or MeOCH$_2$CH$_2$ or Me$_3$Si
R = Ph, R' = H or MeOCH$_2$CH$_2$

BINOL ligands

Scheme 5-35   LLB-catalyzed asymmetric hydrophosphonylation of arylaldehydes using special BINOL ligands; LLB = La/Li/BINOL

useful. After optimization, this catalyst was used for reactions of several aldehydes (Scheme 5-36). No *ee* was observed for the aliphatic aldehyde, with *ee*s in the range 35–74% for the other examples [27].

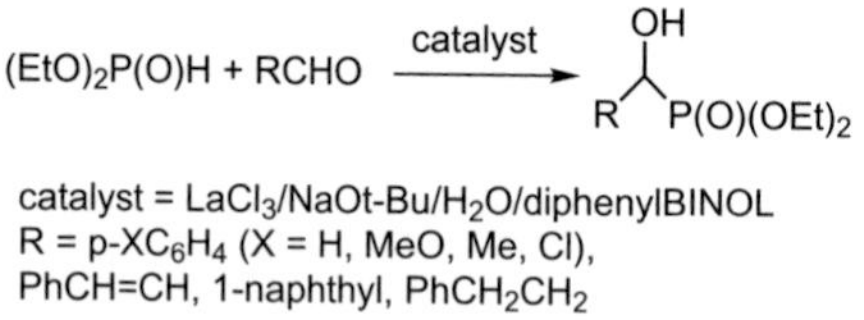

$$(EtO)_2P(O)H + RCHO \xrightarrow{\text{catalyst}} \underset{R}{\overset{OH}{\bigwedge}} P(O)(OEt)_2$$

catalyst = LaCl$_3$/NaOt-Bu/H$_2$O/diphenylBINOL
R = p-XC$_6$H$_4$ (X = H, MeO, Me, Cl),
PhCH=CH, 1-naphthyl, PhCH$_2$CH$_2$

Scheme 5-36   La/Li/(diphenylBINOL)-catalyzed asymmetric hydrophosphony-lation of aldehydes

Shibasaki showed that an aluminum-lithium-BINOL complex (ALB) also catalyzes the asymmetric addition of dialkyl phosphites to aldehydes, with *ee*s ranging from 55 to 90% for aryl or unsaturated aldehydes (Scheme 5-37).

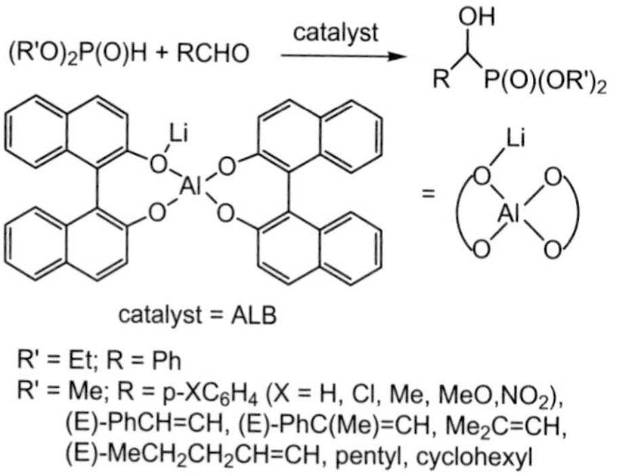

$$(R'O)_2P(O)H + RCHO \xrightarrow{\text{catalyst}} \underset{R}{\overset{OH}{\bigwedge}} P(O)(OR')_2$$

catalyst = ALB

R' = Et; R = Ph
R' = Me; R = p-XC$_6$H$_4$ (X = H, Cl, Me, MeO,NO$_2$),
     (E)-PhCH=CH, (E)-PhC(Me)=CH, Me$_2$C=CH,
     (E)-MeCH$_2$CH$_2$CH=CH, pentyl, cyclohexyl

Scheme 5-37   Al/Li/BINOL (ALB)-catalyzed asymmetric hydrophosphonylation of aldehydes

In contrast, much lower *ee*s were obtained for hexanal and cyclohexanecarbox-aldehyde.

In the proposed mechanism (Scheme 5-38), the lithium naphthoxide acts as a Brønsted base to deprotonate the phosphite, and the aluminum is a Lewis acid [28].

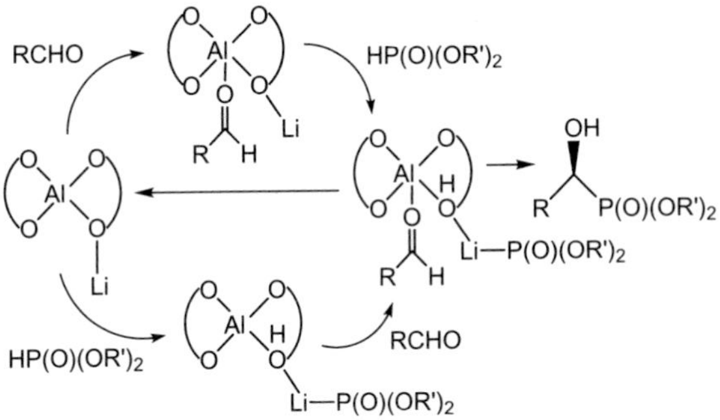

Scheme 5-38 Proposed mechanism for ALB-catalyzed asymmetric hydrophosphonylation of aldehydes; ALB = Al/Li/BINOL

More recently, Shibuya used the ALB catalyst for addition of methyl phosphinate to aldehydes (Scheme 5-39).

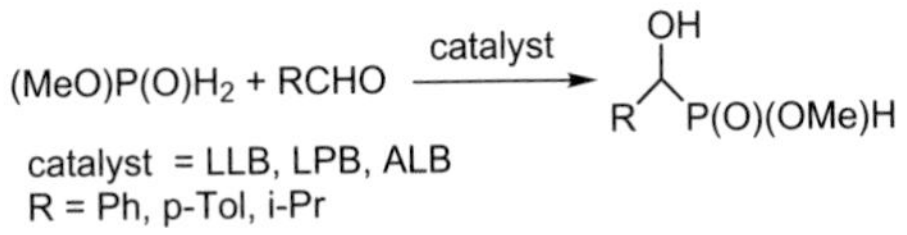

catalyst = LLB, LPB, ALB
R = Ph, p-Tol, i-Pr

Scheme 5-39 Asymmetric addition of methyl phosphinate to aldehydes using heterobimetallic BINOL catalysts; LLB = La/Li/BINOL, LPB = La/K/BINOL, ALB = Al/Li/BINOL

With benzaldehyde, no product was obtained with LLB, and LPB gave no *ee*, but ALB catalysis gave the product in 85% *ee* and 62% yield. When excess aldehyde was used, bis-hydrophosphinylation was possible (Scheme 5-40).

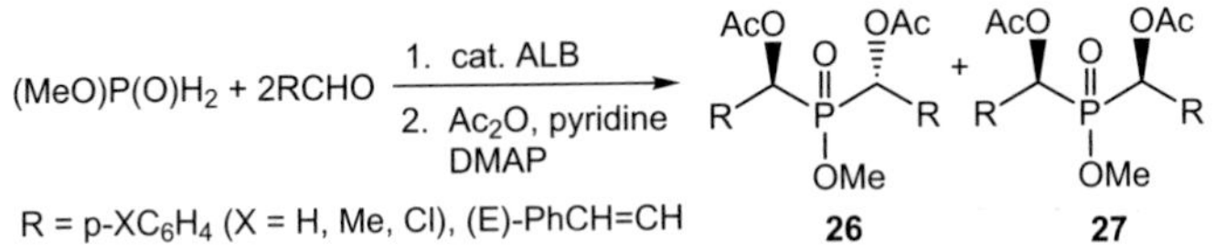

R = p-XC$_6$H$_4$ (X = H, Me, Cl), (E)-PhCH=CH

Scheme 5-40 ALB-catalyzed double asymmetric addition of methyl phosphinate to aldehydes; ALB = Al/Li/BINOL

The ratio of products **26** and **27** depended on the aldehyde, and **26** was formed in 61–82% *ee* [29].

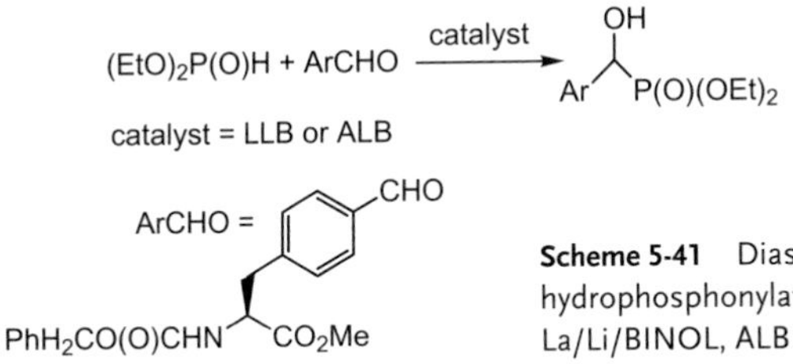

catalyst = LLB or ALB

Scheme 5-41 Diastereoselective LLB- or ALB-catalyzed hydrophosphonylation of a chiral aldehyde; LLB = La/Li/BINOL, ALB = Al/Li/BINOL

The functionalized benzaldehyde derivative in Scheme 5-41 underwent diastere-oselective addition of diethyl phosphite catalyzed either by LLB (75:25 ratio of diastereomers) or ALB (80:20 ratio) [30].

5.3.3.4 Zinc and Aluminum Catalysts

Kee reported using zinc catalysts for the addition of dimethyl phosphite to benzaldehyde (Scheme 5-42).

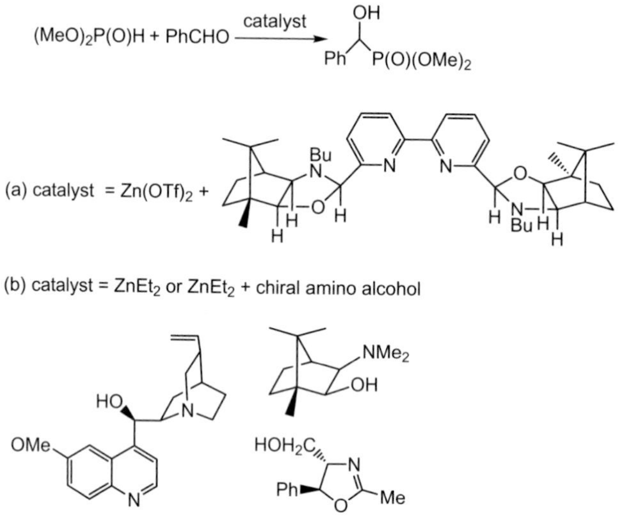

Scheme 5-42 Zinc-catalyzed asymmetric hydrophosphonylation of benzaldehyde

A combination of the Lewis acid zinc triflate and the bases NEt_3 or pyridine acted as an achiral catalyst for this reaction. Instead, using a chiral base which incorporates a bipy ligand to bind zinc gave 26% *ee* of the product (Scheme 5-42a). Alternatively, diethylzinc was an active precatalyst, but attempts to use chiral amino alcohols as ligands in this system gave low *ees* (Scheme 5-42b) [31].

Chiral aluminum SALEN complexes have been used by Kee for asymmetric addition of dimethyl phosphite to benzaldehyde derivatives (Scheme 5-43).

The Al-Me complexes **28a-b** were catalyst precursors for the reaction, which was not affected by air or water and did not require dry or degassed reagents. This system gave high yields but ees ranged only from 10 to 54% with **28a**. In contrast, the t-Bu-substituted SALEN complex **28b** gave racemic products at a slower rate [32].

A Hammett plot for *para*-substituted benzaldehydes showed that electron-rich aldehydes gave higher *ees* (r = –0.4). As in Shibuya's related results (Section 5.3.3.1 above), this indicates that aldehyde coordination is important in enantiodifferentiation, but the lower *r* value (compared to Shibuya's r = –1.30) suggests a weaker electronic influence, probably due to the relative Lewis acidities of Al and La. For ortho-substituted aldehydes, lower ees were observed, presumably due to steric effects. Although Al–Cl and Al-triflate complexes **29–30a-b** did not catalyze the reaction, they

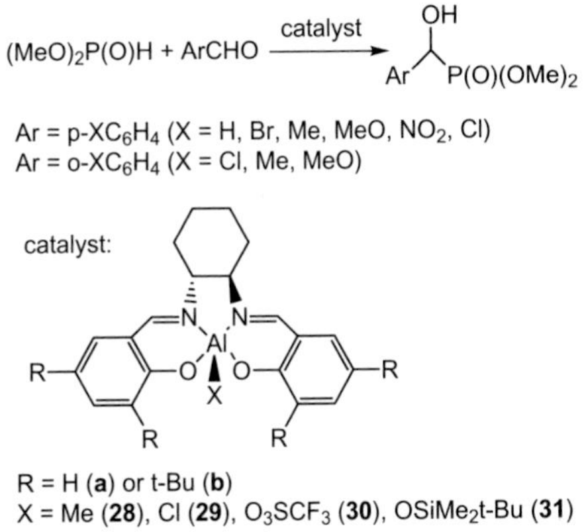
(MeO)2P(O)H + ArCHO
catalyst
OH
Ar
P(O)(OMe)2
Ar = p-XC6H4 (X = H, Br, Me, MeO, NO2, Cl)
Ar = o-XC6H4 (X = Cl, Me, MeO)
catalyst:
R = H (a) or t-Bu (b)
X = Me (28), Cl (29), O3SCF3 (30), OSiMe2t-Bu (31)

Scheme 5-43 Aluminum(SALEN)-catalyzed asymmetric hydrophosphonylation of arylaldehydes

were excellent co-catalysts with **28a.** For example, with equal catalyst loadings, a mixture of **28a** and **29a** gave faster rates than **28a** alone, with similar enantioselectivity. Siloxide complex **31** was not reported to act as a catalyst.

A proposed mechanism (Scheme 5-44) begins with deprotonation of dimethyl phosphite to give an Al-phosphito complex (**32**) which can react with the aldehyde *via* either a chelate or open transition state, the latter possibly involving cooperative action of two aluminum centers, consistent with the observation of co-catalysis. Following P–C bond formation, several possible rearrangements could regenerate the active catalyst and form the product.

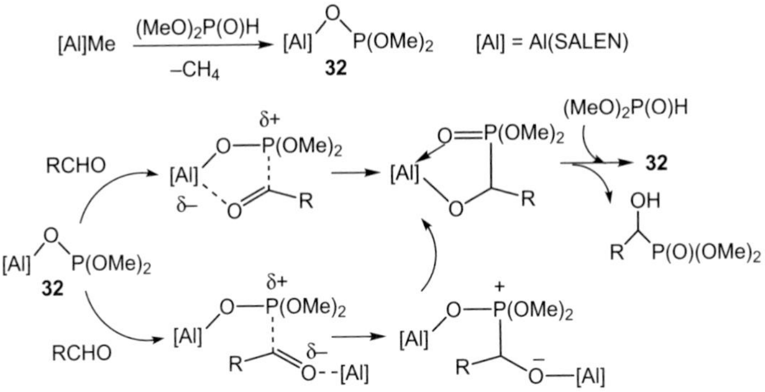
[Al]Me
(MeO)2P(O)H
–CH4
P(OMe)2
32
[Al] = Al(SALEN)
RCHO
(MeO)2P(O)H
32
OH
P(O)(OMe)2

Scheme 5-44 Proposed mechanism for aluminum(SALEN)-catalyzed asymmetric hydrophosphonylation of arylaldehydes

5.3.3.5 **Imines**

Shibasaki reported the first catalytic asymmetric hydrophosphonylation of imines in 1995 (Scheme 5-45) using heterobimetallic LLB-type catalysts.

Reaction rates, yields, and *ee*s depended on the imine, the catalyst, the solvent and the temperature; ees from 49 to 96% were obtained. Scheme 5-46 shows the mechanism proposed for these reactions.

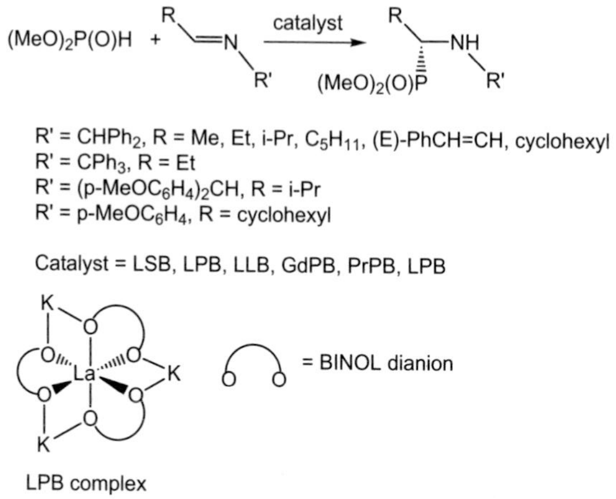

$$(MeO)_2P(O)H \; + \quad \overset{R}{\underset{R'}{\diagup}}\hspace{-0.5em}=N \quad \xrightarrow{\text{catalyst}} \quad \overset{R}{\underset{(MeO)_2(O)\overset{|}{P}}{\diagup}}\hspace{-0.5em}\overset{|}{\underset{R'}{\diagdown}}NH$$

R' = CHPh₂, R = Me, Et, i-Pr, C₅H₁₁, (E)-PhCH=CH, cyclohexyl
R' = CPh₃, R = Et
R' = (p-MeOC₆H₄)₂CH, R = i-Pr
R' = p-MeOC₆H₄, R = cyclohexyl

Catalyst = LSB, LPB, LLB, GdPB, PrPB, LPB

= BINOL dianion

LPB complex

Scheme 5-45 Asymmetric hydrophosphonylation of imines catalyzed by heterobimetallic rare earth/alkali metal/BINOL complexes. Catalyst nomenclature: first letter = rare earth (L = La); second letter = alkali (L = lithium, S = sodium, P = potassium)

First, deprotonation of dimethyl phosphite accompanied by coordination of oxygen to the oxophilic lanthanide gives **33**. Nucleophilic attack of P on the imine carbon along with N-coordination gives **34**; proton transfer followed by product decomplexation regenerates the catalyst [33].

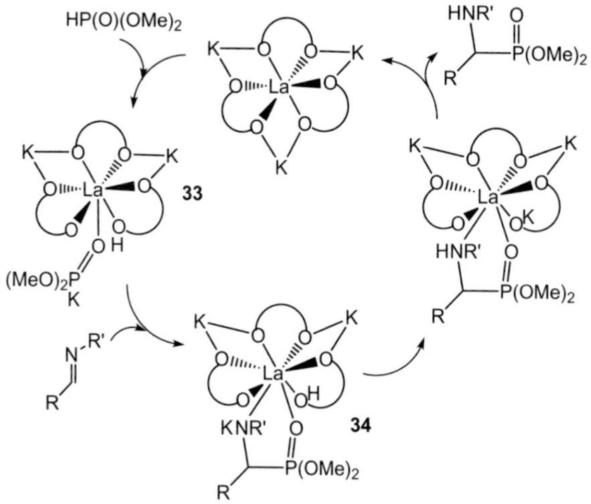

Scheme 5-46 Proposed mechanism for LPB-catalyzed asymmetric hydrophosphonylation of imines

The same group extended this work to a cyclic imine (Scheme 5-47); better results were obtained with heterobimetallic lanthanide catalysts than with chiral titanium alkoxides.

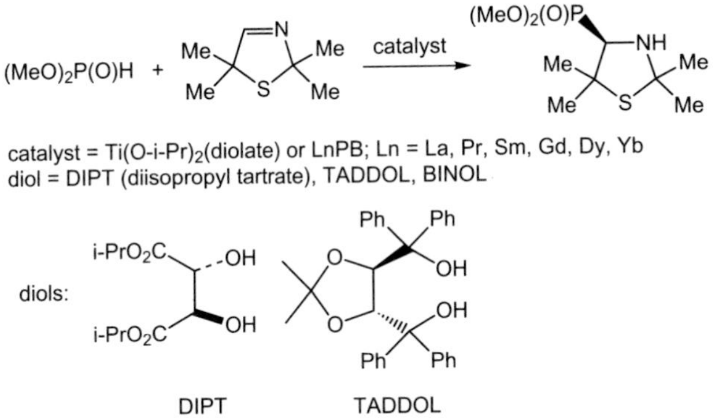

catalyst = Ti(O-i-Pr)₂(diolate) or LnPB; Ln = La, Pr, Sm, Gd, Dy, Yb
diol = DIPT (diisopropyl tartrate), TADDOL, BINOL

diols:

DIPT   TADDOL

Scheme 5-47 Asymmetric hydrophosphonylation of a cyclic imine catalyzed by heterobimetallic rare earth/alkali metal/BI-NOL complexes or by chiral titanium alkoxide complexes

With a YbPB catalyst at room temperature, 86% yield and 98% *ee* were obtained. After extensive optimization of the catalyst, solvent, temperature, pressure, and catalytic loading, 98% yield and 98% *ee* was achieved using YbPB (5 mol%) at 50°C for 48 h in 1:7 THF:toluene. The active catalyst was isolated; its structure is similar to that shown in Scheme 5-46, and a similar mechanism was proposed. Additional spectroscopic studies suggested that complexation of the phosphite to the lanthanide center was a plausible first step, and that the P–C bond is formed by nucleophilic attack of phosphorus on an N-complexed imine [34].

Similar chemistry with diphenylphosphine oxide also resulted in high yields (50–98%) and *ees* (75–93%) (Scheme 5-48). For this chemistry, the PrPB catalyst was the optimum choice [35].

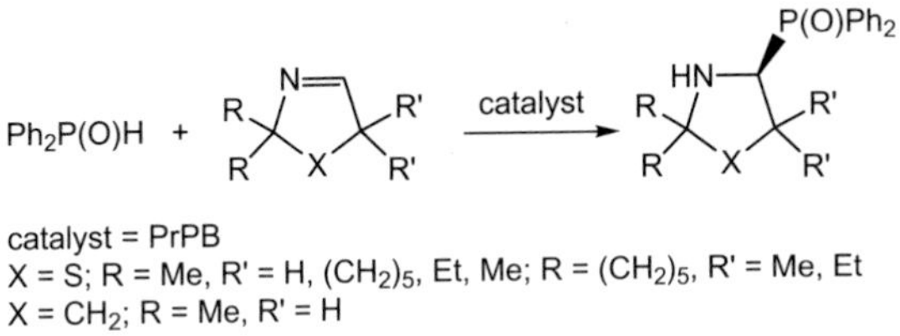

catalyst = PrPB
X = S; R = Me, R' = H, (CH₂)₅, Et, Me; R = (CH₂)₅, R' = Me, Et
X = CH₂; R = Me, R' = H

Scheme 5-48 PrPB-catalyzed asymmetric hydrophosphinylation of cyclic imines

5.3.4
Summary

In comparison to related P(III) chemistry, metal-catalyzed additions of P–H bonds in P(V) compounds to unsaturated substrates have been studied in more detail, and several synthetically useful processes have been developed. In particular, the use of heterobimetallic BINOL-based catalysts allows asymmetric hydrophosphonylation of aldehydes and imines in high yield and enantiomeric excess.

5.4
Outlook

Metal-catalyzed additions of P(III)–H and P(V)–H bonds to unsaturated substrates have been studied much less than related additions of, for example, B–H or Si–H bonds [36]. Already, some synthetically useful processes have been developed, and further work is likely to produce additional useful transformations as well as more fundamental information on the mechanisms of these reactions.

Acknowledgements   We thank the National Science Foundation, Petroleum Research Fund, Exxon Education Foundation, Union Carbide, DuPont, and Cytec Canada for support of our studies in this area.

References

1 D. G. GILHEANY, C. M. MITCHELL in *The Chemistry of Organophosphorus Compounds*, F. R. HARTLEY (ed.), John Wiley and Sons, Chichester, England, **1990**, Vol. 1, pp. 151-190.

2 (a) H. R. HUDSON in *The Chemistry of Organophosphorus Compounds*; F. R. HARTLEY (ed.), John Wiley and Sons: Chichester, England, **1990**, Vol. 1, pp 399-401. (b) M. DANKOWSKI in *The Chemistry of Organophosphorus Compounds*; F. R. HARTLEY (ed.), John Wiley and Sons: Chichester, England, 1990, Vol. 1, pp 496-499. (d) W. WOLFSBERGER, *Chem. Ztg.* **1988**, *112*, 215-221. (e) W. WOLFSBERGER, *Chem. Ztg.* **1988**, *112*, 53-68. (f) R. ENGEL, *Synthesis of Carbon-Phosphorus Bonds*, CRC Press: Boca Raton, **1988**.

3 For examples see: (a) M. M. RAUHUT, I. HECHENBLEIKNER, H. A. CURRIER, F. C. SCHAEFER, V. P. WYSTRACH, *J. Am. Chem. Soc.* **1959**, *81*, 1103-1107. (b) R. B. KING, J. C. CLOYD JR., P. N. KAPOOR, *J. Chem. Soc., Perkin Trans. I* **1973**, 2226-2229. (c) M. HABIB, H. TRUJILLO, C. A. ALEXANDER, B. N. STORHOFF, *Inorg. Chem.* **1985**, *24*, 2344-2349.

4 For examples see: (a) D. L. DuBOIS, W. H. MYERS, D. W. MEEK, *J. Chem. Soc., Dalton Trans.* **1975**, 1011-1015. (b) W. WOLFSBERGER, *Chem. Ztg.* **1990**, *114*, 353-354. (c) M. M. RAUHUT, H. A. CURRIER, A. M. SEMSEL, V. P. WYSTRACH, *J. Org. Chem.* **1961**, *26*, 5138-5145.

5 (a) M. REUTER, E. WOLF, German patent 1,078,574 (to Hoechst), **1960**. See also *Chem. Abstr.* **1961**, *55*, 16427c. (b) M. REUTER, L. ORTHNER, German patent 1,035,135 (to Hoechst), **1958**. See also *Chem. Abstr.* **1960**, *54*, 14124i.

6 (a) A. P. KHARDIN, O. I. TUZHIKOV, L. I. GREKOV, R. K. VALETDINOV, V. I. PANKOV, E. V. MATVEEVA, G. V. NAZAROVA, B. N. POPOV, D. D. CHUVASHOV, *Otkrytiya, Izobret.* **1985**, *10*, 79. (b) USSR patent SU 1,145,022. See also: *Chem. Abstr.* **1985**, *103*, 10371510x. (c) Y. A. DORFMAN, L. V. LEVINA, L. GREKOV, A. V. KOROLEV, *Kinet. Katal.* **1990**, *30*, 662-667. See also: *Chem. Abstr.* **1990**, *112*, 56107p.

7 (a) P. A. T. HOYE, P. G. PRINGLE, M. B. SMITH, K. WORBOYS, *J. Chem. Soc., Dalton Trans.* **1993**, *74*, 269-274. (b) J. W. ELLIS, K. N. HARRISON, P. A. T. HOYE, A. G. ORPEN, P. G. PRINGLE, M. B. SMITH, *Inorg. Chem.* **1992**, *31*, 3026-3033. (c) K. N. HARRISON, P. A. T. HOYE, A. G. ORPEN, P. G. PRINGLE, M. B. SMITH, *J. Chem. Soc., Chem. Commun.* **1989**, 1096-1097.

8 (a) P. G. PRINGLE, M. B. SMITH, *J. Chem. Soc., Chem. Commun.* **1990**, 1701-1702. (b) E. COSTA, P. G. PRINGLE, M. B. SMITH, K. WORBOYS, *J. Chem. Soc., Dalton Trans.* **1997**, 4277-4282. (c) A. G. ORPEN, P. G. PRINGLE, M. B. SMITH, K. WORBOYS, *J. Organomet. Chem.* **1998**, *550*, 255-266.

9 (a) D. K. WICHT, I. V. KOURKINE, B. M. LEW, J. M. NTHENGE, D. S. GLUECK, *J. Am. Chem. Soc.* **1997**, *119*, 5039-5040. (b) D. K. WICHT, I. V. KOURKINE, I. KOVACIK, D. S. GLUECK, T. E. CONCOLINO, G. P. A. YAP, C. D. INCARVITO, A. L. RHEINGOLD, *Organometallics* **1999**, *18*, 5381-5394.

10 U. NAGEL, B. RIEGER, A. BUBLEWITZ, *J. Organomet. Chem.* **1989**, *370*, 223-239.

11 E. COSTA, P. G. PRINGLE, K. WORBOYS, *Chem. Commun.* **1998**, 49-50.

12 P. G. PRINGLE, D. BREWIN, M. B. SMITH, K. WORBOYS, in *Aqueous Organometallic Chemistry and Catalysis*, I. T. HORVATH and F. JOO (eds.), Kluwer: Dordrecht, **1995**, Vol. 5, pp 111-122.

13 I. KOVACIK, D. K. WICHT, N. S. GREWAL, D. S. GLUECK, C. D. INCARVITO, I. A. GUZEI, A. L. RHEINGOLD, *Organometallics* **2000**, *19*, 950-953.

14 (a) M. R. DOUGLASS, T. J. MARKS, *J. Am. Chem. Soc.* **2000**, *122*, 1824-1825. (b) M. A. GIARDELLO, W. A. KING, S. P. NOLAN, M. PORCHIA, C. SISHTA, T. J. MARKS in *Energetics of Organometallic Species*, J. A. MARTINHO SIMOES (ed.), Kluwer: Dordrecht, **1992**, pp 35-51.

15 L.-B. HAN, M. TANAKA, *J. Am. Chem. Soc.* **1996**, *118*, 1571-1572.

16 A. ALLEN JR., D. R. MANKE, W. LIN, *Tetrahedron Lett.* **2000**, *41*, 151-154.

17 L.-B. HAN, N. CHOI, M. TANAKA, *Organometallics* **1996**, *15*, 3259-3261.

18 L.-B. HAN, R. HUA, M. TANAKA, *Angew. Chem. Int. Ed. Engl.* **1998**, *37*, 94-96.

19 For reviews, see: (a) H. GROGER, B. HAMMER, *Chem. Eur. J.* **2000**, *6*, 943-948. (b) D. F. WIEMER, *Tetrahedron* **1997**, *53*, 16609-16644. (c) M. C. MITCHELL, T. P. KEE, *Coord. Chem. Rev.* **1997**, *158*, 359-383.

20 T. YOKOMATSU, T. YAMAGISHI, S. SHIBUYA, *Tetrahedron: Asymmetry* **1993**, *4*, 1401-1404.

21 T. YOKOMATSU, T. YAMAGISHI, S. SHIBUYA, *Tetrahedron: Asymmetry* **1993**, *4*, 1779-1782.

22 T. YOKOMATSU, T. YAMAGISHI, S. SHIBUYA, *Tetrahedron: Asymmetry* **1993**, *4*, 1783-1784.

23 N. P. RATH, C. D. SPILLING, *Tetrahedron Lett.* **1994**, *35*, 227-230.

24 T. YOKOMATSU, T. YAMAGISHI, S. SHIBUYA, *J. Chem. Soc., Perkin Trans. 1* **1997**, 1527-1533.

25 M. D. GROANING, B. J. ROWE, C. D. SPILLING, *Tetrahedron Lett.* **1998**, *39*, 5485-5488.

26 H. SASAI, M. BOUGAUCHI, T. ARAI, M. SHIBASAKI, *Tetrahedron Lett.* **1997**, *38*, 2717-2720.

27 C. QIAN, T. HUANG, C. ZHU, J. SUN, *J. Chem. Soc., Perkin Trans. 1* **1998**, 2097-2103.

28 T. ARAI, M. BOUGAUCHI, H. SASAI, M. SHIBASAKI, *J. Org. Chem.* **1996**, *61*, 2926-2927.

29 T. YAMAGISHI, T. YOKOMATSU, K. SUEMUNE, S. SHIBUYA, *Tetrahedron* **1999**, *55*, 12125-12136.

30 T. YOKOMATSU, T. YAMAGISHI, K. MATSUMOTO, S. SHIBUYA, *Tetrahedron* **1996**, *52*, 11725-11738.

31 S. R. DAVIES, M. C. MITCHELL, C. P. CAIN, P. G. DEVITT, R. J. TAYLOR, T. P. KEE, *J. Organomet. Chem.* **1998**, *550*, 29-57.

32 J. P. DUXBURY, A. CAWLEY, M. THORNTON-PETT, L. WANTZ, J. N. D. WARNE, R. GREATREX, D. BROWN, T. P. KEE, *Tetrahedron Lett.* **1999**, *40*, 4403-4406.

33 H. SASAI, S. ARAI, Y. TAHARA, M. SHIBASAKI, *J. Org. Chem.* **1995**, *60*, 6656-6657.

34 (a) H. GROGER, Y. SAIDA, S. ARAI, J. MARTENS, H. SASAI, M. SHIBASAKI, *Tetrahedron Lett.* **1996**, *37*, 9291-9292. (b) H. GROGER, Y. SAIDA, H. SASAI, K. YAMAGUCHI, J. MARTENS, M. SHIBASAKI, *J. Am. Chem. Soc.* **1998**, *120*, 3089-3103.

35 K. YAMAKOSHI, S. J. HARWOOD, M. KANAI, M. SHIBASAKI, *Tetrahedron Lett.* **1999**, *40*, 2565-2568.

36 L.-B. HAN, M. TANAKA, *Chem. Commun.* **1999**, 395-402.

6
O–H Activation and Addition to Unsaturated Systems

Kazuhide Tani and Yasutaka Kataoka

6.1
Introduction

Hydrido(hydroxo) and hydrido(alkoxo) complexes of late transition metals produced by activation of the O–H bond is postulated to be involved as an important step of various transition metal-mediated catalytic transformations such as water gas shift reactions, Wacker-type reactions, hydration and alcoholysis, transfer hydrogenation of unsaturated compounds, etc. [1]. However, isolation of such complexes is rare. Studies on the O–H bond activation by transition metal complexes are also much less developed compared to those on C–H bond activation by transition metal complexes [2], and review articles dealing with the activation of the O–H bond by transition metals are rare [1]. The chemistry of alkoxo complexes of late transition metals does not appear to have received much attention [3]. A widely accepted explanation for the scarcity of alkoxides or hydroxides of late transition metals was that such metal-oxygen bonds are characteristically weak because of a mismatch of these hard basic ligands with soft late transition metals, and they have propensity to easily decompose presumably by β-hydride elimination to metal hydrides [4]. The bond energy of an Ir–OMe bond estimated recently, however, was revealed to be higher than that of an Ir–Me bond (see below). Thus, the intrinsic bond strength between oxygen and late transition metals is not as weak as was previously thought. There is no inherent instability attached to transition metal alkoxides. Thus, considerable attention has recently been focused on the activation of O–H bonds by transition metals and the reactivity of M–OR (R = H, alkyl, aryl, acyl group) bonds, and examples of the isolation of hydrido(hydroxo), hydrido(alkoxo), etc., complexes are increasing. In this review article we survey the chemistry of mainly late transition metal complexes resulting from oxidative addition of O–H bonds to transition metals as well as stoichiometric and catalytic reactions involving activation of the O–H bond by late transition metal complexes as an important reaction step.

6.2
Transition Metal Complexes Resulting from the Activation of O–H Bonds and Reaction with Related Complexes

Oxidative addition of the O–H bond to transition metal complexes gives hydrido(hydroxo), hydrido(alkoxo) or hydrido(carboxylato) complexes (Eq. 6.1), but well-characterized complexes obtained as primary products from the reaction of the compound, XO–H (XO–H = water, alcohol, and carboxylic acid) with late transition metals are quite rare [1]. Furthermore, the crystal structures of very few complexes of this type have been reported. In this section we will survey late transition metal complexes resulting from activation of water, alcohol, and carboxylic acid.

$$\text{XO–H} \quad + \quad L_nM \quad \rightleftharpoons \quad L_nM\overset{\text{OX}}{\underset{\text{H}}{<}} \quad \cdots \quad (6.1)$$

$$X = H, R, C(O)R$$

6.2.1
Preparation of Hydrido(hydroxo), Hydrido(alkoxo), and Hydrido(carboxylato) Complexes by Metathesis

Before discussing the preparation of late transition metal complexes resulting from the activation of O–H bonds by late transition metal complexes, we will describe metathesis methods for the preparation of hydrido(hydroxo), hydrido(alkoxo), and hydrido(carboxylato) complexes. Though many methods of preparation of transition metal hydroxides, alkoxides, etc. by a metathesis reaction have been reported [1], only a limited number of examples of the preparation of hydrido(hydroxo), hydrido(alkoxo) complexes etc. by metathesis are available.

In 1977, the mononuclear hydrido(hydroxo) complexes, $[RuH(OH)(PPh_3)_2(S)]$ (S = THF, H_2O) (1) were prepared by the metathesis of $[RuHCl(PPh_3)_3]$ with NaOH or KOH [5]. Depending on the reaction conditions, the reaction leads to hydroxo-bridged dimers, $[\{RuH(PPh_3)_2(S)\}_2(\mu\text{-OH})_2]$ (S = Me_2CO, H_2O, or ^tBuOH) and a tetranuclear complex of stoichiometry $[Ru_4H_4(OH)_2(PPh_2)_2(CO)_2(PPh_3)_6(Me_2CO)_2]$ (Eq. 6.2). Another five-coordinate hydrido(hydroxo) complex $[OsH(OH)(CO)(P^iPr_3)_2]$ (2) was recently prepared by metathesis reaction of $[OsH(Cl)(CO)(P^iPr_3)_2]$ with KOH (Eq. 6.3) [6].

The hydrido(ethoxo) complex carrying an electron-donating η^5-C_5Me_5 (= Cp*) ligand, $[Cp*IrH(OEt)(PPh_3)]$ (4), was prepared by a metathesis reaction between $[Cp*IrCl_2(PR_3)]$ (3) and NaOEt followed by β-H elimination from the intermediate diethoxide complex (Eq. 6.4) [7]. Several other iridium alkoxide analogs $[Cp*IrH(OR)$

$$RuH(OH)(PPh_3)_2(S) \quad (1)$$

or

$$RuH(Cl)(PPh_3)_3 + \text{NaOH or KOH} \quad \xrightarrow{\text{solvent}} \quad \{RuH(PPh_3)_2(S)\}_2(\mu\text{-OH})_2 \quad (6.2)$$

or

tetranuclear species

$$S = \text{solvent}$$

(PPh$_3$)] (5) (R = C$_2$D$_5$, ^{n}Pr, ^{i}Pr, Ph) were prepared by treatment of the ethoxide 4 with an excess of the corresponding alcohol (Eq. 6.5). The hydroxo(isopropoxo) complex 5c was characterized by X-ray diffraction [8].

$$\text{OsH(Cl)(CO)(P}^i\text{Pr}_3)_2 \quad + \quad \text{KOH} \quad \xrightarrow{-\text{KCl}} \quad \text{OsH(OH)(CO)(P}^i\text{Pr}_3)_2 \qquad (6.3)$$

2

$$\text{3} \quad \underset{\text{PPh}_3}{\overset{\text{Cl}}{\text{Cp*Ir}}}\text{Cl} \quad +2\,\text{NaOEt} \quad \xrightarrow{-2\,\text{NaCl}} \quad \left[\underset{\text{PPh}_3}{\overset{\text{OEt}}{\text{Cp*Ir}}}\text{OEt} \right] \quad \xrightarrow{-\text{MeCHO}} \quad \underset{\text{PPh}_3}{\overset{\text{H}}{\text{Cp*Ir}}}\text{OEt} \qquad (6.4)$$

3   **4**

$$\underset{\text{PPh}_3}{\overset{\text{H}}{\text{Cp*Ir}}}\text{OEt} \quad + \quad \text{ROH} \quad \longrightarrow \quad \underset{\text{PPh}_3}{\overset{\text{H}}{\text{Cp*Ir}}}\text{OR} \quad + \quad \text{EtOH} \qquad (6.5)$$

4

a : R = C$_2$D$_5$
b : R = ^{n}Pr
c : R = ^{i}Pr
d : R = Ph

5

Several hydrido(phenoxo) complexes of nickel, *trans*-[NiH(OPh)L$_2$] (6) (**a:** L = P^{i}Pr$_3$; **b:** L = PCy$_3$; **c:** L = PBn$_3$), have been prepared by the metathesis reaction of NaOPh with *trans*-[NiHClL$_2$] (Eq. 6.6). The complex 6c was obtained as the phenol-solvated complex whose structure was determined by X-ray analysis [9]. An analogous platinum complex *trans*-[PtH(OPh)(PEt$_3$)$_2$] (7) was prepared by the reaction of *trans*-[PtH(NO$_3$)(PEt$_3$)$_2$] with NaOPh (Eq. 6.7). The complex 7 is air-stable but thermally sensitive and decomposes at room temperature. The structure was elucidated by X-ray analysis [10].

$$\textit{trans}\text{-NiH(Cl)L}_2 \quad + \quad \text{NaOPh} \quad \xrightarrow{-\text{NaCl}} \quad \textit{trans}\text{-NiH(OPh)L}_2 \qquad (6.6)$$

6

a : L = P^{i}Pr$_3$; **b** : L = PCy$_3$; **c** : L = PBn$_3$

$$\textit{trans}\text{-PtH(NO}_3)(\text{PEt}_3)_2 \quad + \quad \text{NaOPh} \quad \xrightarrow{-\text{NaNO}_3} \quad \textit{trans}\text{-PtH(OPh)(PEt}_3)_2 \qquad (6.7)$$

7

Seven-coordinate hydrido(carboxylato)osmium complexes, [OsH$_2$(Cl)(κ^2-O$_2$CMe)(P^{i}Pr$_3$)$_2$] (9) and [OsH$_2$(κ^2-O$_2$CMe)(κ^1-OC(O)Me)(P^{i}Pr$_3$)$_2$] (10) were prepared by metatheses of OsH$_2$Cl$_2$(P^{i}Pr$_3$)$_2$ (8) with K[O$_2$CMe] and Ag[O$_2$CMe] respectively (Scheme 6-1) [11–13]. The reaction with even excess K[O$_2$CMe] gave only mono(acetato) complex 9. The formation of bis(acetato) complex 10 needs the addition of 2 equiv. of Ag[O$_2$CMe] to a dichloromethane solution of 8. The complex 10 was characterized by an X-ray analysis. Cationic hydrido(acetato) complexes 11 and 12 were obtained by treating the complex 10 with HBF$_4$.

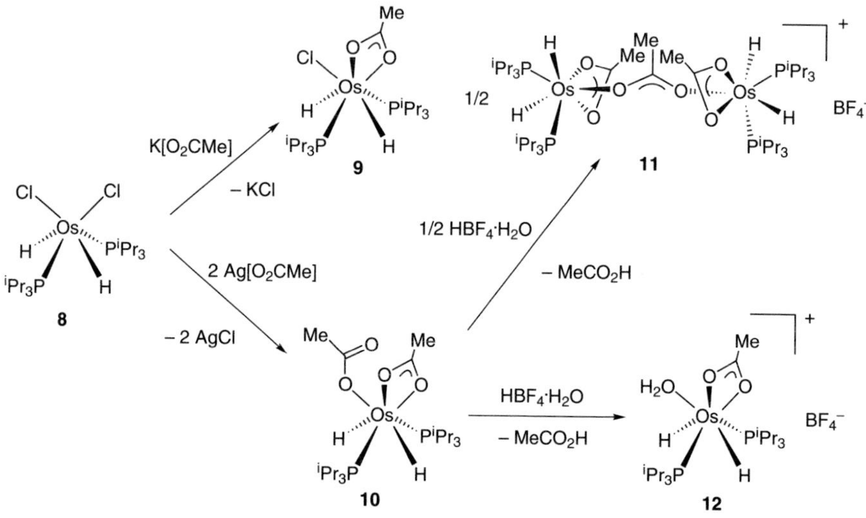

Scheme 6-1 Preparation of Hydrido(acetato) complexes by metathesis reactions

6.2.2
Complexes Resulting from Activation of Water: Hydrido(hydroxo) Complexes

Early in the 1970s, Muetterties reported very briefly that a trialkylphosphine complex of platinum, [Pt(PEt$_3$)$_3$] (**26a**) reacted with water reversibly to form an unstable species, [PtH(PEt$_3$)$_3$]OH (**27**), which was not isolated (Scheme 6-2) [14]. The first hydroxoplatinum complex prepared by O–H bond activation was reported by Bennet in 1973. The platinum alkyne complex **13** reacted with water in toluene under reflux to afford hydroxoplatinum(II) complex **14a** in good yield, which is probably formed *via* a hydrido(hydroxo) complex (Eq. 6.8) [15, 16]. Similar examples of activation of water with the electron-rich benzyne complex **15** [17] or the cyclometalated complex [OsH(η^2-CH$_2$PMe$_2$)(PMe$_3$)$_3$] (**17a**) [18] to form a hydroxo(phenyl) complex (**16**) or a neutral hydrido(hydroxo) complex, *cis*-[OsH(OH)(PMe$_3$)$_4$] (**18a**), were reported (Eqs. 6.9 and 6.10). In the latter case, the hydrido(hydroxo) complex was obtained, but its hydrido ligand did not come from water but was already present as a hydrido ligand in the starting complex **17a**. The ruthenium analog, *cis*-[RuH(OH)(PMe$_3$)$_4$] (**18b**), was also prepared by a similar method from the ruthenium complex [RuH(η^2-CH$_2$PMe$_2$) (PMe$_3$)$_3$] (**17b**) but in low yield [19]. The methyl analog of the cyclometalated complex, [RuMe(η^2-CH$_2$PMe$_2$)(PMe$_3$)$_3$] (**19**), reacted with a slight excess of water at room temperature to give the methyl(hydroxo) complex **20** in a better yield (Eq. 6.11) [19].

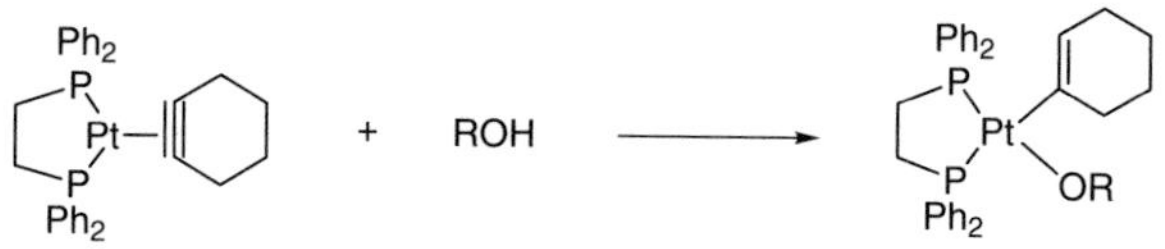

$$(6.8)$$

13   a : R = H   14
            b : R = Me

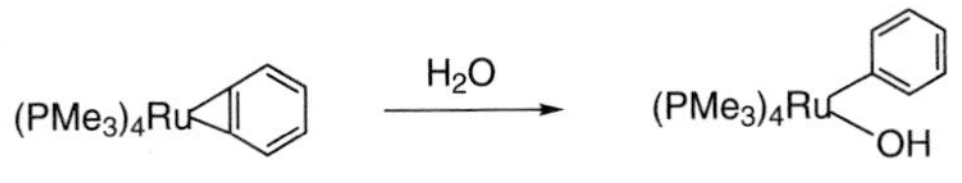

$$(PMe_3)_4Ru \quad \xrightarrow{H_2O} \quad (PMe_3)_4Ru\text{—OH} \qquad (6.9)$$

15   **16**

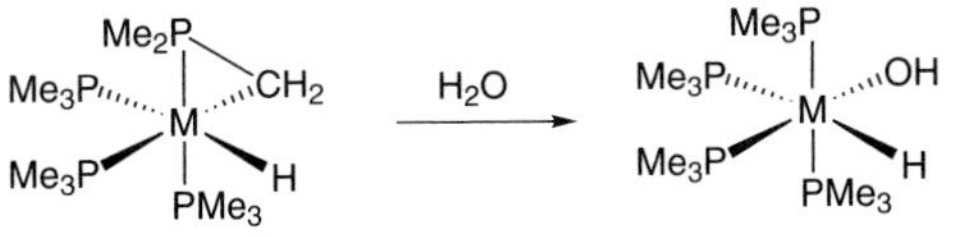

$$17 \quad \xrightarrow{H_2O} \quad 18 \qquad (6.10)$$

a : M = Os
b : M = Ru

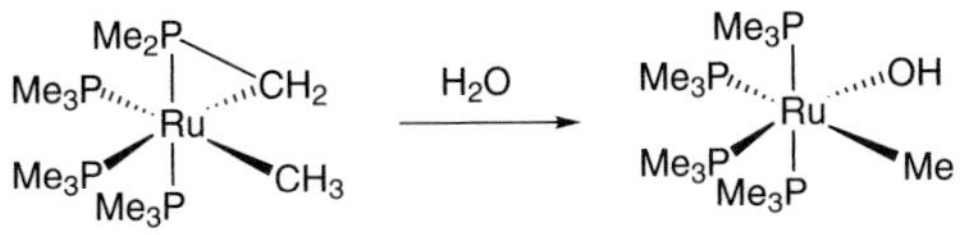

17   **18**

$$19 \quad \xrightarrow{H_2O} \quad 20 \qquad (6.11)$$

19   **20**

A pyrolylnickel complex **21** also could activate water to afford a bridging hydroxo complex **22** accompanying cleavage of the Ni-carbon bond (Eq. 6.12) [20]. The hydrido(hydroxo) species was not observed in the reaction mixture.

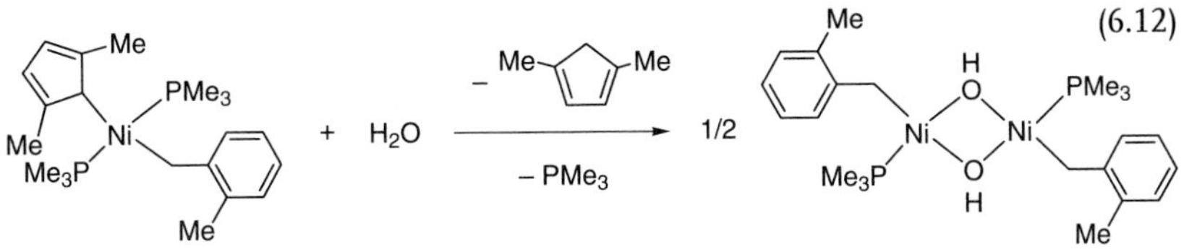

$$(6.12)$$

21   **22**

Herberhold reported activation of water with $Re_2(CO)_{10}$. The reaction proceeded at 200°C to give a tetranuclear complex $[Re(CO)_3(\mu^3\text{-OH})]_4$ (**23**) in quantitative yield, and evolution of dihydrogen and CO was observed (Eq. 6.13) [21]. Complex **23** has a pseudo-cubane structure without metal-metal bonds in which the $Re(CO)_3$ groups are linked by triply-bridging OH ligands. Also in this case, no presumed intermediate hydrido(hydroxo) species was detected.

$$2\ Re_2(CO)_{10} \quad + 4\ H_2O \quad \xrightarrow{200\ °C} \quad [Re(CO)_3(\mu_3\text{-OH})]_4 \quad + 8\ CO \quad + \ 2\ H_2 \qquad (6.13)$$

23

In 1979, the first isolation of the hydrido(hydroxo) complex by oxidative addition of water to an electron-rich platinum(0) complex was accomplished by Yoshida and Otsuka [22]. Highly coordinatively unsaturated bis(triisopropylphosphine)platinum (**24b**) can activate water very easily at room temperature to give the hydrido(hydroxo) complex, *trans*-[PtH(OH)(P^{i}Pr$_3$)$_2$] (**25**), as an extremely air-sensitive and thermally unstable colorless compound (Scheme 6-2). Addition of water to the tris(phosphine)platinum complexes PtL$_3$ (**26**) (**a**: L = PEt$_3$; **b**: L = P^{i}Pr$_3$), however, generates strong hydroxy bases, [PtH(PEt$_3$)$_3$]OH (**27**) or *trans*-[PtH(P^{i}Pr$_3$)$_2$(S)]OH (S = solvent) (**28**), depending on the phosphine ligand used, instead of neutral hydrido(hydroxo) complexes. The complexes **27** and **28** were detected in the reaction mixture of the PtL$_3$-catalyzed water gas shift reaction and could be isolated as BF$_4^-$ or BPh$_4^-$ salts [22, 23].

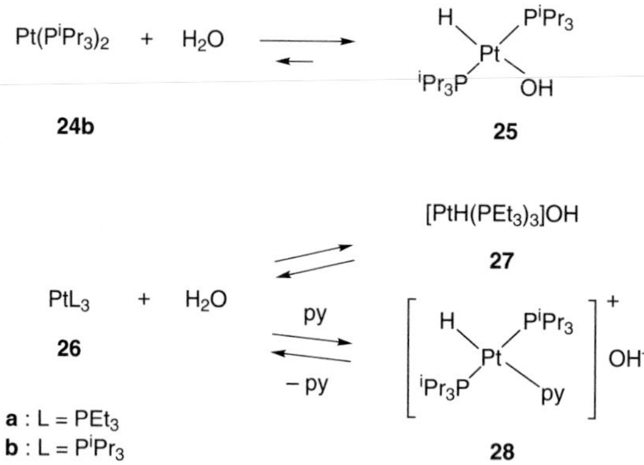

Scheme 6-2 Preparation of hydrido(hydroxo) and hydrido (hydroxide) complexes by oxidative addition of water

The ruthenium hydrido(hydroxo) complex, *cis*-[RuH(OH)(PMe$_3$)$_4$] (**18b**), was prepared in good yield by oxidative addition of water to an ethylene complex **29** (Eq. 6.14) [19]. The reaction of the ethylene complex carrying chelate diphosphine ligands, [Ru(C$_2$H$_4$)(dmpe)$_2$] (**30**) [dmpe = 1,2-bis(dimethylphosphino)ethane], with water at 90°C, however, yielded a binuclear hydrido(hydroxo) water complex *trans*-[RuH(OH)(dmpe)$_2$•H$_2$O]$_2$ (**31**) in which two hydrido(hydroxo)ruthenium fragments are linked together by two water molecules *via* bridging hydrogen bonds (Eq. 6.15). The structure of **31** was determined by an X-ray analysis. Elimination of the hydrogen-bound water by molecular sieves yielded the monomeric hydrido(hydroxo) complex **32**.

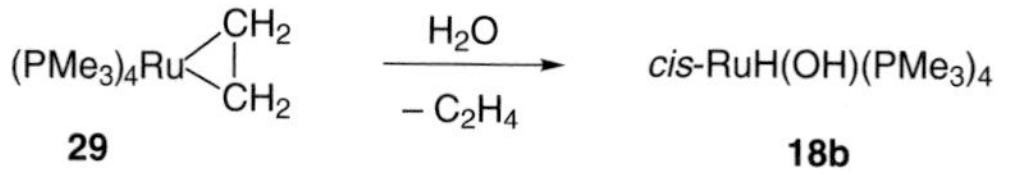

$$(PMe_3)_4Ru{\overset{CH_2}{\underset{CH_2}{\big\langle}}} \quad \xrightarrow[-\,C_2H_4]{H_2O} \quad cis\text{-RuH(OH)(PMe}_3)_4 \qquad (6.14)$$

$$\textbf{29} \qquad\qquad\qquad\qquad\qquad \textbf{18b}$$

Oxidative addition of water to [RhH(P^{i}Pr$_3$)$_3$] (**33**) and [Rh$_2$H$_2$(μ-N$_2$)(PCy$_3$)$_4$] (Cy = *c*-C$_6$H$_{11}$) (**34**) took place at room temperature in pyridine to give [RhH$_2$(py)$_2$(P^{i}Pr$_3$)$_2$]OH (**35**) and [RhH$_2$(py)$_2$(PCy$_3$)$_2$]OH (**36**) respectively, which were both characterized by ^{1}H NMR [24]. Although the complexes are stable in aqueous pyridine, a facile reductive elimination of water occurs in a dry solvent to regenerate the starting complexes indicating reversibility of the H$_2$O addition. The formation of the hydrido(hy-

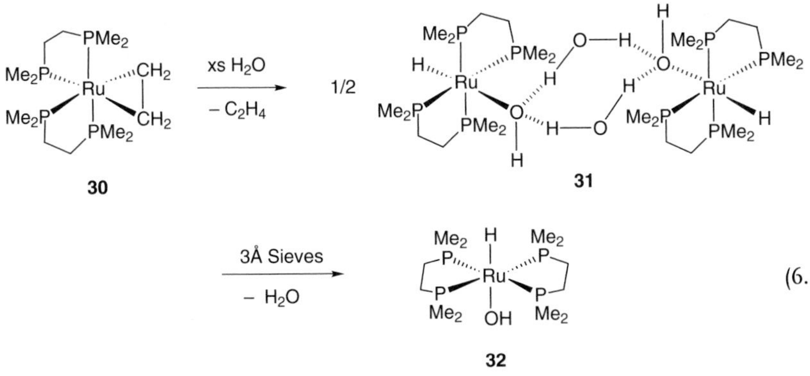

$$\text{30} \quad \xrightarrow[- C_2H_4]{xs\ H_2O} \quad 1/2 \quad \text{31}$$

$$\text{31} \quad \xrightarrow[- H_2O]{3\text{Å Sieves}} \quad \text{32} \qquad (6.15)$$

droxide) complexes was confirmed by isolation of the BPh$_4^-$ salt as colorless crystals (Scheme 6-3).

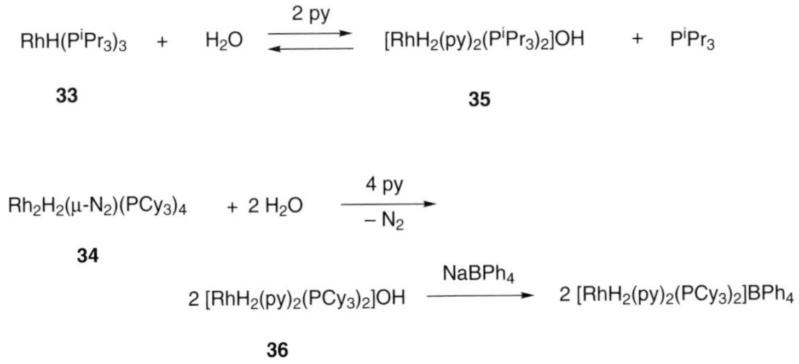

$$\text{RhH(P}^i\text{Pr}_3)_3 \; + \; H_2O \; \underset{}{\overset{2\ py}{\rightleftharpoons}} \; [\text{RhH}_2(\text{py})_2(\text{P}^i\text{Pr}_3)_2]\text{OH} \; + \; \text{P}^i\text{Pr}_3$$

$$\text{33} \qquad\qquad\qquad\qquad \text{35}$$

$$\text{Rh}_2\text{H}_2(\mu\text{-N}_2)(\text{PCy}_3)_4 \; + \; 2\ H_2O \; \xrightarrow[- N_2]{4\ py}$$

$$\text{34}$$

$$2\ [\text{RhH}_2(\text{py})_2(\text{PCy}_3)_2]\text{OH} \; \xrightarrow{\text{NaBPh}_4} \; 2\ [\text{RhH}_2(\text{py})_2(\text{PCy}_3)_2]\text{BPh}_4$$

$$\text{36}$$

Scheme 6-3 Oxidative addition of water to rhodium phosphine complexes

Milstein reported the facile oxidative addition of water to a cationic iridium per-alkylphosphine complex, [Ir(PMe$_3$)$_4$]PF$_6$ (37), to give the hydrido(hydroxo) complex, cis-[IrH(OH)(PMe$_3$)$_4$]PF$_6$ (38), as air-stable and thermally stable colorless crystals, which does not undergo reductive elimination of water even at 100°C (Scheme 6-4) [25]. The structure of the hydrido(hydroxo) complex 38 was confirmed by an X-ray analysis as well as by a neutron diffraction study [26]. In contrast to 38, the strongly basic trans-hydrido(hydroxide)platinum complexes 27, 28 and hydridorhodium complexes 35, 36 containing outer-sphere hydroxide eliminate water readily and are stable only in the presence of a large excess of water (see above, cf. Schemes 6-2 and 6-3). The complex 38 is a relatively weak base. However, 38 undergoes exchange reactions involving the OH group. Reaction of 38 with D$_2$O yields exclusively the hydrido(deuteroxo) complex 38b, not the deutero(deuteroxo) complex, indicating that a reductive elimination-oxidative addition sequence is not involved. Reaction of 38 with MeOH gave the hydrido(methoxo) complex 54. Similar but neutral iridium phosphine complexes, [IrCl(PR$_3$)$_3$] (39) (a: R = Me; b: R = Et), reacted also easily with water at low temperature (–30°C) to give mer-cis-[HIr(OH)Cl(PR$_3$)$_3$] (40) (a: R = Me;

b: R = Et) (Scheme 6-5). The trimethylphosphine complex **40a** was also prepared from the reaction of the cyclooctene complex **41**, which is completely dissociated into **39a** and cyclooctene in C_6D_6, with water. [27] Different from the cationic complex **38**, these hydrido(hydroxo) complexes **40** undergo reductive elimination of water at room temperature generating the starting complex **39**.

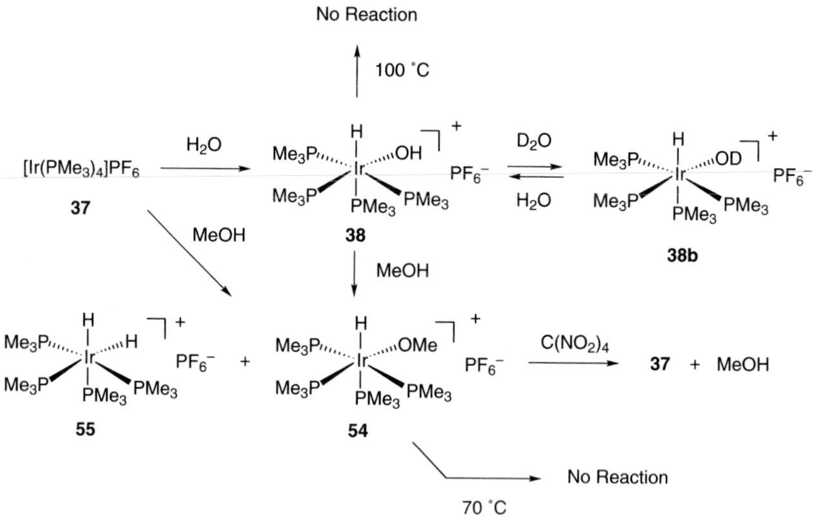

Schemes 6-4 Oxidative addition of water and methanol to cationic iridium phosphine complex **37**

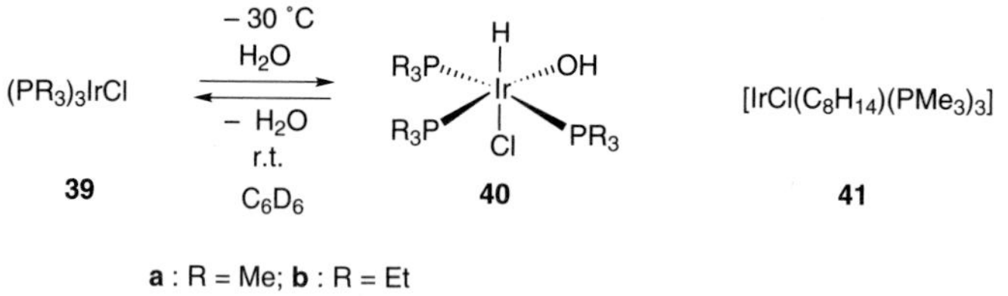

a : R = Me; b : R = Et

Schemes 6-5 Oxidative addition of water to neutral iridium phosphine complexes

Very recently we [28] and Togni's group [29] have found independently that similar dinuclear iridium diphosphine complexes, [IrCl(diphosphine)]₂ (**42a**: diphosphine = BPBP; **42b**: diphosphine = BINAP; **44**: diphosphine = Josiphos) carrying an aryldiphosphine, which is not very electron-donating, can activate water easily to give the stable corresponding hydrido(μ-hydroxo) complexes (Scheme 6-6). Reaction of the extremely air-sensitive iridium-peraryldiphosphine complexes **42a** and **42b** with excess water in THF at ambient temperature gave cationic hydrido(hydroxo) complexes **43a** and **43b** respectively, as air-stable pale yellow powders in almost quantitative yields. The hydrido(hydroxo) complexes **43** are cationic, having a triply bridged $Ir_2(\mu_2\text{-OH})(\mu_2\text{-Cl})$ core and are characterized by comparison of the spectral data with those of the hydrido(methoxo) complex **69b** (Fig. 6-1), whose structure was confirmed by X-ray analysis. The hydrido(hydroxo) complexes **43** show catalytic activity in the hydration of nitriles (see below). Another iridium diphosphine complex **44**, which exists as two geometrical isomers, reacted with a slight excess of water in toluene at room temperature to give a mixture of two geometrical isomers of neutral hydrido(hydroxo) complexes, *syn-trans*-**45** and *anti-cis*-**45**, in good yield. The *anti-cis* isomer isomerizes in THF to the *syn-trans* one and the structure of the latter isomer, *syn-trans*-**45** was confirmed by X-ray analysis. Different from the above-mentioned hydrido(hydroxo) complexes **43**, *syn-trans*-**45** is a neutral complex and has a distorted octahedral structure at the Ir(III) centers, each of which carries two bridging hydroxo ligands, one terminal hydrido, and one chloro donor. The complex shows C2 symmetry, with the 2-fold axis running through the center of the Ir_2O_2 plane. Interatomic distances are normal and the Ir–Ir distance of 3.209(2) Å does not imply a metal-metal interaction. The complex **45** appears to be the first compound of this type having only the OH-fragments acting as bridges. When the complex **45** in THF-d_8 was heated at 100°C in a sealed NMR tube for several hours, re-appearance of **44** was observed in the ^{31}P NMR spectrum, but the elimination was not a clean reaction.

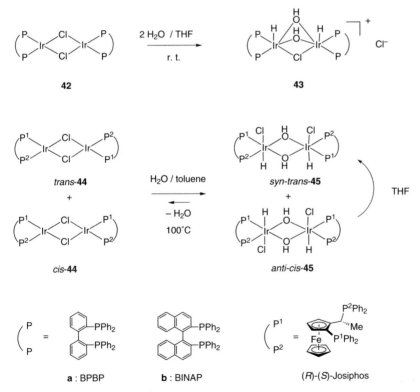

Schemes 6-6 Oxidative addition of water to Ir(I) diphosphine complexes **42** and **44**

Platinum(II) complexes bearing nitrogen donor ligands, [PtMe$_2$(N–N)] (46) (N–N = 2,2′-bipyridyl or 1,10-phenanthroline) can also activate water (Eq. 6.16) [30, 31]. Complex 46 reacted with an excess of water in acetone at room temperature to give the platinum(IV) hydroxo complex, [PtMe$_2$(OH)(N–N)(OH$_2$)]$^+$OH$^-$ (47a) in good yield. Elemental analysis, IR, NMR, and conductivity characterized the complexes. Oxidative addition of water to platinum(II) complex 46 is remarkably different from that to platinum(0) complexes such as 24 or 26 (see Scheme 6-2). The possibility that an oxidative addition occurred first to produce [PtMe$_2$(OH)(H)(N–N)], followed by rapid reaction of the Pt–H group with water, was suggested.

$$\begin{array}{ccc} & \xrightarrow[\text{r.t.}]{\text{ROH}} & \end{array}$$

$$\tag{6.16}$$

46

    a : R = H
    b : R = Me
    c : R = Et
    d : R = ^{i}Pr

47

N   N   = 2,2′-bipyridyl or 1,10-phenanthroline

6.2.3
Complexes Resulting from Activation of Alcohols: Hydrido(alkoxo) Complexes

Hydrido(alkoxo) complexes of late transition metals are postulated as intermediates in the transition metal-catalyzed hydrogenation of ketones (Eq. 6.17), the hydrogenation of CO to MeOH, hydrogen transfer reactions and alcohol homologation. However, the successful isolation of such complexes from the catalytic systems was very rare [32–37].

$$\begin{array}{ccc} \diagdown\!\!\diagup C=O & \xrightarrow[\text{H}_2]{\text{L}_n\text{M}} & \left[\diagup_{\text{CH}}\!\!\diagdown\!\!\text{O}\diagdown\!\!^{\text{ML}_n}_{\text{H}} \right] & \longrightarrow & \diagdown\!\!\diagup \text{CHOH} + \text{L}_n\text{M} \end{array}$$

$$\tag{6.17}$$

The hydrido(alkoxo)ruthenium complex [RuH{OCH(CF$_3$)Ph}(PPh$_3$)$_3$] (48) has been prepared by insertion of a fluoroketone, PhCOCF$_3$, into one of the Ru–H bonds in the dihydridoruthenium complex [RuH$_2$(PPh$_3$)$_4$] (Eq. 6.18) [38]. Hydrogenolysis of complex 48 with H$_2$, ethanol, or isopropanol at room temperature liberates the fluoro alcohol, PhCH(OH)CF$_3$, in good yields, producing ruthenium hydride complexes. Complex 48 serves also as an active catalyst precursor for the hydrogenation of trifluoroacetophenone at 60–80°C. We have also reported the isolation of a hydrido(alkoxo)iridium complex (50) as a model complex of an intermediate complex for the very efficient catalytic hydrogenation of ketone substrates with rhodium(I)-peralkyldiphosphine complexes. Complex 50 was isolated in good yield as air-stable yellow crystals from the reaction of [Ir(cod){^{i}Pr$_2$P(CH$_2$)$_3$P^{i}Pr$_2$}]ClO$_4$ (49) and PhC(O)C(O)NHCH$_2$Ph under an atmospheric pressure of dihydrogen (Eq.

6.19) [33]. Although the crystallographic structure of complex **50** was not available, the complex was well characterized by elemental analysis as well as spectroscopic studies. The rhodium analog of **49** showed high catalytic activity for the hydrogenation of various kinds of ketone substrates under 1 atm of dihydrogen at ambient temperature, but the rhodium hydrido(alkoxo) complex analogous to **50** could not be detected in the reaction mixture. In contrast, complexes **49** and **50** did not show catalytic activity under similar conditions.

$$\text{RuH}_2(\text{PPh}_3)_4 \quad + \quad \underset{\text{Ph}}{\overset{\overset{\displaystyle O}{\|}}{\bigwedge}}\text{CF}_3 \quad \xrightarrow{60\ ^\circ\text{C}} \quad \text{RuH}\{\text{OCH}(\text{CF}_3)\text{Ph}\}(\text{PPh}_3)_3 \tag{6.18}$$

48

$$\xrightarrow[\text{r.t.}]{\text{H}_2,\ \text{EtOH, or }^i\text{PrOH}} \quad \underset{\text{Ph}}{\overset{\overset{\displaystyle OH}{|}}{\bigwedge}}\text{CF}_3 \quad + \quad \text{RuH}_m(\text{PPh}_3)_n$$

$$[\text{Ir(cod)}\{^i\text{Pr}_2\text{P(CH}_2)_3\text{P}^i\text{Pr}_2\}]\text{ClO}_4 \quad + \quad \underset{\text{Ph}}{\overset{\overset{\displaystyle O}{\|}}{\bigwedge}}\underset{\underset{\displaystyle O}{\|}}{\bigwedge}\text{NHCH}_2\text{Ph} \quad \xrightarrow[\substack{\text{r.t.}\\-\text{COD}}]{\text{H}_2\ (1\text{atm})}$$

49

50

(6.19)

The first preparation of alkoxo complexes by oxidative addition of an alcohol was reported for the preparation of methoxoplatinum(II) complexes, e.g., **14b**, which was prepared by the oxidative addition of MeOH to the platinum(0) cycloalkyne complex **13** (Eq. 6.8) [15, 16]. Yoshida and Otsuka [39] reported that reaction of two coordinate platinum complexes, [PtL$_2$] (**24**) (**a:** L = P^{i}Pr$_3$, **b:** L = PCy$_3$, **c:** L = PPh^{t}Bu$_2$) with MeOH at room temperature gave stable *trans*-[PtH$_2$L$_2$] (**51**), formed by β-H elimination from the corresponding oxidative addition products, hydrido(methoxo) complexes (Eq. 6.20). The tris(phosphine) complex, [Pt(^{i}PPr$_3$)$_3$] (**26b**) also gave the dihydrido complex **51a** by reaction with MeOH. The corresponding palladium(0) complexes *trans*-[PdL$_2$] (L = P^{t}Bu$_3$, PPh^{t}Bu$_2$, PCy$_3$), however, were inert toward methanol.

$$\underset{\textbf{24}}{\text{PtL}_2} \quad \xrightarrow[\substack{-\ [\text{CH}_2\text{O}]_x\\ \text{r. t.}}]{\text{MeOH}} \quad \underset{\textbf{51}}{trans\text{-PtH}_2(\text{L})_2} \qquad \begin{array}{l}\textbf{a}: \text{L} = \text{P}^i\text{Pr}_3\\ \textbf{b}: \text{L} = \text{PCy}_3\\ \textbf{c}: \text{L} = \text{PPh}^t\text{Bu}_2\end{array} \tag{6.20}$$

The dimethylplatinum(II) complexes having chelate nitrogen ligands, [PtMe$_2$(N–N)] (N–N = 2,2'-bipyridyl or 1,10-phenanthroline) (**46**) reacted with alcohols (ROH, R = Me, Et, ^{i}Pr) to give hygroscopic and basic platinum(IV) complexes,

$[PtMe_2(OR)(N–N)(OH_2)]^+OH^-$ (47) (Eq. 6.16) [30, 31]. These complexes, likely resulting from an oxidative ROH addition, were characterized by elemental analysis, IR and NMR spectroscopy, conductivity measurements and conversion to derivatives containing weakly coordinating bulky anions. These reactions are of interest because they represent the first examples of oxidation of platinum(II) complexes with alcohols and provide the first stable alkoxoplatinum(IV) complexes. The alkoxo–platinum(IV) bond is inert against solvolysis by alcohols, water and even dilute perchloric acid.

The formation of other stable hydrido(alkoxo) complexes of platinum(IV) formulated as $[PtH(OR)(Cl)(SnCl_3)(PPh_3)_2]$ (R = Et, ^nPr, and Bu) (52) from the reaction of $Pt(PPh_3)_4$ with the corresponding alcohol in the presence of $SnCl_2•2H_2O$ or $SnCl_4•5H_2O$ was reported (Scheme 6-7) [40]. The formulation of the hydrido(alkoxo) complexes was proposed based on elemental analyses as well as spectroscopic results, but the exact structures were not established. Formation of slightly different hydrido(alkoxo) complexes depending on using either $SnCl_2•2H_2O$ or $SnCl_4•5H_2O$ was claimed. Similar reactions using isopropyl alcohol provided a different type of reaction products (53). Note that MeOH could not be activated by these systems.

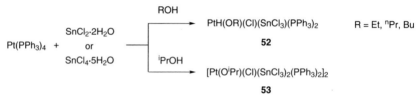

Scheme 6-7 Reaction of $Pt(PPh_3)_4$ with alcohols in the presence of tin chlorides

The cationic iridium complex bearing electron-donating trimethylphosphine ligands, $[Ir(PMe_3)_4]PF_6$ (37) reacted with MeOH to give the hydrido(methoxo) complex, cis-$[IrH(OMe)(PMe_3)_4]PF_6$ (54), as a mixture with cis-$[IrH_2(PMe_3)_4]PF_6$ (55) (Scheme 6-4) [25]. Complex 54 was also obtained quantitatively by dissolution of the hydrido(hydroxo) complex, cis-$[IrH(OH)(PMe_3)_4]PF_6$ (38), in MeOH and subsequent removal of the solvent in vacuo. The structure of 54 was confirmed crystallographically. The hydrido(methoxo) complex 54 is air-stable and thermally stable and does not change upon heating at 70°C. However, it does undergo reductive elimination of MeOH upon oxidation with $C(NO_2)_4$.

Neutral hydrido(methoxo) complexes of iridium, mer-cis-$[IrH(OMe)Cl(PEt_3)_3]$ (56) and $[IrH(OMe)Cl(PMe_3)_3]$ (58) were obtained by oxidative addition of MeOH to $[IrCl(PEt_3)_3]$ (39b) and $[IrCl(PMe_3)_3]$ (39a) at –30°C, respectively (Scheme 6-8) [27]. In contrast to the cationic hydrido(methoxo) complex 54, the neutral complexes 56 and 58 undergo MeOH reductive elimination in benzene at room temperature, generating the starting complexes 39b and 39a. This result contradicts the expectations that reductive elimination is facilitated by lower electron density or by higher steric hindrance (the cationic complexes are less electron-rich than the neutral complexes and the PMe_3 ligand is bulkier than the chloro ligand). For the hydrido(methoxo) complex, O–H reductive elimination is preferred over β-H elimination. Irreversible β-H elimination of the equilibrium mixture of [IrCl(PEt₃)₃] (39b) and *mer-cis*-[IrH(OMe)Cl(PEt₃)₃] (56) in benzene to give the dihydride, *mer-cis*-[IrH₂Cl(PEt₃)₃] (57), occurred slowly at room temperature. From a kinetic study, the strength of the bond between the methoxo group and iridium was estimated, D(Ir–OMe) = 61–68 kcal mol⁻¹, which is higher than those between late transition metals and the methyl group, e.g., D(Ir–Me) = 46 kcal mol⁻¹, D(Pt–Me) = 38 kcal mol⁻¹ [41]. Relative M–X bond strengths (X⁻ = H⁻, Me⁻, MeO⁻, etc.) were experimentally determined for [Cp*RuX(PMe₃)₂] and for [PtMe(X){Ph₂P(CH₂)₂PPh₂}], and the general order of LₙM–O > LₙM–C(sp³) was reported [42]. From a best fit to an ETC model, the values of the LₙM–OMe bond strength were found to be 33.9 kcal mol⁻¹ (for M = Ru) and 24.7 kcal mol⁻¹ (for M = Pt) [43].

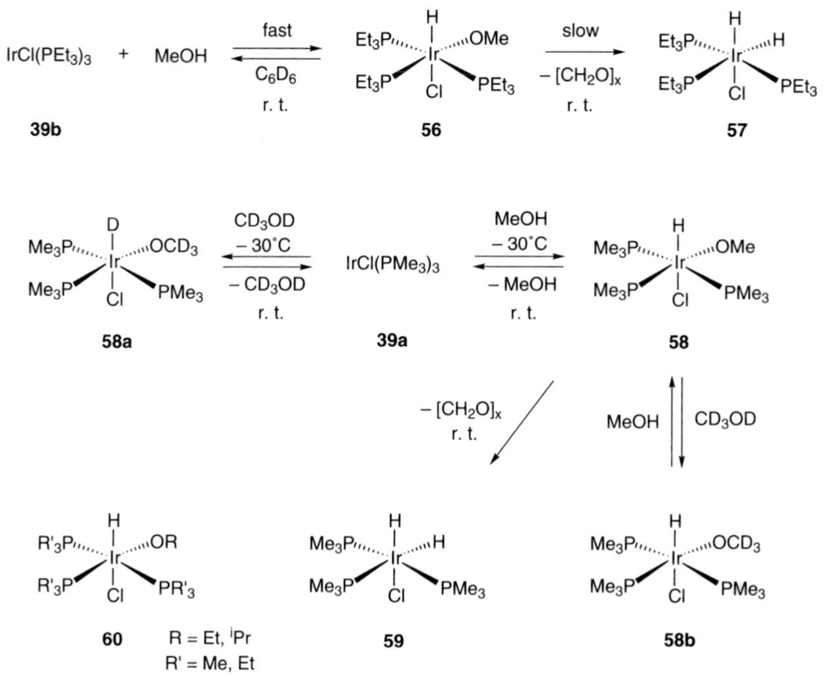

Schemes 6-8 Oxidative addition of alcohols to neutral iridium phosphine complexes

The detailed decomposition (β-H elimination) mechanism of the hydrido(alkoxo) complexes, *mer-cis*-[Ir(H)(OR)Cl(PR′₃)₃] (R = Me, Et, ⁱPr; R′ = Me, Et; H *trans* to Cl) (56, 58, 60), forming the dihydrides *mer-cis*-[Ir(H)₂Cl(PR′₃)₃] (57, 59) along with the corresponding aldehyde or ketone was examined (Scheme 6-8). The hydrido(ethoxo) as well as the hydrido(isopropoxo) complexes 60 could also be prepared by oxidative addition of ethanol and isopropanol to the phosphine complexes 39 [44]. In the initial stage of the β-H elimination, a pre-equilibrium is assumed in which an unsaturated pentacoordinated product is generated by an alcohol-assisted dissociation of the chloride. From this intermediate the transition state is reached, and the rate-determining step is an irreversible scission of the β-C–H bond. This process has a low activation energy ($\Delta H^{\ddagger}_{obs}$ =24.1 + 1.8 kcal mol^{-1}; $\Delta S^{\ddagger}_{obs}$ = 0.6 + 5.9 e.u.). The reaction rate is of first order in the iridium complex and of 1.33 order in the alcohol, which serves as a catalyst. The rate depends on the nature of the phosphine (PEt$_3$ > PMe$_3$), on the alkyl substituent of the alkoxide (Me > Et >> ^{i}Pr), and on the solvent (benzene > N-methylpyrrolidone), but is not affected by excess phosphine.

Bergman reported dramatic solvent effects of liquid xenon on the reactivity of [Cp*Ir(H)$_2$(PMe$_3$)] (61) for C–H vs O–H activation (Scheme 6-9) [45]. Irradiation of 61 in liquid xenon containing MeOH or EtOH gave the oxidative addition products of O–H bonds, [Cp*IrH(OMe)(PMe$_3$)] (62a) or [Cp*IrH(OEt)(PMe$_3$)] (62b) respectively, whereas the reactions in the alcohol solvents produced [Cp*Ir(CO)(PMe$_3$)] (63) or [Cp*IrH(CH=CH$_2$)(PMe$_3$)] (64) resulting from initial oxidative addition of the iridium center into a C–H bond in the alcohol. Irradiation of the dihydride 61 in xenon with ^{i}PrOH or ^{t}BuOH as a substrate, however, gave the C–H insertion product, [Cp*IrH(CH$_2$CH(Me)OH)(PMe$_3$)] (65) or [Cp*IrH(CH$_2$(Me)$_2$OH)(PMe$_3$)] (66) respectively. The abrupt change in reactivity between EtOH and ^{i}PrOH is remarkable.

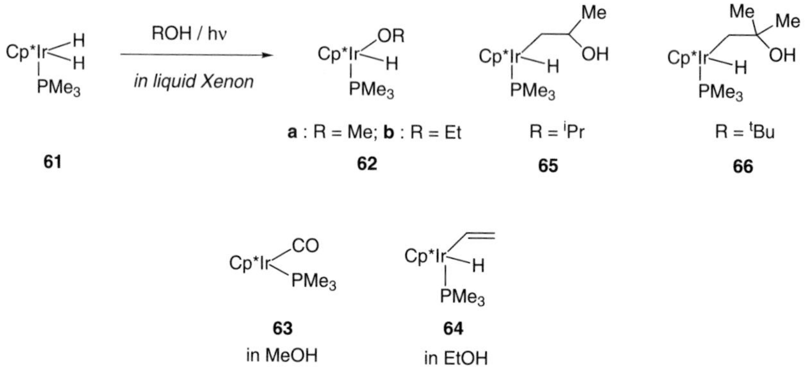

Scheme 6-9 Reaction of Cp*IrH$_2$(PMe$_3$) with alcohols

Merola examined the reactions of [Ir(cod)(PMe$_3$)$_3$]Cl (67) with several alcohols (Scheme 6-10) [46]. Primary and secondary alcohols gave only a complex mixture containing hydrido- and hydrido(carbonyl)iridium(III) compounds. The tertiary alcohol ^{t}BuOH, which has no β-hydrogen, reacted with 67 at 70–80°C to form isobutylene and mer-[IrH(OH)(Cl)(PMe$_3$)$_3$] (40a), which could also be prepared by the reaction between 67 and water. Phenols reacted with [Ir(cod)(PMe$_3$)$_3$]Cl (67) to give cleanly O–H addition products, mer-[IrH(OAr)(Cl)(PMe$_3$)$_3$] (68). One such compound (Ar = 3,5-dimethylphenyl) was crystallographically characterized.

Very air-sensitive iridium diphosphine complexes carrying a peraryldiphosphine ligand, [IrCl(diphosphine)]$_2$ (42a, 42b)(a: diphosphine = BPBP; b: diphosphine = BINAP [47, 48]) can also activate MeOH in addition to H$_2$O at room temperature very easily. Reaction of 42 with excess MeOH in toluene at room temperature gave air-stable and thermally stable colorless hydrido(methoxo) complexes, [{IrH(diphosphine)}$_2$(μ-OMe)$_2$(μ-Cl)]Cl (69) quantitatively (Eq. 6.21) [49]. The structure of 69b,

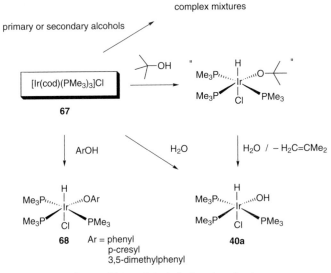

Schemes 6-10 Oxidative addition of alcohol, phenol, and water to cationic iridium phosphine complex **67**

which is a cationic binuclear triply bridged complex (one chloride bridge and two methoxo bridges) was confirmed by X-ray analysis (Fig. 6-1). Complex **69b** contains a solvated MeOH. The presence of strong hydrogen bonding between the solvated MeOH and the chloride counter anion was suggested by ^{1}H NMR and verified by X-ray analysis [MeO⋯Cl = 2.62(2)Å]. Exchange between the methoxo and hydrido ligands and the solvated MeOH was not detected below 40°C. Thermolysis of complex **69** in toluene/MeOH (1:1 v/v) at 80°C in a sealed tube gave mainly the β-H elimina-

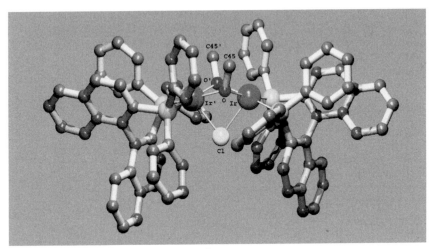

Fig 6-1 Crystal structure of the cationic part of [{IrH((R)-binap)}$_2$(μ-OMe)$_2$(μ-Cl)]Cl•MeOH (**69b**)

tion product, [{IrH(diphosphine)}$_2$(μ-H)$_2$(μ-Cl)]Cl (**70**) (Eq. 6.21) [50]. In contrast to the above-mentioned complexes, capable of activating O–H bonds, complexes **42** carry a peraryldiphosphine, which does not have a strong electron-donating property, as an auxiliary ligand. Nevertheless, complexes **42** can easily activate an O–H bond. Reaction of **42** with EtOH and ^{i}PrOH did not give the corresponding hydrido(alkoxo) complexes, but gave a complex mixture comprising several hydrides, including complexes **70**, and carbonyl complexes. The hydrido(methoxo) complexes **69** can serve as active catalysts for transfer hydrogenation of alkyne using MeOH as a hydrogen donor (see Scheme 6-30).

[Ir(μ-Cl)(diphosphine)]$_2$ + MeOH $\xrightarrow[\text{toluene}]{\text{r.t.}}$

42

[{IrH(diphosphine)}$_2$(μ-OMe)$_2$(μ-Cl)]Cl $\xrightarrow[\substack{\text{MeOH / toluene} \\ (1 / 1)}]{80\,°C}$ [{IrH(diphosphine)}$_2$(μ-H)$_2$(μ-Cl)]Cl

69 **70**

a : diphosphine = BPBP   (6.21)
b : diphosphine = BINAP

The formation of the first osmium hydrido(alkoxo) complex, a yellow air-stable and thermally stable hydrido(methoxo)osmium(II) complex, *trans*-[OsH(OMe)(Cl)(NO) (P^{i}Pr$_3$)$_2$] (**72**), by the oxidative addition of MeOH to a 16-electron complex, *trans*-[OsCl(NO)(P^{i}Pr$_3$)$_2$] (**71**) was briefly reported (Eq. 6.22) [51].

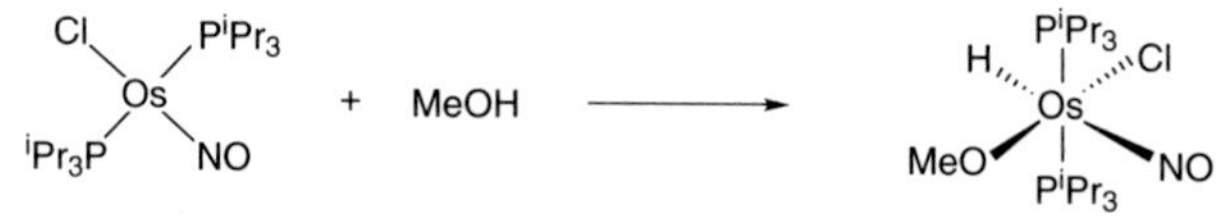

[structure 71: Cl, P^{i}Pr$_3$, Os, ^{i}Pr$_3$P, NO] + MeOH ⟶ [structure 72: H, P^{i}Pr$_3$, Cl, Os, MeO, NO, P^{i}Pr$_3$]   (6.22)

71 **72**

Several hydrido(phenoxo) complexes of late transition metals have been prepared by the oxidative addition of phenol which has no β-hydrogen. The oxidative addition of phenol or pentafluorophenol to the highly coordinatively unsaturated platinum [52] or palladium [53] bis(phosphine)complexes, [M(PCy$_3$)$_2$] (**24b**: M = Pt; **73**: M = Pd) gave the corresponding hydrido(phenoxo) complexes, *trans*-[MH(OAr)(PCy$_3$)$_2$] (**74**) (a: M = Pt, Ar = Ph or C$_6$F$_5$; b: M = Pd, Ar = Ph) (Scheme 6-11). The structure of the palladium complex, *trans*-[PdH(PhO)(PCy$_3$)$_2$]•PhOH, which contains a solvated phenol molecule bonded *via* a hydrogen bridge to the oxygen center of the metal-coordinated phenoxy group, was confirmed by X-ray crystallography. Similar hydrido(phenoxo) complexes of nickel and platinum, *trans*-[MH(OPh)(PBn$_3$)$_2$] (**6c**: M = Ni, **76**: M = Pt) were also prepared by oxidative addition of phenol to platinum(0) and nickel(0) ethylene complexes, [M(C$_2$H$_4$)(PBn$_3$)$_2$] (**75**) (a: M = Ni, b: M = Pt) [9]. The reaction of [Pd(C$_2$H$_4$)(PBn$_3$)$_2$] with phenol, however, yielded only [Pd(PBn$_3$)$_3$]. The reaction of *cis*-[Pt(C$_2$H$_4$)(dmpe)] (**77**) with phenol gave [PtEt(OPh)(dmpe)] (**78**) and not the corresponding *cis*-hydrido(phenoxo) complex. Addition of phenol to [PdMe$_2$(dmpe)] (**79**) yielded [PdMe(OPh)(dmpe)] (**80**). Some of these phenoxo complexes were characterized crystallographically also.

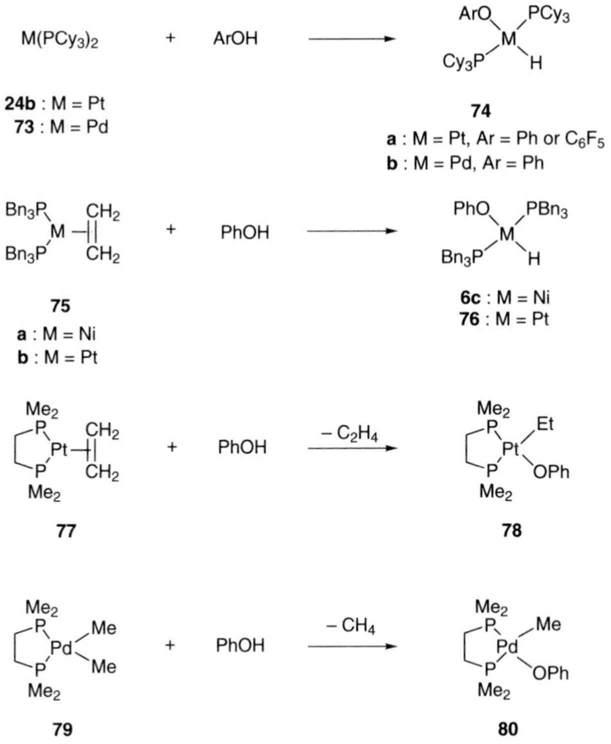

M(PCy₃)₂ + ArOH ⟶ [structure 74: ArO–M(PCy₃)(Cy₃P)–H]

$$M(PCy_3)_2 \ + \ ArOH \ \longrightarrow \ \begin{array}{c} ArO \quad PCy_3 \\ M \\ Cy_3P \quad H \end{array}$$

24b : M = Pt
73 : M = Pd

74

a : M = Pt, Ar = Ph or C₆F₅
b : M = Pd, Ar = Ph

$$\begin{array}{c} Bn_3P \quad CH_2 \\ M-\!\!\shortparallel \\ Bn_3P \quad CH_2 \end{array} \ + \ PhOH \ \longrightarrow \ \begin{array}{c} PhO \quad PBn_3 \\ M \\ Bn_3P \quad H \end{array}$$

75

a : M = Ni
b : M = Pt

6c : M = Ni
76 : M = Pt

$$\begin{array}{c} Me_2 \\ P \quad CH_2 \\ Pt-\!\!\shortparallel \\ P \quad CH_2 \\ Me_2 \end{array} \ + \ PhOH \ \xrightarrow{-\,C_2H_4} \ \begin{array}{c} Me_2 \\ P \quad Et \\ Pt \\ P \quad OPh \\ Me_2 \end{array}$$

77

78

$$\begin{array}{c} Me_2 \\ P \quad Me \\ Pd \\ P \quad Me \\ Me_2 \end{array} \ + \ PhOH \ \xrightarrow{-\,CH_4} \ \begin{array}{c} Me_2 \\ P \quad Me \\ Pd \\ P \quad OPh \\ Me_2 \end{array}$$

79

80

Schemes 6-11 Preparation of hydrido(phenoxo) complexes by oxidative addition of phenols

The reaction of phenols and fluorinated alcohols with dialkyl bis(phosphine) complexes of nickel and palladium has been examined in detail [5]. trans-[PdR₂L₂] (R = Me, Et; L = PMe₃, PEt₃) (**81**) and trans-[NiMe₂(PMe₃)₂] (**82**) reacted with 2 equiv. of fluorinated alcohols and para-substituted phenols R′OH [R′ = CH(CF₃)Ph, Ph, p-CH₃C₆H₄, etc.] to give complexes trans-[PdR(OR′)(HOR′)L₂] (**83**) and trans-[NiMe(OR′)(HOR′) (PMe₃)₂] (**84**) respectively, in which the presence of strong O-H⋯O hydrogen bonding between the alkoxide (or aryloxide) ligand and the alcohol both in the solid state and in solution was confirmed (Scheme 6-12).

6.2.4
Complexes Resulting from Activation of Carboxylic Acids: Hydrido(carboxylato) Complexes

Examples of the isolation of hydrido(carboxylato) complexes by oxidative addition of carboxylic acid are very rare. Wilkinson [55] examined oxidative addition of fluoro carboxylic acids to Vaska's complex, trans-[IrCl(CO)(PPh₃)₂]. They isolated colorless hydrido(carboxylato) complexes of stoichiometry, [IrCl(H)(OCORF)(CO)(PPh₃)₂] (RF = CF₃, C₂F₅), but did not completely succeed in characterizing the products of the oxidative addition. The complex is claimed to be a mixture of four closely similar

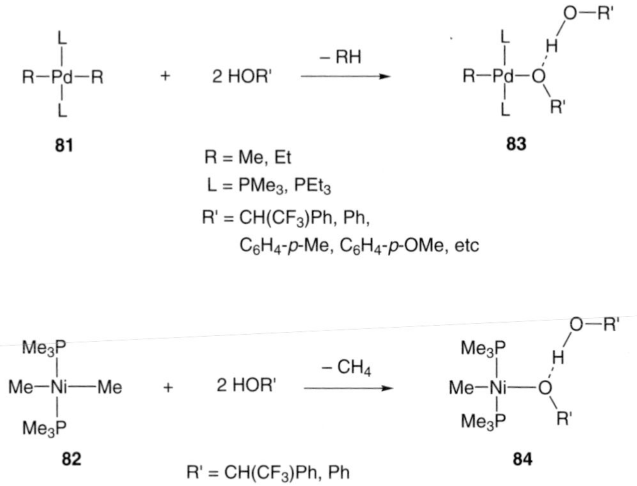

Schemes 6-12 Reactions of dialkyl complexes of Pd and Ni with alcohols

species by NMR. From the isolated colorless complex, carboxylic acid easily dissociates in solution or even in the solid state upon heating. Wilke et al. reported the preparation of hydrido(acetato)nickel complex **86a** by the oxidative addition of acetic acid to the nickel dinitrogen complex **85** (Eq. 6.23) [56]. The preparation of a series of analogous hydrido(carboxylato) complexes of nickel **86** by the same procedure was subsequently also reported by another group, and the properties of the hydrido(carboxylato)nickel complexes were examined (Eq. 6.23) [57]. Kubota et al. [58] reported the preparation of hydrido(carboxylato)iridium complexes, $[IrH(Cl)(RCOO)(PPh_3)_2]$ (**88**) (R = H, Me, Et, ^nPr, Ph, CF_3), by oxidative addition of several carboxylic acids to the iridium dinitrogen complex **87** (Eq. 6.24). In these complexes the car-

$$[Ni(PCy_3)_2]_2N_2 + 2 RCO_2H \xrightarrow[\text{toluene / ether}]{- N_2} 2 NiH(O_2CR)(PCy_3)_2 \qquad (6.23)$$

85   **86**

a : R = Me; **b** : R = H; **c** : R = Ph; **d** : R = CF_3

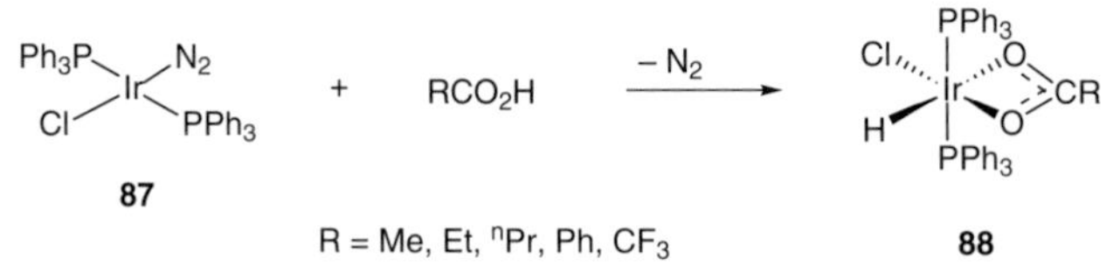

$$(6.24)$$

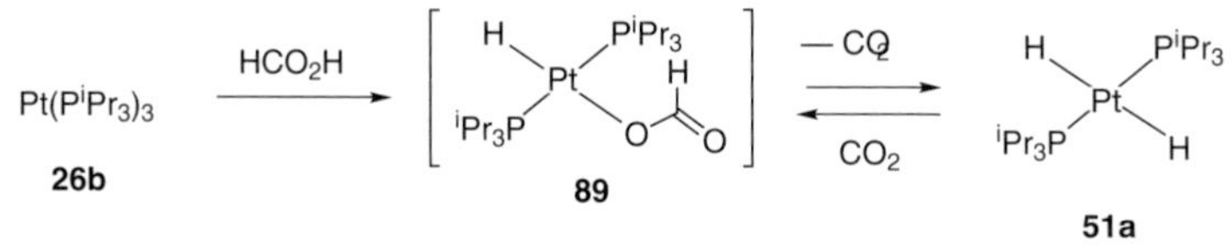

87

R = Me, Et, ^nPr, Ph, CF_3 **88**

$$(6.25)$$

26b   **89**   **51a**

boxylato ligands coordinate as bidentate ligands. Yoshida and Otsuka observed rapid catalytic decomposition of formic acid into CO_2 and H_2 with [Pt(P^{i}Pr$_3$)$_3$] (26b) [23]. They claimed that the decomposition proceeded through oxidative addition of HCO_2H to give trans-[PtH(O$_2$CH)(P^{i}Pr$_3$)$_2$] (89), which is followed by β-H elimination affording trans-[PtH$_2$(P^{i}Pr$_3$)$_2$] (51a) and CO_2 (Eq. 6.25). Although the hydrido(carboxylato) complex 89 could not be isolated from the reaction mixture, the complex 89 was separately prepared by CO_2 insertion into the Pt–H bond of trans-[PtH$_2$(P^{i}Pr$_3$)$_2$] (51a) and used as a catalyst for the decomposition of formic acid.

Merola reported the preparation of hydrido(carboxylato)iridium(III) complexes, mer-[IrCl(OC(O)R)(H)(PMe$_3$)$_3$] (90) (R = Ph, Me), by oxidative addition of acetic acid or benzoic acid to [Ir(cod)(PMe$_3$)$_3$]Cl (67) [46]. The structure of 90 (R = Ph) in which the carboxylato ligand coordinates as an η^1-ligand, was confirmed by X-ray analysis. The reaction of 67 with salicylic acid yielded the product 91, which resulted from activation of the O–H bond of the carboxylato but not of the hydroxo group (Scheme 6-13).

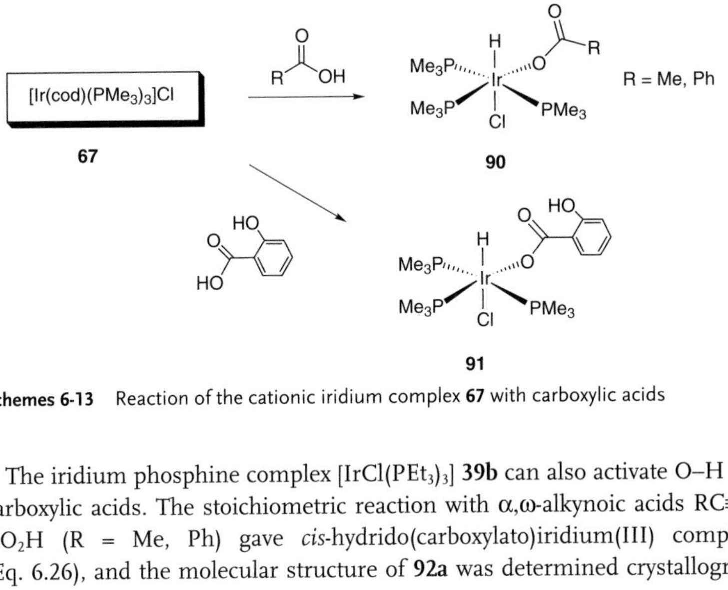

Schemes 6-13 Reaction of the cationic iridium complex 67 with carboxylic acids

The iridium phosphine complex [IrCl(PEt$_3$)$_3$] 39b can also activate O–H bonds of carboxylic acids. The stoichiometric reaction with α,ω-alkynoic acids RC≡C(CH$_2$)$_2$ CO$_2$H (R = Me, Ph) gave cis-hydrido(carboxylato)iridium(III) complexes 92 (Eq. 6.26), and the molecular structure of 92a was determined crystallographically [59].

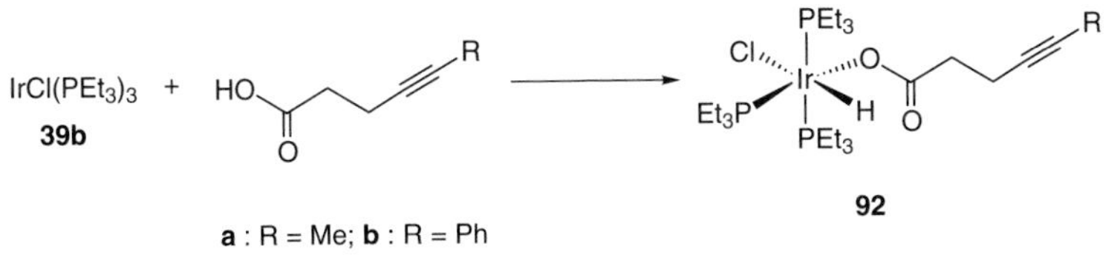

(6.26)

a : R = Me; b : R = Ph

The dinuclear iridium(I) diphosphine complexes 42 can also activate carboxylic acids easily. For example, the reaction of [IrCl(binap)]$_2$ (42b) with an excess of acetic acid or benzoic acid in toluene at room temperature gave the corresponding (hydrido)carboxylato complexes [{IrH(binap)}$_2$(μ-OC(O)R)$_2$(μ-Cl)]Cl (**93**) (**a**: R = Me; **b**: R = Ph) [50]. Complexes **93** were characterized by elemental analyses, IR and NMR spectroscopy, and conductivity measurements. For these compounds, triply bridging structures similar to that of the hydrido(methoxo) complex **69b** were proposed (Eq. 6.27).

$$[IrCl(binap)]_2 \quad + \quad RCO_2H \quad \longrightarrow \quad [\{IrH(binap)\}_2(\mu\text{-}OC(O)R)(\mu\text{-}Cl)]Cl \qquad (6.27)$$

42b   **a** : R = Me; **b** : R = Ph   **93**

6.2.5
Reaction of Hydrido(hydroxo), Hydrido(alkoxo) and Hydrido(carboxylato) Complexes

Reactions of the hydrido(hydroxo) complex **2** with several substrates were examined (Scheme 6-14) [6]. The reactions are fairly complicated and several different types of reactions are observed depending on the substrate. Methyl acrylate and small Lewis bases such as CO, P(OMe)$_3$, ^{t}BuNC coordinate to the five-coordinated complex **2** affording the corresponding six-coordinate complexes. In reactions with the unsaturated bonds in dimethylacetylenedicarboxylate, carbon dioxide, phenylisocyanate indications for the addition across the O–H bond but not across the Os–OH bond were obtained. In reactions with olefins such as methyl vinyl ketone or allyl alcohol, elimination of a water molecule was observed to afford a hydrido metallacyclic compound or a hydrido(ethyl) complex. No OH insertion product was obtained.

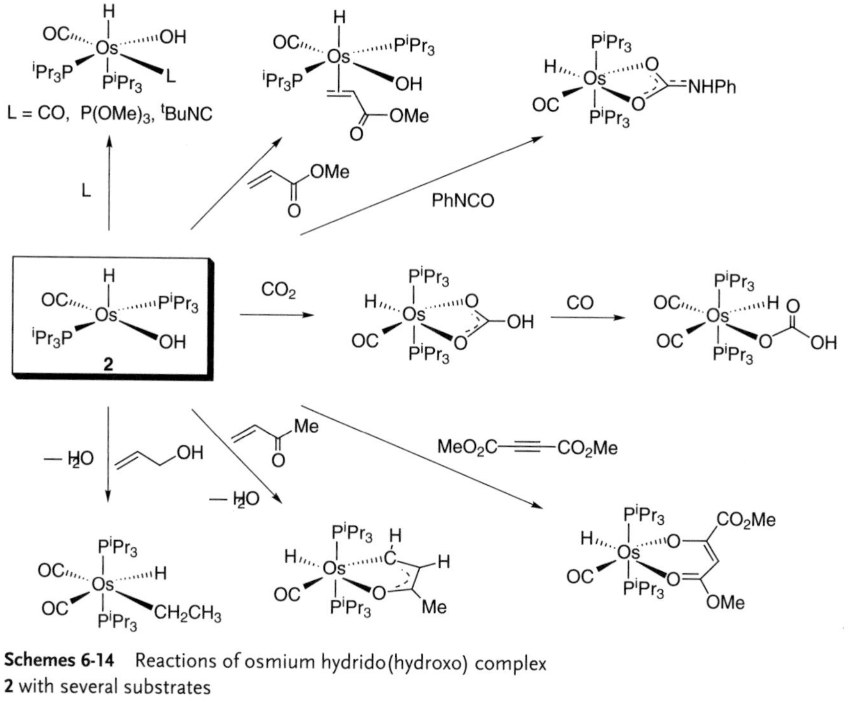

Schemes 6-14   Reactions of osmium hydrido(hydroxo) complex **2** with several substrates

The reactivity of hydrido(ethoxo) complex **4** was examined (Scheme 6-15) [8]. Metatheses similar to those postulated for alcohol exchange (Eq. 6.5) occurred between HCl, LiCl, phenyl acetate or primary amines and yielded complexes **94**. The reaction of **4** with cyclic anhydrides proceeded similarly to give iridium-assisted ring opening products **95**. Heterocumulenes afforded the insertion products **96** into the Ir–O bond.

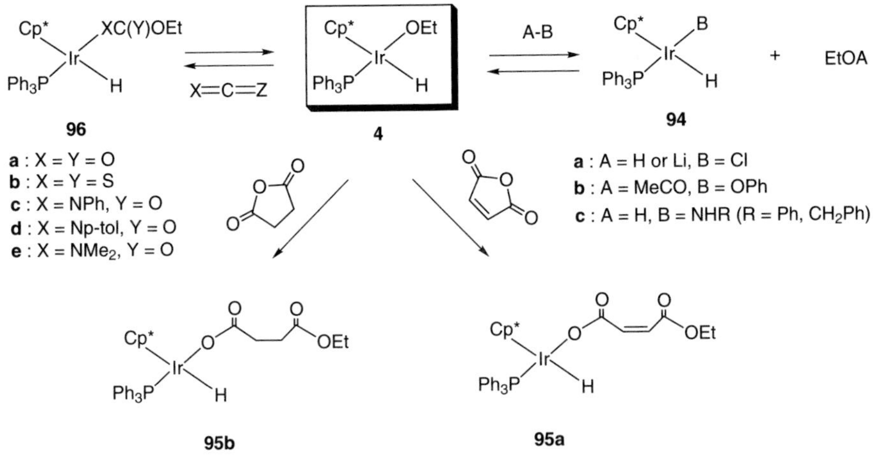
Cp*
XC(Y)OEt
Ir
Ph3P
H
X=C=Z
OEt
A-B
B
+
EtOA
96
4
94
a : X = Y = O
b : X = Y = S
c : X = NPh, Y = O
d : X = Np-tol, Y = O
e : X = NMe2, Y = O
a : A = H or Li, B = Cl
b : A = MeCO, B = OPh
c : A = H, B = NHR (R = Ph, CH2Ph)
95b
95a

Schemes 6-15 Reactivity of hydrido(ethoxo)iridium complex **4**

The Pt–OMe bond of complex **14b** is reactive, giving either the corresponding hydroxo derivative **14a** upon reaction with water or the methoxycarbonyl derivative **97** with CO (Eq. 6.28) [16].

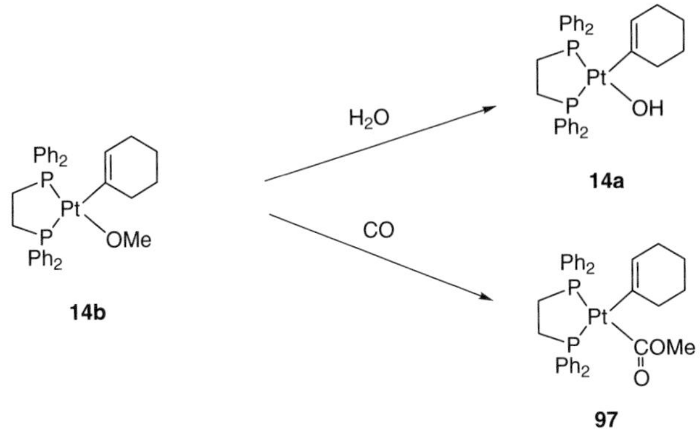
Ph2
P
Pt
OMe
14b
H2O
CO
OH
14a
COMe
O
97

(6.28)

The reactivities of hydrido(phenoxo) complexes of *trans*-[MH(OPh)L_2] (**6**: M = Ni; **7**: M = Pt) (L = phosphine) were examined (Eqs. 6.29, 6.30; Scheme 6-16), and a high nucleophilicity for the metal-bound phenoxide was suggested [9, 10]. Reaction with methyl iodide produced anisole and *trans*-[MH(I)L_2] for both Ni and Pt complexes. Phenyl isocyanate also provided the insertion products into the metal–phenoxo bond, *trans*-[MH{N(Ph)C(O)OPh}L$_2$], but the nickel complexes have a higher reactivity. The reaction of nickel complexes **6** with olefins results in reductive elimination of phenol in contrast to the insertion chemistry for the platinum analog **7**.

$$\text{\textit{trans}-NiH(OPh)(L)}_2 \ + \ \text{MeI} \ \xrightarrow[\text{r.t.}]{\text{C}_6\text{D}_6} \ \text{\textit{trans}-NiH(I)(L)}_2 \ + \ \text{MeOPh} \qquad (6.29)$$

$$\mathbf{6}$$

$$L = \text{P}^i\text{Pr}_3, \text{PCy}_3, \text{PBn}_3$$

$$\text{\textit{trans}-NiH(OPh)(L)}_2 \ + \ \text{PhNCO} \ \underset{\text{r.t.}}{\overset{\text{C}_6\text{D}_6}{\rightleftharpoons}} \ \text{\textit{trans}-NiH(PhNC(O)OPh)(L)}_2 \qquad (6.30)$$

$$\mathbf{6}$$

$$L = \text{P}^i\text{Pr}_3, \text{PCy}_3, \text{PBn}_3$$

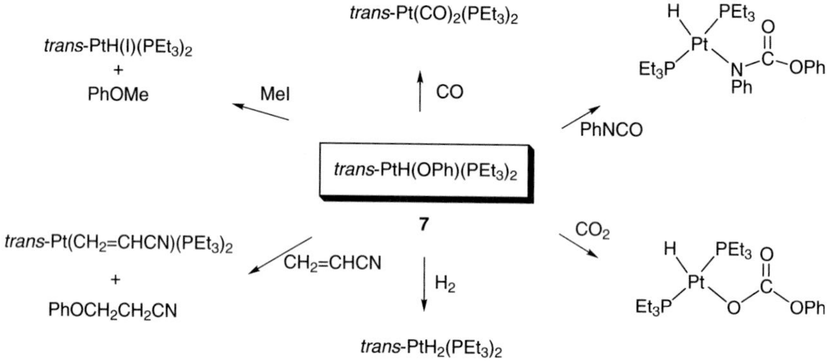

Schemes 6-16 Reactivity of *trans*-PtH(OPh)(PEt$_3$)$_2$ **7**

The reactivities of several hydrido(carboxylato) complexes of iridium have been studied [58]. Complexes **88** reacted with carbon monoxide to afford carbonyl complexes, [IrCl(H)(OCOR)(CO)(PPh$_3$)$_2$] (**98**), in which the carboxylato ligands changed to monodentate ligands and exist as a mixture of isomers (Eq. 6.31).

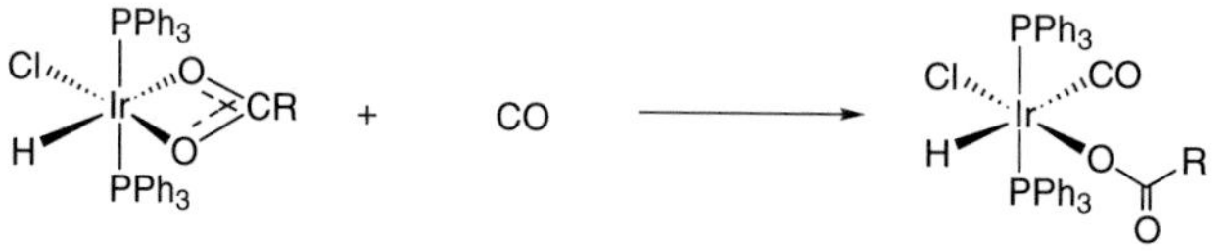

$$\mathbf{88} \qquad R = \text{Me, Et, }^n\text{Pr, Ph, CF}_3 \qquad \mathbf{98} \qquad (6.31)$$

The reactions of a neutral **10** as well as a cationic dihydrido(acetato)osmium complex **12** with acetylenic compounds were examined (Scheme 6-17) [11–13]. A vinylidene **99**, an osmacyclopropene **100**, or a carbyne complex **101** were obtained, depending on the starting hydrido(acetato) complexes or the kind of acetylene used. In any case, the reaction proceeded by insertion of a C≡C triple bond into one of the two Os–H bonds, but the acetato ligands do not take part in the reaction and act as stabilizing ligands.

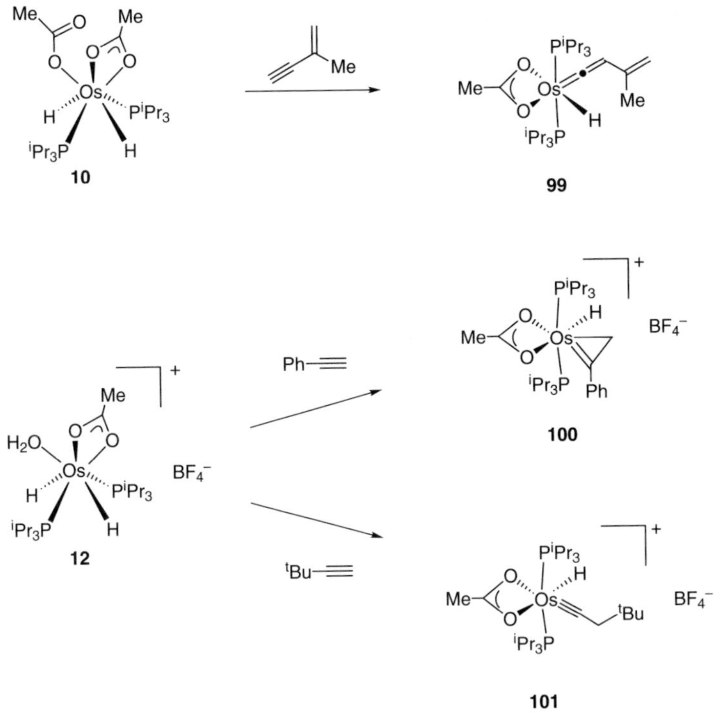

Schemes 6-17 Reaction of hydrido(carboxylato)osmium(IV) complexes with alkynes

6.3
Catalytic Reactions Involving Activation of O–H Bonds

In this section we deal with reactions in which in one step, formally an O–H bond activation, is involved. Although the precise reaction mechanisms have not been elucidated, some of these reactions are considered to proceed by nucleophilic attack of water, an alcohol, etc. to a substrate activated by a transition metal. We choose to emphasize examples coming from our own research activities in this field.

6.3.1
Water Gas Shift Reactions

Yoshida and Otsuka found that platinum(0) complexes [PtL$_3$] (**26**) (**a**: L = PEt$_3$; **b**: L = P^{i}Pr$_3$) and rhodium hydrido complexes such as [RhHL$_3$] (L = P^{i}Pr$_3$ **33**, PEt$_3$), [Rh$_2$H$_2$(μ-N$_2$)(PCy$_3$)$_4$] (**34**), *trans*-[RhH(N$_2$)(PPh^{t}Bu$_2$)$_2$], and [RhH(P^{t}Bu$_3$)$_2$], all of which carry electron-donating alkylphosphine ligands, can catalyze the water gas shift reaction under fairly mild conditions (100–150°C; CO 20 kg/cm^{-2}) (Eq. 6.32) [23, 60]. Among these complexes, [RhH(P^{i}Pr$_3$)$_3$] (**33**) was the most active catalyst precursor. Several complexes were isolated from or detected in the reaction mixture (see above) and, based on these, reaction mechanisms were proposed in which oxidative addition of water plays an important role. Taqui Khan et al. reported that $K[Ru^{II}(Hedta)(CO)]$ can catalyze the water gas shift reaction under mild conditions at 20–80°C and 1–35 atm of CO [61]. Based on kinetic studies, they proposed a mechanism involving the oxidative addition of H_2O to the carbonyl complex to form hydrido(hydroxo) species, *cis*- and *trans*-$[Ru^{IV}(Hedta)(CO)H(OH)]^-$ as the rate determining step. The formation of a Ru^{IV} hydrido(hydroxo) complex was suggested by ^1H NMR and UV spectroscopy. The maximum turnover number of 350 mol of CO_2/H_2 per mol of catalyst per hour was achieved when the reaction was carried out at 50°C and 15 atm of CO.

$$CO \quad + \quad H_2O \quad \overset{\text{Catalyst}}{\rightleftharpoons} \quad CO_2 \quad + \quad H_2 \qquad (6.32)$$

Catalyst : PtL_3 (**26**) (**a** : L = PEt_3; **b** : L = P^iPr_3)

$RhH(L)_3$ (L = P^iPr_3 (**33**), PEt_3), $Rh_2H_2(\mu\text{-}N_2)(PCy_3)_4$ (**34**),

trans-$RhH(N_2)(PPh^tBu_2)_2$, $RhH(P^tBu_3)_2$

$K[Ru(Hedta)(CO)]$

6.3.2
Wacker-type Reactions

Wacker olefin oxidation, which is depicted in its simplest form in Eq. (6.33), contains palladium(II)-catalyzed hydration of olefin in its important step (Eq. 6.34) and is discussed extensively [62]. In this review article we introduce two asymmetric Wacker type reactions.

$$H_2C=CH_2 \quad + \quad 1/2\ O_2 \quad \xrightarrow{\overset{PdCl_2}{\underset{}{CuCl_2}}} \quad CH_3CHO \qquad (6.33)$$

$$Pd^{(2+)}(H_2C=CH_2) \quad + \quad H_2O \quad \longrightarrow \quad Pd^{(2+)}(CH_2CH_2OH) \quad + \quad H^+ \qquad (6.34)$$

Hosokawa and Murahashi reported the intramolecular cyclization of *o*-allylphenols catalyzed by a chiral π-allylpalladium complex in the presence of cupric acetate under an oxygen atmosphere, although the optical yields were rather low (Eq. 6.35) [63, 64].

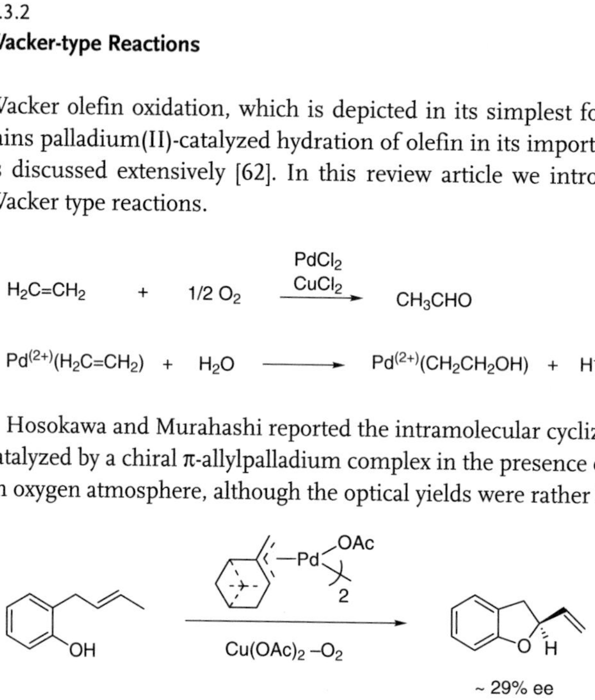

$$(6.35)$$

~ 29% ee

In 1997, Uozumi and Hayashi found high enantioselective Wacker-type cyclization of *o*-allylphenols or *o*-homoallylphenols by using Pd(II) catalysts coordinated with chiral bis(oxazoline) ligands based on the 1,1′-binaphthyl backbone (Eq. 6.36)

[65]. Both chemical yield and enantioselectivity are strongly affected by the anionic part of the catalyst, and the trifluoroacetate anion is considered to play a key role in the catalysis. Excess benzoquinone is required for high yield of the product, and use of methanol as solvent is essential for the successful cyclization.

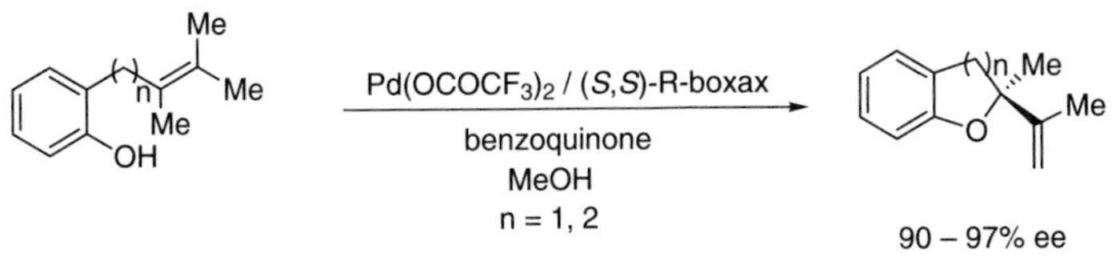

$$ (6.36) $$

90 – 97% ee

(S,S)-R-boxax : R = ^{i}Pr, CH$_2$Ph

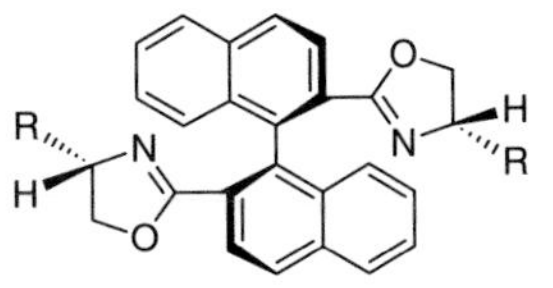

6.3.3
Hydration, Alcoholation and the Related Reactions of Unsaturated Compounds

Catalytic hydration and alcoholation of unsaturated compounds such as alkenes or alkynes would be a high value-adding step in the synthesis of compounds of complicated structure as well as in the large-scale production of industrially useful simple compounds. The activation of the O–H bond of water, alcohol, or carboxylic acid by transition metals is relevant to a variety of such catalytic processes.

6.3.3.1 Hydration and Methanolysis of Nitrile
Several methods of hydration of nitriles with strong acid or base to produce the corresponding carboxamides have been proposed (Eq. 6.37), but generally the reactions are slow and the selectivity of carboxamide is not good enough [66]. Therefore, hydration of nitriles using transition metal complexes under neutral conditions has been intensively studied. Several hydroxo transition metal complexes and their precursors have been proposed as homogeneous catalyst systems showing high turnovers. Representative examples of effective catalysts are [Pt(C$_6$H$_8$)(L$_2$)] (L$_2$ = 2PPh$_3$, diphos (13); C$_6$H$_8$ = cyclohexyne) [67], trans-[M(OH)(CO)(PPh$_3$)$_2$] (M = Rh, Ir) [67], [PdCl(OH)(bipy)(H$_2$O)] [68–70], [PtL$_2$] (24) (b: L = PCy$_3$; c: L = PPh^{t}Bu$_2$) [22], [PtL$_3$] (26) (a: L = PEt$_3$; b: L = P^{i}Pr$_3$) [22], [PtR(OH)L$_2$] (R = Ph, L = P^{t}Bu$_2$Me, P^{t}BuMe$_2$, PEt$_3$, PMe$_2$Ph, PMePh$_2$; R = Me, L = P^{t}Bu$_2$Me, P^{i}Pr$_3$) [71], trans-[PtH(Cl)(PR$_3$)$_2$] (R = Me, Et)/1 equiv. NaOH [72], etc.

$$ \text{RCN} \quad + \quad \text{H}_2\text{O} \quad \xrightarrow{\text{Catalyst}} \quad \text{RCONH}_2 \qquad (6.37) $$

Among these catalysts, systems obtained by the reaction between trans-[PtH(Cl)(PR$_3$)$_2$] (R = Me and Et) and 1 equiv. of NaOH are the most active [72]. For example, acetonitrile was hydrated at 80°C to acetamide with a turnover frequency [TOF = mol substrate/mol catalyst · h] of 178 for the R = Me derivative. These catalysts also exhibit appreciable reactivity even at 25°C and remain active for days giving turnover numbers (TON) of about 6000. The catalytic intermediates, such as *trans*-[PtH(OH)(PEt₃)₂], [PtH(H₂O)(PEt₃)₂]⁺, [PtH(NCMe)(PEt₃)₂]⁺, and [PtH(NHC(O)Me)(PEt₃)₂] have been intercepted and spectroscopically characterized, and a reaction mechanism in which proton transfer from solvating water to coordinated *N*-carboxamide is the rate-determining step is proposed (Scheme 6-18). A mechanistic study for the hydration of acrylonitrile was also performed.

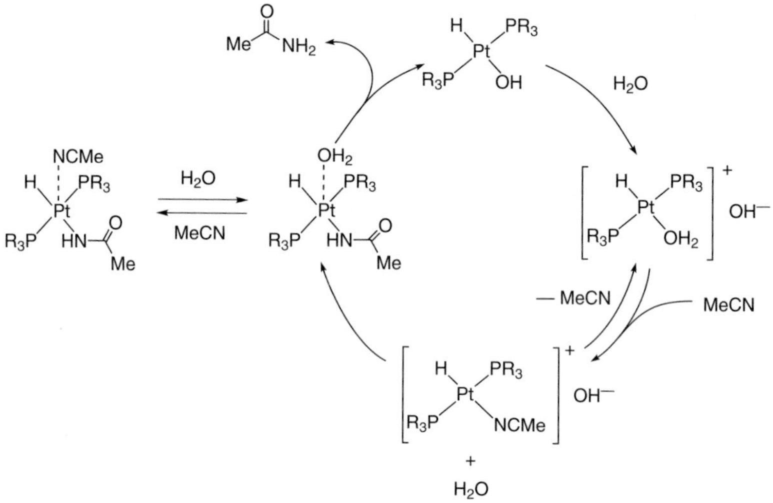

Schemes 6-18 A plausible mechanism of hydration of acetonitrile catalyzed by unhindered hydridobis(phosphine)platinum(II) complexes

Some Pt complexes bearing electron-donating phosphines can also catalyze the hydration of the C=C double bond of acrylonitrile or crotonitrile and yielded β-hydroxypropionitrile or β-hydroxybutyronitrile respectively besides the corresponding amide (Eq. 6.38) [22, 75]. Among the platinum phosphine complexes examined, [Pt(PEt₃)₃] (**26a**), carrying less bulky ligands, was the most effective for the hydration of the olefinic bond. The present catalyst system was ineffective for hydration of other olefins, however.

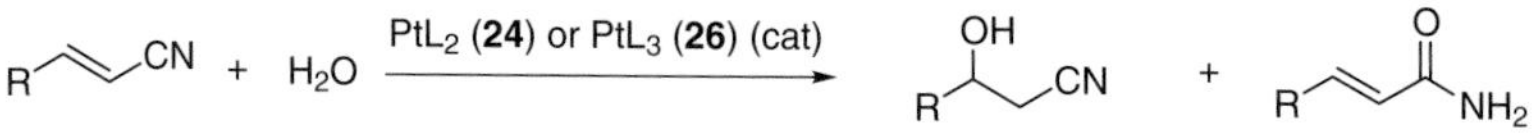

$$R = H, Me; L = PEt_3, P^iPr_3, PCy_3, PPh^tBu_2 \tag{6.38}$$

Several cationic palladium(II) aqua complexes, [Pd(H₂O)₄]⁺, *cis*-[PdL(H₂O)₂]⁺ (L = en, methionine methyl ester, 1,5-dithiacycloocta-3-ol), and [Pd(dien)(H₂O)]⁺, serve as the active catalyst for the selective hydration of various nitriles to the corresponding carboxamides, e.g., CHCl₂CN was hydrated to CHCl₂C(O)NH₂ in the presence of

$[Pd(H_2O)_4]^+$ at 40°C [73]. A kinetic study indicated that internal attack on the Pd-co-ordinated nitrile ligand by the aqua (not hydroxide) ligand and external attack on the nitrile ligand by solvent water occur at a similar rate.

Murahashi et al. also reported on the catalytic hydration of nitriles promoted by $[RuH_2(PPh_3)_4]$ [74]. Upon treatment with only 1–2 equiv. of water under neutral conditions at 120°C, hydration of various aliphatic and aromatic nitriles proceeds smoothly to give the corresponding amides in good yield. It is unique that the hydration proceeds with one equivalent of water, and thus other functional groups are tolerated in the hydrolysis. The authors proposed a reaction mechanism in which nucleophilic attack of water on the nitrile coordinated to the ruthenium center is the key step (Scheme 6-19). The hydration can be extended to the catalytic transformation of γ-cyano ketones to ene-lactams (Eq. 6.39). Cyanoketones bearing ester groups were converted into the corresponding ene-lactams without hydrolysis of the ester groups (Eq. 6.40), and thus the present catalytic hydration of nitriles was applied to the synthesis of several natural products.

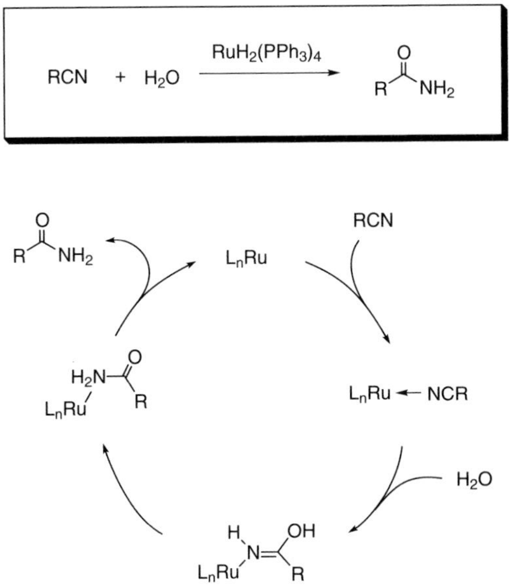

Schemes 6-19   Catalytic cycle of hydration of nitriles catalyzed by $RuH_2(PPh_3)_4$

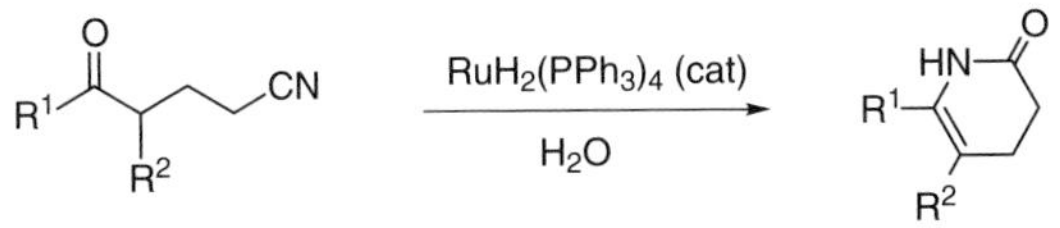

(6.39)

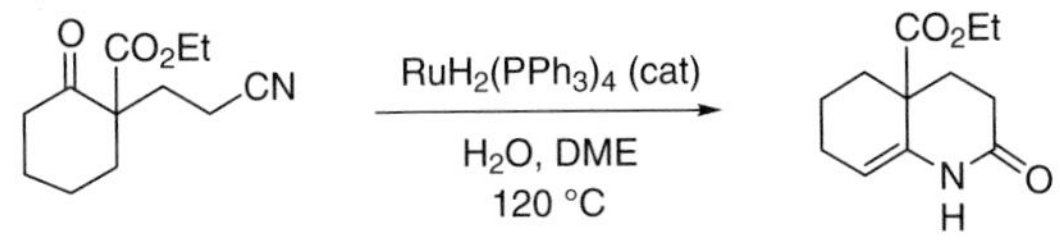

(6.40)

We also found that iridium hydrido(hydroxo) complexes like [{IrH(diphosphine)}$_2$(μ-OH)$_2$(μ-Cl)]Cl (**43**) and the precursor diphosphine complexes **42** can also catalyze the hydration of nitriles. In the presence of catalytic amounts of these complexes, heating acetonitrile and benzonitrile with excess water at 120°C gave the corresponding amides [47, 50].

Very recently Chin et al. reported Ir complex-catalyzed hydration as well as the first example of methanolysis of nitriles [75]. Hydration of RCN (R = Me, MeCH=CH, Ph) is catalyzed by [Cp*Ir(η^3-CH$_2$CHCHPh)(NCR)]OTf (R = Me, MeCH=CH, Ph) in the presence of Na$_2$CO$_3$ (Eq. 6.41). The methanolysis of benzonitrile also proceeded to give the imino ether in the presence of [Cp*Ir(η^3-CH$_2$CHCHPh)(NCPh)]OTf and Na$_2$CO$_3$ (Eq. 6.42). The methanolysis of benzonitrile proceeded much more rapidly than the hydration. Based on the isolated or detected intermediate complexes, plausible reaction mechanisms are suggested.

$$\text{RCN} \quad + \quad \text{H}_2\text{O} \xrightarrow[\text{Na}_2\text{CO}_3,\ 70\ ^\circ\text{C}]{\text{(cat)}} \quad \underset{\text{R}}{\overset{\text{O}}{\|}}\text{NH}_2 \qquad (6.41)$$

R = Me, —HC=CHMe, Ph

$$\text{PhCN} \quad + \quad \text{MeOH} \xrightarrow[\text{Na}_2\text{CO}_3,\ 70\ ^\circ\text{C}]{\text{(cat)}} \quad \underset{\text{Ph}}{\overset{\text{OMe}}{\|}}\text{NH} \qquad (6.42)$$

6.3.3.2 Addition of O–H Bonds across Alkenes and Related Reactions

From the beginning of the 1970s until the mid 1980s, several examples of the telomerization of dienes with water [76, 77] or methanol [78, 79] to isomeric mixtures of dienols or dienol ethers catalyzed by palladium-phosphine complexes in the presence of carbon dioxide have been reported. Neither the yield nor the selectivity were very high. However, when allene was employed as a 'diene', 3-methyl-2-methylene-3-buten-1-ol was obtained with fairly good selectivity (up to 98%) (Eq. 6.43) [78].

$$2 \ \ =\!\bullet\!= \ \ + \ \text{H}_2\text{O} \xrightarrow[\text{THF}]{\overset{\text{CO}_2}{\text{Pd (dba)}_2-\text{PPh}_3}} \qquad \text{OH} \qquad (6.43)$$

The Co$_2$(CO)$_8$/pyridine system can catalyze carbomethoxylation of butadiene to methyl 3-pentenoate (Eq. 6.44) [80]. The reaction mechanism of the cobalt-catalyzed carbalkoxylation of olefins was investigated and the formation of a methoxycarbonylcobalt species, 'MeOC(O)Co' from a cobalt carbonyl complex with methanol as an intermediate is claimed [81, 82].

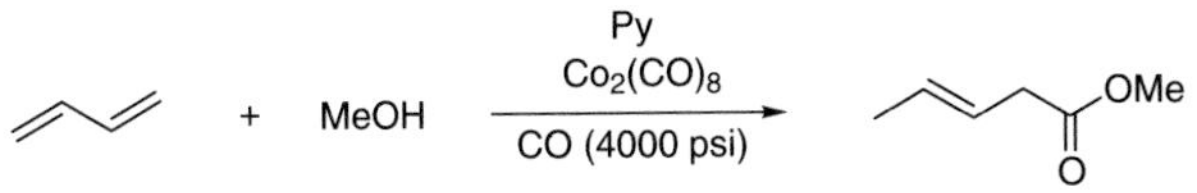

$$\text{(6.44)}$$

Efficient anti-Markonikov addition of water to terminal olefins producing primary alcohols would be one of the most desirable catalytic processes (Eq. 6.45). As one example of such a reaction, Jensen and Trogler reported the anti-Markonikov hydration of terminal olefins catalyzed by a platinum(II) trimethylphosphine complex producing primary alcohols [83]. The report, however, was claimed to be of doubtful reproducibility [84].

$$\text{(6.45)}$$

6.3.3.3 Addition of O–H Bonds across Alkynes

Hydration of unactivated alkynes is an important method for functionalizing this plentiful hydrocarbon source. Therefore, a variety of metal ions have been proposed as catalysts for this reaction, and almost all of the reported additions of water to terminal alkynes follow the Markonikov rule. The hydration of 1-alkynes with Hg(II) salts in sulfuric acid [85], $RuCl_3$/aq.HCl [86, 87], $K[Ru^{III}(edta-H)Cl]\cdot 2H_2O$ [88], $RhCl_3\cdot 3H_2O$/aq. HCl [89], $RhCl_3/NR_4^+$ [90], Zeise-type Pt(II) complexes [91–93], and $NaAuCl_4$ [94] produced exclusively methyl ketones (Eq. 6.46).

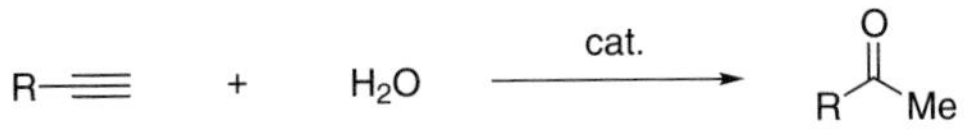

$$\text{(6.46)}$$

R = H, alkyl, Ph

cat. : $RuCl_3$ / HCl, $RhCl_3$ / HCl, $[Pt(C_2H_4)Cl_2]_2$
$K[Ru(III)(edta-H)Cl]_2\cdot H_2O$
$[MeN(C_8H_{17})_3]^+[RhCl_4(H_2O)_2]^-$

Several mechanistic studies have been performed, and similar mechanisms involving formation of an acetylenic π-complex coordinated by H_2O as an intermediate have been proposed (Scheme 6-20) [86, 88–90].

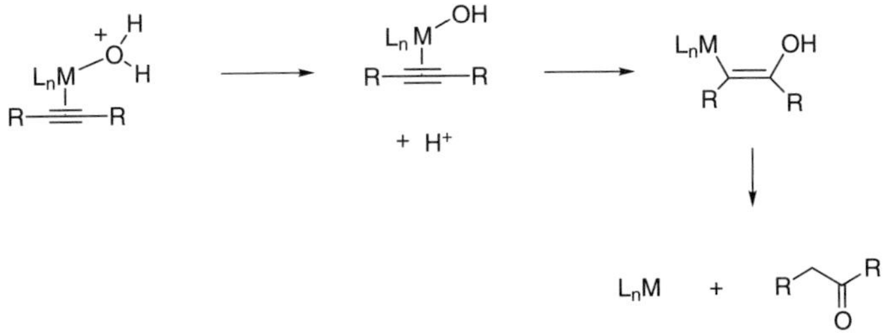

Schemes 6-20 Mechanism for transition metal-catalyzed hydration of acetylene

In 1998, Wakatsuki et al. reported the first anti-Markonikov hydration of 1-alkynes to aldehydes by an Ru(II)/phosphine catalyst. Heating 1-alkynes in the presence of a catalytic amount of [RuCl$_2$(C$_6$H$_6$)(phosphine)] {phosphine = PPh$_2$(C$_6$F$_5$) or P(3-C$_6$H$_4$SO$_3$Na)$_3$} in 2-propanol at 60–100°C leads to predominantly anti-Markovnikov addition of water and yields aldehydes with only a small amount of methyl ketones (Eq. 6.47) [95]. They proposed the attack of water on an intermediate ruthenium vinylidene complex. The C–C bond cleavage or decarbonylation is expected to occur as a side reaction together with the main reaction leading to aldehyde formation. Indeed, olefins with one carbon atom less were always detected in the reaction mixtures (Scheme 6-21).

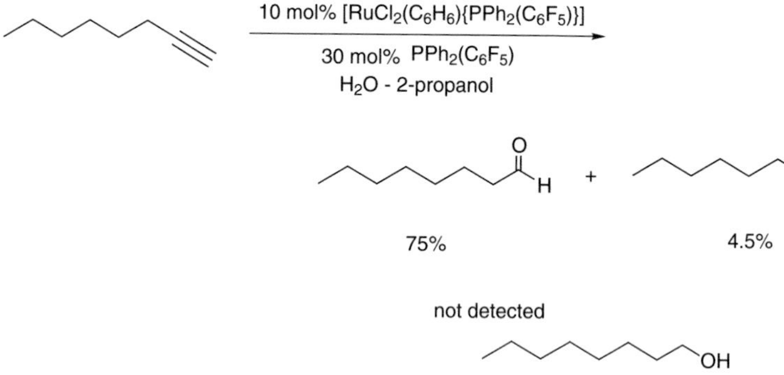

(6.47)

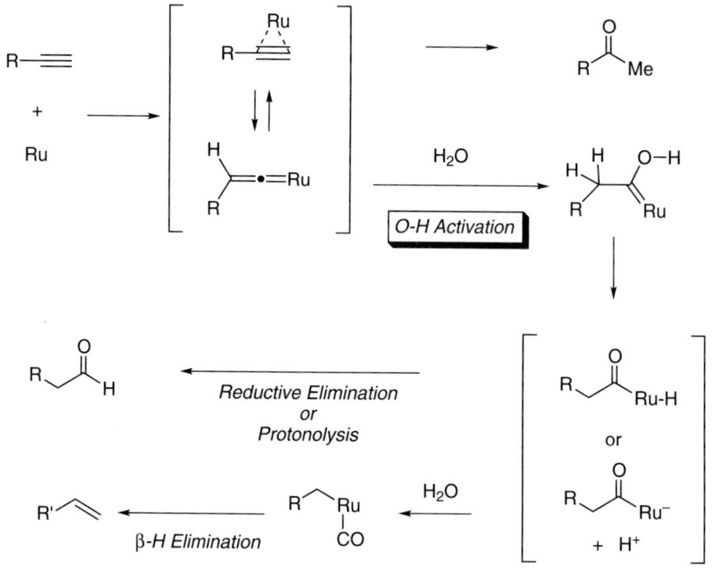

Schemes 6-21 Proposed reaction path for Ru-catalyzed hydration of terminal alkynes

Several Pt(II) complexes are also effective for the hydration of internal alkynes (Eq. 6.48). Internal alkynes were converted to the corresponding internal ketones in good yields by refluxing in THF with a slight excess of water in the presence of a catalytic amount of Zeise's dimer, $[PtCl_2(C_2H_4)]_2$ [91–93]. Electron-donating substituents in the alkyne give better results in this reaction than electron-withdrawing substituents. Unsymmetrically substituted alkynes are hydrated to yield both ketones. The formation of the regioisomer in which the carbonyl moiety is in the α-position to the bulkier substituent is slightly perferred. A mechanism involving nucleophilic attack of water on the Pt(II)-coordinated alkyne is postulated. When substituents have heteroatoms such as oxygen or sulfur, a chelation control process seems to govern the regiochemistry, and in some cases fairly high regioselectivity was achieved. The best results are obtained with substrates having the heteroatom in the β- or γ-position to the alkyne functionality (Eq. 6.49). An alkyne having electron-withdrawing substituents, e.g., dimethyl acetylenedicarboxylate, however, was not a good substrate. This reaction is specific for water, and not effective for the addition of alcohol, e.g., MeOH.

$$R-\!\!\!\equiv\!\!\!-R' + H_2O \xrightarrow[\text{THF}]{[PtCl_2(C_2H_4)]_2} \quad \underset{R}{\overset{O}{\|}}\!\!\!\diagup\!R' + R\diagup\!\!\underset{O}{\overset{R'}{\|}} \qquad (6.48)$$

R, R' = alkyl, Ph

$$\text{(alkyne-OMe)} + H_2O \xrightarrow[\text{H}_2\text{O/THF}]{[PtCl_2(C_2H_4)]_2} \text{(ketone-OMe)} + \text{(ketone-OMe)} \qquad (6.49)$$

94%   6%

Gold(III) complexes can catalyze hydration of unactivated alkynes. Treatment of 1-alkynes with 2 mol% of NaAuCl$_4$ in refluxing aqueous methanol afforded the corresponding methyl ketones. Internal acetylenes were also hydrated to mixtures of ketones (Eq. 6.50) [94]. Although no regioselective hydration of simple internal alkynes was observed, a highly regioselective hydration was achieved with methyl propargyl ethers 102, probably because of the interaction of the ether oxygen center with the Au(III) center (Eq. 6.51); internal methyl propargyl ethers of tertiary alcohols 102 (R^1, R^2 = alkyl, R^3 = alkyl or Ph) afforded α,β-unsaturated ketones 103 by regioselective hydration followed by elimination of methanol. A propargyl methyl ether of a primary alcohol 102 (R^1, R^2 = H; R^3 = alkyl) afforded α-methoxy ketone 104 in excellent yields. In contrast, a methyl ether of an α-ethynyl alcohol 102 (R^1 = alkyl; R^2, R^3 = H) afforded α-methoxyketone 105 [96]. The direct formation of dimethyl acetals from alkynes by addition of 2 equiv. of methanol catalyzed by NaAuCl$_4$ in anhydrous methanol is also achieved (Eq. 6.52) [96] while the above-mentioned platinum(II) catalysts were only effective for hydration [91]. The gold catalyst, however, showed the same disadvantage as the mercury catalysts. It was quickly reduced to inactive metallic gold; so more than 50 mol of product could not be synthesized per mol of the gold complex [97].

$$R-\!\!\!\equiv\!\!\!-R' \xrightarrow[\text{H}_2\text{O - MeOH}]{2 \text{ mol\% NaAuCl}_4} \underset{R}{\overset{O}{\|}}\!\!\!\diagup\!\!\!\diagup\!R' \qquad \begin{array}{l} R = C_6H_{13}, R' = H : 91\% \\ R = Ph, R' = H : 91\% \\ R = R' = C_6H_{13} : 94\% \end{array} \qquad (6.50)$$

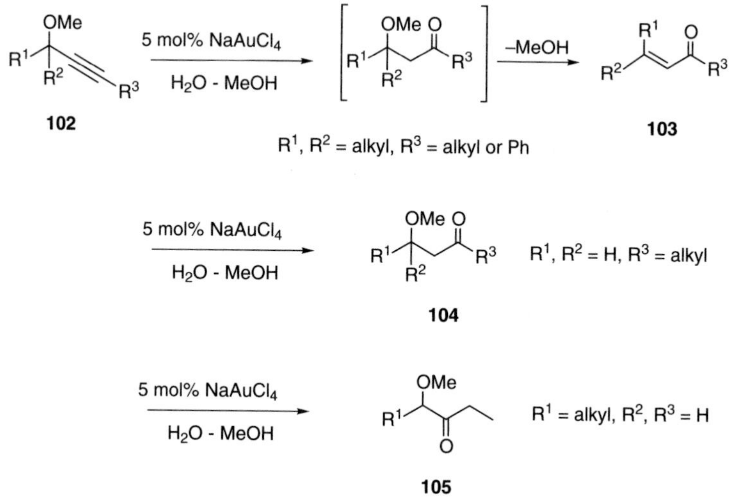

$$R^1 \overset{OMe}{\underset{R^2}{\diagdown}} R^3 \quad \xrightarrow[\text{H}_2\text{O - MeOH}]{\text{5 mol\% NaAuCl}_4} \quad \left[R^1 \overset{OMe}{\underset{R^2}{\diagdown}} \overset{O}{\diagdown} R^3 \right] \quad \xrightarrow{-\text{MeOH}} \quad R^2 \overset{R^1}{\diagdown} \overset{O}{\diagdown} R^3 \quad (6.51)$$

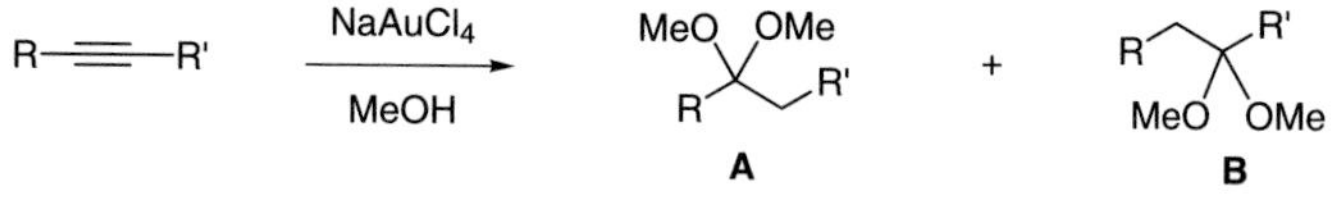

102   **103**

R^1, R^2 = alkyl, R^3 = alkyl or Ph

$$\xrightarrow[\text{H}_2\text{O - MeOH}]{\text{5 mol\% NaAuCl}_4} \quad R^1 \overset{OMe}{\underset{R^2}{\diagdown}} \overset{O}{\diagdown} R^3 \qquad R^1, R^2 = \text{H}, R^3 = \text{alkyl}$$

104

$$\xrightarrow[\text{H}_2\text{O - MeOH}]{\text{5 mol\% NaAuCl}_4} \quad R^1 \overset{OMe}{\diagdown} \overset{}{\underset{O}{\diagdown}} \qquad R^1 = \text{alkyl}, R^2, R^3 = \text{H}$$

105

$$R \overline{\quad\equiv\quad} R' \quad \xrightarrow[\text{MeOH}]{\text{NaAuCl}_4} \quad R \overset{MeO \quad OMe}{\diagdown\diagdown} R' \quad + \quad R \overset{R'}{\underset{MeO \quad OMe}{\diagdown}} \quad (6.52)$$

                                         A   **B**

	Product		*Product*
$R = C_6H_{13}$, $R' = H$	**A** (91%)	$R = \overset{OH}{\diagup}(CH_2)_6-$, $R' = H$ 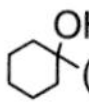	**A** (91%)
$R = C_{10}H_{21}$, $R' = H$	**A** (94%)	$R = R' = C_6H_{13}$	**A** = **B** (94%)
$R = Ph$, $R' = H$	**A** (91%)	$R = C_6H_{13}$, $R' = Me$	**A** / **B** = 40 / 60 (91%)

The RuCl$_3$/aq. HCl system is effective for hydration of phenylpropiolic acid to give acetophenone and carbon dioxide *via* the β-keto acid (Eq. 6.53) [86].

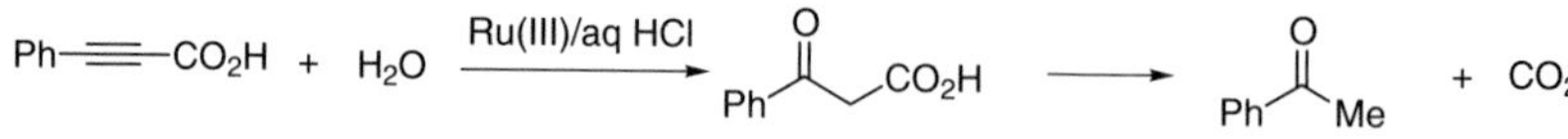

$$Ph \overline{\quad\equiv\quad} CO_2H + H_2O \quad \xrightarrow{\text{Ru(III)/aq HCl}} \quad Ph \overset{O}{\diagup} CO_2H \quad \longrightarrow \quad Ph \overset{O}{\diagup} Me + CO_2$$

$$(6.53)$$

Regioselective intermolecular hydration of alkynones by Pd(II) species under neutral conditions was reported [98]. Depending on the structure of the alkynones, 1,4- or 1,5-diketones were produced selectively (Scheme 6-22). Reaction of 3-alkynyl methyl ketone or 2(2-alkynyl)cyclohexanone with excess water at room temperature in the presence of a catalytic amount of [PdCl$_2$(MeCN)$_2$] in acetonitrile under ultrasonic wave irradiation afforded the corresponding 1,4-diketone. In similar reactions with 2-(2-alkynyl)cyclopentanone, 2-(3-alkynyl)cyclohexanone, or 4-alkynyl methyl ketone, the 1,5-diketones were obtained. Simple alkynes such as 1-octyne and 5-decyne were not hydrated under comparable conditions. It is proposed that the spatial distance between the carbonyl oxygen and any of the acetylenic carbons determines the regioselectivity, i.e. oxypalladation followed through **A** gives a 1,4-diketone and reactions through **B** provide a 1,5-diketone.

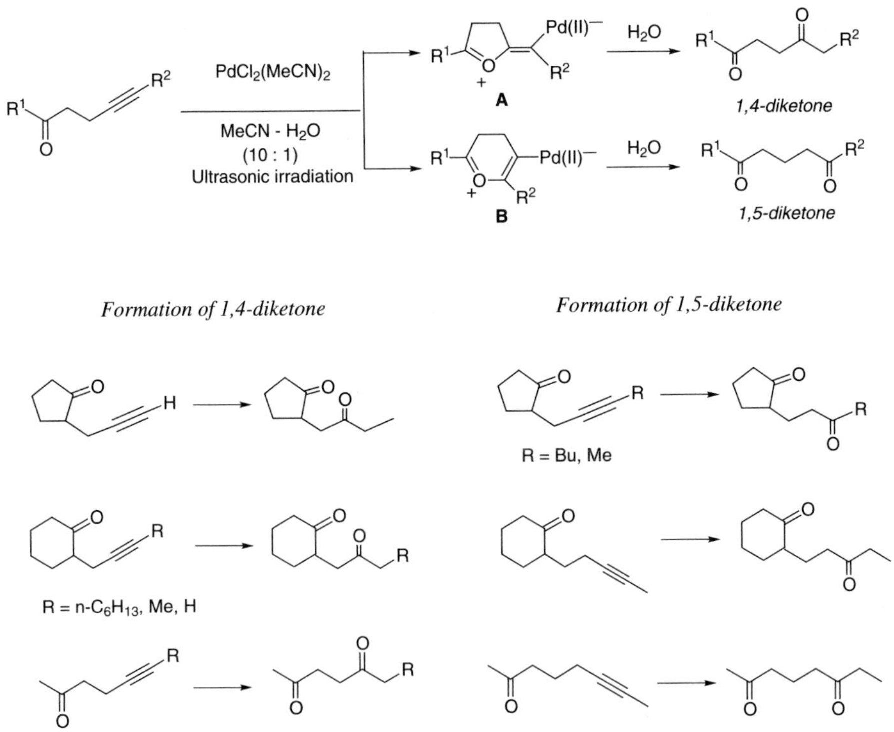

Schemes 6-22 Pd-catalyzed regioselective hydration of alkynones

The stereoselective catalyzed addition of water or methanol to dimethyl acetylenedicarboxylate (DMAD) was reported to yield oxalacetic acid dimethylester or dimethyl methoxyfumarate. The catalyst precursor *cis*-[Pd(PMe$_2$Ph)$_2$(solvent)$_2$] [BF$_4$]$_2$ was prepared from *cis*-[PdCl$_2$(PMe$_2$Ph)$_2$] and AgBF$_4$ (Eq. 6.54). The analogous platinum complex was not effective, however [99].

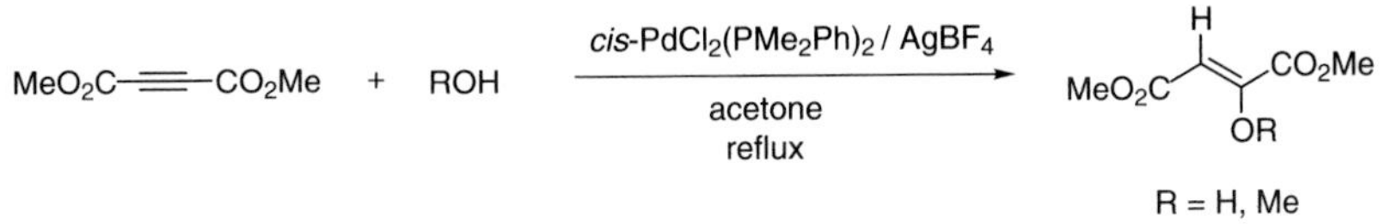

$$R = H, Me$$

$$(6.54)$$

We also reported that the MCl$_2$(ligand)–AgPF$_6$ system [M = Pd, Pt; ligand = 2PPh$_3$, Ph$_2$P(CH$_2$)$_n$PPh$_2$ (n=2,5)] gives efficient catalysts for the addition of methanol to DMAD under very mild conditions, i.e. at room temperature [100]. In our case, the platinum catalysts were more active than the palladium analogs and even active for the addition of methanol to non-activated acetylenes such as 6-dodecyne or diphenylacetylene [101]. The catalytic activity is influenced by the phosphine ligand and by the anion as well as by the ratio of the dichloroplatinum complex to the silver salt (Eq. 6.55). Stoichiometric reactions of DMAD and MeOH with *cis*-[PtCl$_2$(PPh$_3$)$_2$]

and 1 equiv. of AgPF$_6$ gave trans-[PtCl(PPh$_3$)$_2$\{η^1-(E)-MeO$_2$CC=C (OMe)CO$_2$Me\} **106** in good yield (Eq. 6.56) [101]. This complex corresponds to an intermediate in the catalytic cycle. Similar cationic methoxyvinyl palladium complexes **107** carrying a tridentate phosphine ligand were also prepared, and the structure of **107a** was characterized crystallographically [102]. We have further found that the stereoselective trans addition of an alcohol to DMAD is effectively catalyzed by a silver(I) salt (Eq. 6.57) [103]. Besides methanol, ethanol, iso-propanol and benzyl alcohol can also be added to give the corresponding alcoholation products in good yield. Reaction of a terminal alkyne, like methyl propyonate, with MeOH in the presence of AgOTf gave the trans addition product, methyl Z-3-methoxyacrylate, in good yield. An internal alkyne which has one ester group, e.g. ethyl 2-tridecynoate, also reacted with methanol to give a mixture composed of vinyl ethers, an acetal and a ketone, which were all obtained by regioselective addition of methanol (Eq. 6.58). The MeO group was introduced exclusively in the β-position to the ester group. The addition of methanol to non-activated alkynes, however, was not catalyzed by silver salts.

Recently Teles found that cationic gold(I) complexes of the general type [L-Au$^+$] (where L is a phosphine, a phosphite or an arsine) were excellent catalysts for the ad-

$$R\!\!\equiv\!\!R' \quad \xrightarrow[\text{CH}_2\text{Cl}_2 \text{ / MeOH}]{\overset{\text{PtCl}_2(\text{ligand})}{\underset{\text{AgOTf or AgPF}_6}{\big|}}} \quad \overset{R}{\underset{\text{a}}{\text{C(=O)}\text{R}'}} \quad + \quad \overset{R}{\underset{\text{b}}{\text{C(=O)}\text{R}'}} \tag{6.55}$$

ligand : 2 PPh$_3$, Ph$_2$P(CH$_2$)$_n$PPh$_2$ (n = 2, 5)

R = n-C$_{10}$H$_{21}$, R' = Cy : 89 % (**a** / **b** = 75 / 25)
R' = Ph : 91 % (**a** / **b** = 33 / 67)
R' = H : 90 % (**a** only)

$$\text{cis-PtCl}_2(\text{PPh}_3)_2 \quad + \quad \text{MeCO}_2\!\!\equiv\!\!\text{CO}_2\text{Me} \quad \xrightarrow[\text{MeOH / CH}_2\text{Cl}_2]{\text{AgPF}_6} \quad \textbf{106} \tag{6.56}$$

107

P O X = a:
b : Ph$_2$PC$_2$H$_4$OC$_2$H$_4$PPh$_2$

$$\text{MeCO}_2\!\!\equiv\!\!\text{CO}_2\text{Me} \quad + \quad \text{ROH} \quad \xrightarrow{\text{1 mol\% AgOTf}} \quad \begin{array}{c}\text{MeO}_2\text{C}\\ \text{H}\end{array}\!\!=\!\!\begin{array}{c}\text{OR}\\ \text{CO}_2\text{Me}\end{array} \tag{6.57}$$

R = Me, Et, ^{i}Pr, CH$_2$Ph

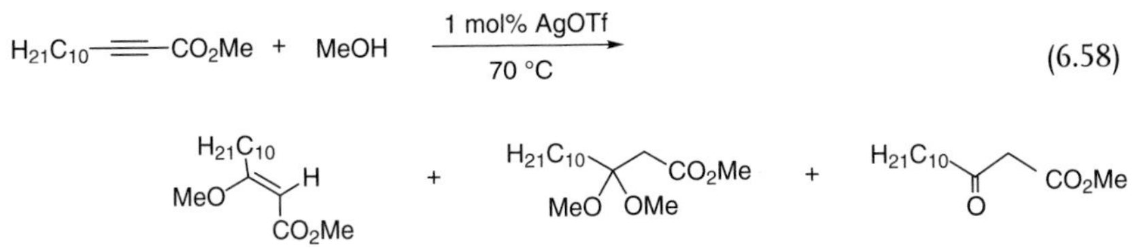

$$H_{21}C_{10}\text{---}\!\!\equiv\!\!\text{---}CO_2Me \; + \; MeOH \quad \xrightarrow[70\ °C]{1\ mol\%\ AgOTf} \qquad\qquad (6.58)$$

dition of alcohols to alkynes under mild conditions [97]. The addition of alcohols to both non-activated terminal and internal alkynes proceeded smoothly to give the corresponding acetals and/or enol ethers depending on the substrate. In addition to methanol, other alcohols like ethanol, isopropanol and allyl alcohol added also smoothly to terminal alkynes (Scheme 6-23). Propargyl alcohols can also be used as substrates. These catalysts achieved total turnover numbers of up to 10^5 mol of product per mol of catalyst, with turnover frequencies of up to 5400 h^{-1}. They are neither water nor air sensitive, and the reaction can be conducted without solvent. The nature of the ligand L in the cationic gold complex had a considerable influence upon catalytic activity ($AsPh_3 < PEt_3 < PPh_3 < P(4\text{-}F\text{-}C_6H_4)_3 < P(OMe)_3 < P(OPh)_3$). Anions present in the solution are also a very important factor with respect to catalytic activity. Based on the experimental results and *ab initio* calculations, Teles proposed the mechanism shown in Scheme 6-24 for the catalytic addition of alcohols to alkynes. *Ab initio* calculations predict that mono- and disubstituted alkynes are better ligands than MeOH and favor the formation of a π-coordinated alkyne-Au complex **A**, which is then attacked by MeOH in an associative manner to give the intermediate **B**.

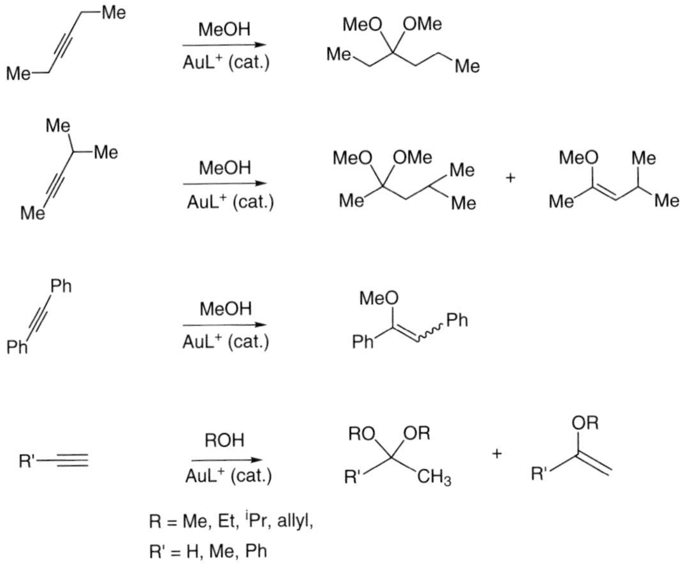

Schemes 6-23 Au-catalyzed alcoholation of alkynes

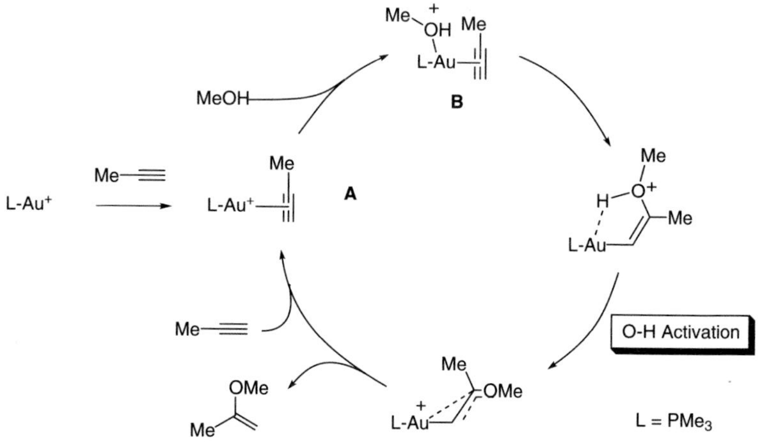

Schemes 6-24 Proposed mechanism for the addition of MeOH to Propyne by Au(PMe$_3$)$^+$

The intramolecular addition of the O–H bond to alkynes catalyzed by palladium complexes has been developed by K. Utimoto et al. (Eq. 6.59) [104]. An alkynyl alcohol can be converted to a cyclic alkenyl ether in the presence of a catalytic amount of [PdCl$_2$(PhCN)$_2$ or [PdCl$_2$(MeCN)$_2$] in ether or THF at room temperature. When the reaction was carried out in MeCN–H$_2$O under reflux in the presence of a catalytic amount of PdCl$_2$, hydration of the acetylenic alcohol occurred and the ketoalcohol was obtained in good yield instead.

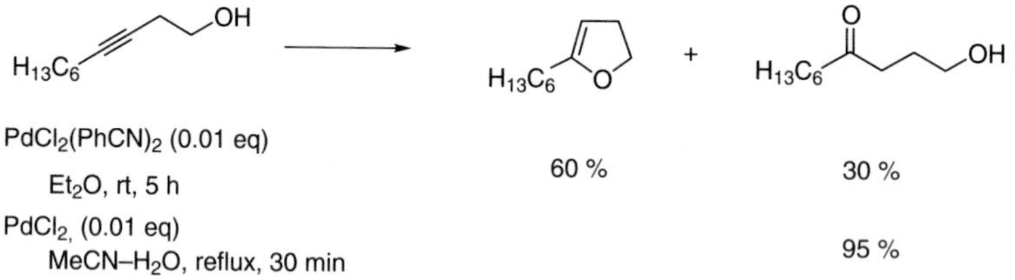

PdCl$_2$(PhCN)$_2$ (0.01 eq)
Et$_2$O, rt, 5 h

PdCl$_2$, (0.01 eq)
MeCN–H$_2$O, reflux, 30 min

60 %

30 %

95 %

(6.59)

6.3.3.4 Addition of O–H Bonds to Methylenecyclopropane

Recently, Y. Yamamoto reported a palladium-catalyzed hydroalkoxylation of methylene cyclopropanes (Scheme 6-25) [105]. Curiously, the catalysis proceeds under very specific conditions, i.e. only a 1 : 2 mixture of [Pd(PPh$_3$)$_4$] and P(o-tolyl)$_3$ leads to an active system. Other combinations using Pd(0 or II) precursors with P(o-tolyl)$_3$ or 1,3-bis(diphenylphosphino)propane, the use of [Pd(PPh$_3$)$_4$] without P(o-tolyl)$_3$ or with other phosphine ligands were all inefficient for the hydroalkoxylation. The authors assumed a mechanism in which oxidative addition of the alcohol to a Pd(0) center yields a hydrido(alkoxo) complex which is subsequently involved in hydropalladation of methylenecyclopropane.

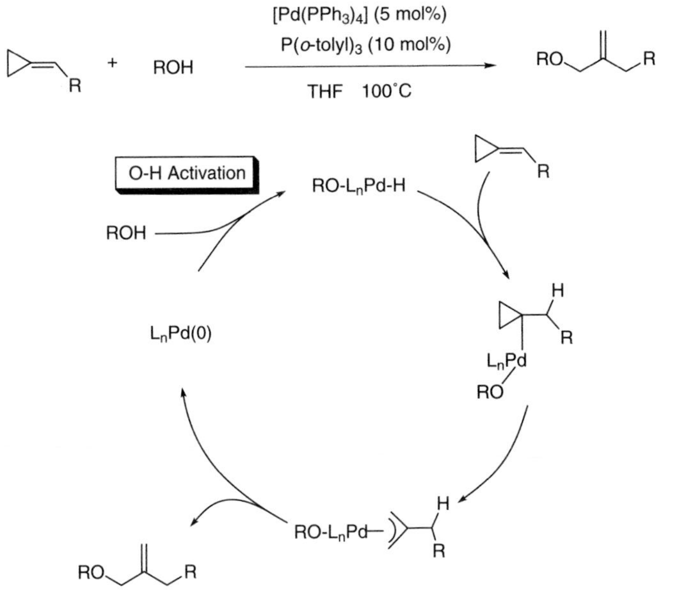

Schemes 6-25 Pd-catalyzed hydroalkoxylation of methylenecyclopropanes and the proposed mechanism

6.3.4
Addition of Carboxylic Acids to Unsaturated Compounds

Utimoto and Nozaki reported on the palladium(II)-catalyzed regio- and stereoselective cyclization of alkynoic acid [106]. Refluxing alkynoic acid in the presence of a
catalytic amount of a palladium(II) complex, e.g. [PdCl$_2$(PhCN)$_2$], in THF or acetonitrile produced the corresponding cyclization product selectively. While the cyclization of 3-alkynoic acids proceeds in 5-*Endo-Dig* manner affording 3-alken-4-olides
(Eq. 6.60), 4- and 5-alkynoic acids cyclized in 5- and 6-*Exo-Dig* manner to afford γ-
alkylidene butyrolactones and δ-alkylidene valerolactones respectively, in excellent
to good yields (Eq. 6.61). In case of the cyclization of 2-substituted 4-alkynoic acid,
the reaction proceeded stereoselectively to afford the Z-isomer preferentially. The
authors proposed the reaction to take place *via* the organopalladium intermediate
108 and succeeded in trapping it with an electrophile to afford a γ-butyrolactone
bearing a stereodefined alkylidene group in γ-position [107]. Chan et al. also reported Rh-catalyzed intramolecular cyclizations of alkynoic acids to alkylidene lactones
[108]. Several alkynoic acids can be effectively cyclized to the corresponding exocyclic enol lactones with [RhCl(Cy$_2$PC$_2$H$_4$PCy$_2$)]$_2$ in methylene chloride at room temperature. Exclusive formation of the Z-olefinic isomers was observed, i.e., *trans* addition of the carboxylate O–H across the triple bond (Eq. 6.62). There is also a strong
preference for the formation of five- *vs* six-membered rings in the substituted pen-
tynoic acid systems. A proposed mechanism involves initial OH activation by the
metal center followed by nucleophilic attack of the carboxylate anion on the metal-

coordinated alkyne of the intermediate (Scheme 6-26). For the model complexes of the catalytic intermediates **A** and **B**, analogous iridium complexes **92** and **109** were prepared by stoichiometric reactions between [IrCl(PR$_3$)$_3$] and an α,ω-alkynoic acid [59].

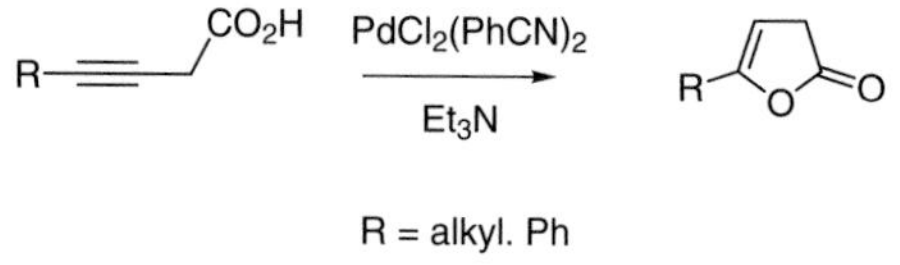

$$\text{(6.60)}$$

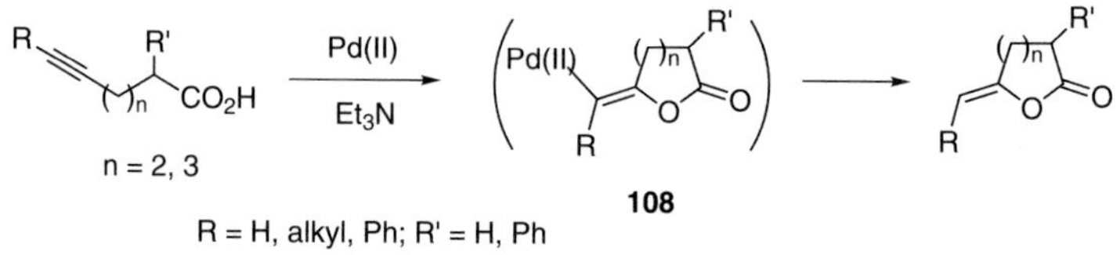

R = alkyl. Ph

$$\text{(6.61)}$$

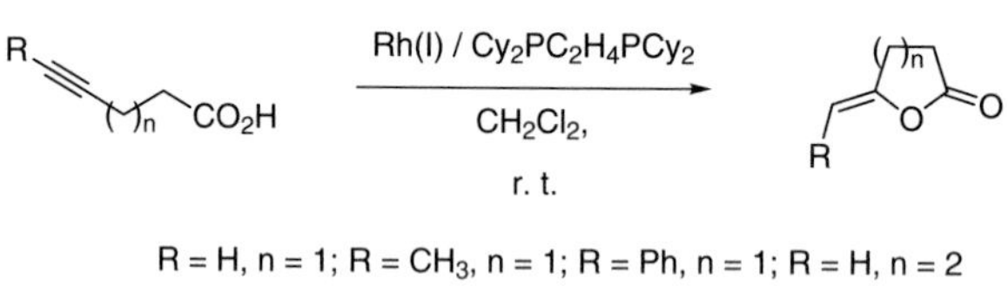

n = 2, 3

108

R = H, alkyl, Ph; R' = H, Ph

$$\text{(6.62)}$$

R = H, n = 1; R = CH$_3$, n = 1; R = Ph, n = 1; R = H, n = 2

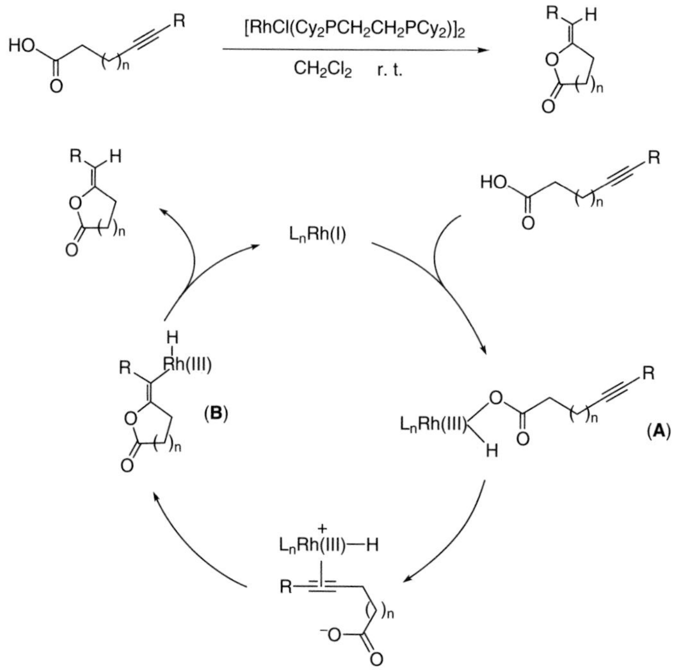

Schemes 6-26 A proposed reaction pathway for Rh(I)-catalyzed cyclization of α,ω-alkynoic acids to alkylidene lactones

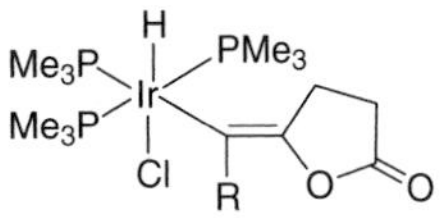

109

In 1998, the first example of the palladium-catalyzed hydrocarboxylation of allenes was reported (Eq. 6.63) [109]. Various kinds of carboxylic acids reacted
smoothly with monoarylallenes in the presence of a catalytic amount of the
$Pd_2(dba)_3 \cdot CHCl_3$/dppf complex, affording the corresponding (*E*)-allyl esters regio-
and stereoselectively in high yield. Initial formation of hydrido(carboxylato) Pd(II)
species by oxidative addition of carboxylic acid to Pd(0) and the subsequent hy-
dropalladation of an allene affording π-allylpalladium carboxylate are postulated as
the reaction mechanism (Scheme 6-27).

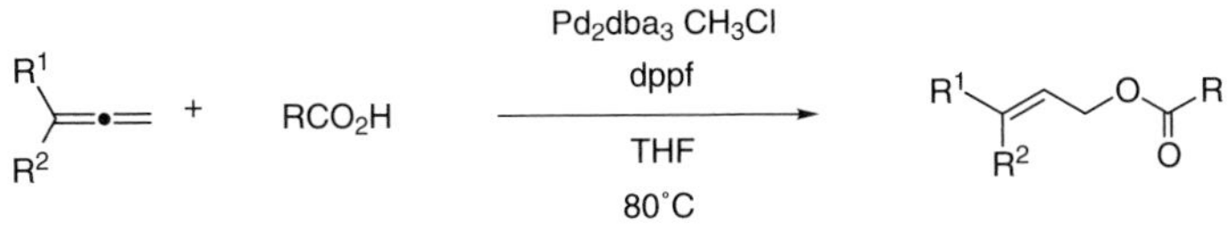

(6.63)

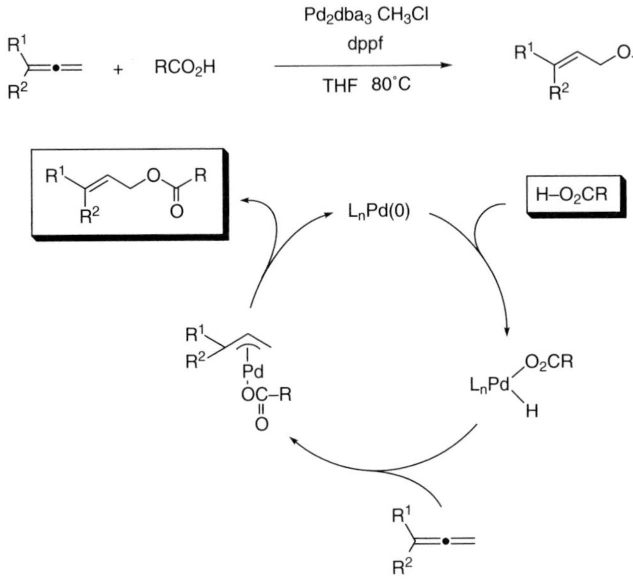

Schemes 6-27 Pd(0)-catalyzed hydrocarboxylation of allene

6.3.5
H–D Exchange Reaction

Hydrido(hydroxo) complexes derived from the oxidative addition of H_2O to
$[Pt(PR_3)_3]$ **(26)** are stronger bases than aqueous alkali in organic media. Platinum(0)

peralkylphosphine complexes such as [PtL₃] (26) (a: L = PEt₃, b: L = PⁱPr₃), and [Pt(PCy₃)₂] (24b) can catalyze H–D exchange of activated C–H bonds such as α-hydrogen atoms of ketones, aldehydes, sulfones, sulfoxides, and nitroalkanes with D₂O [22]. For example, 86% of the methyl hydrogens of acetophenone were exchanged for deuterium by heating with a large excess of D₂O in the presence of a catalytic amount of [Pt(PⁱPr₃)₃] (26b) at 80°C for 20 h in THF. [Pt(PEt₃)₃] (26a) is a more active catalyst than aqueous NaOH. A mechanism for this H–D exchange reaction is proposed in Scheme 6-28.

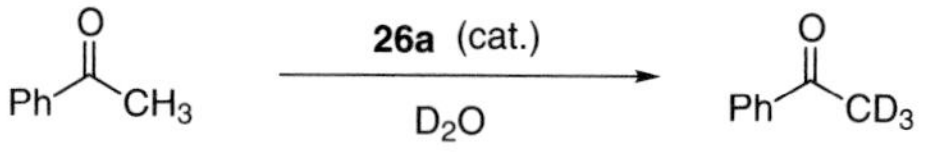

$$\text{Ph}\overset{O}{\underset{}{\|}}\text{CH}_3 \xrightarrow[\text{D}_2\text{O}]{\textbf{26a (cat.)}} \text{Ph}\overset{O}{\underset{}{\|}}\text{CD}_3$$

$$\text{Pt(PEt}_3)_3 + \text{D}_2\text{O} \rightleftharpoons [\text{PtD(PEt}_3)_3]\text{OD} \qquad (1)$$
$$\textbf{26a}$$

$$[\text{PtD(PEt}_3)_3]\text{OD} + \text{PhCOCH}_3 \rightleftharpoons \text{PtD(CH}_2\text{COPh)(PEt}_3)_3 + \text{DHO} \qquad (2)$$

$$\text{PtD(CH}_2\text{COPh)(PEt}_3)_3 + \text{D}_2\text{O} \rightleftharpoons [\text{PtD(PEt}_3)_3]\text{OD} + \text{PhCOCH}_2\text{D} \qquad (3)$$

$$\text{PtD(CH}_2\text{COPh)(PEt}_3)_3 \longrightarrow \text{Pt(PEt}_3)_3 + \text{PhCOCH}_2\text{D} \qquad (4)$$

Schemes 6-28   Mechanism for Pt(PEt₃)₃-catalyzed H–D exchange of acetophone with D₂O

The [PtL₃]/D₂O system, however, was incapable of exchanging aromatic hydrogens. Rh(I) hydrides, [RhH(PⁱPr₃)₃] (33) and [Rh₂H₂(μ-N₂)(PCy₃)₄] (34), on the contrary, are active catalysts for H–D exchange between a wide range of aromatic hydrogens and D₂O [24]. Heating a mixture of pyridine and D₂O in the presence of a catalytic amount of 33 at 80°C for 20 h gave deuterated pyridine without positional preference, deuteration at 2,6-, 3,5- and 4-positions being 62, 59, and 58% respectively (Eq. 6.64). A wide range of aromatic compounds, e.g., PhMe₃, PhOMe, PhF, naphthalene, etc. were deuterated with D₂O by the Rh(I) hydrides. Deuterium incorporation into the Me group was also observed for PhMe and PhOMe. It is proposed that the H–D exchange of aromatic hydrogens with D₂O catalyzed by 33 proceeds via initial formation of [RhD(PⁱPr₃)₃] through addition of D₂O to 33 followed by reductive elimination of DHO from the adduct [RhH(D)(py)₂(PⁱPr₃)₂]OD. H–D exchange of H₂ with D₂O can also be catalyzed by 33 and 34 (Eq. 6.65).

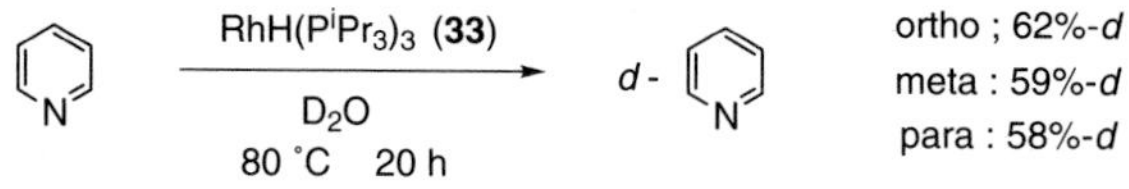

$$\underset{\text{N}}{\bigcirc} \xrightarrow[\substack{\text{D}_2\text{O}\\80\,°\text{C}\quad20\,\text{h}}]{\text{RhH(PⁱPr}_3)_3\ (33)} d\text{-}\underset{\text{N}}{\bigcirc} \qquad \begin{array}{l}\text{ortho ; 62\%-}d\\\text{meta : 59\%-}d\\\text{para : 58\%-}d\end{array} \qquad (6.64)$$

$$\underset{\textbf{33}}{\text{RhH(PⁱPr}_3)_3} + \text{D}_2\text{O} \rightleftharpoons \text{RhD(PⁱPr}_3)_3 + \text{DHO} \qquad (6.65)$$

6.3.6
Miscellaneous

6.3.6.1 Reductive Dimerization of Alkynes

The cationic Rh(I) complex [Rh(MeOH)$_2$(binap)]ClO$_4$ catalyzes selectively the reductive dimerization of dialkyl acetylenedicarboxylates to give 1,2,3,4-tetrakis(alkoxycarbonyl)-1,3-butadienes in MeOH (Eq. 6.66) [110]. The source of the hydrogen was found by a deuterium-labeling experiment to be not only the hydroxy hydrogen but also the methyl hydrogen of MeOH. Based on the experimental results, a plausible reaction pathway for this reductive dimerization was proposed (Scheme 6-29). Consistently, in the reaction mixture, formaldehyde was detected by the chromotropic acid test. In the first step, coordinated MeOH in [Rh(MeOH)$_2$(binap)]ClO$_4$ is activated by the rhodium to afford a cationic hydrido(methoxo) complex (not isolated). The O–H bond activation is a key step in the production of the active species for this dimerization.

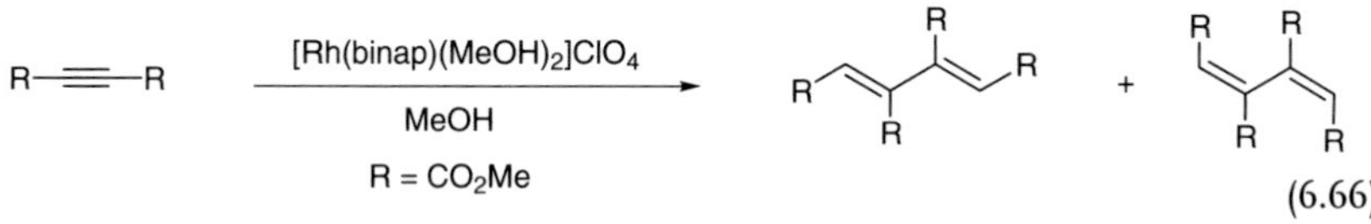

$$(6.66)$$

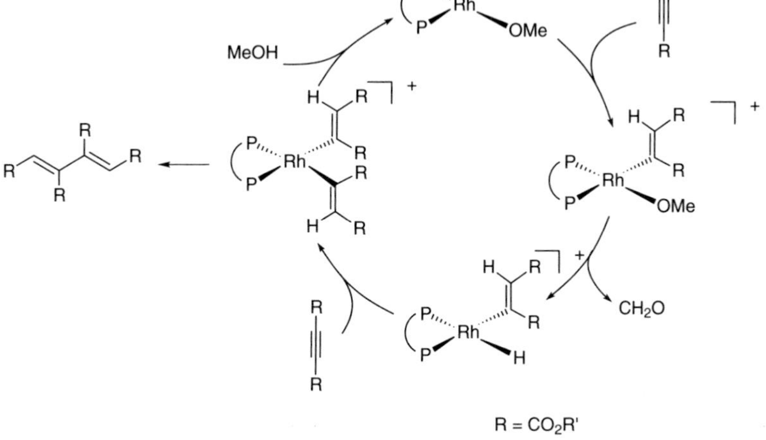

Schemes 6-29 A plausible reaction pathway for Rh(I)-catalyzed reductive dimerization of DMAD

6.3.6.2 Transfer Hydrogenation of Unsaturated Compounds

Recently we have shown that methanol can be utilized as an efficient hydrogen source for the hydrogenation of alkynes as well as alkenes; hydrido(methoxo)iridium(III) complexes, [{IrH(diphosphine)}$_2$(μ-OMe)$_2$(μ-Cl)]Cl (69) [a: diphosphine = BPBP or b: diphosphine = BINAP] can catalyze efficiently the transfer hydrogenation of alkynes with methanol to give *trans*-alkenes selectively. Small amounts of alkanes are also produced (Eq. 6.67) [49]. The hydrogenation of several inner alkynes proceeded effectively at 80°C in a 1:1 mixture of toluene and methanol in the presence of a catalytic amount of the hydrido(methoxo) complex 69. The initial reduction product was the *cis* olefin, but this isomerized easily to the *trans* olefin under the reaction conditions (Scheme 6-30). As the hydrogen source, methanol was far more efficient than ethanol or even isopropanol. This provides a rare example of catalytic reduction using methanol as a hydrogen donor.

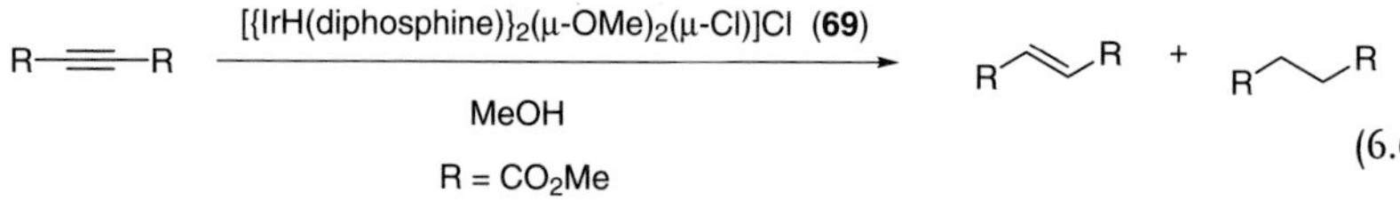

$$R-\!\!\!\equiv\!\!\!-R \xrightarrow[\substack{\text{MeOH} \\ R = CO_2Me}]{\text{[{IrH(diphosphine)}}_2(\mu\text{-OMe})_2(\mu\text{-Cl})]Cl \ (69)}} R\diagup\!\!\!\diagdown R \ + \ R\diagdown\!\!\!\diagup R$$

(6.67)

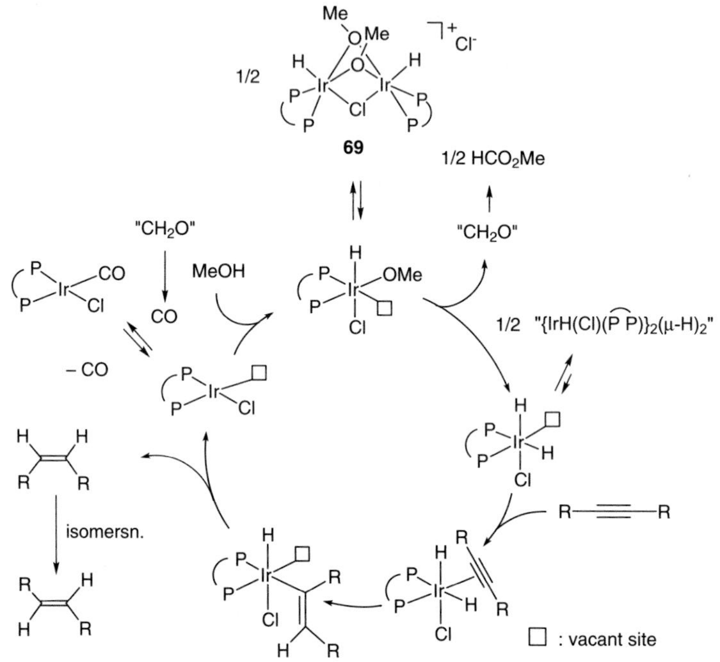

Schemes 6-30 A plausible reaction pathway for Ir(I)-catalyzed reduction of alkynes with MeOH

6.3.6.3 Transition Metal-Catalyzed Silanone Generation

The platinum complex [{Pt(dmpe)}$_2$(μ-H)$_2$](OTf)$_2$ can catalyze the generation of silanones from secondary silanols under very mild conditions (at room temperature) in acetone (Eq. 6.68) [111]. The silanol $^i\text{Pr}_2\text{SiHOH}$ was converted to the cyclic trimer of diisopropylsilanone $(^i\text{Pr}_2\text{SiO})_3$ in 75% yield. A β-hydrogen elimination from Pt-Si–O–H of $[\text{Pt(dmpe)}\{\text{Si}(^i\text{Pr})_2\text{OH}\}(\text{OTf})]$ which was obtained by initial Si–H oxidative addition was postulated (Scheme 6-31).

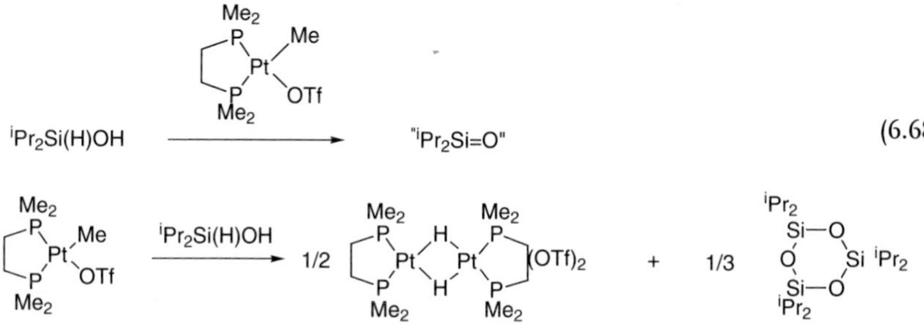

$$^i\text{Pr}_2\text{Si(H)OH} \longrightarrow {}^{"}\text{Pr}_2\text{Si=O}{}^{"} \tag{6.68}$$

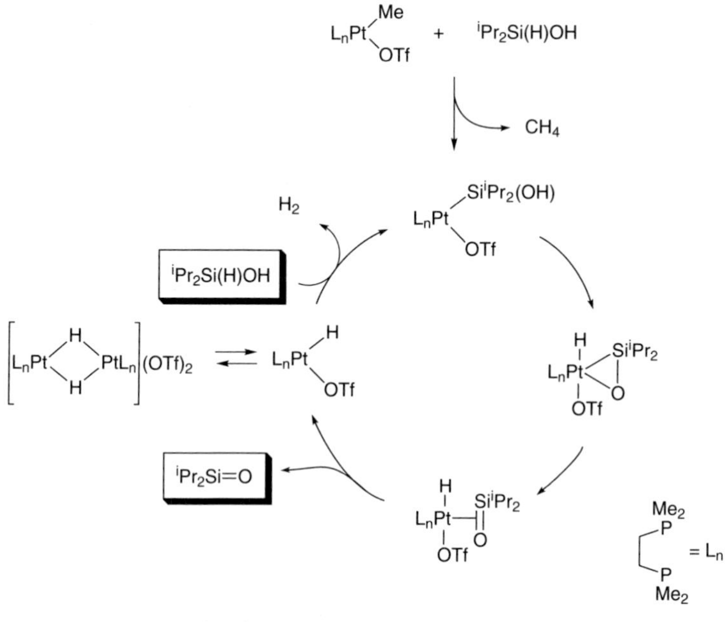

Schemes 6-31 Pt(II)-catalyzed generation of silanones and the proposed mechanism

References

1 BRYNDZA, H. E.; TAM, W. *Chem. Rev.* **1988**, *88*, 1163-1188.

2 SHILOV, A. E.; SHUL'PIN, G. B. *Chem. Rev.* **1997**, *97*, 2879-2932.

3 MEHROTRA, R. C.; AGARWAL, S. K.; SINGH, Y. P. *Coord. Chem. Rev.* **1985**, *68*, 101-130.

4 HARTLEY, F. R. *The Chemistry of Platinum and Palladium*; Wiley: New York, **1973**.

5 CHAUDRET, B. N.; COLE-HAMILTON, D. J.; NOHR, R. S.; WILKINSON, G. *J. Chem. Soc. Dalton Trans.* **1977**, 1546-1557.

6 EDWARDS, A. J.; ELIPE, S.; ESTERUELAS, M. A.; LAHOZ, F. J.; ORO, L. A.; VALERO, C. *Organometallics* **1997**, *16*, 3828-3836.

7 NEWMAN WINSLOW, L. J.; BERGMAN, R. G. *J. Am. Chem. Soc.* **1985**, *107*, 5314-5315.

8 GLUECK, D. S.; NEWMAN WINSLOW, L. J.; BERGMAN, R. G. *Organometallics* **1991**, *10*, 1462-1479.

9 SELIGSON, A. L.; COWAN, R. L.; TROGLER, W. C. *Inorg. Chem.* **1991**, *30*, 3371-3381.

10 COWAN, R. L.; TROGLER, W. C. *J. Am. Chem. Soc.* **1989**, *111*, 4750-4761.

11 CROCHET, P.; ESTERUELAS, M. A.; LÓPEZ, A. M.; MARTÍNEZ, M.-P.; OLIVÁN, M.; OÑATE, E.; RUIZ, N. *Organometallics* **1998**, *17*, 4500-4509.

12 BUIL, M. L.; EISENSTEIN, O.; ESTERUELAS, M. A.; GARCÍA-YEBRA, C.; GUTIÉRREZ-PUEBLA, E.; OLIVÁN, M.; OÑATE, E.; RUIZ, N.; TAJADA, M. A. *Organometallics* **1999**, *18*, 4949-4959.

13 BUIL, M. L.; ESTERUELAS, M. A.; GARCÍA-YEBRA, C.; GUTIÉRREZ-PUEBLA, E.; OLIVÁN, M. *Organometallics* **2000**, *19*, 2184-2193.

14 GERLACH, D. H.; KANE, A. R.; PARSHALL, G. W.; JESSON, J. P.; MUETTERTIES, E. L. *J. Am. Chem. Soc.* **1971**, *93*, 3543-3544.

15 BENNETT, M. A.; ROBERTSON, G. B.; WHIMP, P. O.; YOSHIDA, T. *J. Am. Chem. Soc.* **1973**, *95*, 3028-3030.

16 BENNETT, M. A.; YOSHIDA, T. *J. Am. Chem. Soc.* **1978**, *100*, 1750-1759.

17 HARTWIG, J. F.; ANDERSEN, R. A.; BERGMAN, R. G. *J. Am. Chem. Soc.* **1989**, *111*, 2717-2719.

18 GOTZIG, J.; WERNER, R.; WERNER, H. *J. Organomet. Chem.* **1985**, *285*, 99-114.

19 BURN, M. J.; FICKES, M. G.; HARTWIG, J. F.; HOLLANDER, F. J.; BERGMAN, R. G. *J. Am. Chem. Soc.* **1993**, *115*, 5875-5876.

20 CARMONA, E.; MARIN, J. M.; PANEQUE, M.; POVEDA, M. *Organometallics* **1987**, *6*, 1757-1765.

21 HERBERHOLD, M.; SÜB, G.; ELLERMANN, J.; GÄBELEIN, H. *Chem. Ber.* **1978**, *111*, 2931-2941.

22 YOSHIDA, T.; MATSUDA, T.; OKANO, T.; KITANI, T.; OTSUKA, S. *J. Am. Chem. Soc.* **1979**, *101*, 2027-2038.

23 YOSHIDA, T.; UEDA, Y.; OTSUKA, S. *J. Am. Chem. Soc.* **1978**, *100*, 3941-3942.

24 YOSHIDA, T.; OKANO, T.; SAITO, K.; OTSUKA, S. *Inorg. Chim. Acta* **1980**, *44*, L135-L136.

25 MILSTEIN, D.; CALABRESE, J. C.; WILLIAMS, I. D. *J. Am. Chem. Soc.* **1986**, *108*, 6387-6389.

26 STEVENS, R. C.; BAU, R.; MILSTEIN, D.; BLUM, O.; KOETZLE, T. F. *J. Chem. Soc., Dalton Trans.* **1990**, 1429-1432.

27 BLUM, O.; MILSTEIN, D. *Angew. Chem. Int. Ed. Engl.* **1995**, *34*, 229-231.

28 Tani, K.; Iseki, A.; Yamagata, T. *Angew. Chem. Int. Ed. Engl.* **1998**, *37*, 3381-3383.

29 Dorta, R.; Togni, A. *Organometallics* **1998**, *17*, 3423-3428.

30 Monaghan, P. K.; Pudderphatt, R. J. *Inorg. Chem. Acta* **1982**, *65*, L59-L61.

31 Monaghan, P. K.; Puddehatt, R. J. *Organometallics* **1984**, *3*, 444-449.

32 Tani, K.; Suwa, K.; Tanigawa, E.; Yoshida, T.; Okano, T.; Otsuka, S. *Chem. Lett.* **1982**, 261-264.

33 Tani, K.; Tanigawa, E.; Tatsuno, Y.; Otsuka, S. *J. Organometal. Chem.* **1985**, *279*, 87-101.

34 Milstein, D. *J. Am. Chem. Soc.* **1986**, *108*, 3525-3526, and references therein.

35 Thompson, J. S.; Bernard, K. A.; Rappoli, B. J.; Atwood, J. D. *Organometallics* **1990**, *9*, 2727-2731.

36 Thompson, J. S.; Randall, S. L.; Atwood, J. D. *Organometallics* **1991**, *10*, 3906-3910.

37 Bryndza, H. E.; Calabrese, J. C.; Marsi, M.; Roe, D. C.; Tam, W.; Bercaw, J. E. *J. Am. Chem. Soc.* **1986**, *108*, 4805-4813.

38 Hayashi, Y.; Komiya, S.; Yamamoto, T.; Yamamoto, A. *Chem. Lett.* **1984**, 1363-1366.

39 Yoshida, T.; Otsuka, S. **1977**, *J. Am. Chem. Soc.*, 2134-2140.

40 Akl, N. S.; Tayim, H. A. *J. Organomet. Chem.* **1985**, *297*, 371-374.

41 Mondal, J. U.; Blake, D. M. *Coord. Chem. Rev.* **1982**, *47*, 205-238.

42 Bryndza, H. E.; Fong, L. K.; Paciello, R. A.; Tam, W.; Bercaw, J. E. *J. Am. Chem. Soc.* **1987**, *109*, 1444-1456.

43 Drago, R. S.; Wong, N. M.; Ferris, D. C. *J. Am. Chem. Soc.* **1992**, *114*, 91-98.

44 Blum, O.; Milstein, D. *J. Am. Chem. Soc.* **1995**, *117*, 4582-4594.

45 Sponsler, M. B.; Weiller, B. H.; Stoutland, P. O.; Bergman, R. G. *J. Am. Chem. Soc.* **1989**, *111*, 6841-6843.

46 Ladipo, F. T.; Kooti, M.; Merola, J. S. *Inorg. Chem.* **1993**, *32*, 1681-1688.

47 Yamagata, T.; Iseki, A.; Tani, K. *Chem. Lett.* **1997**, 1215-1216.

48 Dorta, R.; Egli, P.; Zürcher, F.; Togni, A. *J. Am. Chem. Soc.* **1997**, *119*, 10857-10858.

49 Tani, K.; Iseki, A.; Yamagata, T. *Chem. Commun.* **1999**, 1821-1822.

50 Tani, K.; Yamagata, T.; Iseki, A. , Unpublished results.

51 Werner, H.; Michenfelder, A.; Schulz, M. *Angew. Chem. Int. Ed. Engl.* **1991**, *30*, 596-598.

52 Fornies, J.; Green, M.; Spencer, J. L.; Stone, F. G. A. *J. Chem. Soc., Dalton Trans* **1977**, 1006-1009.

53 Braga, D.; Sabatino, P.; Bugno, C. D.; Leoni, P.; Pasquali, M. *J. Organomet. Chem.* **1987**, *334*, C46-C48.

54 Kim, Y.-J.; Osakada, K.; Takenaka, A.; Yamamoto, A. *J. Am. Chem. Soc.* **1990**, *112*, 1096-1104.

55 Singer, H.; Wilkinson, G. *J. Chem. Soc. (A)* **1968**, 2516-2520.

56 Jonas, K.; Wilke, G. *Angew. Chem.* **1969**, *81*, 534.

57 Darensbourg, M. Y.; Ludwig, M.; Riordan, C. G. *Inorg. Chem.* **1989**, *28*, 1630-1534.

58 Smith, S. A.; Blake, D. M.; Kubota, M. *Inorg. Chem.* **1972**, *11*, 660-662.

59 Marder, T. B.; Chan, D. M.-T.; Fultz, W. C.; Calabrese, J. C.; Milstein, D. *J. Chem. Soc., Chem. Commun.* **1987**, 1885-1887.

60 Yoshida, T.; Okano, T.; Ueda, Y.; Otsuka, S. *J. Am. Chem. Soc.* **1981**, *103*, 3411-3422.

61 Taqui Khan, M. M.; Halligudi, S. B.; Shukla, S. *Angew. Chem. Int. Ed. Engl.* **1988**, *27*, 1735-1736.

62 Tsuji, J. *Palladium Reagents and Catalysis*; John Wiley and Sons: Chichester, **1995**.

63 Hosokawa, T.; Uno, T.; Inui, S.; Murahshi, S.-I. *J. Am. Chem. Soc.* **1981**, *103*, 2318-2323.

64 Hosokawa, T.; Okuda, C.; Murahashi, S.-I. *J. Org. Chem.* **1985**, *50*, 1282-1287.

65 Uozumi, Y.; Kato, K.; Hayashi, T. *J. Am. Chem. Soc.* **1997**, *119*, 5063-5064.

66 Larock, R. C. *Comprehensive Organic Transformations*; VCH New York, Weinheim, Cambridge **1989**.

67 Bennett, M. A.; Yoshida, T. *J. Am. Chem. Soc.* **1973**, *95*, 3030-3031.

68 Villain, G.; Kalck, P.; Gaset, A. *Tetrahedron Lett.* **1980**, *21*, 2901-2904.

69 Villain, G.; Constant, G.; Gaset, A.; Kalck, P. *J. Mol. Catal.* **1980**, *7*, 355-364.

70 VILLIAN, G.; GASET, A.; KALCK, P. *J. Mol. Catal.* **1981**, *12*, 103-111.

71 ARNOLD, D. P.; BENNETT, M. A. *J. Organomet. Chem.* **1980**, *199*, 119-135.

72 JENSEN, C. M.; TROGLER, W. C. *J. Am. Chem. Soc.* **1986**, *108*, 723-729.

73 KAMINSKAIA, N. V.; KOSTIC, N. M. *J. Chem. Soc., Dalton Trans.* **1996**, 3677-3686.

74 MURAHASHI, S.-I.; NAOTA, T. *Bull. Chem. Soc. Jpn.* **1996**, *69*, 1805-1824 and references cited therein.

75 CHIN, C. S.; CHONG, D.; LEE, B.; JEONG, H.; WON, G.; DO, Y.; PARK, Y. *J. Organometallics* **2000**, *19*, 638-648.

76 ATKINS, K. E.; WALKER, W. E.; MANYIK, R. M. *J. Chem. Soc., Chem. Commun.* **1971**, 330.

77 INOUE, Y.; SATO, M.; SATAKE, M.; HASHIMOTO, H. *Bull. Chem. Soc. Jpn.* **1983**, *56*, 637-638.

78 INOUE, Y.; OHTSUKA, Y.; HASHIMOTO, H. *Bull. Chem. Soc. Jpn.* **1984**, *57*, 3345-3346.

79 YAGI, H.; TANAKA, E.; ISHIWATARI, H.; HIDAI, M.; UCHIDA, Y. *Synthesis* **1977**, 334-335.

80 MATSUDA, A. *Bull. Chem. Soc. Jpn.* **1973**, *46*, 524-530.

81 MILSTEIN, D.; HUCKABY, J. L. *J. Am. Chem. Soc.* **1982**, *104*, 6150-6152.

82 MILSTEIN, D. *Acc. Chem. Res.* **1988**, *21*, 428-434.

83 JENSEN, C. M.; TROGLER, W. C. *Science (Washington, D. C.)* **1986**, *233*, 1069-1071.

84 RAMPRASAD, D.; YUE, H. J.; MARSELLA, J. A. *Inorg. Chem.* **1988**, *27*, 3151-3155.

85 LAROCK, R. C.; LEONG, W. W. *Addition of H–X Reagents to Alkenes and Alkynes*; TROST, B. M., FLEMMING, I. and SEMMELHOCK, M. F., (eds.); Pergamon Press: Oxford, **1991**; Vol. 4, pp 269.

86 HALPERN, J.; JAMES, B. R.; KEMP, A. L. W. *J. Am. Chem. Soc.* **1961**, *83*, 4097-4098.

87 HALPERN, J.; JAMES, B. R.; KEMP, A. L. W. *J. Am. Chem. Soc.* **1966**, *88*, 5142-5147.

88 TAQUI KHAN, M. M.; HALLIGUDI, S. B.; SHUKLA, S. *J. Mol. Cat.* **1990**, *58*, 299-305.

89 JAMES, B. R.; REMPEL, G. L. *J. Am. Chem. Soc.* **1969**, *91*, 863-865.

90 BLUM, J.; HUMINER, H.; ALPER, H. *J. Mol. Cat.* **1992**, *75*, 153-160.

91 HISCOX, W.; JENNINGS, P. W. *Organometallics* **1990**, *9*, 1997-1999.

92 HARTMAN, J. W.; HISCOX, W. C.; JENNINGS, P. W. *J. Org. Chem.* **1993**, *58*, 7613-7614.

93 JENNINGS, P. W.; HARTMAN, J. W.; HISCOX, W. C. *Inorg. Chim. Acta* **1994**, *222*, 317-322.

94 FUKUDA, Y.; UTIMOTO, K. *J. Org. Chem.* **1991**, *56*, 3729-3731.

95 TOKUNAGA, M.; WAKATSUKI, Y. *Angew. Chem. Int. Ed.* **1998**, *37*, 2867-2869.

96 FUKUDA, Y.; UTIMOTO, K. *Bull. Chem. Soc. Jpn.* **1991**, *64*, 2013-2015.

97 TELES, J. H.; BRODE, S.; CHABANAS, M. *Angew. Chem. Int. Ed.* **1998**, *37*, 1415-1417.

98 IMI, K.; IMAI, K.; UTIMOTO, K. *Tetrahedron Lett.* **1987**, *28*, 3127-3130.

99 AVSHU, A.; O'SULLIVAN, R. D.; PARKINS, A. D. *J. Chem. Soc., Dalton Trans.* **1983**, 1619-1624.

100 KATAOKA, Y.; MATSUMOTO, O.; OHASHI, M.; YAMAGATA, T.; TANI, K. *Chem. Lett.* **1994**, 1283-1284.

101 KATAOKA, Y.; MATSUMOTO, O.; TANI, K. *Organometallics* **1996**, *15*, 5246-5249.

102 KATAOKA, Y.; TSUJI, Y.; MATSUMOTO, O.; OHASHI, M.; YAMAGATA, T.; TANI, K. *Chem. Commun.* **1995**, 2099 - 2100.

103 KATAOKA, Y.; MATSUMOTO, O.; TANI, K. *Chem. Lett.* **1996**, 727-728.

104 UTIMOTO, K. *Pure Appl. Chem.* **1983**, *55*, 1845-1852.

105 CAMACHO, D. H.; NAKAMURA, I.; SAITO, S.; YAMAMOTO, Y. *Angew. Chem. Int. Ed.* **1999**, *38*, 3365-3367.

106 LAMBERT, C.; UTIMOTO, K.; NOZAKI, H. *Tetrahedron Lett.* **1984**, *25*, 5323-5326.

107 YANAGIHARA, N.; LAMBERT, C.; IRITANI, K.; UTIMOTO, K.; NOZAKI, H. *J. Am. Chem. Soc.* **1986**, *108*, 2753-2754.

108 CHAN, D. M. T.; MARDER, T. B.; MILSTEIN, D.; TAYLOR, N. J. *J. Am. Chem. Soc.* **1987**, *109*, 6385-6388.

109 AL-MASUM, M.; YAMAMOTO, Y. *J. Am. Chem. Soc.* **1998**, *120*, 3809-3810.

110 TANI, K.; UEDA, K.; ARIMITSU, K.; YAMAGATA, T.; KATAOKA, Y. *J. Organomet. Chem.* **1998**, *560*, 253-255.

111 GOIKHMAN, R.; AIZENBERG, M.; SHIMON, L. J. W.; MILSTEIN, D. *J. Am. Chem. Soc.* **1996**, *118*, 10894-10895.

7
Sulfur (and Related Elements)–X Activation

Hitoshi Kuniyasu

7.1
Introduction

Although the transition metal-catalyzed heterofunctionalization of unsaturated organic compounds with heteroatom compounds represented by hydrosilylation of alkene and alkyne has been investigated for more than three decades [1–3], the study of the activation of S–X and Se–X bonds by transition metals is a relatively new research area. One reason for such tardiness seems to stem from the widespread misconception that 'sulfur compounds poison the catalyst' [4]. It is time to consider the meaning of these words more carefully. Indeed, studies dealing with the synthesis of 'stable' transition metal compounds with thiolate (chalcogenate) as ligands are ubiquitous, and made us feel that a catalytic transformation involving repeated M–S bond formation and cleavage was not likely to be promising in terms of reactivity of the metal complexes. Suppose some sulfur compound was added into some transition metal-catalyzed reaction system, like hydrogenation, Heck reaction or similar; the sulfur compound could potentially poison the catalyst by sticking to the coordination site required for the desired catalytic reaction. However, this concept does not mean that catalytic transformations of organic sulfur compounds are themselves not catalyzed by transition metal complexes. Even if an M–S bond cleavage is a thermodynamically unfavorable process and not confirmed by stoichiometric reaction, the 'stable' M–S bond could be regenerated in the next catalytic cycle. We need to bear in mind that the minimum requirements to make a reaction system catalytic are: (1) equilibrium in each elementary process in a whole catalytic cycle, and (2) total thermodynamic advantage of the reaction system in question. Moreover, according to a recent DFT calculation, the actual strength of an M–S bond is not strong compared with that of an M–H or an M–C bond. (For instance, 48.4 kcal/mol for a Pd–SCH3 bond at best [5].)

The progress on S–C bond activation, which covers the reduction of a C–S bond to a C–H bond, cross coupling reaction of sulfides with main group organometallic nucleophiles, ring opening reactions of thietanes and thiiranes, and desulfurization of thiols, sulfides, and thiophenes has already been reviewed elsewhere [6–10], and these topics are beyond the subject of this book. Here we focus on catalytic S (and related elements)–X bond activation using unsaturated organic compounds with emphasis on their reaction mechanisms. Finally, some stoichiometric reactions related to the mechanisms of these catalytic reactions will be introduced [11].

7.2
Catalytic S–H Activation

Pioneering work involving the combination of an organosulfide, a C–C unsaturated organic compound and a transition metal catalyst was reported by Reppe in 1953, in which Ni(CO)$_4$-catalyzed 'hydrothiocarboxylation' of alkyne or ethylene by thiol and CO was briefly described to give the corresponding thioesters **1** or **2** (an application of one of the so-called Reppe reactions) (Eqs. 7.1 and 7.2) [12, 13].

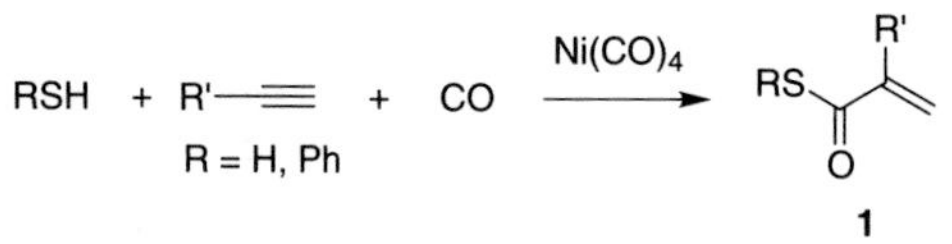

$$RSH + R'\!\!-\!\!\!\equiv\!\!\!- + CO \xrightarrow{Ni(CO)_4} \underset{\mathbf{1}}{RS\!\!\overset{R'}{\underset{O}{\diagup\!\!\diagdown}}} \tag{7.1}$$

$$R = H, Ph$$

$$RSH + \;=\!= \;+ CO \xrightarrow{Ni(CO)_4} \underset{\mathbf{2}}{\overset{SR}{\underset{O}{\diagdown\!\!\diagup}}} \tag{7.2}$$

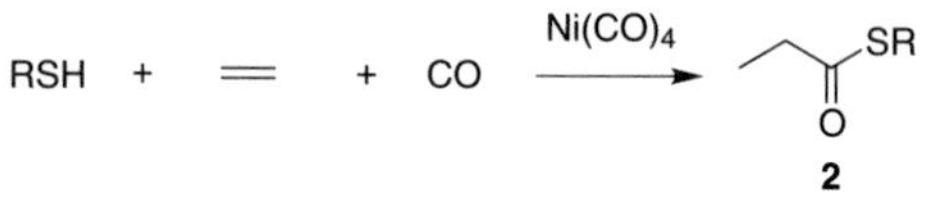

Although these reactions indicated the great potential of transition metal-catalyzed reactions of organosulfides with C–C unsaturated organic compounds, little attention has been paid to such a combination of reactions for many years since then. In 1960, Holmquist and Carnaham reported the Co-catalyzed reaction of thiol with CO (1000 atm) at 250–300 °C to afford thioester **3** in up to 46% yield with ca. 1–5 wt% of catalyst (Eq. 7.3) [14].

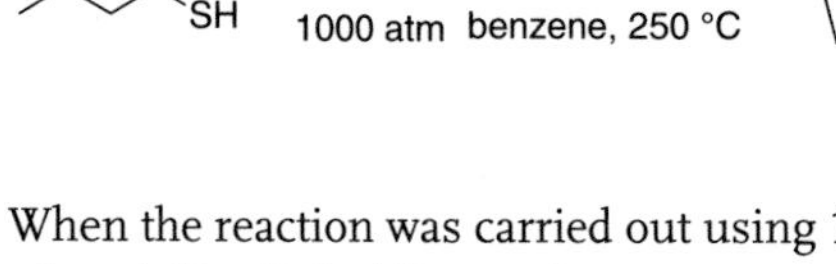

$$\underset{\text{100-1000 atm}}{RSH + CO} \xrightarrow[\text{C}_6\text{H}_6 \text{ or toluene}]{Co_2(CO)_8 \text{ (1~5 wt\%)}} \underset{\substack{\mathbf{3} \\ \text{up to 46\%}}}{\overset{O}{\underset{}{R\diagup\!\!\overset{\|}{\diagdown}SR}}} \tag{7.3}$$

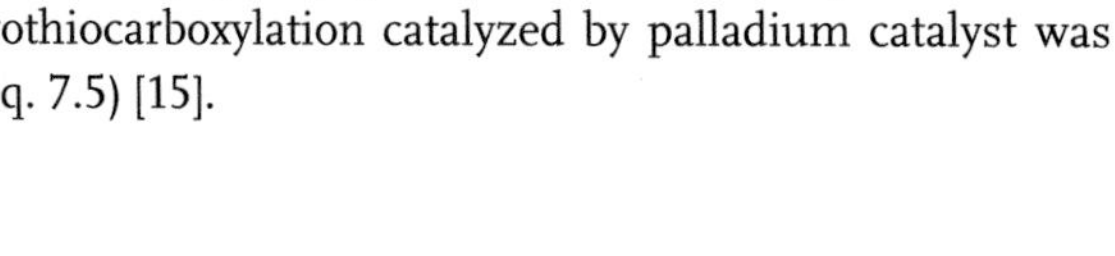

$$\underset{}{\overset{SH}{\diagup\!\!\diagdown\!\!\diagdown_{SH}}} + CO \xrightarrow[\text{1000 atm benzene, 250 °C}]{Co_2(CO)_8 \text{ (1~5 wt\%)}} \underset{\mathbf{4}}{\overset{O}{\diagup\!\!\diagdown_S}} \text{ or } \underset{\mathbf{5}}{\overset{O}{\diagdown\!\!\diagup_S}} \tag{7.4}$$

When the reaction was carried out using 1,3-butanedithiol, thiolactone **4** or **5** was produced (Eq. 7.4). (The product was not specified in the paper.) In 1973, the hydrothiocarboxylation catalyzed by palladium catalyst was reported in a US Patent (Eq. 7.5) [15].

$$
\begin{array}{c}
\text{Pd(PPh}_3)_2\text{Cl}_2 \\
\text{SnCl}_2
\end{array}
$$

RSH + R'—≡ + CO $\xrightarrow[\substack{\text{solvent} \\ 120\ ^\circ\text{C, 6 h}}]{}$ R'—CH=CH–C(O)–SR + R'–CH(SR)–C(O)–   (7.5)

R' = Et, Me,   R = C$_5$H$_{11}$ 2000 psi   up to 61%   minor
Bu, Ph

50 mmol   **6**   **7**

4 mmol

t-BuSH + R—≡ + CO $\xrightarrow[\substack{\text{THF (20 mL)} \\ \text{AcOH (0.1 mL)} \\ 100\ ^\circ\text{C, 21 h}}]{\text{Pd(OAc)}_2/i\text{-Pr}_3\text{P}}$ R–CH$_2$CH$_2$–C(O)–SBu-t + R–CH(SBu-t)–   (7.6)

220 mmol   R = C$_4$H$_9$ 750 psi   up to 79%   7%

300 mmol   **8**   **9**

Unlike the case of the Ni-catalyzed reaction, which afforded the branched thioester (Eq. 7.1), the PdCl$_2$(PPh$_3$)$_3$/SnCl$_2$-catalyzed reaction with 1-alkyne and 1-alkene predominantly provided terminal thioester **6** in up to 61% yield in preference to **7**. In 1983, a similar hydrothiocarboxylation of an alkene was also documented by using a Pd(OAc)$_2$/P(^{i}-Pr)$_3$ catalyst system with t-BuSH to form **8** in up to 79% yield (Eq. 7.6) [16]. It was mentioned in the patent that the Pt-complex also possessed catalytic activity for the transformation, although the yield of product was unsatisfactory. In 1984, the hydrothiocarboxylation of a 1,3-diene catalyzed by Co$_2$(CO)$_8$ in pyridine was also reported in a patent [17]. In 1986, Alper et al. reported that a similar transformation to the one shown in Eq. (7.3) can be realized under much milder reaction conditions in the presence of a 1,3-diene [18], and the carboxylic ester **10** was produced using an aqueous alcohol as solvent (Eq. 7.7) [19].

RSH + R'OH + CO $\xrightarrow[\substack{\text{H}_2\text{O} \\ 190\ ^\circ\text{C, 24 h}}]{\substack{5\ \text{mol\%} \\ \text{Co}_2(\text{CO})_8}}$ R–C(O)–OR' + H$_2$S   (7.7)

R = Ar, ArCH$_2$ 850 psi   **10**

Co(R)(CO)$_4$ $\xleftarrow{\text{CO, R'OH}}$ ... $\xrightarrow[\text{- OH}^-]{\text{H}_2\text{O}}$ ...
+
SH$^-$

They proposed a reaction mechanism including: (1) reaction of RSH with Co(CO)$_4^-$ to provide CoR(CO)$_4$ and SH$^-$, (2) insertion of CO into the Co–R bond, and (3) alcoholysis of the resultant Co[C(O)R](CO)$_4$ to afford **10**.

BuSH + CH$_2$=CH–CH=CH$_2$ $\xrightarrow[\text{THF, 80 }^\circ\text{C}]{\text{Pd(acac)}_2/\text{PPh}_3/\text{Et}_3\text{Al}}$ BuS–CH$_2$CH=CHCH$_3$ + BuS–CH$_2$CH$_2$CH=CH$_2$ (7.8)

3 equiv   **11**   **12**

                                     7   :   3
                                           95%

To my knowledge, the first clear example of a transition metal-catalyzed addition of a thiol to a carbon-carbon unsaturated compound, 'hydrothiolation', was reported by Dzhemilev et al. in 1981 (Eq. 7.8) [20].

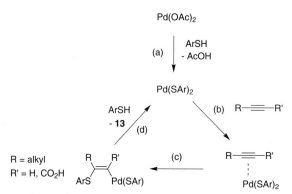

$$\text{ArSH} \quad + \quad \text{R}\!\!\equiv\!\!\text{R'(H)} \quad \xrightarrow{\text{cat.}} \quad \underset{\underset{\textbf{13}}{\overset{}{\text{ArS}}}}{\overset{\text{R} \quad \text{R'(H)}}{}}\!\!\diagdown\!\!\underset{\text{H}}{} \tag{7.9}$$

~87%

cat; Pd(OAc)$_2$, Pd(PPh$_3$)$_4$, Pt(PPh$_3$)$_4$, and NiCl$_2$(PPh$_3$)$_2$

R = n-C$_6$H$_{13}$, R' = H; R = TMS, R' = H; R = R' = Pr; R= n-C$_5$H$_{11}$, R' = CO$_2$H etc.

The addition of BuSH to 1,3-butadiene was catalyzed by Pd(acac)$_2$/PPh$_3$/Et$_3$Al in THF at 80°C to provide a mixture of **11** and **12** (7:3) in 95% combined yield. In 1992, it was reported that a number of commercially available metal complexes such as Pd(OAc)$_2$, Pd(PPh$_3$)$_4$, Pt(PPh$_3$)$_4$, and NiCl$_2$(PPh$_3$)$_2$ had the catalytic activity for the hydrothiolation of alkynes (Eq. 7.9) [21]. The Pd(OAc)$_2$-catalyzed reaction carried out in THF exhibited the best results to provide the Markovnikov adducts **13** when terminal alkynes were employed. The results of the hydrothiolation of internal alkynes and the deuteriothiolation of 1-octyne demonstrated that the hydrothiolation took place in a *cis*-fashion. The proposed reaction mechanism (Scheme 7-1) includes: (1) ligand exchange between Pd(OAc)$_2$ and ArSH to afford Pd(SAr)$_2$ with AcOH, (2) coordination of alkyne to the Pd center, (3) regioselective *cis*-insertion of the alkyne into the Pd–S bond with Pd bound to the carbon carrying R' (R' = H or CO$_2$H), and (4) protolysis of the vinyl palladium complex by ArSH and/or AcOH with the stereochemistry of vinyl moiety retained.

Scheme 7-1 A proposed reaction path of Pd(OAc)$_2$-catalyzed hydrothiolation of alkyne

Although the path (a), which is the initiation stage of the catalytic reaction, was actually confirmed by a stoichiometric reaction, no direct evidence has been provided about paths (b)–(d). When the reaction was carried out by using PdCl$_2$(PhCN)$_2$ as a catalyst precursor, the Markovnikov adducts formed *in situ* isomerized into internal vinyl sulfide **14** (Eq. 7.10) [22].

$$\text{ArSH} + \underset{R}{\overset{R}{\equiv}} \quad \xrightarrow[\substack{\text{benzene, 80 °C} \\ \text{20 h}}]{\substack{\text{5 mol\%} \\ \text{PdCl}_2\text{(PhCN)}_2}} \quad \underset{\substack{\text{up to 77\%} \\ \textbf{14}}}{R \diagup \diagdown \text{SAr}} \qquad (7.10)$$

$$\text{PhSH} + \text{PdCl}_2\text{(PhCN)}_2 \quad \xrightarrow{\hspace{3cm}} \quad \underset{\textbf{15}}{[\text{PdCl(SPh)(PhSH)}]_n} \qquad (7.11)$$

The authors proposed the complex $[\text{PdCl(SPh)(PhSH)}]_n$ **15** as an active catalyst, with thiol acting as a ligand of Pd(II) (Eq. 7.11). On the other hand, the reaction using RhCl(PPh$_3$)$_3$ as a catalyst precursor furnished the anti-Markovnikov *cis*-adducts **16** when a terminal acetylene was employed as a substrate (Eq. 7.12).

$$\text{ArSH} + R\text{—}\!\!\equiv \quad \xrightarrow{\text{RhCl(PPh}_3\text{)}_3} \quad \underset{\textbf{16}}{\overset{R}{\underset{H}{\diagup}\!\!=\!\!\underset{SAr}{\diagdown}}} \qquad (7.12)$$

The proposed reaction mechanism (Scheme 7-2) comprises: (1) oxidative addition of ArSH to RhCl(PPh$_3$)$_3$ to give Rh(H)(Cl)(SPh)(PPh$_3$)$_n$, (2) coordination of alkyne to the Rh complex, (3) *cis*-insertion of alkyne into the Rh–H bond with Rh positioned at terminal carbon and H at internal carbon, (4) reductive elimination of **16** from the Rh(III) complex to regenerate the Rh(I) complex.

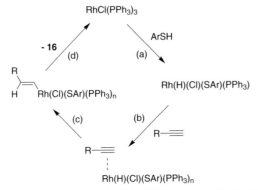

Scheme 7-2 A proposed reaction path of RhCl(PPh$_3$)$_3$-catalyzed hydrothiolation of alkyne

Although the path (a) has been verified by a stoichiometric reaction [23], the details of exact reaction mechanism remain unsettled. Triggered by this publication [and the Pd-catalyzed doublethiolation of alkynes described in Eq. (7.7) in Section 7-3], a number of transition metal-catalyzed additions of S–X or Se–X bonds to C-C unsaturated organic compounds started to be published. In 1994, Bäckvall et al. applied the Pd(OAc)$_2$-catalyzed hydrothiolation to conjugated enynes and obtained **17**, precursor of 2-(phenylsulfinyl) and 2-(phenylsulfonyl) 1,3-diene, **18** and **19**, both of which are useful synthetic building blocks for organic reactions (Eq. 7.13) [24].

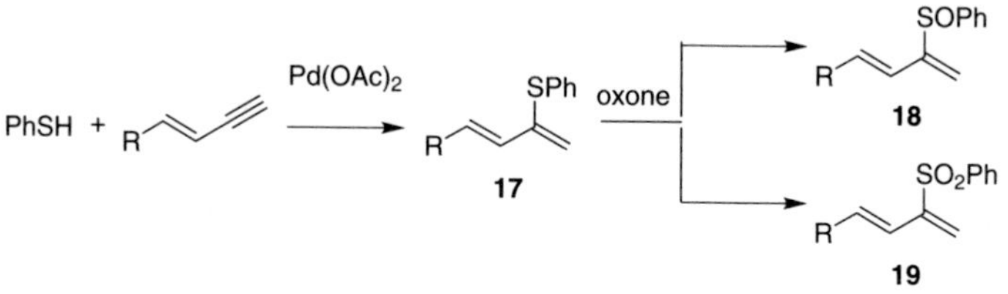

$$ \text{PhSH} + \underset{R}{\diagdown\!\!\!\equiv} \xrightarrow{\text{Pd(OAc)}_2} \underset{\mathbf{17}}{R\diagdown\!\!\diagup^{\text{SPh}}} \xrightarrow{\text{oxone}} \begin{cases} R\diagdown\!\!\diagup^{\text{SOPh}} \quad \mathbf{18} \\ R\diagdown\!\!\diagup^{\text{SO}_2\text{Ph}} \quad \mathbf{19} \end{cases} \tag{7.13}$$

In 1995, Koelle et al. reported the hydrothiolation of activated alkynes catalyzed by [Cp*Ru(SR)]$_2$ to provide **20** as a mixture of stereoisomers (Eq. 7.14) [25].

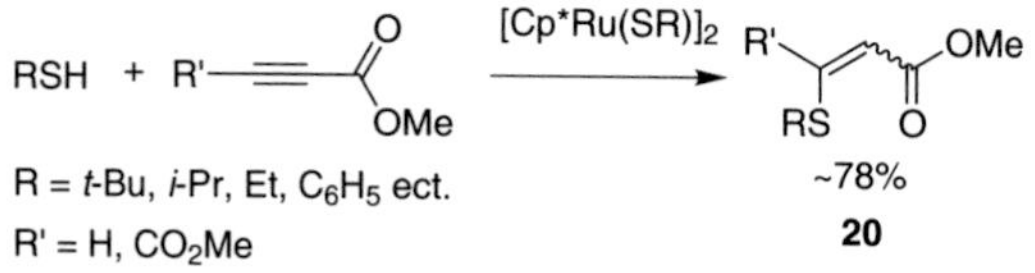

$$ \text{RSH} + R'\!-\!\!\!\equiv\!\!\!-\overset{O}{\underset{\text{OMe}}{\diagup\!\!\!\diagdown}} \xrightarrow{\text{[Cp*Ru(SR)]}_2} \underset{\underset{\mathbf{20}}{\sim 78\%}}{\underset{RS\ \ O}{R'\diagdown\!\!\diagup\!\!\diagdown\!\!\diagup^{\text{OMe}}}} \tag{7.14}$$

R = *t*-Bu, *i*-Pr, Et, C$_6$H$_5$ ect.
R' = H, CO$_2$Me

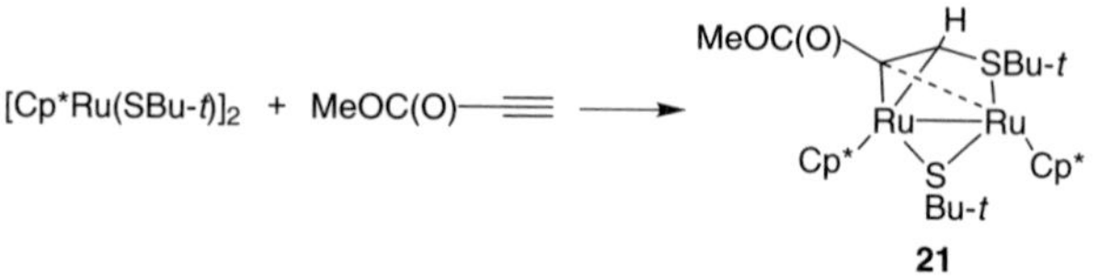

$$ \text{[Cp*Ru(SBu-}t)]_2 + \text{MeOC(O)}\!-\!\!\!\equiv \longrightarrow \underset{\mathbf{21}}{\begin{array}{c} \text{MeOC(O)} \quad\quad\text{H} \\ \diagup\!\!\!\diagdown\!\!\!\text{SBu-}t \\ \text{Ru}\!-\!\!-\!\text{Ru} \\ \text{Cp*} \diagup\ \ \diagdown \text{Cp*} \\ \text{S} \\ \text{Bu-}t \end{array}} \tag{7.15}$$

The authors confirmed the formation of vinyl Ru-complex **21** by the reaction of [Cp*Ru(SBu-*t*)]$_2$ with methyl propiolate (Eq. 7.15). To my knowledge, this is the first observation of the insertion of an alkyne into the M–S bond within a catalytically active metal complex. In 2000, Gabriele et al. reported the Pd-catalyzed cycloisomerization of (Z)-2-en-4-yne-1-thiol affording a thiophene derivative **22** (Eq. 7.16) [26].

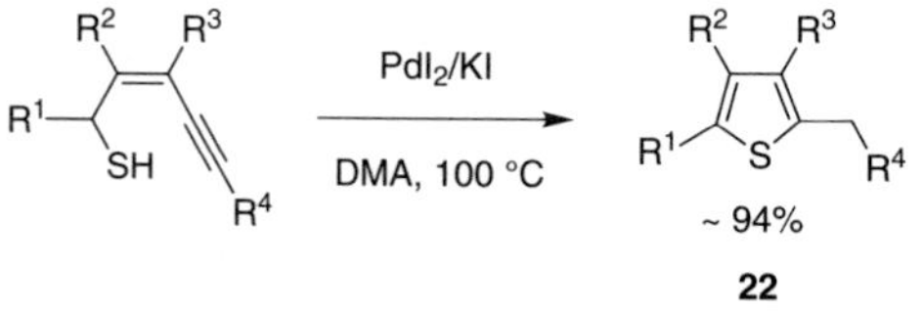

$$ \underset{\mathbf{}}{\begin{array}{c} R^2 \quad R^3 \\ R^1\diagdown\!\!\!\diagup\!\!\!\diagdown \\ \text{SH} \quad \diagdown R^4 \end{array}} \xrightarrow[\text{DMA, 100 °C}]{\text{PdI}_2/\text{KI}} \underset{\underset{\mathbf{22}}{\sim 94\%}}{\begin{array}{c} R^2 \quad R^3 \\ R^1\diagdown\!\!\!\diagup\!\!\!\diagdown\!\!\!\diagup R^4 \\ \text{S} \end{array}} \tag{7.16}$$

The authors proposed the conventional nucleophilic attack by an SH group to the triple bond of the alkyne coordinating to Pd(II) [path (a) in Scheme 7-3].

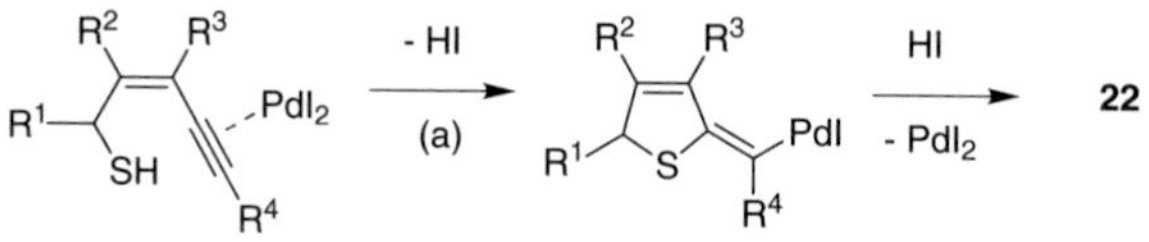

$$ \underset{\mathbf{}}{\begin{array}{c} R^2 \quad R^3 \\ R^1\diagdown\!\!\!\diagup\!\!\!\diagdown\!\!\!\cdot\text{PdI}_2 \\ \text{SH} \quad \diagdown R^4 \end{array}} \xrightarrow[\text{(a)}]{- \text{HI}} \underset{\mathbf{}}{\begin{array}{c} R^2 \quad R^3 \\ R^1\diagdown\!\!\!\diagup\!\!\!\diagdown\!\!\!\diagup\!\!\!\cdot\text{PdI} \\ \text{S} \quad\quad R^4 \end{array}} \xrightarrow[- \text{PdI}_2]{\text{HI}} \mathbf{22} $$

Scheme 7-3 A proposed reaction path of Pd-catalyzed cycloisomerization of (Z)-2-en-4-yne-1-thiol

However, a mechanism including intramolecular *cis*-thiopalladation is also possible (path (b) in Scheme 7-4).

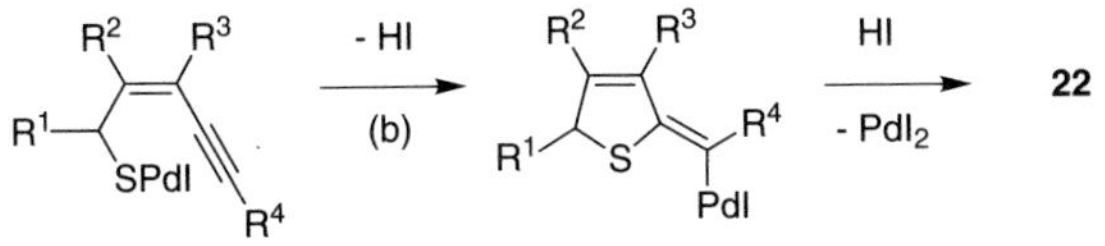

Scheme 7-4 An alternative reaction path of Pd-catalyzed cycloisomerization of (Z)-2-en-4-yne-1-thiol

In 1995, Ogawa et al. reported the Rh-catalyzed 'thioformylation' of alkyne, that is, addition of RS and formyl groups onto adjacent carbons to provide 23 (Eq. 7.17) [27].

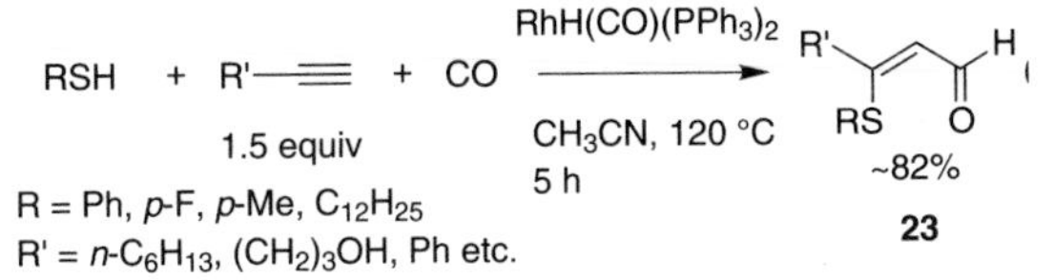

$$RSH + R'\!\!-\!\!\equiv\!\! + CO \xrightarrow[\substack{CH_3CN, 120\ °C \\ 5\ h}]{RhH(CO)(PPh_3)_2} \quad (7.17)$$

1.5 equiv

R = Ph, *p*-F, *p*-Me, $C_{12}H_{25}$
R' = *n*-C_6H_{13}, $(CH_2)_3OH$, Ph etc.

~82%

23

This reaction exemplified that the regiochemistry of RS and H introduced by carbonylative addition differed from that of those by simple hydrothiolation. In the Rh-catalyzed hydrothiolation, the ArS group added to the terminal carbon and H to the internal carbon (Eq. 7.12). On the other hand, in the Rh-catalyzed thioformylation, RS was placed at internal carbon and formyl at the terminal carbon in spite of using the same catalyst precursor, [RhCl(PPh₃)₃], which was also active for the thioformylation shown in Eq. (7.17). In 1997, the Pt-catalyzed hydrothiocarboxylation using RSH, alkyne and CO was reported to furnish 24 (Eq. 7.18), which showed the same regiochemistry as the Ni-catalyzed reaction shown in Eq. (7.1) [28].

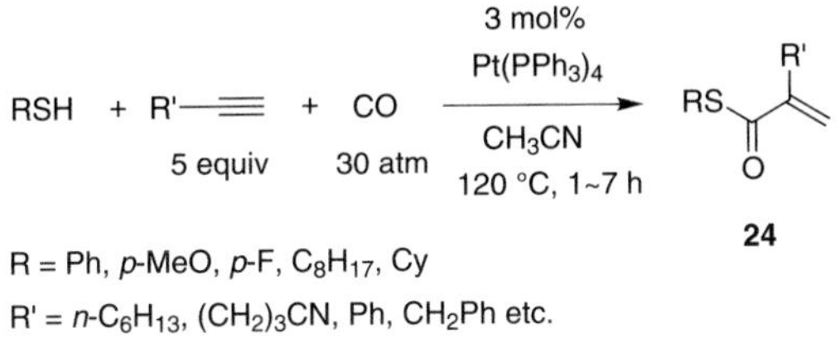

$$RSH + R'\!\!-\!\!\equiv\!\! + CO \xrightarrow[\substack{CH_3CN \\ 120\ °C,\ 1\sim7\ h}]{\substack{3\ mol\% \\ Pt(PPh_3)_4}} \quad (7.18)$$

5 equiv   30 atm

24

R = Ph, *p*-MeO, *p*-F, C_8H_{17}, Cy
R' = *n*-C_6H_{13}, $(CH_2)_3CN$, Ph, CH_2Ph etc.

Two possible reaction mechanisms, (A) and (B), are shown in Schemes 7-5 and 7-6. The former includes: (a) the oxidative addition of RSH to Pt(PPh₃)ₙ to give Pt(H)(SR)(PPh₃)₂, (b) insertion of CO into the Pt–S bond, (c) insertion of the alkyne into the Pt–C(O)SR bond with Pt bound to the terminal carbon and RSC(O) to the internal carbon, and (c) reductive elimination of the product 24 under regeneration of Pt(PPh₃)ₙ or (d′) protolysis by RSH to furnish 24 and Pt(H)(SR)(PPh₃)₂.

The step (a) was observed in a stoichiometric reaction [29]; however, nothing is known about the other routes (b)–(d). An alternative mechanism (B) includes: (a)

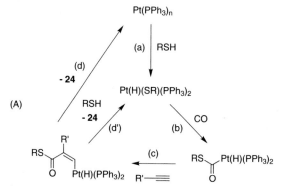

Scheme 7-5 A possible path of Pt(PPh₃)₄-catalyzed hydrothio-carboxylation of alkyne

oxidative addition of RSH to Pt(PPh₃)ₙ to give Pt(H)(SR)(PPh₃)₂; (b) insertion of the alkyne into the Pt–H bond with Pt bound to the internal carbon and H to the terminal carbon, (c) insertion of CO into the vinyl-C–Pt bond, and (d) reductive elimination of the product **24** under regeneration of Pt(PPh₃)ₙ.

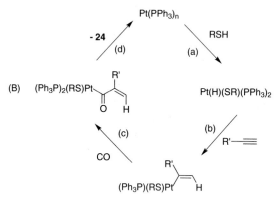

Scheme 7-6 Another possible reaction path of Pt(PPh₃)₄-catalyzed hydrothiocarboxylation of alkyne

When the platinum-catalyzed hydrothiolation was performed for acetylenic alcohols, intramolecular cyclization took place to afford α-methylene lactone **25** in up to 67% yield (Eq. 7.19) [30].

$$(7.19)$$

Thiol is acting as a co-catalyst in this transformation. In 1997, Alper et al. reported the reaction of thiols, propargyl alcohol, and CO catalyzed by Pd(PPh₃)₄ (Eq. 7.20) [31].

$$
\begin{array}{l}
\text{RSH} + \quad \text{(propargyl alcohol + R}^3\text{)} + \text{CO} \xrightarrow[\substack{\text{THF} \\ \text{400 psi, 100 °C}}]{\substack{\text{Pd(PPh}_3)_4 \\ \text{cat. } p\text{-TsOH}}} \quad \textbf{26} \\
\text{R = Ar, n-C}_8\text{H}_{17} \qquad \qquad \qquad \qquad \qquad \sim 85\%
\end{array}
\tag{7.20}
$$

$$
\xrightarrow[\substack{\text{toluene} \\ \text{600 psi, 110 °C}}]{\substack{\text{Pd(PPh}_3)_4 \\ \text{cat. } p\text{-TsOH}}} \quad \textbf{27} \quad \sim 77\%
$$

$$
\xrightarrow[\substack{\text{DME} \\ \text{400 psi, 100 °C}}]{\text{Pd(PPh}_3)_4} \quad \textbf{28} \quad \sim 88\%
$$

Depending on the reaction conditions, the major products were different: (1) formation of **26** in the presence of catalytic amount of p-TsOH in THF, (2) formation of **27** under higher pressure of CO at higher temperature in toluene, (3) formation of **28** in DME. Although no comment was made about the reaction mechanism, the formation of a propargyl or allenyl palladium complex may be involved in the catalytic cycle. In 1999, Alper et al. also reported the hydrothiocarboxylation of conjugated enynes to form **29** in up to 76% yield. The same regio- and stereochemistry as in the Pt-catalyzed reaction (Eq. 7.18) was realized using a palladium catalyst (Eq. 7.21) [32].

$$
\begin{array}{l}
\text{RSH} + \quad \text{(enyne, R}^1, R^2\text{)} + \text{CO} \xrightarrow[\substack{\text{THF} \\ 110\,°C}]{\substack{\text{Pd(OAc)}_2 \\ \text{dppp}}} \quad \textbf{29} \\
\qquad \qquad \quad 3 \sim 5 \text{ equiv} \qquad 400 \text{ psi} \qquad \quad \text{up to 76\%} \\
\text{R = Ph, PhCH}_2, \text{C}_8\text{H}_{17} \\
R^1 = R^2 = \text{Me; } R^1 = \text{H, } R^2 = \text{Ph etc.}
\end{array}
\tag{7.21}
$$

It was described that the use of the combination of Pd(OAc)₂/dppp /THF resulted in the highest yield. It is not clear whether such a combination of catalyst, ligand and solvent is restricted to the hydrothiocarboxylation of conjugated enynes or also may be applied to other alkynes. Furthermore, it remains to be explored whether, in the Pt-catalyzed reaction performed in CH₃CN [see Eq. (7.18)], conjugated enynes can be used.

$$
\text{Pd(OAc)}_2 + n\text{L} + \text{CO} \longrightarrow \text{Pd(0)L}_n + \text{Ac}_2\text{O} + \text{CO}_2
\tag{7.22}
$$

The oxidative addition of thiol to Pd(0), which can be generated *in situ* from Pd(OAc)$_2$ and excess phosphine (Eq. 7.22) [33], provides Pd(H)(SR)L$_n$ and was proposed as the initial step to enter the catalytic cycle. After the oxidative addition (path (a)), two routes, (A) in Scheme 7-7 and (B) in Scheme 7-8, can be envisioned.

The first one, (A), includes: (b) insertion of CO into the Pd–S bond; (c) insertion of the C≡C triple bond of the enyne into the Pd–C(O)SR bond whereby Pd binds to the terminal carbon and the RSC(O) group to the internal carbon, and (d) C–H bond-forming reductive elimination or protolysis by the thiol to form **29** (Scheme 7-7).

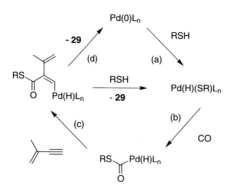

Scheme 7-7 A proposed reaction path of Pd-catalyzed hydrothiocarboxylation of enyne (R^1 = H, R^2 = Me)

The other mechanism (B) includes: (b) insertion of the C≡C triple bond of the enyne into the Pd–H bond with Pd bound to the internal carbon and H to the terminal carbon to give **30**, (c) insertion of CO into the Pd–C bond, and (d) reductive elimination of **29** under regeneration of Pd(0)L$_n$ (Scheme 7-8).

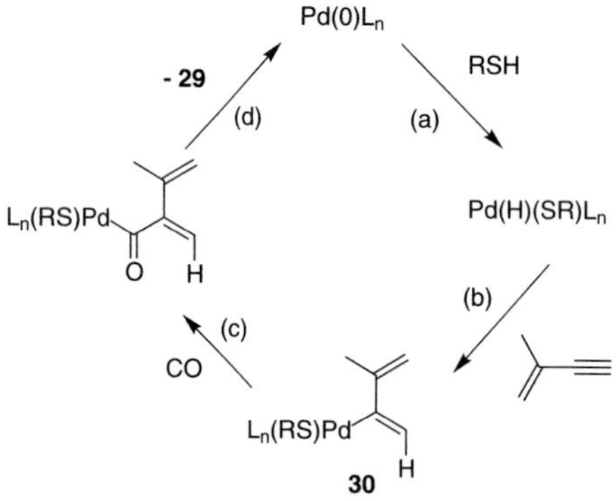

Scheme 7-8 Another possible route of Pd-catalyzed hydrothiocarboxylation of enyne

As Pd(OAc)$_2$ reacts with RSH to provide Pd(SR)$_2$, which could react with phosphine to form Pd(SPh)$_2$L$_n$ (Eq. 7.23) [34]; another mechanism, not involving the formation of Pd(0), can also be envisioned.

$$Pd(OAc)_2 + RSH + nL \longrightarrow Pd(SR)_2L_n + AcOH \qquad (7.23)$$

The path shown in Scheme 7-9 includes: (a) insertion of CO into the Pd–S bond of $Pd(SR)_2L_n$, (b) insertion of the C≡C triple bond of the enyne to afford **31**, and (c) protolysis of **31** by RSH to provide **29** with regeneration of $Pd(SR)_2L_n$.

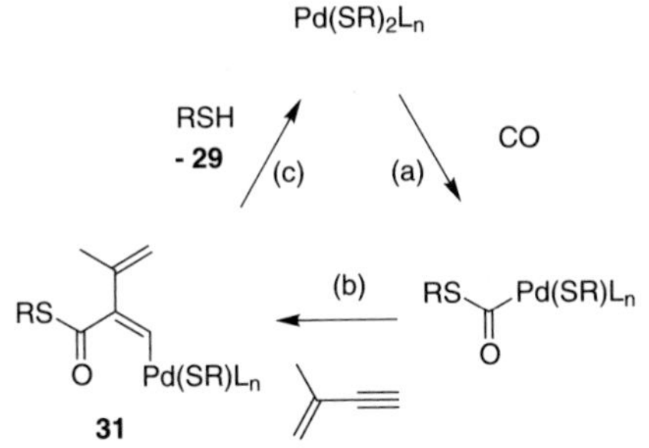

Scheme 7-9 The other possible route of Pd-catalyzed hydrothio-carboxylation of enyne

In 1996, Ogawa et al. reported the hydrothiolation of an allene catalyzed by $Pd(OAc)_2$ to provide **32**, whose formation can be explained as follows: (1) insertion of the allene into the Pd–S bond of $Pd(SPh)_2$ to give a π-allyl palladium thiolate **33**, and (2) protolysis by PhSH to form **32** and $Pd(SPh)_2$ (Eq. 7.24) [35]. The authors proposed the direct reaction of a σ-allyl palladium with PhSH before the formation of **33**.

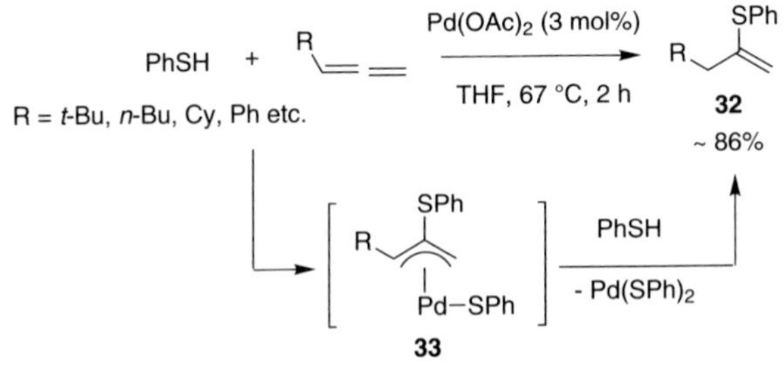

$$(7.24)$$

In 1998, Alper et al. reported the hydrothiocarboxylation of allenes using the same catalyst precursor and PPh_3 to provide **34** (Eq. 7.25) [36]. Again, the regiochemistry

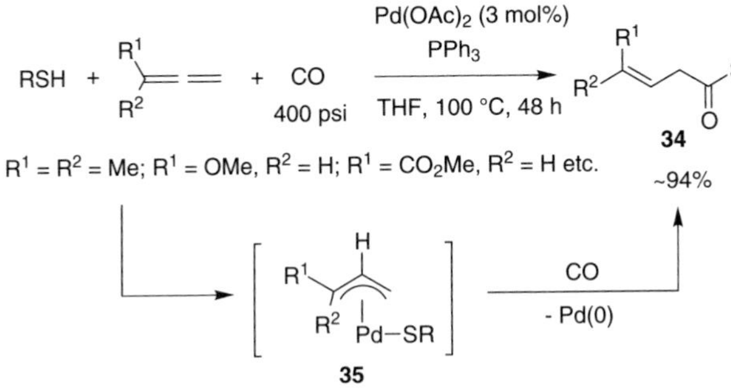

$$(7.25)$$

differed according to whether RS and H were incorporated by the hydrothiolation or the carbonylative reaction. A possible reaction path includes: (1) insertion of the allene into the Pd–H bond to form the π-allyl palladium complex **35**, (2) insertion of CO into the Pd–C bond, and (3) reductive elimination of **34** with the re-formation of Pd(0).

When the reaction of 1,1-dimethyl allene with o-BrC$_6$H$_4$SH was carried out in the presence of Pd(OAc)$_2$/dppf/i-Pr$_2$HN/CO in benzene, hydrothiolation of the allene took place (Eq. 7.26) [37]. However, the regiochemistry of the adduct **36** was different from that obtained by the Pd(OAc)$_2$-catalyzed hydrothiolation of mono-substituted allenes (cf. Eq. 7.24), showing that the regiochemistry of the hydrothiolation of allenes can be controlled by the reaction conditions even when the same metal(Pd) catalyst is used.

$$ \text{ArSH} \quad + \quad \text{(allene)} \quad \xrightarrow[\substack{\text{dppf} \\ i\text{-Pr}_2\text{NH, CO (400 psi)} \\ \text{PhH, 100 °C}}]{\text{Pd(OAc)}_2 \text{ (3 mol\%)}} \quad \underset{\substack{\textbf{36} \\ 72\%}}{\text{(product with SAr)}} \tag{7.26} $$

Ar = o-Br

On the other hand, when the reaction was performed with o-IC$_6$H$_4$SH, the cyclization product **37** was formed in up to 92% yield (Eq. 7.27).

$$ \text{(aryl-I, SH, X)} + \text{(alkene R}^1\text{R}^2\text{)} + \text{CO} \xrightarrow[\substack{\text{dppf} \\ i\text{-Pr}_2\text{N, benzene} \\ 100 °C}]{\text{Pd(OAc)}_2 \text{ (3 mol\%)}} \underset{\substack{\textbf{37} \\ \sim92\%}}{\text{(product)}} \tag{7.27} $$

400 psi

R^1 = R^2 = H; R^1 = H, R^2 = Cy, etc.
X = H, Cl, Me

A possible reaction mechanism shown in Scheme 7-10 includes: (a) oxidative addition of the S–H bond to Pd(0), (b) insertion of the allene into the Pd–H bond to form the π-allyl palladium **38**, (c) reductive elimination of allyl sulfide, (d) oxidative addition of the I-aryl bond into the Pd(0), (e) insertion of CO into the Pd–C bond, (f) insertion of the tethered C=C into the Pd–C(O) bond, and (g) β-elimination to form **37** followed by the formation of [baseH]I and Pd(0).

Kurosawa et al. have reported that the relative stability of the π-allyl palladium thiolate **39** and the allyl sulfide/Pd(0) was highly ligand dependent. In the presence of PPh$_3$ or P(OMe)$_3$ the stability was in favor of reductive elimination (Eq. 7.28), while in the presence of olefin or in the absence of any additional ligand the stability was in favor of oxidative addition (Eq. 7.29) [38]. This can explain the reactivity of the π-allyl palladium thiolate **33** and **38** proposed in Eq. (7.24) and path (c) of Scheme 7-10. The complex **33** should react with PhSH, but C–S bond-forming reductive elimination has to be suppressed in order to obtain the desired product **32**. On the other hand, the complex **38** requires the phosphine ligand to promote the C–S bond-forming reductive elimination.

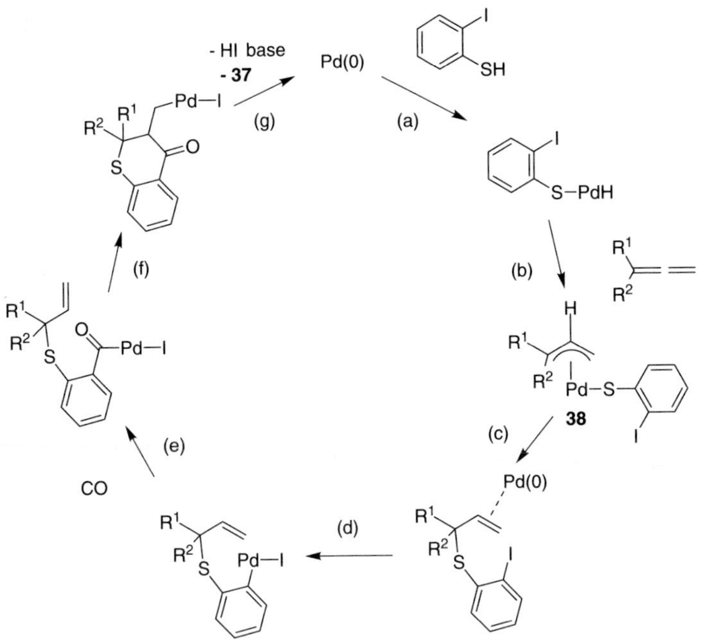

Scheme 7-10 A proposed reaction path of Pd-catalyzed carboylative heteroannulation of *o*-iodothiophenol with allene

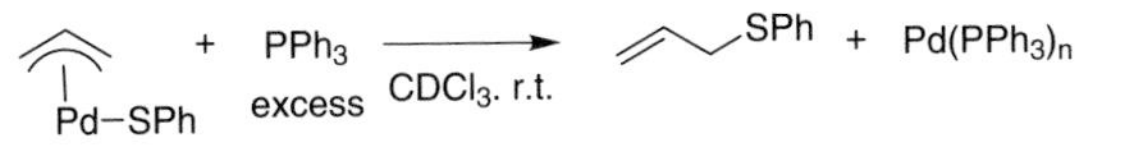

$$\text{(allene)} \quad + \quad PPh_3 \quad \xrightarrow[\text{CDCl}_3.\ \text{r.t.}]{\text{excess}} \quad \text{(allyl-SPh)} \quad + \quad Pd(PPh_3)_n \tag{7.28}$$

39

$$\text{(allyl-SPh)} \quad + \quad Pd_2(dba)_3 \quad \xrightarrow[\text{CDCl}_3.\ \text{r.t.}]{} \quad \textbf{39} \tag{7.29}$$

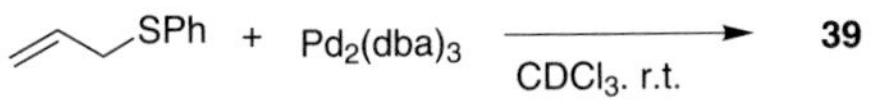

The addition of selenol to acetylene and allene, 'hydroselenation' is also catalyzed by Pd- Pt-, Rh-, and Ni-complexes to give **40** and **41/42** (Eqs. 7.30, 7.31) [39, 40].

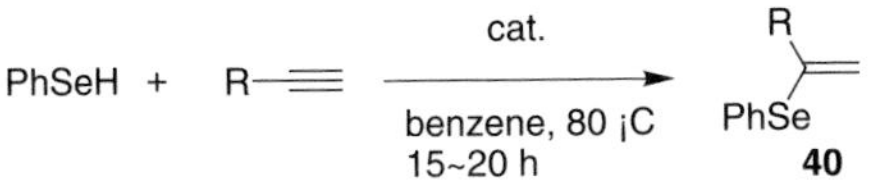

$$PhSeH \quad + \quad R\!\!\equiv\!\! \quad \xrightarrow[\substack{\text{benzene, 80 °C} \\ 15\sim20\ \text{h}}]{\text{cat.}} \quad \underset{\substack{PhSe \\ \textbf{40}}}{\overset{R}{\diagup\!\!=}} \tag{7.30}$$

$$\sim 72\%$$

cat; $Pd(OAc)_2$, $Pd(PPh_3)_4$, $PdCl_2(PhCN)_2$, $Pt(PPh_3)_4$, $RhCl(PPh_3)_3$, and $NiCl_2(PPh_3$

$R = n\text{-}C_6H_{13}$, $(CH_2)_3OH$, Ph

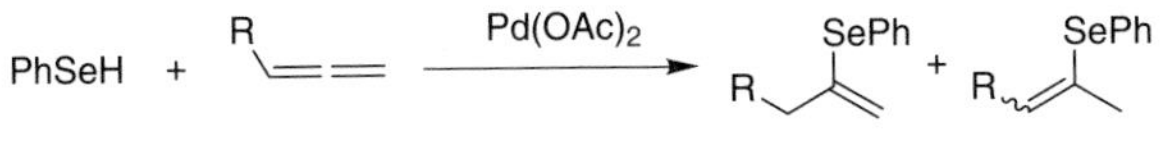

$$PhSeH \quad + \quad R\!\!\diagdown\!\!=\!\!= \quad \xrightarrow{Pd(OAc)_2} \quad \underset{\textbf{41}}{\overset{SePh}{R\diagup\!\!\diagup}} \quad + \quad \underset{\textbf{42}}{\overset{SePh}{R\diagdown\!\!\diagup}} \tag{7.31}$$

These reactions could proceed either *via:* (1) insertion of the alkyne or the allene into the M–Se bonds, and (2) C–H bond-forming reductive elimination (or protolysis by selenol) or *via:* (1) insertion of the alkyne or the allene into the M–H bond, and (2) C–Se bond-forming reductive elimination.

In 1998, the Pd-catalyzed formation of thioester **43** by the reaction of a thiol, an allyl alcohol, and CO in the presence of *p*-TsOH was reported (Eq. 7.32) [41].

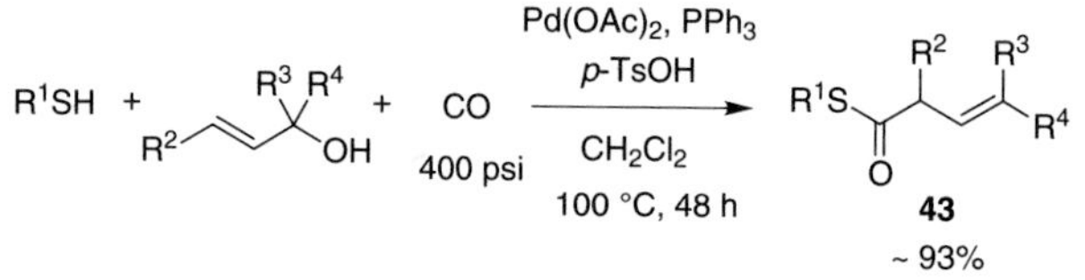

$$R^1SH + \underset{R^2}{\overset{R^3 \ R^4}{\diagdown}}\!\!\!\!\!\diagup\!\!\!\!\diagdown OH + CO \xrightarrow[\substack{CH_2Cl_2 \\ 100\,°C,\ 48\ h}]{\substack{Pd(OAc)_2,\ PPh_3 \\ p\text{-}TsOH}} \underset{\underset{O}{\parallel}}{R^1S}\diagdown\!\!\!\!\diagup\overset{R^2 \quad R^3}{\diagdown}\!\!\!\!\diagup R^4 \qquad (7.32)$$

400 psi

43
~ 93%

The proposed mechanism (Scheme 7-11) includes: (a) oxidative addition of a protonated alcohol to Pd(0) to provide the π-allyl palladium complex **44**, (b) nucleophilic replacement of H_2O by PhSH, (c) insertion of CO into the Pd–C bond, and (d) re-

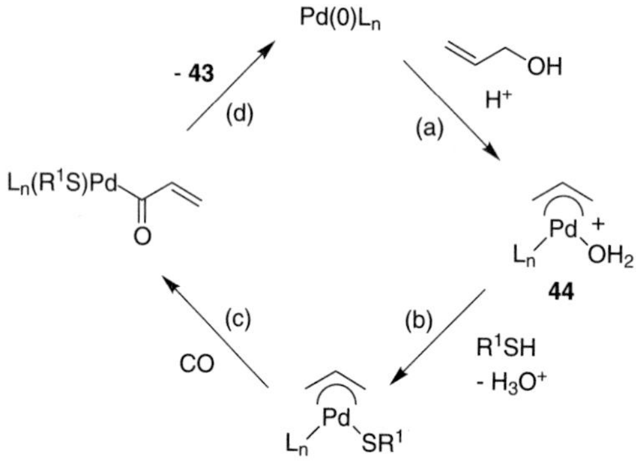

Scheme 7-11 A proposed reaction path of Pd-catalyzed dehydroxythiocarboxylation of allylic alcohol ($R^2 = R^3 = R^4 = H$)

ductive elimination of the product **43**. The paths (c) and (d) are likely to be similar to the processes involved in the hydrothiocarboxylation of allene (Eq. 7.25). Kondo et al. have reported the Ru-catalyzed formation of allyl sulfide **45** from alkyl carbonates and RSH in acetonitrile at r.t. after 1 h (Eq. 7.33) [42].

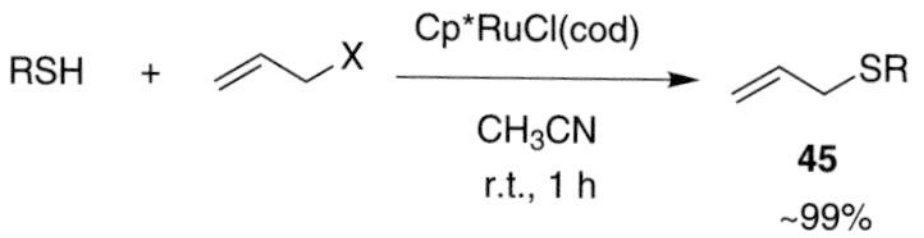

$$RSH + \diagup\!\!\!\!\diagdown\!\!\!\!\diagup X \xrightarrow[\substack{CH_3CN \\ r.t.,\ 1\ h}]{Cp^*RuCl(cod)} \diagup\!\!\!\!\diagdown\!\!\!\!\diagup SR \qquad (7.33)$$

45
~99%

X = OCO_2Me, OCO_2Et, $OCOCF_3$, OAc, OH

R = Ph, *n*-C_5H_{11}, *c*-Hex, C_2H_4OH etc.

Allyl acetate and allyl alcohol can be employed as substrates, too. The reaction using **46** took place with retention of the stereochemistry (Eq. 7.34), indicating that the

$$n\text{-}C_5H_{11}SH \quad + \quad \overset{\text{46}}{\underset{\text{'''CO}_2\text{Me}}{\bigcirc}} \quad \xrightarrow[\substack{\text{CH}_3\text{CH}_2\text{CN} \\ \text{reflux, 10 h}}]{\text{Cp*RuCl(cod)}} \quad \overset{}{\underset{\text{'''SC}_5\text{H}_{11}\text{-}n}{\bigcirc}} \qquad (7.34)$$

$$\overset{\text{PdSPh}}{\underset{\text{47}}{\bigcirc}}\text{CO}_2\text{Me} \quad \xrightarrow[\text{CDCl}_3,\ 0\ °\text{C, 5 min}]{\text{5 equiv of PPh}_3} \quad \overset{\text{CO}_2\text{Me}}{\underset{\text{PhS}}{\bigcirc}} \qquad (7.35)$$

overall process proceeds under double inversion of the stereochemistry. (It is known that the reductive elimination of the allyl sulfide from π-allyl palladium thiolate **47** proceeds under inversion of the configuration (Eq. 7.35) [38].

The Pd- and Ru-catalyzed carbonylation of allyl sulfide was reported to form **48** in good yield (Eq. 7.36) [43].

$$RS{\diagup\!\!\!\!\diagdown} \quad + \quad CO \quad \xrightarrow[\text{toluene, 140 °C}]{\substack{\text{Pd(OAc)}_2/\text{dppe} \\ \text{or Ru}_3(\text{CO})_{12}}} \quad RS\overset{O}{\diagdown}{\diagup\!\!\!\!\diagdown} \qquad (7.36)$$
$$68\ \text{atm} \qquad\qquad\qquad\qquad \overset{\textbf{48}}{\sim 88\%}$$

$$\text{ArOTf} \quad + \quad \text{BuSH} \quad \xrightarrow[\substack{t\text{-BuONa} \\ \text{toluene, 80 °C, 24 h}}]{\substack{\text{Pd(OAc)}_2 \\ tol\text{-BINAP}}} \quad \begin{matrix}\text{ArSBu} \\ \textbf{49} \\ 60 \sim 93\%\end{matrix} \qquad (7.27)$$

Ar = Ph, 2-naphthyl, p-BuC$_6$H$_4$, p-MeOC$_6$H$_4$ etc.

The Pd-catalyzed coupling reaction of aryl triflate with BuSNa can be nicely achieved by using *tol*-BINAP as ligand to furnish **49** (Eq. 7.37) [44]. No formation of **49** was observed in the presence of P(tol-*p*)$_3$ or P(2-furyl)$_3$, again indicating that the reactivity for the C–S bond-forming reductive elimination from Pd(SR)(Ar)L$_n$ strongly depends on the species of L. The reaction using PhSH did not afford the desired coupling product.

$$\begin{matrix}\text{ArSH} \\ 19.6\ \text{mmol}\end{matrix} \quad + \quad \underset{4.1\ \text{mmol}}{\overset{R\diagdown{}^O}{\underset{N}{\bigwedge}}} \quad \xrightarrow[\text{CH}_2\text{Cl}_2,\ 0\ °\text{C, 14 h}]{\text{Et}_2\text{Zn/L-DIPT}} \quad \underset{\substack{\textbf{50} \\ \text{up to 88\% ee}}}{\overset{R\diagdown{}^O}{\underset{\text{ArS}\diagdown\ \diagup\text{NH}}{\bigcirc}}} \qquad (7.38)$$

So far, progress on the late transition metal-catalyzed reactions utilizing S–H bond activation has been surveyed. Finally, the recent advancement of chiral Lewis acid-catalyzed reactions using thiol as nucleophile will be briefly described. Oguni et al. reported the asymmetric ring-opening reaction of *n*-acylaziridine with *t*-BuC$_6$H$_4$SH in the presence of Et$_2$Zn/L-(+)-diisopropyl tartarate to give **50** in up to 88% *ee* in 98% yield (Eq. 7.38) [45].

The enantioselectivity was significantly influenced by the steric factor of the thiols employed. When *p*-MeC$_6$H$_4$SH and PhSH were used, the optical yields decreased to 69% and 3%, respectively. Shibasaki et al. have reported that gallium-lithium-binaphthoxide (GLB) **51** became a good catalyst for the enantioselective ring opening reaction of epoxide for the production of **52** (Eq. 7.39) [46].

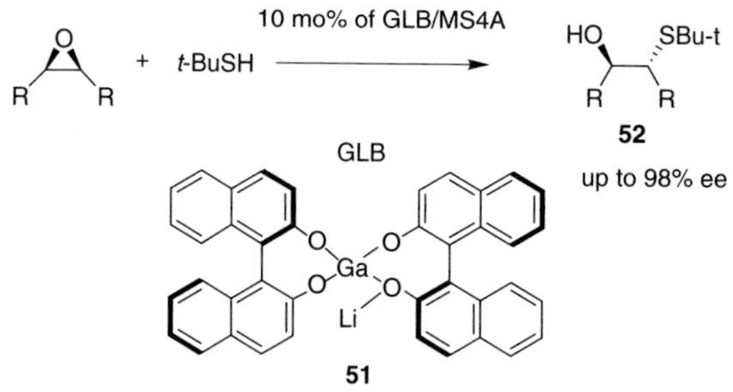

(7.39)

By using 10 mol% of **51**, MS$_4$A, and *t*-BuSH, the desired product **52** was obtained in up to 98% *ee* in 80% yield. A complementary role by two metals (Ga and Li) in activating and positioning both of the substrates has been proposed. The MS4A (sodium aluminosilicate) accelerated the reaction; however, the actual role of this additive was not clearly defined, although the possibilty that MS4A delivers Na$^+$ ions was pointed out. Tomioka et al. reported the asymmetric Michael addition of an aromatic thiol to α,β-unsaturated esters in the presence of 8 mol% of **53** to provide **54** in up to 97% *ee* in 99% yield (Eq. 7.40) [47].

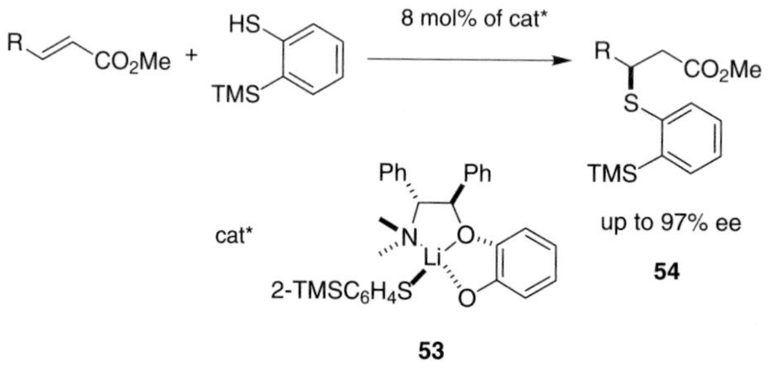

(7.40)

By using LaNa$_3$-tris(binaphthoxide) (LSB) **55**, catalytic asymmetric Michael addition of thiols to cycloalkenones took place to provide the adduct **56** with high *ee*s in good yields (Eq. 7.41) [48].

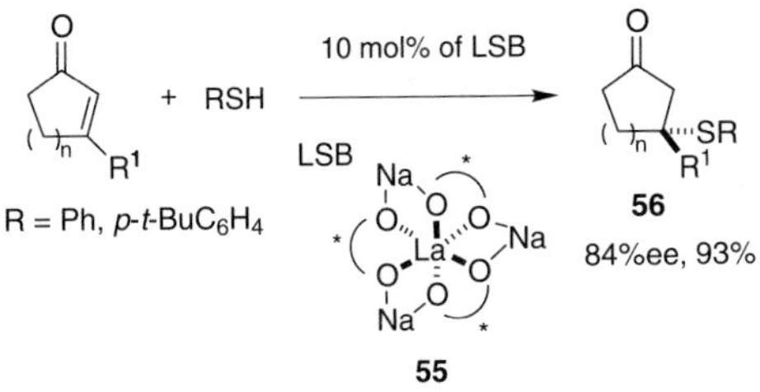

$$(7.41)$$

R = Ph, p-t-BuC$_6$H$_4$

55

56

84%ee, 93%

7.3
Catalytic S–S Activation

To my knowledge, the first transition metal-catalyzed reaction utilizing S–S bond activation was reported by Holmquist el al. in 1960 [14]. The reaction of (PhS)$_2$ with CO (950 atm) in the presence of chromium oxide on Al$_2$O$_3$ at 275°C furnished thioester **57** in 31% yield (Eq. 7.42).

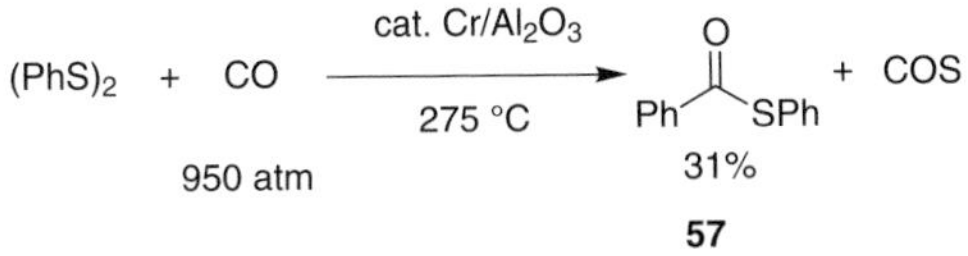

$$(7.42)$$

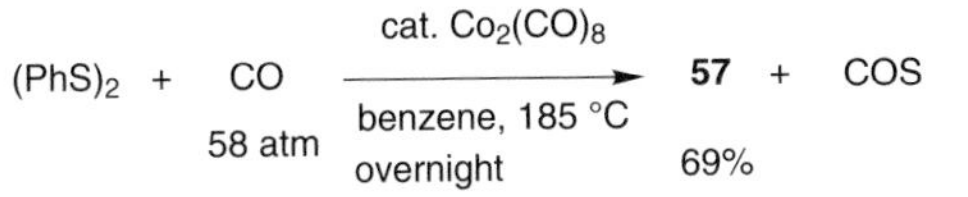

(PhS)$_2$ + CO $\xrightarrow[\text{275 °C}]{\text{cat. Cr/Al}_2\text{O}_3}$ Ph–C(O)–SPh + COS

950 atm

31%

57

$$(7.43)$$

(PhS)$_2$ + CO $\xrightarrow[\substack{\text{benzene, 185 °C}\\ \text{overnight}}]{\text{cat. Co}_2\text{(CO)}_8}$ **57** + COS

58 atm

69%

In 1985, Alper et al. discovered that the same reaction took place under much milder conditions using Co$_2$(CO)$_8$ as a catalyst and benzene as a solvent (58 atm of CO at 185°C overnight) to afford **57** in 69% yield (Eq. 7.43) [49]. The proposed mechanism (Scheme 7-12) includes: (a) oxidative addition of (PhS)$_2$ to Co$_2$(CO)$_8$ to provide Co(SPh)(CO)$_4$, (b) insertion of CO into the Co–S bond; (c) elimination of COS, (d) insertion of CO into the Co–Ph bond, and (e) reaction of (PhS)$_2$ with Co[C(O)Ph](Co)$_4$ to provide **57** under regeneration of Co(SPh)(CO)$_4$.

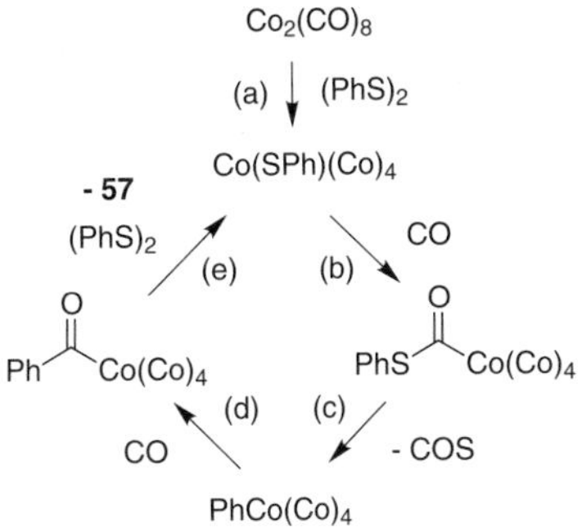

Scheme 7-12 A proposed reaction path of Co-catalyzed dethio-carbonylation of (PhS)$_2$

This exemplified that the oxidative addition of S–S bond to a low-valent metal complex is one prototype to initiate a reaction using a disulfide. In 1987, Uemura et al. reported an analogous transformation using $(PhSe)_2$ instead of $(PhS)_2$ to afford the phenyl selenobenzoate 58 in up to 78% yield under 100 atm of CO in benzene at 200°C (Eq. 7.44) [50].

$$(PhSe)_2 + CO \xrightarrow[\text{benzene, 200 °C} \atop \text{5-100 atm}]{\text{cat. Co}_2\text{(CO)}_8/\text{PPh}_3} \quad \underset{\substack{\mathbf{58} \\ \sim 78\%}}{Ph\overset{O}{\underset{}{\diagup}}SePh} \qquad (7.44)$$

$$(PhSe)_2 + Ph\overset{O}{\underset{}{\diagup}}Co(Co)_4 \xrightarrow[\text{THF, 20 °C}]{} \underset{45\%}{\mathbf{58}} \qquad (7.45)$$

They confirmed that benzoylcobalt tetracarbonyl actually reacted with $(PhSe)_2$ to furnish 58 in 45% yield at 20°C (Eq. 7.45). In 1986, Dzhemilev et al. reported the reaction of $(PhS)_2$ with 1,3-butadiene catalyzed by [Ni(acac)$_2$/(dppe)] in DMF at 150°C to provide a mixture of 59, 60, and 61 in 80% combined yield (Eq. 7.46) [51].

$$(PhS)_2 + \diagup\!\!\!\diagdown\!\!\!\diagup \xrightarrow[\text{DMF, 150 °C}]{\text{Ni(acac)}_2/\text{dppe}} \underset{\mathbf{59}}{PhS\diagup\!\!\!\diagdown\!\!\!\diagup} + \underset{\mathbf{60}}{PhS\diagup\!\!\!\diagdown\!\!\!\diagup\!\!\!\diagdown} \qquad (7.46)$$

$$+ \underset{\substack{\mathbf{61} \\ n = 1\sim4}}{PhS\!\!\left(\!\diagup\!\!\!\diagdown\!\!\!\diagup\!\right)_n\!\!SPh} \quad 80\% \text{ combined yield}$$

Although the selectivity of the reaction was unsatisfactory, this is, to my knowledge, the first documentation describing the transition metal catalyzed 'doublethiolation' of C–C unsaturated organic compounds.

$$(ArY)_2 + R\!\!-\!\!\equiv \xrightarrow[\text{benzene, 80 °C, 12 h}]{\text{Pd(PPh}_3)_4} \underset{\substack{\mathbf{62} \\ \text{up to 98\%}}}{\overset{R}{\underset{ArY\quad YAr}{\diagup\!\!\diagdown}}} \qquad (7.47)$$

$$R = n\text{-}C_6H_{13}, Ph, Me_3Si \text{ etc.}$$

In 1991, the first practical doublethiolation and doubleselenation, that is, the Pd-catalyzed addition of diaryl disulfide and diaryl diselenide to terminal alkynes, were reported to provide cis-adducts 62 in good yields (Eq. 7.47) [52]. The proposed reaction mechanism (Scheme 7-13) comprises: (a) oxidative addition of $(ArY)_2$ (Y = S, Se) to Pd(PPh$_3$)$_n$, (b) coordination of alkyne to the Pd-complex, (c) insertion of the alkyne into the Pd–Y bond to provide vinyl palladium complex 63, in which the Pd binds to the terminal carbon and the YAr group to the internal carbon, and (d) C–Y bond-forming reductive elimination to give 62 under regeneration of Pd(PPh$_3$)$_n$. Although the path (a) is known to proceed at room temperature, there is no direct evidence for the paths (b)–(d). It should be emphasized, however, that this paper clear-

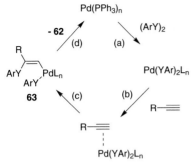

Scheme 7-13 A proposed reaction path of Pd-catalyzed doublethiolation and doubleselenation of alkyne

ly indicated for the first time the generality of the insertion of alkynes into Pd–S and
Pd–Se bonds. Only terminal alkynes could be used in these transformations. When
the reaction was carried out in the presence of CO, 'thio-thiocarboxylation' and 'seleno-selenocarboxylation', that is, introduction of ArY and ArYC(O) groups onto adjacent carbons took place to afford **64** (Eq. 7.48).

$$(ArY)_2 + R\!\!\equiv\!\! + \underset{1\sim60\ atm}{CO} \xrightarrow[\text{benzene, 80 °C, 12 h}]{\text{Pd(PPh}_3)_4 \text{ or} \atop \text{PdCl}_2(\text{PPh}_3)_2} \underset{\underset{\textbf{64}}{\text{up to 89\%}}}{} \tag{7.48}$$

R = alkyl; either Pd(PPh₃)₄ or PdCl₂(PPh₃)₂
R = Ph; must use PdCl₂(PPh₃)₂
R = Me₃Si; both failed

The carbon monoxide was regioselectively attached to the terminal carbon. The
nature of the alkynes dramatically influenced the efficiency of the carbonylation.
Aliphatic terminal alkynes like 1-octylene were successfully carbonylated by both
Pd(PPh₃)₄ and PdCl₂(PPh₃)₂ as catalyst precursors. However, the reaction between
phenyl acetylene, (ArY)₂, and CO catalyzed by Pd(PPh₃)₄ provided only **62**. The use
of PdCl₂(PPh₃)₂ overcame this restriction and **64** was obtained in moderate yields (R
= Ph; Y = S, 29%; Y = Se, 86%). In marked contrast, when trimethylsilylacetylene
was employed as a substrate, only a simple double thiolation took place with
Pd(PPh₃)₄, while no reaction was observed with PdCl₂(PPh₃)₂.

Starting from **63**, the carbonylation may proceed *via* coordination and insertion of
CO into the vinyl-C–Pd bond to provide an α,β-unsaturated acyl complex. This complex reacts with (ArY)₂, and subsequently the C–Y bond is formed by reductive elimination to give **64** (Scheme 7-14). Because the compound **64** could be directly converted into the corresponding enal **65** by the Pd-catalyzed reduction with Bu₃SnH,
this sequence is synthetically equivalent to the regio- and stereoselective thioformylation and selenoformylation of alkynes (Eq. 7.49) [53, 54].

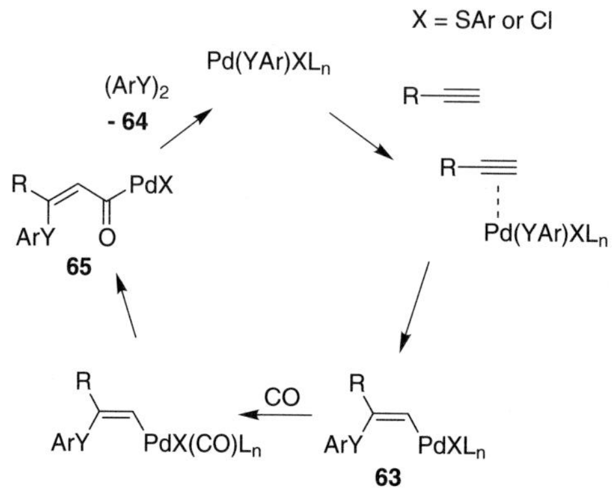

X = SAr or Cl

Scheme 7-14 A possible reaction path of Pd-catalyzed thio-thio-carboxylation and selenoselenocarboxlation of alkyne

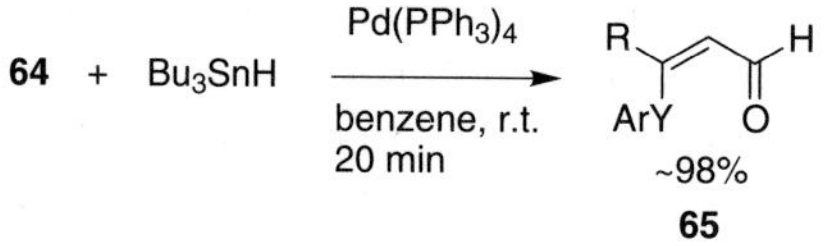

$$ \textbf{64} \ + \ Bu_3SnH \ \xrightarrow[\substack{\text{benzene, r.t.} \\ \text{20 min}}]{Pd(PPh_3)_4} $$

~98%

65

$$(7.49)$$

Two possible reaction mechanisms, A and B, are envisioned as shown in Scheme 7-15 and Scheme 7-16, respectively.

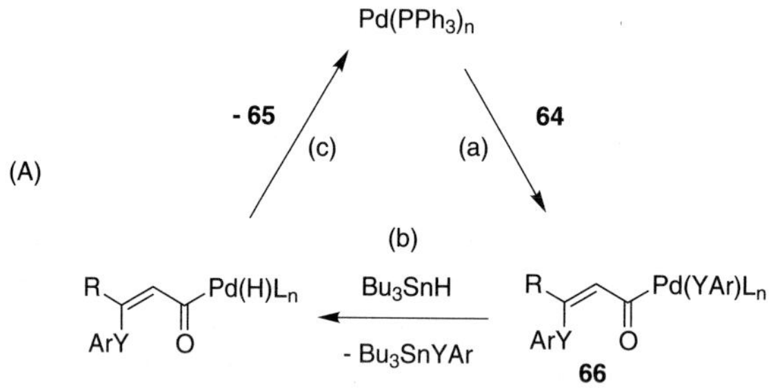

Scheme 7-15 A possible reaction path of Pd-catalyzed reduction of **64** to **65**

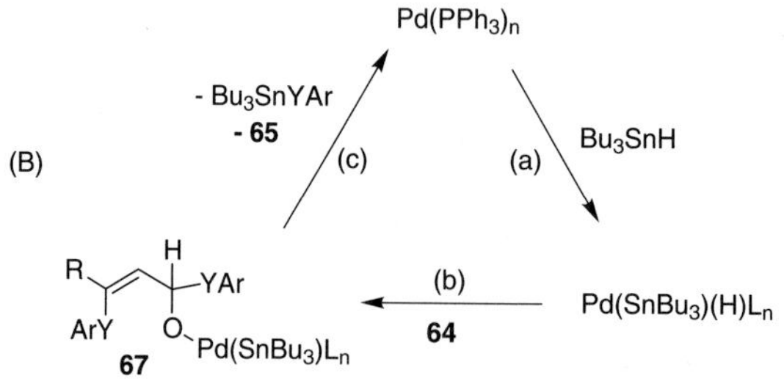

Scheme 7-16 An alternative path of Pd-catalyzed reduction of **64** to **65**

The former, A, includes: (a) oxidative addition of **64** to Pd(PPh₃)ₙ to form **66**; (b) ligand exchange between H of Bu₃SnH and YAr of **66**; (c) reductive elimination of

65 with regeneration of $Pd(PPh_3)_n$. An alternative mechanism B may be formulated: (a) oxidative addition of Bu_3SnH to $Pd(PPh_3)_n$, (b) addition of the Pd–H bond to $–C(O)YAr$ of **64** to form **67**, and (c) β-YAr elimination to produce **65** followed by reductive elimination of Bu_3SnYAr from the resultant Pd(II) complex. When alkynes having an –OH group in propargyl or homopropargyl positions were treated under thio-thiocarboxylation or seleno-selenocarboxylation conditions at 100°C for 50 h, a one-pot lactonization occurred to furnish **68** in up to 70% yield (Eq. 7.50) [55].

$$(PhY)_2 \ + \ \text{HO} \diagdown \cdots \equiv \ + \ CO \xrightarrow[\substack{60 \ \text{atm} \\ 50 \ \text{h}}]{\substack{Pd(PPh_3)_4 \\ \text{benzene, } 100 \ °C}} \quad \text{68} \qquad (7.50)$$

Y = S, Se
n = 1, 2
up to 70%

Gareau et al. reported the Pd-catalyzed doublethiolation of terminal alkyne using $(i\text{-}Pr_3SiS)_2$ to provide cis-adduct **69**, which was in situ converted into vic-di(methylthio)alkene (**70**) by the treatment with TBAF and MeI (Eq. 7.51) [56]. Because the direct Pd-catalyzed addition of dialkyl disulfide to terminal acetylene hardly proceeded, this method provided an effective way to synthesize the equivalent of a dialkyldisulfide adduct.

Some noteworthy studies using other substrates than alkynes have been conducted in S–S and Se–Se bond activation chemistry.

$$(i\text{-}Pr_3SiS)_2 \ + \ R\!\!-\!\!\!\equiv \ \xrightarrow{Pd(PPh_3)_4} \ \underset{\textbf{69}}{\overset{R}{\underset{(i\text{-}Pr)_3SiS \quad SSi(Pr\text{-}i)_3}{\diagup}}} \ \xrightarrow[\text{MeI}]{\text{TBAF}} \ \underset{\textbf{70}}{\overset{R}{\underset{MeS \quad SMe}{\diagup}}} \qquad (7.51)$$

Fukuzawa et al. reported that Pd-catalyzes the reaction of allylacetate, $(PhSe)_2$ and SmI_2 to form allylselenide **71** (Eq. 7.52) [57]. The complex **72**, which can be generated by the contact of $(PhSe)_2$ with Pd(0) in the actual reaction mixture, was hardly active as a catalyst, indicating that the reaction proceeded via oxidative addition of allyl acetate to Pd(0) rather than oxidative addition of $(PhSe)_2$ to Pd(0).

$$(PhSe)_2 \ + \ R\diagdown\diagup\diagdown OAc \ + \ SmI_2 \ \xrightarrow[\text{THF, } 25 \ °C, \ 10\text{-}15 \ h]{PdCl_2/PPh_3} \ R\diagdown\diagup\diagdown SePh \qquad (7.52)$$

71

$$\underset{\textbf{72}}{\overset{Ph}{\underset{Ph}{\overset{Ph_3P \diagdown Se \diagup SePh}{\underset{PhSe \diagup Se \diagdown PPh_3}{Pd \qquad Pd}}}}}$$

hardly active as a catalyst

The Pd-catalyzed reaction of $(ArS)_2$ with isocyanide ArNC **73** (both Ar = p-tol) and its mechanistic study was reported in 1997 (Eq. 7.53) [58]. When the reaction of $(ArS)_2$ with 4 equiv of **73** was performed in the presence of 2 mol% of $Pd(PPh_3)_4$ in benzene at 80°C for 12 h, 81% of the 1:2-adduct **75** was obtained with a concomitant formation of the 1:1-adduct **74** and the 1:3-adduct **76**. A similar reaction using Pt(PPh₃)₄ as a catalyst precursor ended up with the formation of a complicated polymeric product.

$$(ArS)_2 + ArNC \xrightarrow[\substack{\text{benzene} \\ \text{80 °C}}]{Pd(PPh_3)_4} \underset{\textbf{74}}{ArS \overset{NAr}{\diagdown} SAr} + \underset{\textbf{75}}{ArS \overset{NAr}{\diagdown}_2 SAr} \underset{\textbf{76}}{ArS \overset{NAr}{\diagdown}_3 SAr} \qquad (7.53)$$

73

major

A possible reaction path shown in Scheme 7-17 includes: (a) oxidative addition of (ArS)₂ to Pd(PPh₃)ₙ to give Pd(SAr)₂(PPh₃)ₙ **77**, (b) coordination of isocyanide **73** to the palladium center to form Pd(SAr)₂(CNAr)(PPh₃)ₙ **78**, (c) insertion of isocyanide into the Pd–S bond to give the thioimidoyl palladium complex **79**. Then either reductive elimination (d) may occur to give **74** directly or, (e), the insertion of further isocyanide(s) into the Pd–S or the Pd–C bonds provides either **80** or **81**. From these latter complexes, reductive elimination (f) of the 1:m-adduct regenerates Pd(PPh₃)ₙ.

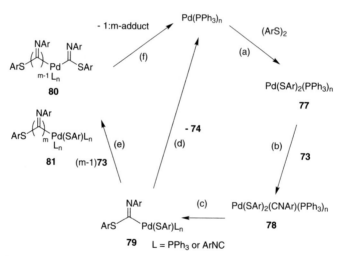

Scheme 7-17 A possible path of Pd-catalyzed reaction of (ArS)₂ with **73**

The stoichiometric reaction (Scheme 7-18) showed the actual formation of **78** via **77′** and **77″** at room temperature. Heating of **78** provided neither **79** nor **74**. However, in the presence of another equiv of (ArS)₂, **74** (73% after 15 h) and **77″** were produced at 80°C.

The reaction of **74** with Pd(PPh₃)₄ afforded **78**, but the intermediate **79** was not confirmed (Eq. 7.54). These simple stoichiometric reactions clearly suggested the relative energy of a reaction system composed of 2 equiv of (ArS)₂, 1 equiv of **73**, and 1 equiv of Pd(PPh₃)₄ (Scheme 7-19).

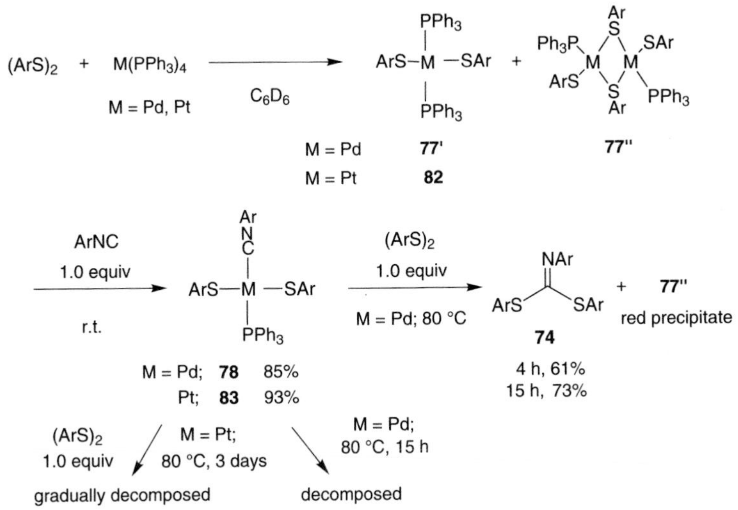

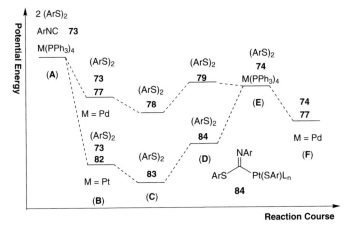

Scheme 7-18 Mechanistic study for the Pd- and Pt-catalyzed reaction of (ArS)$_2$ with **73**

$$74 \quad + \quad M(PPh_3)_4 \quad \xrightarrow{\quad C_6D_6 \quad} \quad 78(83) \qquad\qquad (7.54)$$

The oxidative addition of (ArS)$_2$ to Pd(0) and coordination of **73** to the resultant Pd(II) both lower the total energy [state (**C**) from (**A**) *via* (**B**)]. Both the insertion of isocyanide into Pd–S of **78** giving state (**D**) and the reductive elimination of **74** from **79** affording state (**E**) are reversible. The equilibrium of the insertion and de-insertion of the isocyanide favors the formation of the product of the de-insertion reaction. [State (**C**) is lower than state (**D**).] Although state (**C**) is more stable than state (**E**), the short-lived Pd(0) can be trapped by (ArS)$_2$ to give **77** [state (**F**) from state (**E**)].

Scheme 7-19 Predicted qualitative reaction coordinate diagrams (free PPh$_3$s omitted)

The treatment of (ArS)$_2$ with Pt(PPh$_3$)$_4$ and subsequently with **73** produced Pt(SAr)$_2$(ArNC)(PPh$_3$) (**83**) *via* **82** similarly to the case of Pd (Scheme 7-18). However, in contrast to the case of **78**, **83** did not provide **74** even in the presence of another equiv of (ArS)$_2$ at 80°C for 3 days. The oxidative addition of **74** to Pt(PPh$_3$)$_4$ afforded **83** similarly to the case of the corresponding Pd-complex (Eq. 7.54). Based on these results, the reason for the different catalytic activity of Pd-complex and Pt-complex was also explained using Scheme 7-19. It was suspected that the different activities stem from the different relative energy gaps between the M(0) and M(II) states. Since more energy is gained when Pt(0) is converted into Pt(II) than when Pd(0) is oxidized formally to Pd(II) [from state (**A**) to state (**B**)], reductive elimination from the Pt(II) complex [from state (**D**) to state (**E**)] makes C–S bond forming very difficult. This is why the Pt-catalyzed reaction provides polymeric products by multiple isocyanide incorporation.

76 + **73** (7.55)

Pd(PPh$_3$)$_4$ (3 mol%)

4.7 equiv

80 C°, 36 h

74 + **75** + **76** +

1.6% 5% 21%

ArS⟨⟩$_4$SAr + ArS⟨⟩$_5$SAr

NAr NAr

18%(4%) 8%(4%)

ArS⟨⟩$_6$SAr + ArS⟨⟩$_7$SAr + ArS⟨⟩$_8$SAr + ArS⟨⟩$_9$SAr

NAr NAr NAr NAr

5%(3%) 6%(5%) 5%(5%) 8%(11%)

+ others

yields in ( ) based on **73**

The reaction of the 1,3-adduct **76** with an excess amount (4.7 equiv) of **73** produced a mixture of adducts (1:1 to 1:9), which were separated and isolated by PTLC and HPLC (Eq. 7.55). A similar treatment of (PhSe)$_2$ with **73** in the presence of the Pd-complex [Pd(PPh$_3$)$_4$] provided also the 1;2-adduct **86** and the 1;3-adduct **87** (Eq. 7.56) [59].

(ArSe)$_2$ + **73** → Pd(PPh$_3$)$_4$

ArSe⟨⟩SeAr + ArSe⟨⟩$_2$SeAr + ArSe⟨⟩$_3$SeAr (7.56)

NAr NAr NAr

85 **86** **87**

0% 19% 18%

In marked contrast to the reaction with (ArS)$_2$, the 1:1-adduct **85** was not formed at all. The stoichiometric study also supported this fact, indicating that the formation of **85** from (ArSe)$_2$ with **73** is a thermodynamically unfavorable process.

In 1999, Kondo and Mitudo et al. reported the Ru-catalyzed addition of (PhS)$_2$ to alkenes to give **88**, the first transition metal-catalyzed addition of an S–S bond to an alkene (Eq. 7.57) [60]. When the reaction of (RS)$_2$ (R = Ph, Me, Bu) with

2-norbornene was carried out in the presence of Cp*RuCl(cod), the *cis*-adduct **89** was formed, indicating the inclusion of *cis*-insertion of alkene into the Ru–S bond and following C–S bond-forming reductive elimination (Eq. 7.58). The authors proposed the complex [Cp*RuCl(μ-SPh)]₂ (**90**), formed by the oxidative addition of (PhS)₂ to Cp*RuCl(cod), as one possible candidate for the active catalyst (Eq. 7.59).

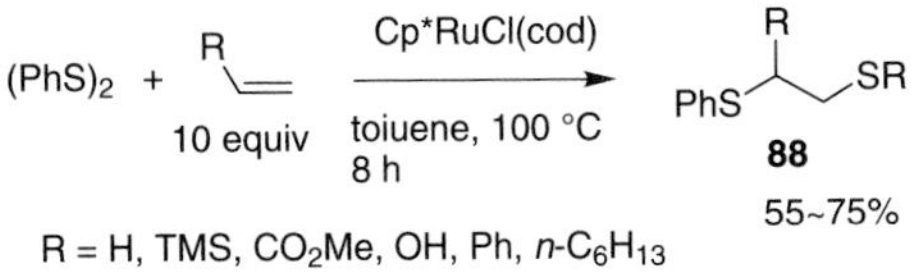

$$
\text{(PhS)}_2 + \underset{\substack{10 \text{ equiv}}}{\overset{R}{\diagdown\!=}} \quad \xrightarrow[\substack{\text{toiuene, 100 °C} \\ 8 \text{ h}}]{\text{Cp*RuCl(cod)}} \quad \underset{\substack{\textbf{88} \\ 55\text{\textasciitilde}75\%}}{\text{PhS}\diagup\overset{R}{\diagup}\!\diagdown\!\text{SR}} \tag{7.57}
$$

R = H, TMS, CO₂Me, OH, Ph, *n*-C₆H₁₃

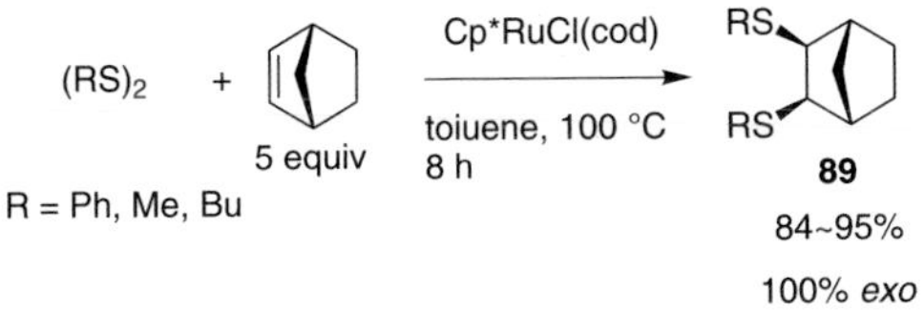

$$
\text{(RS)}_2 + \underset{\substack{5 \text{ equiv}}}{\text{[norbornene]}} \xrightarrow[\substack{\text{toiuene, 100 °C} \\ 8 \text{ h}}]{\text{Cp*RuCl(cod)}} \underset{\substack{\textbf{89} \\ 84\text{\textasciitilde}95\% \\ 100\% \, exo}}{\text{RS···norbornane···RS}} \tag{7.58}
$$

R = Ph, Me, Bu

$$
\text{(PhS)}_2 + \text{Cp*RuCl(cod)} \longrightarrow \underset{\textbf{90}}{\text{[Cp*RuCl(μ-SPh)]}_2} \tag{7.59}
$$

7.4
Catalytic S–X Activation

The activation of S–X bonds, where X is different from H or S, has also attracted much attention, because two different functional groups can be introduced by a single procedure.

The first example was reported in 1989 by Ando et al. and showed the 'germylthiolation' and 'germylselenation', that is, palladium-catalyzed addition of the strained S–Ge bond of thiadigermirane and Se–Ge bond of selenadigermirane to acetylene (C₂H₂) to provide **91** (Eq. 7.60) [61].

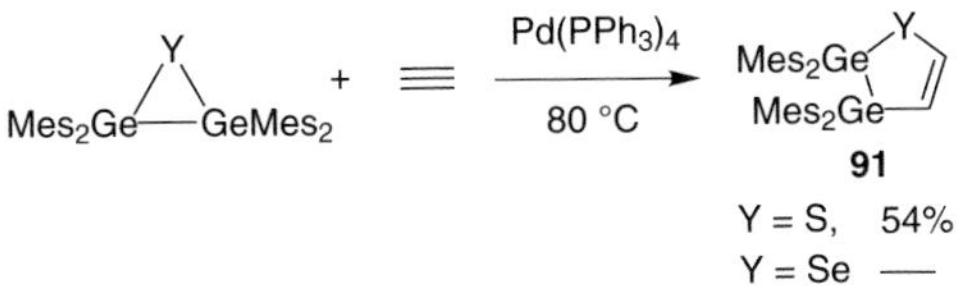

$$
\underset{\text{Mes}_2\text{Ge}\!-\!\text{GeMes}_2}{\overset{Y}{\triangle}} + \equiv \xrightarrow{\substack{\text{Pd(PPh}_3)_4 \\ 80 \text{ °C}}} \underset{\substack{\textbf{91} \\ Y = S, \; 54\% \\ Y = Se \; \text{—}}}{\overset{\text{Mes}_2\text{Ge}\diagdown^{Y}}{\underset{\text{Mes}_2\text{Ge}\diagup}{}}} \tag{7.60}
$$

The proposed mechanism (Scheme 7-20) includes: (a) oxidative addition of Y–Ge (Y = S or Se) bonds to Pd(PPh₃)ₙ, (b) insertion of acetylene into the Pd–Y bond to give **92** or insertion of acetylene into the Pd–Ge bond to form **93**, (c) formation of **91** by either a C–Ge or a C–Y bond-forming reductive elimination with regeneration of Pd(PPh₃)ₙ.

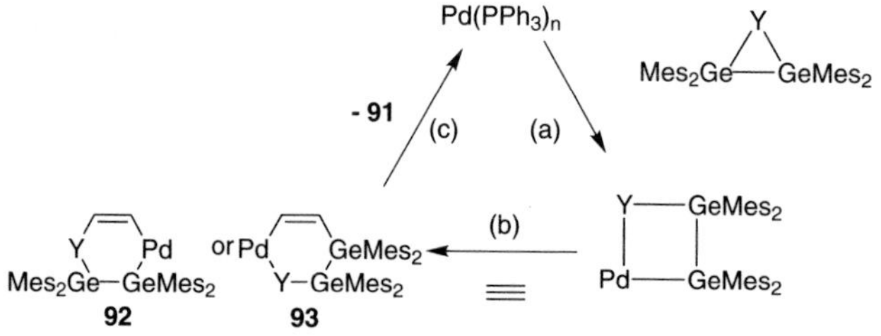

Scheme 7-20 A proposed mechanism of Pd-catalyzed addition of Y-Ge (Y = S, Se) bonds to acetylene

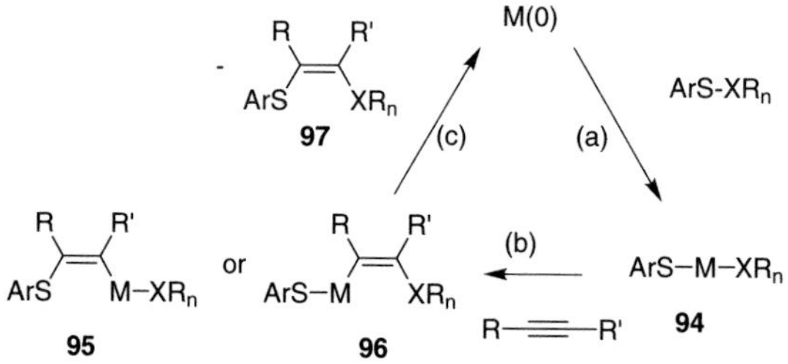

Scheme 7-21 A prototype of the mechanism of M(0)-catalyzed addition of S-X bonds to alkyne

This reaction typifies the two possibilities of reaction routes for M-catalyzed addition of an S–X (or Se–X) bond to alkyne: (a) oxidative addition of the S–X bond to M(0) to form **94**, (b) insertion of alkyne into either the M–S or M–X bond to provide **95** or **96**; (c) C–X or C–S bond-forming reductive elimination to give **97** (Scheme 7-21). Comparable reaction sequences are also discussed when the Chalk-Harrod mechanism is compared with the modified Chalk-Harrod mechanism in hydrosilylations [1,3]. The palladium-catalyzed 'thioboration', that is, addition of an S–B bond to an alkyne was reported by Miyaura and Suzuki et al. to furnish the *cis*-adducts **98** with the sulfur bound to the internal carbon and the boron center to the terminal carbon (Eq. 7.61) [62].

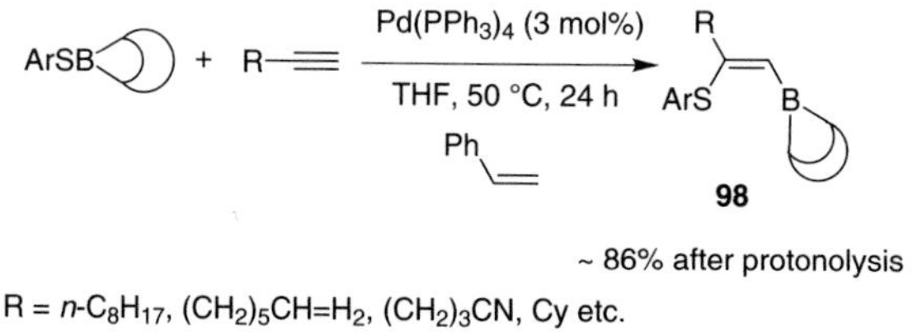

~ 86% after protonolysis

R = *n*-C$_8$H$_{17}$, (CH$_2$)$_5$CH=H$_2$, (CH$_2$)$_3$CN, Cy etc.

Two possible routes are envisioned for X = B in Scheme 7-21. The authors favored a path involving the oxidative addition of the S–B bond to Pd(0), insertion of the alkyne into the Pd–S bond followed by C–B bond-forming reductive elimination. On the other hand, Morokuma et al. studied the mechanism of the addition of HSB(OCH$_2$)$_2$ (**99**) to acetylene (C$_2$H$_2$) using Pd(PH$_3$)$_2$ (**100**) as a catalyst to produce **101** using hybrid density functional calculations (Eq. 7.62) [5].

$$HS-B\overset{O}{\underset{O}{\bigcirc}} + \equiv \quad \xrightarrow[\text{100}]{\text{Pd(PH}_3)_2} \quad HS\overset{\diagup}{\diagdown}B-O\underset{O}{\bigcirc} \qquad (7.62)$$

99   101

They concluded that the oxidative addition of the S–B bond of **99** to **100** is less likely. Their proposed mechanism shown in Scheme 7-22 includes: (a) coordination of C_2H_2 to **100** to form **102**, (b) dissociation of one PH_3 to form **103**, (c) formation of $Pd(SH)[CH=CH(B(OCH_2)_2](PH_3)$ **105**, where the SH and the vinyl group are *trans* through the σ-metathesis-like transition state **104**, (d) isomerization of **105** by SH migration to form **106**, (e) recoordination of the PH_3 to provide **107**, and (f) C–S bond-forming reductive elimination to give **101** with regeneration of **100**. The rate-determining step of the whole catalytic reaction corresponds to the formation of **105** *via* **104** as a transition state, in which the S–B bond attacks the Pd–C bond to produce simultaneously the Pd–S bond and the C–B bond. The Pd–S distance in **104** is 2.474 Å, which is close to the Pd–S covalent bond length of **105** (2.468 Å). The C_2H_2 ligand is strongly bent towards the PH_3 ligand, the C–B bond is formed (1.863 Å, longer than the normal C–B bond length, 1.527 Å), and the S–B bond is broken (2.113 Å from 1.807 Å of **99**). Based on the Pd–B bond distance, some interaction between the B and the Pd center in **104** was assumed. The energy of **104** was calculated to be 18.6 kcal/mol [20.1 kcal/mol for $MeSB(OCH_2)_2$] above that of the reac-

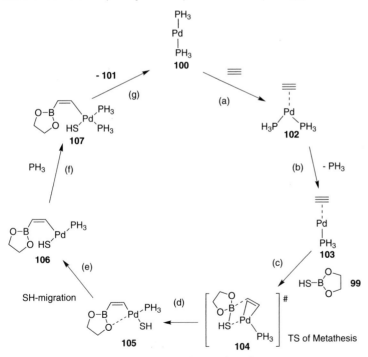

Scheme 7-22 Schematic proposed route obtained from the potential-energy profile of $Pd(PH_3)_2$-catalyzed addition of **99** to C_2H_4

tants (**99** plus C_2H_2). The authors proposed that the ability of the thiolate group, HS, to take a bridging position between the $B(OR)_2$ group and the Pd center significantly stabilizes **104**. On the other hand, the formation of **105** from the reactants was exothermic (18.1 kcal/mol), indicating that the process may not be fast but the equilibrium strongly favors **105**. The barrier height of the C–S bond-forming reductive elimination (**107** to **100**) was 12.5 kcal/mol, and the exothermicity of the total catalytic reaction was calculated to be 35.3 kcal/mol. They also investigated the reason why Pt(0) did not catalyze the thioboration of acetylene. It was found that the large energy barrier (E_a = 27.9 kcal/mol) for the reductive elimination from the Pt(II) thiolate vinylborane complex analogous to **107** is the major obstacle to making the reaction catalytic. However, the estimation of the steric hindrance in **108**, which is the suspected intermediate if Morokuma's result is applied to the reaction of Eq. (7.61), is not clear and the real mechanism seems to be elusive.

108   A possible intermediate for Pd(PPh₃)₄-catalyzed thioboration of 1-alkyne

Tanaka et al. reported on the related palladium-catalyzed 'selenophosphorylation' and 'thiophosphorylation', i.e., additions of Y–P (Y = Se, S) bonds in **109** to an alkyne provides **110** in up to 95% yield (Eq. 7.63) [63, 64]. The oxidative addition of the selenophosphate to $M(PEt_3)_3$ (M = Pd, Pt) (Eq. 7.64) afforded **111**, and the oxidative addition to Pd(PPh₃)₄ was mentioned in the paper. While no information about the insertion process was provided, considering the regio- and sterochemistry of Pd-catalyzed addition of Y–Y and Y–H bonds to alkynes (Eqs. 7.9, 7.30 and 7.47), the insertion of alkyne into Pd–Y (not Pd–P) is more likely.

$$PhYP(O)(OPh)_2 \ + \ R\!\!-\!\!\equiv \quad \xrightarrow[\substack{3 \text{ mo\%}}]{Pd(PPh_3)_4} \quad (7.63)$$

109   Y = Se, THF, 67 ¡C, 15-20 h

                   Y = S, toluene, 100 ¡C, 3-5 h   **110**

R = *n*-C₆H₁₃, H, CH₂OMe, (CH₂)₃OMe, Ph, etc   up to 95%

$$PhSeP(O)(OPh)_2 \ + \ M(PEt_3)_3 \quad \xrightarrow[C_6D_6,\ 25\ °C,\ 0.5\ h]{} \quad PhSe\!-\!M\!-\!P(O)(OPh)_2 \quad (7.64)$$

111

M = Pd, 93%

M = Pt, 95%

The palladium-catalyzed addition of Se–Si ('silylselenation') and Se–Ge bonds to phenyl acetylene was also reported to provide cis-adducts **112** in 25% and 35% yields, respectively (Eq. 7.65) [65].

$$\text{Ph}\!-\!\!\equiv\!\! \quad + \quad \text{PhSeGMe}_3 \quad \xrightarrow[\text{120 ¡C}]{\text{Pd(PPh}_3)_4} \quad \underset{\underset{\textbf{112}}{\text{PhSe}\quad\text{GMe}_3}}{\overset{\text{R}}{\diagdown\!\!=\!\!\diagup}} \tag{7.65}$$

24 h, G =Si, 25%
12 h, G =Ge, 35%

The 'silylthiolation' was more successfully performed using Pt-catalysts, (PhS)$_2$, and (Cl$_3$Si)$_2$ in toluene at 110°C to afford the products **113** after treatment with EtOH in up to 95% yield (Eq. 7.66) [66].

$$\text{R}\!-\!\!\equiv\!\! \quad + \quad (\text{ArS})_2 \quad + \quad (\text{Cl}_3\text{Si})_2 \quad \xrightarrow[\substack{\text{toluene}\\ \text{110 °C, 6 h}}]{\substack{\text{Pt(PPh}_3)_4\\ \text{1.5 mol\%}}} \quad \xrightarrow{\text{EtOH, Et}_3\text{N}} \quad \underset{\underset{\textbf{113}}{\text{ArS}\quad\text{Si(OEt)}_3}}{\overset{\text{R}}{\diagdown\!\!=\!\!\diagup}} \tag{7.66}$$

Ar = Ph, p-Cl, p-Me
R = n-C$_6$H$_{13}$, PhCH$_2$, (CH$_2$)$_3$CN etc.

up to 95%

The formation of ArSSiCl$_3$ **114** from (ArS)$_2$ and (Cl$_3$Si)$_2$ was also catalyzed by the Pt-complex (Eq. 7.67). The equilibrium between Pt(SPh)$_2$(PEt$_3$)$_2$/(Cl$_3$Si)$_2$ and Pt(SPh)(SiCl$_3$)(PPh$_3$)$_2$/PhSSiCl$_3$ is characterized by an equilibrium constant $K = 0.03$ (23°C) (Eq. 7.68). The authors proposed the insertion of alkyne into a Pt–S bond of the S–Pt–Si complex rather than insertions into an S–Pt–S complex or the Pt–Si bond. This assumption is based on: (1) the strong *trans* influence of the silyl ligand, (2) the enhanced bond strength of the Si–Pt bond when Cl serves as an substituent on Si, and (3) the regioselectivity of the insertion reaction.

$$(\text{ArS})_2 \quad + \quad (\text{Cl}_3\text{Si})_2 \quad \xrightarrow[\text{60 °C, 2 h}]{\text{Pt(PPh}_3)_4} \quad \underset{\underset{\textbf{114}}{85\%}}{\text{ArSSiCl}_3} \tag{7.67}$$

$$\underset{\underset{\text{PEt}_3}{|}}{\overset{\overset{\text{PEt}_3}{|}}{\text{PhS-Pt-SPh}}} \quad + \quad (\text{Cl}_3\text{Si})_2 \quad \underset{K = 0.03\ (23\ °\text{C})}{\rightleftharpoons} \quad \underset{\underset{\text{PEt}_3}{|}}{\overset{\overset{\text{PEt}_3}{|}}{\text{PhS-Pt-SiCl}_3}} + \text{PhSSiCl}_3 \tag{7.68}$$

In 1999, it was reported that the palladium catalyzed 'azathiolation', that is, the addition of the S–N bond of sulfenamide **115** to carbon monoxide can be catalyzed by palladium(0) complexes in pyridine to provide the thiocarbamate **116** in good yields (Eq. 7.69) [67]. Contrary to the other S–X bond activations described so far, where X has the same electronegativity as S (i.e. X = S) or lower (X = H, B, Si, Ge, and P), the S–N bond has a strong $S^{\delta+}$–$N^{\delta-}$ bond character and shows unique reactivity.

With regard to the reaction mechanism, a path involving: (a) oxidative addition of

$$\underset{\textbf{115}}{\text{ArSNR}_2} \quad + \quad \underset{\substack{\text{20 atm}}}{\text{CO}} \quad \xrightarrow[\substack{\text{pyridine, 80 °C}\\ \text{1~20 h}}]{\text{Pd(PPh}_3)_4} \quad \underset{\underset{\underset{\text{up to 90\%}}{\textbf{116}}}{\text{ArS}\quad\text{NR}_2}}{\overset{\text{O}}{\underset{\|}{\diagdown\!\!C\!\!\diagup}}} \tag{7.69}$$

the S–N bond of **115** to Pd(0) to give **117**, (b) insertion of CO into the Pd–N bond of **117** [68], and (c) reductive elimination under liberation of **116** can be envisioned (Scheme 7-23). However, this path is unlikely, based on the following observations: (1) The ^{31}P NMR spectrum of the crude reaction mixture of the Pd(PPh$_3$)$_4$-catalyzed reaction of **115** with CO showed the formation of Pd(SAr)$_2$(PPh$_3$)$_n$ (**118**); (2) the complex **118** was actually an active catalyst; (3) no formation of an imine, which may be generated from a palladium amide, Pd–NR$_2$ (R = Et) by β-H elimination was observed during the course of the reaction; (4) no reaction was observed in the stoichiometric reaction of **115** with Pd(PPh$_3$)$_4$.

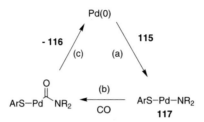

Scheme 7-23 An unlikely path of Pd-catalyzed azathiolation of CO

An alternative proposal for the propagation of this reaction is shown in Scheme 7-24, which includes: (a) insertion of CO into the Pd–S bond of **118** to provide **119**, and (b) reaction of **119** with **115** to give **116** with regeneration of **118** through a σ-metathesis-like transition state **120**. According to this mechanism, the ArS group in **118** and the R$_2$N group of **115** are incorporated simultaneously.

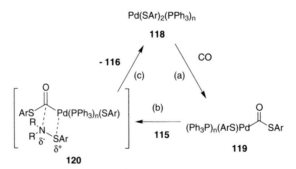

Scheme 7-24 A proposed propagation of Pd-catalyzed azathiolation of CO

Indeed, when [Pd(p-tolS)$_2$(PPh$_3$)]$_2$ **118′** was used as the catalyst in the reaction of PhSNEt$_2$ **115′** with CO, 83% of p-tolS was incorporated into the product **116′** at low conversion of the reaction (Eq. 7.70). Although the formation of **119** was not confirmed, the reactivity of **119** with **115** was examined as follows. When an equiv of PhSC(O)SPh **121** was added into the pyridine solution of Pd(PPh$_3$)$_4$ in the presence of 5 equiv of PhSNMe$_2$ (**115″**) at –30°C over 10 min, the formation of 80% of [Pd(SAr)$_2$(PPh$_3$)]$_2$ (**118″**) and 18% of **116″** was observed (Eq. 7.71). This clearly indicated that some short-lived Pd[C(O)SPh](SPh)(PPh$_3$)$_n$ (**119′**) was generated by the oxidative addition of **121** to Pd(PPh$_3$)$_4$ and subsequently intercepted by **115″** before de-insertion of CO took place. In the proposed σ-metathesis-like transition state (**120**), the S–N bond simultaneously interacts with both the Pd and the electrophilic carbon of the C(O)SAr group, providing the driving force of the reaction.

$$
\text{PhSNEt}_2 \;+\; \text{CO} \xrightarrow[\substack{\text{20 Kg/cm}^2 \;\; 80\,°C, \text{ pyridine, 3 h} \\ \text{PPh}_3\ (15\ \text{mol\%})}]{\substack{[\text{Pd(Stol-}p)_2(\text{PPh}_3)]_2\ \mathbf{118'} \\ (5\ \text{mol\% of Pd})}} \underset{\substack{\mathbf{26\%}}}{\text{PhS-C(O)-NEt}_2} + \underset{\substack{\mathbf{116'} \\ 8.3\% \\ (83\%\ \text{based on }p\text{-tolS})}}{p\text{-tolS-C(O)-NEt}_2}
$$
(7.70)

115'

$$
\underset{\mathbf{121}}{\text{PhS-C(O)-SPh}} + \text{Pd(PPh}_3)_4 \xrightarrow[\substack{-30\,°C \\ \text{pyridine}}]{\substack{\text{PhSNMe}_2 \\ \mathbf{115''} \\ 5\ \text{equiv}}} \underset{\substack{\mathbf{118''} \\ 80\% \\ \text{by }^{31}\text{P NMR}}}{[\text{Pd(SPh)}_2(\text{PPh}_3)]_2} + \underset{\substack{\mathbf{116''} \\ 18\%}}{\text{PhS-C(O)-NMe}_2}
$$
(7.71)

7.5
Related Stoichiometric Reaction

As we have seen, reviewing catalytic S–X bond activations, some reactions complete their catalytic cycles by C–S bond-forming reductive elimination from C–M–S complexes. Hartwig et al. have reported on the mechanism of the C–S bond-forming reductive elimination from Pd(L)(R)(SR′) **122** (Eq. 7.72) [69].

$$
\underset{\mathbf{122}}{\text{Ph}_2\text{P-Pd-SAr}} \longrightarrow \left[\underset{\mathbf{123}}{\text{Ph}_2\text{P-Pd-SAr}} \right] \longrightarrow \text{Ph}_2\text{P-Pd-S-Ar} \longrightarrow \text{Pd(0)} + \text{PhSAr}
$$
(7.72)

They demonstrated that electron-deficient R groups and electron-rich R′ substituents at S accelerated the reductive elimination. They proposed **123** (L$_2$ = DPPE, R = Ph, R′ = Ar) as a transition state, where R acts as an electrophile and thiolate as a nucleophile. The Hammet plot for the reductive elimination showed that the resonance effect of the substituent in R determines the inductive effect of the R group, and the effect in SR′ showed an acceptable linear relationship with the standard σ-values. The relative rate for sulfide elimination as a function of the hybrid valence configuration of the carbon center bonded to palladium followed the trend sp^2 > sp >> sp^3.

It was reported that the oxidative addition of vinyl sulfide **124** to Pt(PPh$_3$)$_2$(C$_2$H$_4$) produced **125** in good yield at room temperature (Eq. 7.73) [70]. Both geminal groups G (Ph, MeOCH$_2$, Me$_3$Si) and ArS in **124** were indispensable to achieve the oxidative addition. This indicates that considerable anionic character is built up at this carbon center in the course of the reaction, pointing toward transition states (or intermediates) like **126** and **127**. The interaction of the p-type orbital at the anionic carbon center with the σ* orbital of the C–S bond (negative hyperconjugation) which is cleaved in **127** was also proposed.

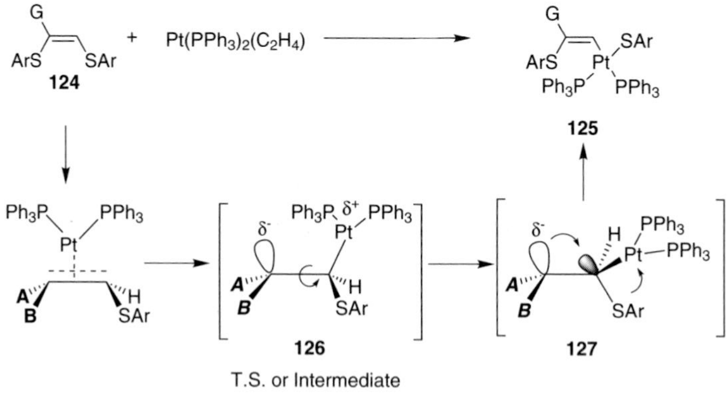

(7.73)

T.S. or Intermediate

Throughout the entire S–X bond activation and the corresponding mechanistic studies, a general feature of the reactivity of S–M bonds was gradually revealed.

Let us consider the general trends of the reactivity of C–C, C–S, and C–Q (Q = Cl, Br, I) bonds towards oxidative addition and reductive elimination (Scheme 7-25). In many cases, either C–C bond-forming reductive elimination from a metal center (a) or the oxidative addition of a C–Q bond to a low-valent metal center is a thermodynamically favorable process (c). On the other hand, the thermodynamics of the C–S bond oxidative addition and reductive elimination (b) lies in between these two cases. In other words, one could more easily control the reaction course by the modulation of metal, ligand, and reactant. Further progress for better understanding of S–X bond activation will be achieved by thorough stoichiometric investigations and computational studies.

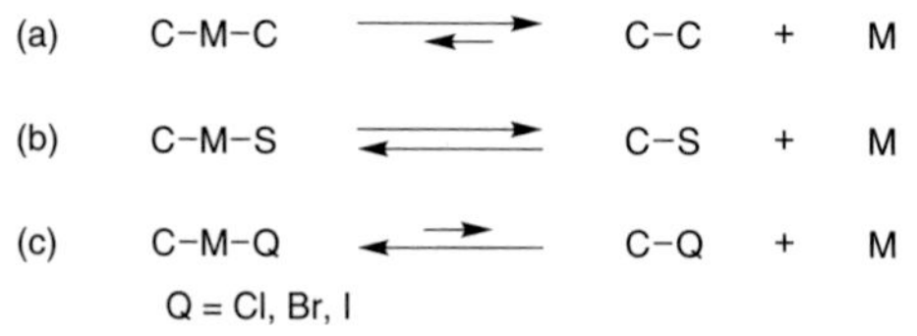

(a)   C–M–C   ⇌   C–C   +   M

(b)   C–M–S   ⇌   C–S   +   M

(c)   C–M–Q   ⇌   C–Q   +   M
       Q = Cl, Br, I

Scheme 7-25   General trend of reactivity of C-C, C-S, and C-Q Bonds toward M: (a) favored reductive elimination, (b) adjustable by metal, ligands and reactand, (c) favored oxidative addition

References

1 (a) A. J. Chalk, J. F. Harrod, *J. Am. Chem. Soc.* **1965**, *87*, 16. (b) J. F. Harrod, A. J. Chalk, *J. Am. Chem. Soc.* **1965**, *87*, 1133.

2 I. Ojima in *The Chemistry of Organic Silicon Compound*;. S. Patai, Z. Rappoport, (eds.); Wiley-Interscience: Chichester, **1989**; Chapter 25.

3 J. P. Collman, L. S. Hegedus, J. R. Norton, R. G. Finke, *Principles and Application of organotransition metal Chemistry*; University Science Books: Mill Valley, CA, **1987**; pp 355.

4 L. L. Hegedus, R. W. McCabe, *Catalyst Poisoning*; Marcel Deckker: New York, **1984**. (b) A. T. Hutton in *Comprehensive Coordination Chemistry*; G. Wilkinson, R. D. Gillard, J. A. McCleverty (eds.); Pergamon Press: Oxford, U. K., **1984**; Vol. 5, p 1151.

5 Q. Cui, D. G. Musaev, K. Morokuma, *Organometallics* **1998**, *17*, 1383.

6 T. Luh, Z. Ni, *Synthesis*, **1990**, 89.

7 M. R. Dubois, *Chem. Rev.* **1989**, *89*, 1.

8 R. D. Adams, S. B. Falloon. *Chem. Rev.* **1995**, *95*, 2587.

9 (a) C. Bianchini, A. Meli. *Acc. Chem. Res.* **1998**, *31*, 109. (b) C. Bianchini; A. Meli. *Synlett* **1997**, 643. (c) B. C. Wiegand, C. M. Friend. *Chem. Rev.* **1992**, *92*, 491.

10 H. Vehrenkamp, *Angew. Chem. Int. Ed. Engl.* **1975**, *14*, 322.

11 (a) L. Han, M. Tanaka, *J. Chem. Soc., Chem. Commun.* **1999**, 395. (b) A. Ogawa, *J. Organomet. Chem.* **2000**, *611*, 463. (c) T. Kondo, T. Mitsudo, *Chem. Rev.* **2000**, *100*, 3205.

12 V. W. Reppe, *Liebigs Ann.* **1953**, *582*, 1.

13 V. W. Reppe, H. Kröper, *Liebigs Ann.* **1953**, *582*, 38.

14 H. E. Holmquist, J. E. Carnahan, *J. Org. Chem.* **1960**, *25*, 2240.

15 J. F. Knifton, US Patent 3,933,884 (1976).

16 P. Foley, US Patent 4,422,977 (1983).

17 J. Kadelka, H. H. Schwarz, Ger. Offen. DE 3,246,149 (1984)

18 S. Antebi, H. Alper, *Organometallics* **1986**, *5*, 596.

19 S. C. Shim, S. Antebi, H. Alper, *J. Org. Chem.* **1985**, *50*, 147.

20 (a) U. M. Dzhemilev, R. V. Kunakova, R. L. Gaisin, *Izv. Akad. Nauk SSSR, Ser. Khim.* **1981**, *11*, 2655.

21 (a) H. Kuniyasu, A. Ogawa, K. Sato, I. Ryu, N. Kambe, N. Sonada, *J. Am. Chem. Soc.* **1992**, *114*, 5902. (b) J. W. McDonald, J. L. Corbin, W. E. Newton, *Inorg. Chem.* **1976**, *15*, 2056.

22 A. Ogawa, T. Ikeda, K. Kimura, T. Hirao, *J. Am. Chem. Soc.* **1999**, *121*, 5108.

23 H. Singer, G. Wilkinson. *J. Chem. Soc. A* **1968**, 2516.

24 J. E. Bäckvall, A. Ericsson, *J. Org. Chem.* **1994**, *59*, 5850.

25 U. Koelle, C. Rietmann, J. Tjoe, T. Wagner, U. Englert, *Organometallics* **1995**, *14*, 703.

26 B. Gabriele, G, Salerno, A. Fazio. *Org. Lett.* **2000**, *2*, 351.

27 A. Ogawa, M. Takeba, J. Kawakami, I. Ryu, N. Kambe, N. Sonoda, *J. Am. Chem. Soc.* **1995**, *117*, 7564.

28 A. Ogawa, J. Kawakami, M. Mihara, T. Ikeda, N. Sonoda, T. Hirao, *J. Am. Chem. Soc.* **1997**, *119*, 12380.

29 (a) A. E. KESKINEN, C. V. SENOFF, *J. Organomet. Chem.* **1972**, *37*, 201. (b) R. UGO, M. LA, S. CENINI, A. SEGRE, F. CONTI, *J. Chem. Soc. A* **1971**, 522.

30 A. OGAWA, K. KAWABE, J. KAWAKAMI, M. MIHARA, T. HIRAO, N. SONODA, *Organometallics* **1998**, *17*, 3111.

31 W. J. XIAO, H. APLER, *J. Org. Chem.* **1997**, *62*, 3422.

32 W. J. XIAO, G. VASAPOLLO, H. ALPER, *J. Org. Chem.* **1999**, *64*, 2080.

33 (a) C. AMATORE, A. JUTAND, M. A. M'BARKI, *Organometallics* **1992**, *11*, 3009. (b) C. AMATORE, E. CARRÉ, A. JUTAND, M. A. M'BARKI, *Organometallics* **1995**, *14*, 1818.

34 T. B. RAUCHFUSS, J. S. SHU, D. M. ROUNDHILL, *Inorg. Chem.* **1976**, *15*, 2096.

35 A. OGAWA, J. KAWAKAMI, N. SONODA, T. HIRAO, *J. Org. Chem.* **1996**, *61*, 4161.

36 W. J. XIAO, G. VASAPOLLO, H. ALPER, *J. Org. Chem.* **1998**, *63*, 2609.

37 W. J. XIAO, H. ALPER, *J. Org. Chem.* **1999**, *64*, 9646.

38 Y. MIYAUCHI, S. WATANABE, H. KUNIYASU, H. KUROSAWA, *Organometallics* **1995**, *14*, 5450.

39 H. KUNIYASU, A. OGAWA, K. SATO, I. RYU, N. SONODA, *Tetrahedron Lett.* **1992**, *33*, 5525.

40 A. OGAWA, A. KUDO, T. HIRAO. *Tetrahedron Lett.* **1998**, *39*, 5213.

41 W. J. XIAO, H. ALPER. *J. Org. Chem.* **1998**, *63*, 7939.

42 T. KONDO, Y. MORISAKI, S. UENOYAMA, K. WADA, T. MITUDO, *J. Am. Chem. Soc.* **1999**, *121*, 8657.

43 C. M. CRUDDEN, H. ALPER, *J. Org. Chem.* **1995**, *60*, 5579.

44 N. ZHENG, J. C. MCWILLIAMS, F. J. FLEITZ, J. D. ARMSTRONG III, R. P. VOLANTE, *J. Org. Chem.* **1998**, *63*, 9606.

45 M. HAYASHI, K. ONO, H. HOSHIMI, N. OGUNI. *J. Chem. Soc., Chem. Commun.* **1994**, 2699.

46 T. IIDA, N. YAMAMOTO, H. SASAI, M. SHIBASAKI. *J. Am. Chem. Soc.* **1997**, *119*, 4783.

47 K. NISHIMURA, M. ONO, Y. NAGAOKA, K. TOMIOKA. *J. Am. Chem. Soc.* **1997**, *119*, 12974.

48 E. EMORI, T. ARAI, H. SASAI, M. SHIBASAKI, *J. Am. Chem. Soc.* **1998**, *120*, 4043.

49 S. ANTEBI, H. ALPER, *Tetrahedron Lett.* **1985**, *26*, 2609.

50 H. TAKAHASHI, K. OHE, S. UEMURA, N. SUGITA, *J. Organomet. Chem.* **1987**, *334*, C43.

51 U. M. DZHEMILEV, R. V. KUNAKOVA, N. Z. BAIBULATOVA, E. M. MUSTAFINA, E. G. GALKIN, G. A. TOLSTIKOV, *Izv. Akad. Nauk SSSR, Ser. Khim.* **1989**, *3*, 747.

52 H. KUNIYASU, A. OGAWA, S. MIYAZAKI, I. RYU, N. KAMBE, N. SONODA, *J. Am. Chem. Soc.* **1991**, *113*, 9796.

53 H. KUNIYASU, A. OGAWA, K. HIGAKI, N. SONODA, *Organometallics* **1992**, *11*, 3937.

54 H. KUNIYASU, A. OGAWA, N. SONODA, *Tetrahedron Lett.* **1993**, *34*, 2491.

55 A. OGAWA, H. KUNIYASU, N. SONODA, T. HIRAO, *J. Org. Chem.* **1997**, *62*, 8361.

56 Y. GAREAU, A. ORELLANA, *Synlett* **1997**, 803.

57 S. FUKUZAWA, T. FUJINAMI, S. SAKAI, *Chem. Lett.* **1990**, 927.

58 H. KUNIYASU, K. SUGOH, S. MOON, H. KUROSAWA, *J. Am. Chem. Soc.* **1997**, *119*, 4669.

59 H. KUNIYASU, A. MARUYAMA, H. KUROSAWA, *Organometallics* **1998**, *17*, 908.

60 T. KONDO, S. UENOYAMA, K. FUJITA, T. MITSUDO, *J. Am. Chem. Soc.* **1999**, *121*, 482.

61 T. TSUMURAYA, W. ANDO, *Organometallics* **1989**, *8*, 2286.

62 T. ISHIYAMA, K. NISHIJIMA, N. MIYAURA, A. SUZUKI, *J. Am. Chem. Soc.* **1993**, *115*, 7219.

63 L. HAN, N. CHOI, M. TANAKA, *J. Am. Chem. Soc.* **1996**, *118*, 7000.

64 L. HAN, M. TANAKA, *Chem. Lett.* **1999**, 863.

65 A. OGAWA, H. KUNIYASU, M. TAKEBA, T. IKEDA, N. SONODA, T. HIRAO, *J. Organomet. Chem.* **1998**, *564*, 1.

66 L. HAN, M. TANAKA, *J. Am. Chem. Soc.* **1998**, *120*, 8249.

67 H. KUNIYASU, H. HIRAIKE, M. MORITA, A. TANAKA, K. SUGOH, H. KUROSAWA, *J. Org. Chem.* **1999**, *64*, 7305.

68 (a) M. Rahim, C. H. Bushweller, K. J.
Ahmed, *Organometallics* **1994**, *13*, 4952.
(b) S. Calet, F. Urso, H. Alper, *J. Am.
Chem. Soc.* **1989**, *111*, 931. (c) M. E. Pi-
otti, H. Alper, *J. Am. Chem. Soc.* **1996**,
118, 111. (d) H. E. Bryndza, W. C.
Fultz, W. Tam, *Organometallics* **1985**, *4*,
939.

69 (a) G. Mann, D. Barañano, J. F.
Hartwig, A. L. Rheingold, I. A.
Guzei, *J. Am. Chem. Soc.* **1998**, *120*,
9205. (b) D, Barañano, J. F. Hartwig,
J. Am. Chem. Soc. **1995**, *117*, 2937.

70 H, Kuniyasu, A. Ohtaka, T. Naka-
zono, M. Kinomoto, H. Kurosawa, *J.
Am. Chem. Soc.* **2000**, *122*, 2375.

8
Hydrozirconation

Alain Igau

8.1
Introduction

Over the last two decades zirconium derivatives have emerged as highly useful intermediates for organic chemistry. The great majority of the known organozirconium derivatives are Zr(IV) [1–7] and Zr(II) [8–10] complexes which contain the zirconocene Cp_2Zr moiety (Cp= η^5-C_5H_5). Hydrozirconation provides a unique and convenient route for preparing alkyl- and alkenyl-zirconocene derivatives of Zr(IV). The early discovery by Wailes and Weigold of the hydrometalation reagent $[Cp_2Zr(H)Cl]_n$ (1) [11] and subsequent investigations by Schwartz and coworkers [5] demonstrated the usefulness and rather broad scope of hydrozirconation. $[Cp_2Zr(H)Cl]_n$ (1), known as zirconocene hydrochloride or Schwartz's reagent, represents probably the first significant usage of a molecular organometallic zirconium compound in organic synthesis. In addition to the unique chemical behavior of Zr(IV)-compounds, subsequent tranformations such as activation by ligand abstraction with silver salts [12–14] or transmetalation with nickel [15, 16], palladium [17, 18], aluminum [19, 20], zinc [21–23] and copper [7, 24, 25] have increased the reactivity of these organozirconocene complexes, which react in the presence of these salts with a large variety of electrophiles. Organozirconocene-Zr(IV) complexes now participate in most of the processes involving the formation of carbon-carbon bonds [12–26].

In deciding on the material to be covered in this chapter, limitations had to be set. The first section will present the synthesis of various zirconocene hydrides. The focus of the subsequent sections is to present a general synopsis of the different aspects of the hydrozirconation reactions using 1 not only on carbon-carbon multiple bonds but also on heteropolar multiple bonds. Those aspects of hydrozirconation that were covered in previous reviews [1–5, 27] and in the excellent chapter by Labinger in *Comprehensive Organic Synthesis* [28] are summarized or briefly mentioned here. For other aspects of organozirconium chemistry not covered by the above-mentioned reviews, the reader is referred to a number of monographs and reviews [29, 30].

Throughout this chapter we will use in the different Schemes the abbreviation [Zr] = Cp$_2$ZrCl–; thus [Zr]–H = [Cp$_2$Zr(H)Cl]$_n$ (**1**) and [Zr]–R = Cp$_2$Zr(R)Cl. Most of the compounds described in this chapter have metal centers with 16 electrons, and it is important to emphasize that any coordinatively unsaturated zirconocene (IV) complexes can potentially exist as dimeric, oligomeric, or polymeric aggregates.

8.2
Zirconium Hydrides

The most commonly employed hydrozirconating reagent is [Cp$_2$Zr(H)Cl]$_n$ (**1**) [11]. This is a white solid polymeric aggregate, sensitive to air and moisture and moderately sensitive to light. It is conveniently prepared on a large scale (80 g) in 70–80% yields and around 95% purity by reduction of Cp$_2$ZrCl$_2$ with a filtered ether solution of LiAlH$_4$ (Scheme 8-1) [31]. The only significant side-product of the reaction, [Cp$_2$Zr(H)$_2$]$_2$, is converted to **1** upon exposure to CH$_2$Cl$_2$ during work-up.

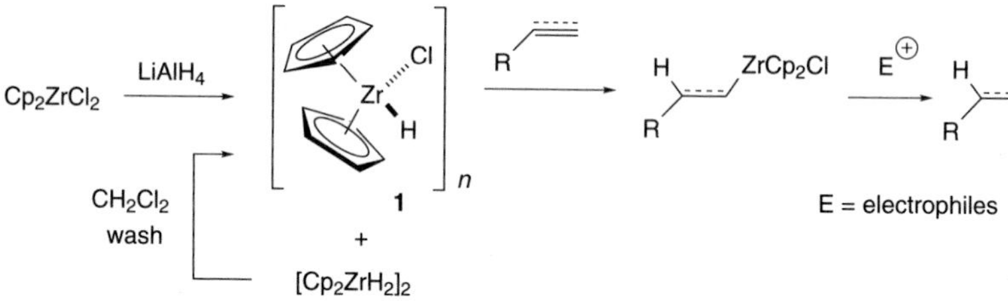

Scheme 8-1 Synthesis of the Schwartz's reagent, **1**

Other hydride sources such as LiAl(O^{t}Bu)$_3$H [32] and Red-Al [19] have been reported for the synthesis of **1**, but work-up did not provide a product free from the presence of contaminants which lower the efficiency of the metal hydride and perturb the reproducibility of the results [33]. It is an aggregate whose molecularity has not yet been established and is sparingly soluble in organic solvents. In standard experimental procedures, the unsaturated substrate is added to a suspension of a stoichiometric amount of **1**, and the mixture is stirred at ambient or slightly elevated temperatures. Formation of the hydrozirconated product is 'visually' controlled by complete dissolution of the white zirconocene reagent **1** ($\approx$ 20 min to 3 h) to give a yellow to orange solution. Generally, the zirconium complex is used directly for subsequent transformation, but it is generally possible to isolate and completely characterize it. The solvent is a crucial component in the hydrozirconation process, as it may considerably affect the reaction rate. Non-polar solvents such as benzene and toluene have been successfully used for hydrozirconation of unsaturated hydrocarbon derivatives. However, from the different classes of solvents used so far in hydrozirconation reactions, ethers appear to be highly efficient [2], and among them THF is the most convenient and inexpensive one. It should also be noted that even though CH$_2$Cl$_2$ is known to be a chlorine atom donor to hydrido zirconocene complexes [11, 31], dichloromethane was reported to be the unique solvent for some hydrozirconation reactions to go to completion [34, 35].

The insolubility of **1** has the major inconvenience that it prevents studying in detail the mechanistic factors which would operate during this hydrometalation process, as in general only the organic substrate and final zirconocene product are observed by NMR spectroscopy.

The vacant orbital in 16e⁻-zirconocene(IV) complexes allows a π-interaction with an incoming alkene or alkyne. However no metal→alkene/alkyne backbonding is possible with the d⁰-Zr-metal center. As a consequence, the metal-olefin interaction is not stabilized, and formation of the thermodynamically favored σ-bound organozirconocene complex (>10 kcal/mol) is then observed [36]. The product is the result of an overall *cis* addition of the zirconocene metal fragment and the hydrogen across the carbon-carbon multiple bonds.

The dynamic olefin insertion process has been modeled using various quantum mechanical methods. A concerted four-center mechanism involving a frontal coplanar attack of the C=C unit on the Zr–H bond of **1** is associated with a low activation energy of 0–15 kcal mol^{-1} and has been proposed for the reaction of ethylene (Scheme 8-2) [37].

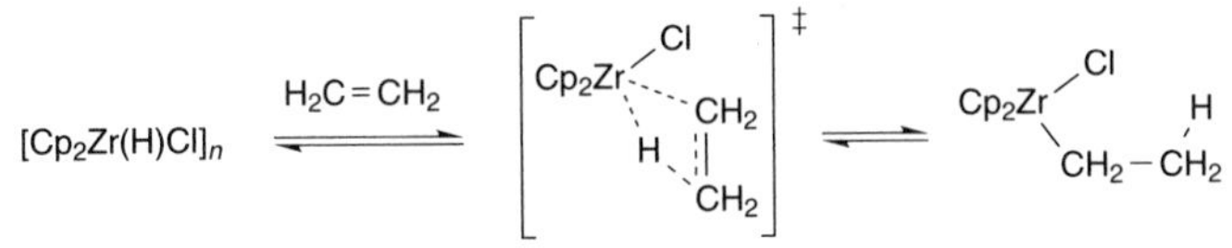

Scheme 8-2 Intermediate in the hydrozirconation of ethylene proposed by Morokuma and coworkers based on *ab initio* MO calculations

To overcome the problem of the insolubility of **1**, different strategies have been developed. Synthesis of alternative soluble hydride reagents, substitution of cyclopentadienyl rings and/or changing the chloro-auxiliary ligand were widely investigated.

The substituted-cyclopentadiene derivatives analogs of [Cp$_2$Zr(H)Cl]$_n$ (**1**), (MeCp)$_2$Zr(H)Cl (**2**) [38], Cp*CpZr(H)Cl (**3**) [39], and [(Me$_2$Si)(C$_5$H$_5$)$_2$Zr(H)Cl] [40] are more soluble in non-polar organic solvents, and consequently they are more reactive. Thus, the direct competition between **1** and **2** for the hydrozirconation reaction of 1-hexene affords predominantly (86%) the product derived from the methylated derivative (Scheme 8-3) [41].

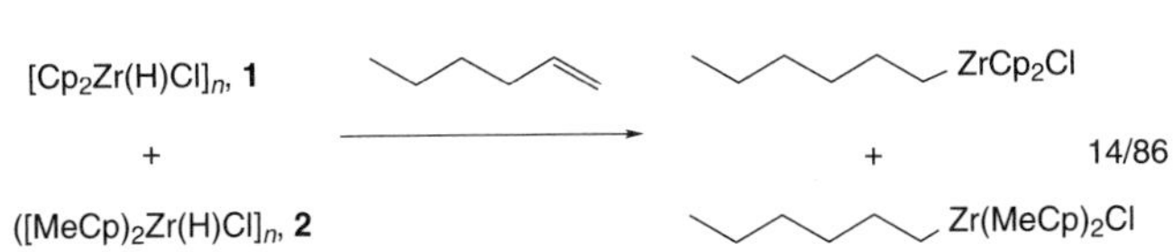

Scheme 8-3 Competition reaction between **1** and **2** in hydrozirkonation

Hydrido zirconocene complexes of the type [Cp*$_2$ZrHX] (X = Cl [42], OR [43], NR$_2$ [44], and PR$_2$ [45]) are rare and considerably more inconvenient and/or expensive to prepare and isolate in pure form than [Cp$_2$Zr(H)Cl]$_n$.

[Cp$_2$Zr(H)Me]$_2$ [46, 47] and [Cp$_2$Zr(H)CH$_2$PPh$_2$]$_n$ [48] are insoluble. Recently, Cp$_2$Zr(H)OTf, soluble in THF up to a concentration of 0.1 M, was prepared by Luinstra and coworkers from the borohydride complex Cp$_2$ZrBH$_4$(OTf) and 1 equiv. of NEt$_3$ in toluene (Scheme 8-4). The larger triflate anion gave a well-defined and very reactive dimeric hydrozirconating reagent, which was successfully applied for the hydrozirconation of alkenes, alkynes, and imines [49, 50].

$$[Cp_2Zr(BH_4)_2] \ + \ [Cp_2Zr(OTf)_2] \ \longrightarrow \ 2\,[Cp_2Zr(BH_4)OTf]$$

$$[Cp_2Zr(BH_4)OTf] \ \xrightarrow[- \ NEt_3 \bullet BH_3]{NEt_3} \ [Cp_2Zr(OTf)(\mu\text{-}H_2)]_2$$

Scheme 8-4 Synthesis of zirconocene hydride Cp$_2$Zr(H)OTf alternative to the Schwartz's reagent,1

A large number of dihydride zirconocene complexes have been prepared, as for example: [Cp$_2$ZrH$_2$]$_2$ [51, 52], [CpCp*ZrH$_2$]$_2$ [53], [Cp*$_2$ZrH$_2$] [54], [(η^5-1,3-C$_5$H$_3$^{t}Bu$_2$)Cp*ZrH$_2$], [(η^5-C$_5$HMe$_4$)Cp*ZrH$_2$], [(η^5-1,3-C$_5$H$_3$^{t}Bu$_2$)$_2$ZrH$_2$], [(η^5-1,3-C$_5$H$_4$^{t}Bu)Cp*ZrH$_2$]$_2$, [(THI)Cp*ZrH$_2$]$_2$, [(η^5-1,3-C$_5$H$_3$^{i}Pr$_2$)$_2$ZrH$_2$], [(η^5-C$_5$HMe$_4$)$_2$ZrH$_2$]$_2$ [55], [(Me$_2$Si)$_2$(η^5-C$_5$H$_3$)$_2$ZrH$_2$]$_2$ [56] [Me$_2$Si(η^5-C$_5$Me$_4$)$_2$ZrH$_2$]$_2$ [57], [(η^5-C$_5$H$_4$R)$_2$ZrH]$_2$(μ-H)$_2$ [58] (R = Me [59], SiMe$_3$ [60], ^{i}Pr, ^{t}Bu [60, 61], PhCH$_2$), [(η^5-1,2,4-C$_5$H$_2$Me$_3$)Cp*ZrH$_2$] [53], [rac-(EBTHI)ZrH$_2$]$_2$ [62], [Cp$_2$Zr(SiPhMeH)]$_2$(μ-H) [63], and [Cp$_2$Zr(SiPh$_2$H)]$_2$ (μ-H) [64]. However most of them are poor hydrozirconating reagents. Moreover, the organohydridozirconocene derivatives of the type [(Cp-substituted)$_2$Zr(R)H] obtained from dihydride complexes generally lead to secondary reactions mainly via reductive elimination [65, 66].

Various procedures for in situ generation of **1** from Cp$_2$ZrCl$_2$ with Red-Al, LiAlH(O^{t}Bu) [67], LiH [68], and LiEt$_3$BH [69] have been reported and used in organic synthesis. Under these experimental conditions the hydrometalating reagents formed in situ must not be considered to be the same as **1**, but they may be expected to react as the soluble monomeric form and therefore improve the reactivity of this metal hydride. Negishi et al. described an in situ hydrogen transfer hydrozirconation with Cp$_2$Zr(^{t}Bu)Cl [67]. It is a clean and high-yield process, but the reaction was initially found to be much slower than with preformed **1** [70, 71]. However the rate of this hydrometalating reagent has been significantly accelerated by addition of catalytic amounts of various Lewis acid metal compounds [72] (Scheme 8-5). Wipf and

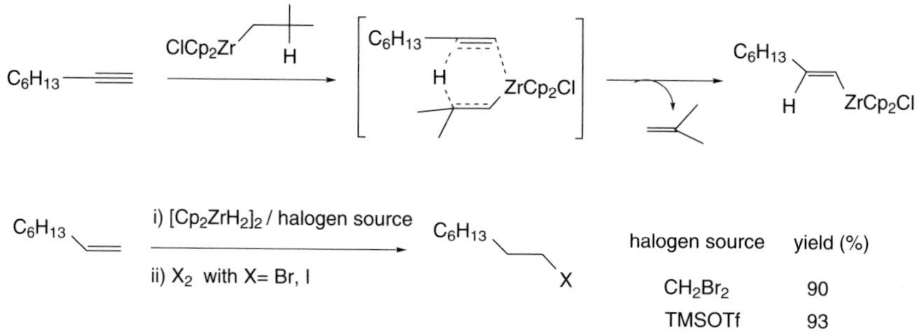

halogen source	yield (%)
CH$_2$Br$_2$	90
TMSOTf	93

Scheme 8-5 Zirconocene hydrides prepared in situ alternative to the Schwartz's reagent,1

coworkers reported that dimeric $[Cp_2ZrH_2]_2$ in the presence of a halogen source provides an efficient and convenient *in situ* alternative to **1** (Scheme 8-5) [73].

Formation of **1** by direct hydrogenation of the dichloride has been disclosed in a patent [74].

Spectroscopic and structural data for some of the hydrido zirconocene complexes described above have been reported and discussed in references cited in this section (see also [75, 76]).

Note that the synthesis and reactivity of anionic [77, 78] and cationic [79] hydrido zirconocene(IV) complexes have been investigated but will not be addressed in this chapter.

Although a large variety of zirconium hydrides have been prepared, only very few reactions are described using these promising alternative complexes to **1**.

8.3
Hydrozirconation across Carbon-Carbon Multiple Bonds

Selective monoalkylation of Cp_2ZrCl_2 to form $Cp_2Zr(R)Cl$ in which R is alkyl or alkenyl is best accomplished by hydrozirconation with $[Cp_2Zr(H)Cl]_n$ (**1**). Most synthetic applications involve the use of terminal organozirconocene complexes. The hydrozirconation reaction on carbon-carbon double or triple bonds with **1** is generally clean and gives high yields of essentially 100% *cis* addition products.

8.3.1
Hydrozirconation across Carbon-Carbon Double Bonds

8.3.1.1 Hydrozirconation of Alkenes

In their early studies, Schwartz and co-workers [5, 80] reported the zirconocene hydrido chloride $[Cp_2Zr(H)Cl]_n$ (**1**) as a reagent capable of reacting under mild conditions with a variey of non-functionalized alkenes to form isolable alkylzirconium(IV) complexes $Cp_2Zr(R)Cl$ in which the zirconium is attached to the least-hindered terminal primary carbon, irrespective of the original location of the double bond in the olefin chain. As an example, at room temperature in benzene, 1-octene, *cis*-4-octene and *trans*-4-octene all yield the *n*-octylzirconocene derivative (Scheme 8-6) [80].

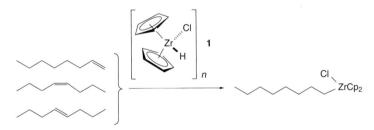

Scheme 8-6 Hydrozirkonation with **1** of *terminal* and *internal* alkenes

The major driving force for this hydrozirconation-isomerization process (which can occur even at low temperature) is steric hindrance, favoring the ultimate formation of the less-hindered terminal alkylzirconium species [81–83]. The rate and efficiency of isomerization is dependent on the substrate as well as the hydrozirconating reagent. As demonstrated by Annby and Karlsson, an increase in the size of the zirconium complex facilitates the migration of the metal fragment (see Scheme 8-11) [41]. This behavior has made hydrozirconation a valuable synthetic tool for a variety of transformations as it allows functionalization of unactivated methyl groups [4, 28, 73]; hydroalumination and hydroboration require considerably higher temperatures for isomerization than hydrozirconation. With cyclic alkenes such as cyclohexene, secondary alkylzirconocene products were observed [5]. Tertiary alkylzirconium compounds appear to be generally inaccessible, even as intermediates. As a consequence, isomerization does not proceed beyond a tertiary position, but, with increased temperature, this limitation can be overcome. Isomerization of the internal double bond is observed when the acyclic derivative is terminated at both ends by a hindered position as shown in Scheme 8-7. Tetrasubstituted alkenes are not hydrozirconated at all [5].

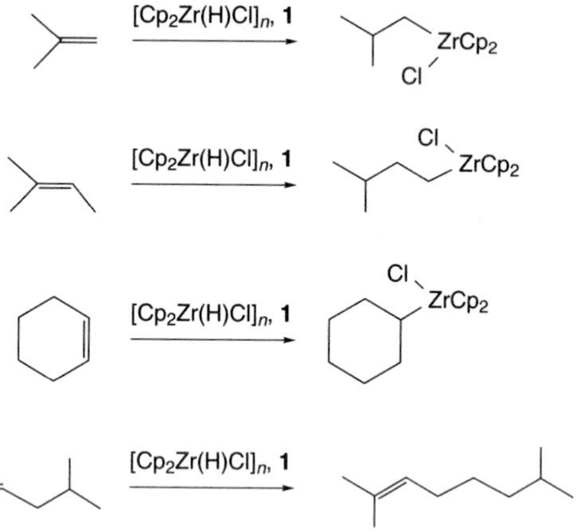

Scheme 8-7 Hydrozirkonation with 1 of substituted alkenes

General reactivity trends for alkenes were established for hydrozirconation by way of qualitative studies: terminal alkene > internal alkene > exocyclic alkene > cyclic alkene ≈ trisubstituted alkene. The rate of hydrozirconation decreases with increasing substitution on the alkene. This property was used for selective monohydrozirconation of conjugated and non-conjugated polyene derivatives (Scheme 8-8) [84–86].

8.3.1.1.1 Isomerization Process

The regioselective hydrozirconation of *internal* unsymmetrical alkenes remains a challenge, as it could considerably expand the use of zirconocene complexes. Little is known about the mechanism of zirconium migration along an alkyl chain.

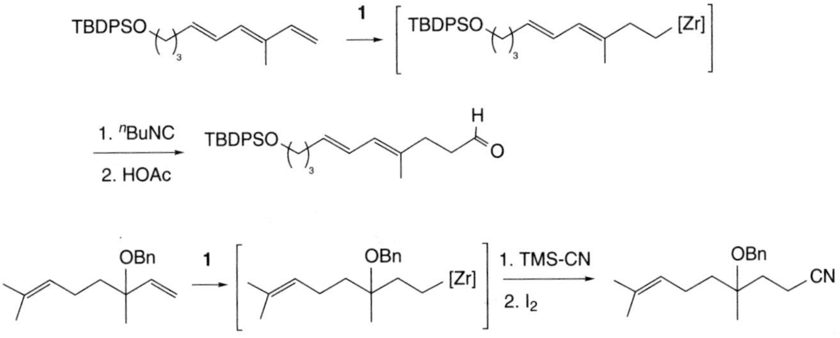

Scheme 8-8 Hydrozirkonation with **1** of polyenes

It has long been established that alkene hydrometalation is basically a reversible process and can lead to migration of the double bond. However, hydrozirconation is not sufficiently reversible to liberate the organic group as an alkene by displacement, suggesting that the regioisomerization process must be non-dissociative. Formation of free terminal alkenes from alkylzirconium complexes has been achieved *via* hydride abstraction with trityl ion in one case [87]. Attempts to quench an internal alkyl adduct following a hydrozirconation/deuteration process resulted in the isolation of only deuterated product with the starting alkene reagent. Very recently Labinger, Bercaw and co-workers [88] made a strikingly important advance in the comprehension of the factors which govern the alkyl isomerization process in organozirconocene complexes. Using appropriate conditions the authors succeeded in observing, by NMR spectroscopy, not only starting material and terminal zirconocene, but also internal alkyl complexes. The rate of isomerization of internal alkyl complexes depends on the length of the alkyl group. Moreover, in the hydrozirconation reaction of *cis*-2-butene, the presence of *trans*-2-butene during the isomerization of **4** to **5** was in favor of a free olefin pathway rather than one involving olefin-hydride intermediates that undergo intramolecular olefin rotation. Addition of 1-pentene to a mixture of **4** to **5** and identification of the corresponding pentyl chloride complex **6** support a facile olefin exchange process (Scheme 8-9).

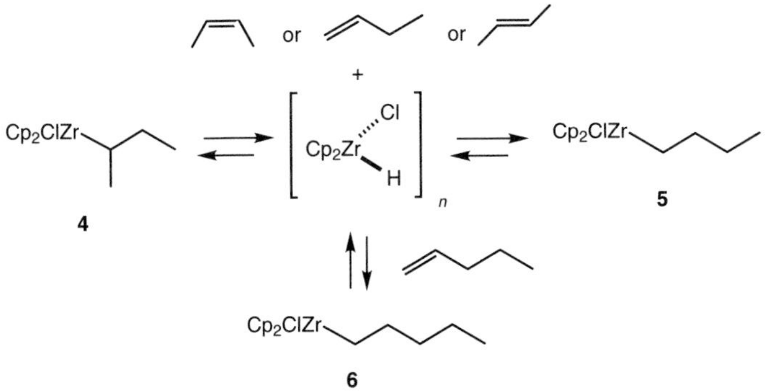

Scheme 8-9 Alkyl isomerization process in alkylzirconocenes

Then, contrary to what was reported previously, the olefin dissociates from the zirconium metal complex. This conclusion was further supported by other experimental observations. However, it cannot be completely excluded that competition between dissociative and direct rearrangement pathways could occur with the different isomerization processes studied up to now. Note that with cationic zirconocene complexes [Cp$_2$Zr-alkyl]$^+$, DFT studies suggest that Zr-alkyl isomerizations occur by the classical reaction route, i.e. β-H transfer, olefin rotation, and reinsertion into the Zr–H bond; the olefin ligand appears to remain coordinated to the Zr metal center [89].

Note that multiple hydrozirconation reactions can be conducted on the same substrate in a one-pot reaction (Scheme 8-10) [90–93].

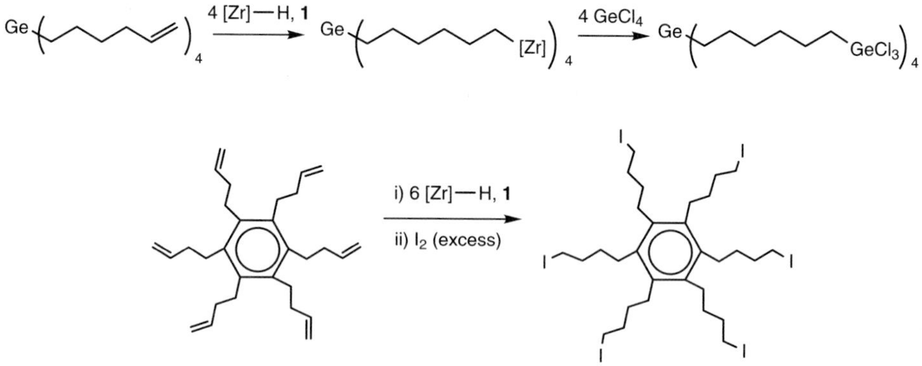

Scheme 8-10 Multiple hydrozirkonation and subsequent transformations

8.3.1.1.2 Regiochemistry

Formation of the least-hindered alkylzirconium complex during the course of the hydrozirconation-isomerization process of alkenes can be altered by the presence of aromatic rings, additional conjugative double bonds, chelating heteroatoms [81, 94] or groups that can be lost by elimination [95, 96].

Styrene hydrozirconation led to a ratio for terminal and benzylic zirconocene products of around 85:15 [97]. Experimental evidence on alkyl-substituted styrene suggested that both electronic [98, 99] and steric effects [41, 86] are important for the formation of the benzylic and terminal zirconium isomers. Migration of the metal fragment during the reaction of the zirconocene moiety might occur, perturbing the terminal/benzyl regioisomer ratio of the isolated products (Scheme 8-11) [67, 83, 98–102].

Hydrozirconation of the sterically less hindered double bond in conjugated dienes is generally observed [103]. However, the presence, as the minor regioisomer, of the allylic product has been reported [5, 86]. Depending on the reaction conditions and the reagents used in the subsequent transformation, the allylic isomer did not form (Scheme 8-12) [2].

The presence of functional groups containing oxygen-, nitrogen-, and phosphorus-chelating heteroatoms does frequently have regiochemical consequences in the

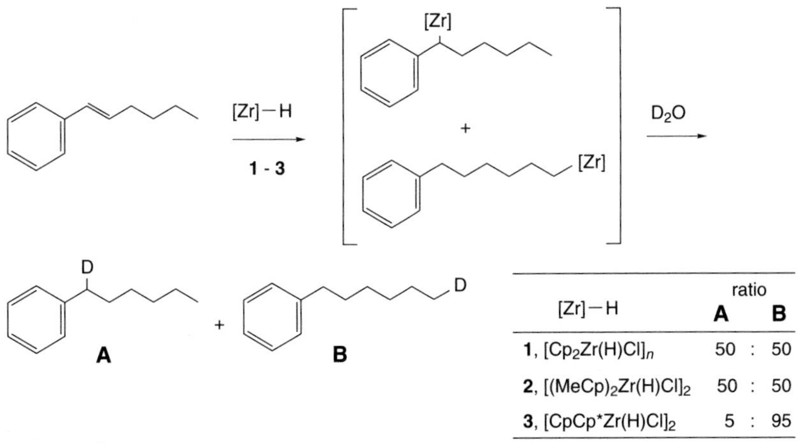

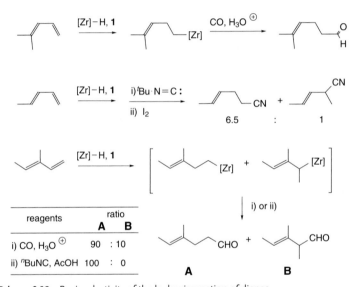

	ratio	
[Zr]–H	**A**	**B**
1, [Cp₂Zr(H)Cl]ₙ	50	50
2, [(MeCp)₂Zr(H)Cl]₂	50	50
3, [CpCp*Zr(H)Cl]₂	5	95

Scheme 8-11   Regioselectivity of the hydrozirconation of substituted styrene

reagents	ratio	
	A	**B**
i) CO, H₃O⁺	90	: 10
ii) ⁿBuNC, AcOH	100	: 0

Scheme 8-12   Regioselectivity of the hydrozirconation of dienes

hydrozirconation reactions on alkenes. Hydrozirconation of internal alkenes containing oxazoline and oxazine functionalities yielded after deuterolysis several positional isomers indicating the formation of O- or N-coordinated zirconocene species as possible products of the hydrozirconation reaction [104–106]. A five-membered Zr←N chelate ring was postulated in the synthesis of 3-hydroxystearate in the reaction of [Cp₂Zr(H)Cl]ₙ (**1**) with the corresponding internal alkene (Scheme 8-13) [107].

Definitive evidence on the role of heteroelements in the regiochemical control of the hydrozirconation may be exemplified with the reaction of phospholene-1 (**7**) with **1**. The *gem*-phosphazirconocene compound **8** is selectively formed with P-coordination on the metal fragment. The β-zirconated product **10** was prepared with **1**

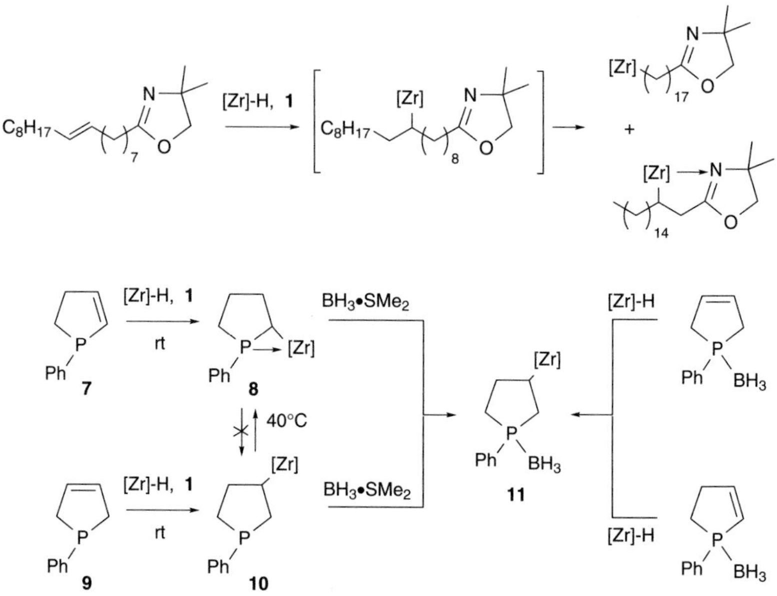

Scheme 8-13 Heteroatom chelating effect in hydrozirconation

and phospholene-2 (**9**). This compound isomerizes at 40°C to the α-zirconated product **8**. Complexation of the lone pair of the phosphorus atom in phospholene-1 with borane gives, upon addition of **1**, the β-zirconated product **11** [108] (Scheme 8-13).

Interestingly, hydrozirconation of vinyl phosphine **12** and vinyl ethoxy derivatives **13** give the expected β-zirconated complexes **14** and **15** respectively ; an O- or P-chelating effect did not occur to form the reverse regioisomer. However addition of chlorodiphenylphosphine to **14** leads to the isomerized 1,1-diphosphine **16**, which demonstrated that subsequent transformation of alkylzirconocene may induce the formation of isomers different from those observed for the corresponding starting zirconium derivatives [109] (Scheme 8-14). The primary zirconocene **15** rearranges by a rapid β-elimination to give alkoxide ROZrCp$_2$Cl and ethylene (Scheme 8-14) [110].

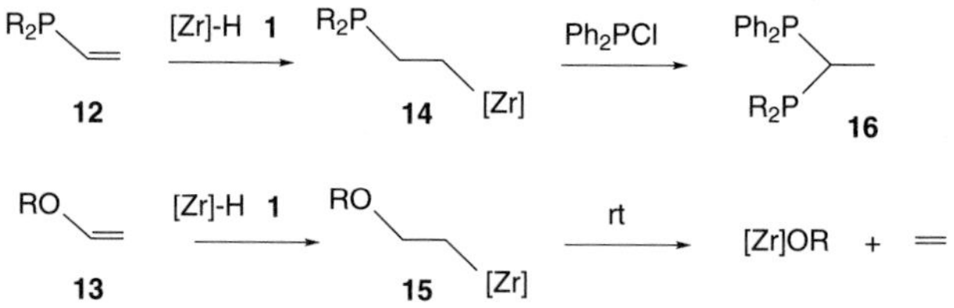

Scheme 8-14 Chemical behavior of β-heterosubstituted zirconocenes

Migration of the Cp$_2$ZrCl metal fragment along an alkyl chain onto a heteroatom was first reported by Annby and Karlsson [81, 95, 96, 111–113]. This rearrangement was used synthetically in the hydrozirconation/transmetalation-promoted ring-opening reactions of phosphorus-, nitrogen- or oxygen-containing unsaturated heterocycles to give a variety of acyclic unsaturated phosphines, alcohols, ethers and esters (Scheme 8-15) [114–117].

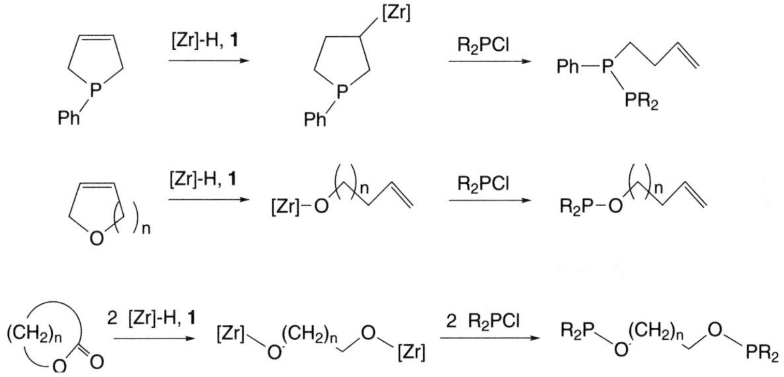

Scheme 8-15 Ring opening reaction *via* hydrozirconation/exchanges reactions (n = 1,2)

Additional coordination upon hydrozirconation (see Scheme 8-13) but also during the subsequent transformations (see Schemes 8-14 and 8-15) has to be taken into account in the analysis of the regiochemistry of the hydrozirconation/transmetalation reaction sequences.

Note that hydrozirconation of 2-vinylfuran gives only the internal product [86] (Scheme 8-16) which probably is the result of the combination of the effects described in this section: (i) O-coordination, (ii) aromatic stabilization, (iii) reduced steric effect of the flat furan ring, which favors the reverse expected regiochemistry in the hydrozirconation reaction of alkenes with $[Cp_2Zr(H)Cl]_n$ (**1**).

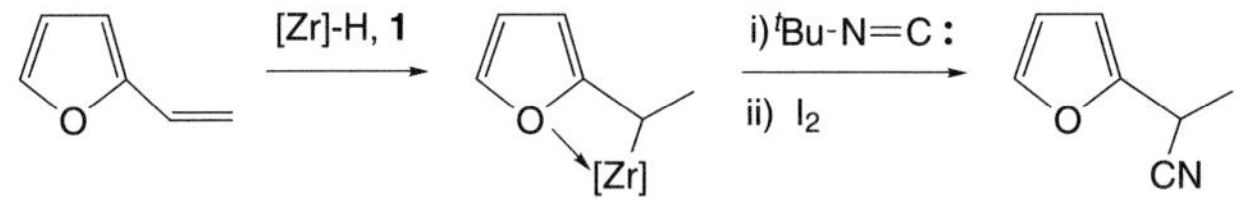

Scheme 8-16 Regioselective formation of internal zirconocene upon heteroatom chelating effect

Hydrozirconation of various alkenylboranes [118–121] and alkenylzinc halides [34] with **1** provides the corresponding 1,1-bimetalloalkanes, which can be selectively converted to functionalized organic compounds [122–125]. Interestingly, alkenylzinc halides (RCH=CHZnX) show remarkable chemoselectivity, and functional groups such as chloride, cyanide, or ester functionality are tolerated [126].

8.3.1.2 Hydrozirconation of Allenes

Allylzirconium complexes are conveniently obtained by the regio- and stereoselective hydrozirconation of allene [127–133] and can be, for example, used subsequently for the MAO-catalyzed allylzirconation of alkynes to prepare enyne derivatives [132]. Alternatively, the preparation of (E)-1,3-dienes from aldehydes and the appropriate allylstannane zirconocene derivative (R = SnBu$_3$) is accomplished (Scheme 8-17) [131]. Note that addition of $[Cp_2Zr(H)Cl]_n$ (**1**) on the allenyl reagent with the functional phosphonic acid dimethyl ester group [R = P(O)(OMe)$_2$] gave, after hydrolysis, the corresponding propynyl compound (Scheme 8-17) [134].

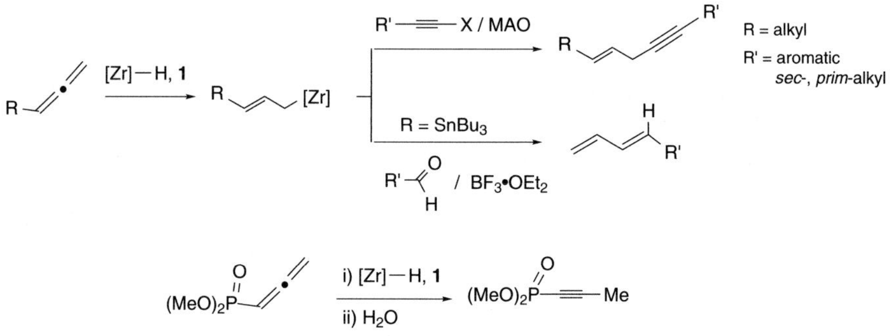

Scheme 8-17   Hydrozirconation on allene derivatives

8.3.2
Hydrozirconation of Alkynes

The hydrozirconation reaction of alkynes tolerates the presence of a large variety of functional groups. As a consequence, this protocol has become very popular in organic synthesis [135–142].

8.3.2.1   Hydrozirconation of Terminal Alkynes

Hydrozirconation of terminal alkynes R–C≡CH (R= aryl, alkyl) with **1** affords terminally (*E*)-Zr-substituted alkenes with high efficiency and excellent stereochemical and regiochemical control (>98%). These alkenylzirconocene complexes are of particular interest for synthetic use [136, 143, 144]. Moreover, beside the electropositive halogen sources [145] and heteroatom electrophiles [3] used in the pioneering studies to directly cleave the Zr–C bond, (*E*)-vinyl-Zr complexes were recently transformed into a number of other *trans*-functionalized alkenes such as (*E*)-vinyl-sulfides[146], vinylic selenol esters [147], vinyl-sulfones [148], vinyl-iodonium [149], vinyl-(RO)$_2$P(O) [150], and vinilic tellurides [143].

Terminal heterosubstituted alkynes X–C≡CH such as RO–C≡CH [16, 17, 151–153], RSe–C≡CH (R= alkyl, aryl) [154–157], Me$_3$Si–C≡CH [158] and R$_2$P–C≡CH [159] were regioselectively converted into the (*E*)-2-ethenylzirconium complexes *via* hydrozirconation using **1**. In marked contrast, with terminal ^{n}BuTe–C≡CH [154] and Bu$_3$Sn–C≡CH [160], the reverse 1,1-X,Zr-dimetallo isomers are formed (Scheme 8-18).

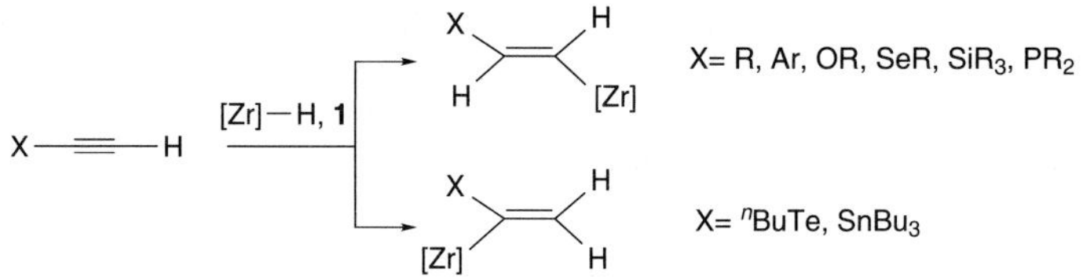

Scheme 8-18   Regio- and stereo-selectivity in hydrozirconation of terminal alkynes

Terminal diynes could not be cleanly monohydrozirconated; it was necessary to protect one of the two acetylenic forms with EtMgBr/TMSCl to get single addition [85].

8.3.2.2 Hydrozirconation of Internal Alkynes

Hydrozirconation of unsymmetrical disubstituted alkynes Me–C≡C-Alkyl produces regioselectively a mixture of the two possible regio-isomers [161]. Excess of 1 favored the formation of the thermodynamically preferred alkenylzirconocene with the metal fragment on the sterically less demanding substituent, with retention of stereoisomeric purity (cis) and with a regioselectivity over 90% [102, 161, 162]. A bis-zirconated intermediate has been suggested to account for the isomerization of the vinylzirconium species formed from the addition of an excess of 1 to acetylene. Lately it was also found that adequate experimental conditions led to the expected single regioisomer (Scheme 8-19) [140, 141, 163–166].

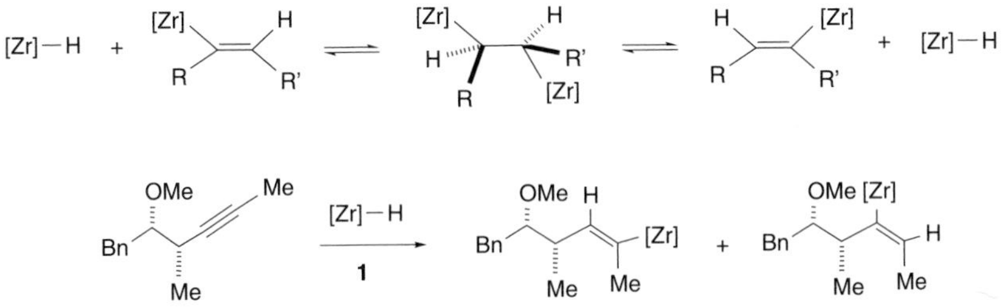

1 (equiv)	Solvent	T (°C)	time (h)	A	regioselectivity	B
2.5	C₆H₆	rt	3.3	72	:	28
2.0	C₆H₆	50	4.0	100	:	0
2.5	THF	rt	4.0	54	:	46
2.5	THF	50	1.0	100	:	0

Scheme 8-19 Regioselectivity in the hydrozirconation of internal alkynes

As for alkenes, the rate of hydrozirconation of alkynes decreases with increasing substitution on the alkyne. An unsymmetrical diyne reacts with **1** preferentially at the less-substituted triple bond [85].

The regiochemistry of the hydrozirconation of disubstituted stannyl- [24, 167–170] and silyl- [171] acetylenes and boron- [118, 172–175] and zinc- [34, 126] alkynyl derivatives result in the formation of 1,1-dimetallo compounds. Hydrozirconation of alkynyliodonium salts affords alkenylchlorozirconocenes with the Zr–C bond geminal to the iodonium moiety [176]. These zirconocene complexes allowed the preparation of (E)-trisubstituted olefins (Scheme 8-20).

Hydrozirconation is useful for regiospecific and/or stereospecific deuterium labeling [177]. Formation of diastereomerically pure dideuterated 3,3-dimethyl-butylzirconium(IV) complexes by successive hydrozirconation sequences showed

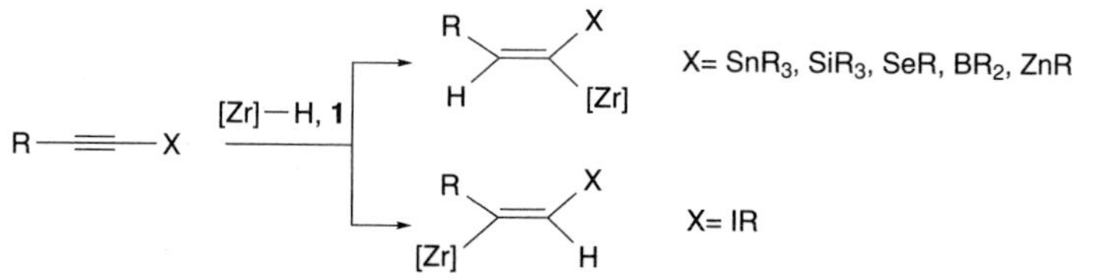

Scheme 8-20 Regio- and stereo-selectivity of the hydrozirconation of disubstituted alkynes

that Zr–H addition on carbon-carbon multiple bonds proceeds stereospecifically *cis* (Scheme 8-21) [5]. These compounds have been used to get some insight in the reaction mechanism of carbon-X bond cleavage processes [178].

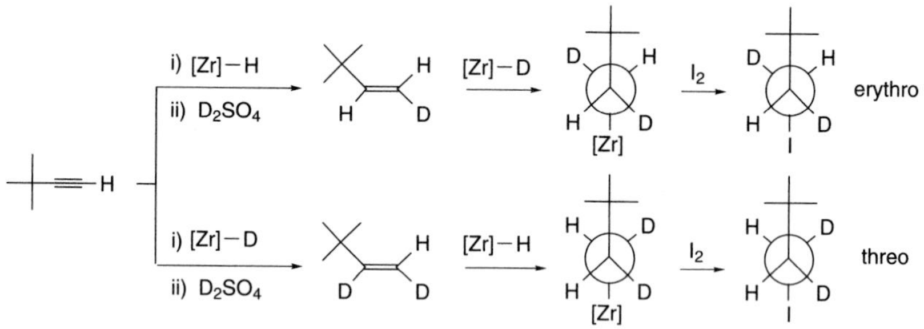

Scheme 8-21 Diastereoselectivity of the hydrozirconation/exchange reactions

8.4
Hydrozirconation across Heteropolar Multiple Bonds

Reduction with $[Cp_2Zr(H)Cl]_n$ (**1**) of common C=X functional groups such as aldehydes, ketones, nitriles and related functionality is also possible.

Carbonyl compounds, such as aldehydes [103, 179], (thio)ketones [31, 94, 180–183], carboxylic acids, and esters [183, 184] with **1** are reduced to alcohols after hydrolysis [5], except in sterically hindered cases (see Section 8.5) [185, 186]. Under the same experimental conditions the regioselective reduction of the oxirane ring with **1** gives also the corresponding alcohol [183, 187].

Hydrozirconation of nitriles RC≡N with **1** led to the formation of aldimido $[RCH=NZrCp_2Cl]$ derivatives [183, 188–191]. Erker and coworkers reported the first X-ray structure analysis of these complexes, which shows a nearly linear Zr–N=CHPh unit indicating substantial Zr–N π character. Moreover, this complex exhibits the stereochemical properties of a heteroallene system [189, 190]. Mono- and di-hydrozirconation of nitriles $R_2C(C≡N)_2$ with **1** [192, 193] have been also successfully accomplished to give $[R_2C(C≡C(N)(CH=NZrCp_2Cl)]$ and $[R_2C(C≡CH=NZr-Cp_2Cl)_2]$, respectively in quantitative yield (Scheme 8-22). Extension of this protocol to phosphinonitrile compounds $R_2P-C≡N$ [194] and $RP(C≡C(N)_2$ allowed the preparation of new imines N-substituted with different heteroelements. Interestingly, hydrozirconation of malononitrile $H_2C(C≡N)_2$ followed by exchange reaction with chlorophosphines allowed the synthesis of new β-diiminate systems [195] (Scheme 8-22).

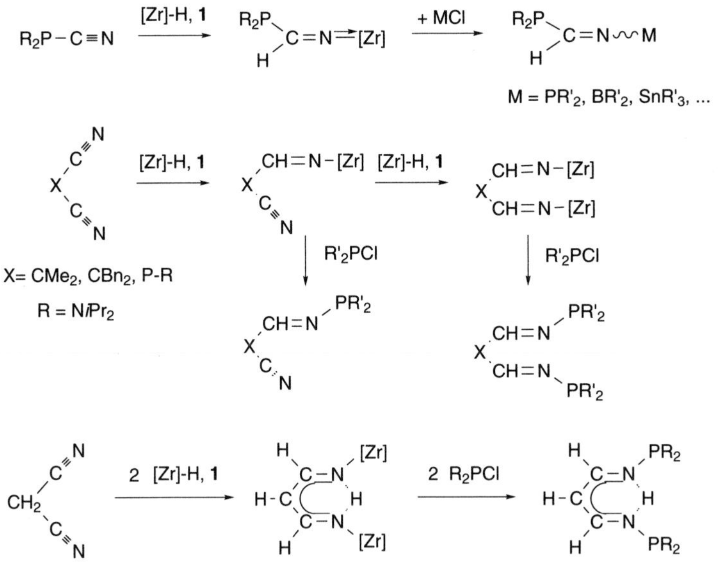

Scheme 8-22 Hydrozirconation on nitrile derivatives

Another route for the formation of imines is the treatment of secondary carboxamides with an excess of **1** [196].

In marked contrast to the other hetero atom multiple bond functions cited in this section the >C=N– imine bond was not found to undergo hydrometalation. Imines with neighboring C–H bonds as in PhCH=NMe do react with **1** *via* imine/enamine tautomerism, but not by hydrozirconation [197].

The reaction of **1** with diphenyldiazomethane resulted in the hydrogenation of the >C=N=N unit to form the hydrazonato ligand, which is found η^2-N,N' bonded to the metal [198, 199].

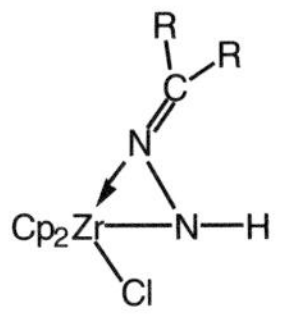

Note that carbon monoxide inserts into the Zr–H bond of **1** (2 equiv.) to afford an η^2-formaldehydo-type complex [(Cp$_2$ZrCl)]$_2$(μ-CH$_2$O) [200–202]. Iminoacyl zirconocene complexes are formed after addition of **1** to isonitriles [203]. Carbon dioxide [183, 202] is reduced to formaldehyde with **1** (2 equiv.). CO$_2$-like molecules such as isocyanates RNCO [204], isothiocyanates RNCS [205], and carbodiimides RNCNR [204] are readily converted to the corresponding bidentate form-amido ligands.

Various new phosphonium salts were prepared in a one-pot reaction of carbon disulfide, phosphines, and **1** followed by an exchange reaction of the zirconated metal fragment with 2 equiv. of electrophiles (Scheme 8-23) [206].

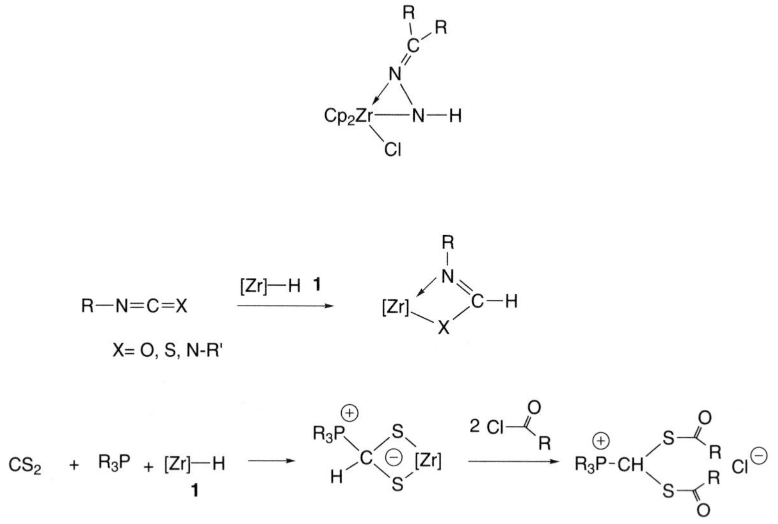

Scheme 8-23 Hydrozirconation of heterocumulenes

Phosphaalkenes –P=C<, and phosphaimines –P=N– react with **1** to give secondary zirconated aminoalkyl or diamino phosphines, respectively, with P-coordination to the metal fragment (Scheme 8-24) [207]. An unexpected methylene-transfer reaction was observed upon reaction of **1** with $Ph_3P=CH_2$ (Scheme 8-24) [208].

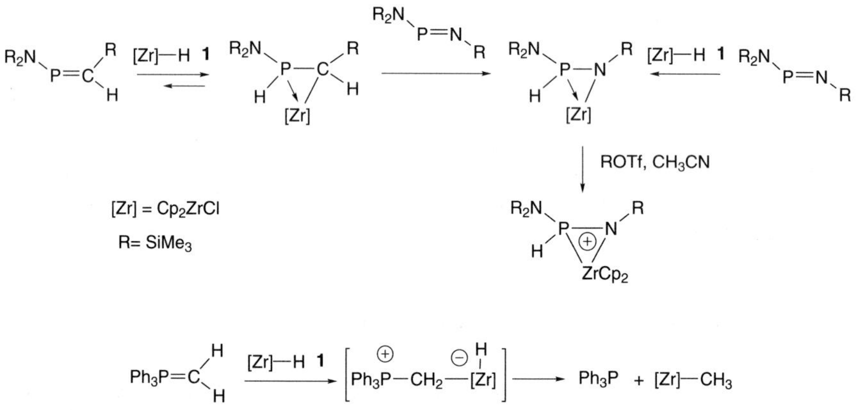

Scheme 8-24 Hydrozirconation of phosphorus-X double bond derivatives

Hydrozirconation of iminoboranes –B≡N–R leads to a four-membered ring with a Zr–H–B hydride bridge [209].

8.5
Chemoselectivity and Functional Group Compatibility

The chemoselectivity of Schwartz's reagent (1) toward alkynes, alkenes, nitriles, and carbonyl groups, and thus its general functional group compatibility, can be modulated. However, it is important to keep in mind that the presence of functional groups may have regiochemical consequences on the hydrozirconation reaction.

8.5.1
Relative Reactivity of C≡C versus C=C

As predicted from the comparative rates for C≡C over C=C hydrozirconation cited earlier, a (poly)enyne is selectively hydrozirconated at the alkyne moiety, whatever the position of the alkene function [138, 210] in the molecule. It can be exemplified by the chemoselective hydrozirconation of 1,3-butenyne. One exception to this chemoselectivity has been reported, which showed the terminal alkene to react with 1 but leaving the TMS-substituted alkyne function intact (Scheme 8-25).

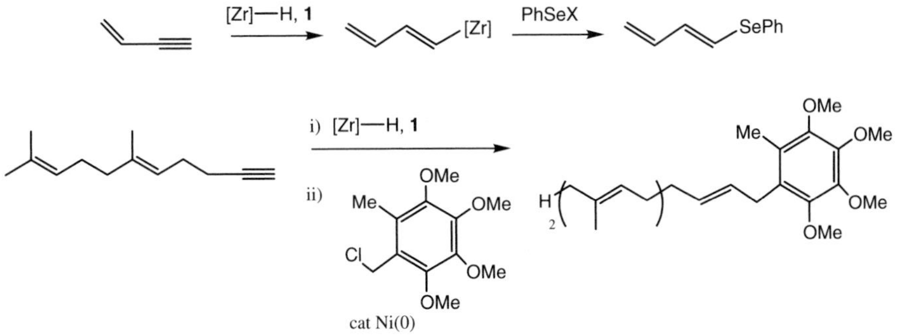

Scheme 8-25   Hydrozirconation of enyne derivatives

The synthetic potential of alkenylzirconium complexes is partially due to the fact that the hydrozirconation of alkynes can be carried out in the presence of some synthetically useful functional groups such as halide [80, 153, 211, 212], acetals, amides, imides, carbamates, sulfides [186], ester, cyano [95, 213] and chiral propargyl amino functionalities [214].

8.5.2
Relative Reactivity of C≡C and C=C versus C≡N

Nitriles [153, 211] are tolerated by an equivalent of reagent in the presence of a particularly reactive site of unsaturation, such as a terminal alkyne or vinyl function, whereas methacrylonitrile gives only C≡N hydrozirconation (Scheme 8-26) [215].

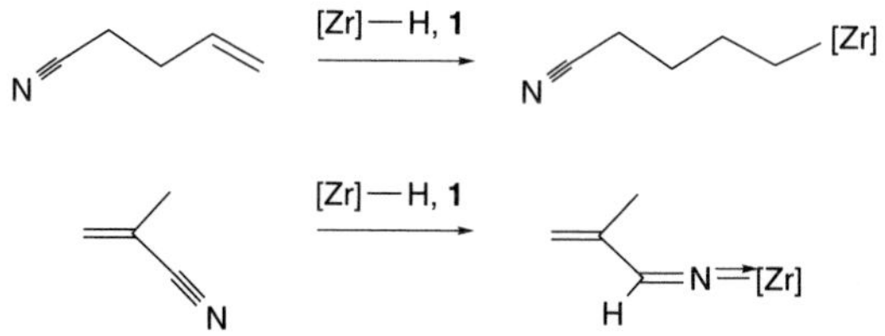

Scheme 8-26 Hydrozirconation of alkenes with nitrile function

8.5.3
Relative Reactivity of C≡C and C=C versus carbonyl functions

Enones and enoates undergo 1,2-reduction [115, 191]. Lipshutz et al. reported the effective protection of carbonyl functions by the triisopropylsilyl acyl silane group (TIPS), which allowed the selective conversion of alkenes or alkynes to the corresponding zirconocene complexes [24]. The aldehyde could subsequently be regenerated by desilylation with TBAF [186].

Carboxylic acids can be protected as oxazolines [96, 105–107, 186, 191] or as ester functions. Alkynic esters such as silyl esters [153, 211], *tert*-butyl esters [216], and even benzyl esters [153, 211] have been successfully hydrozirconated when the reactive site was a terminal alkyne or vinyl group (Scheme 8-27).

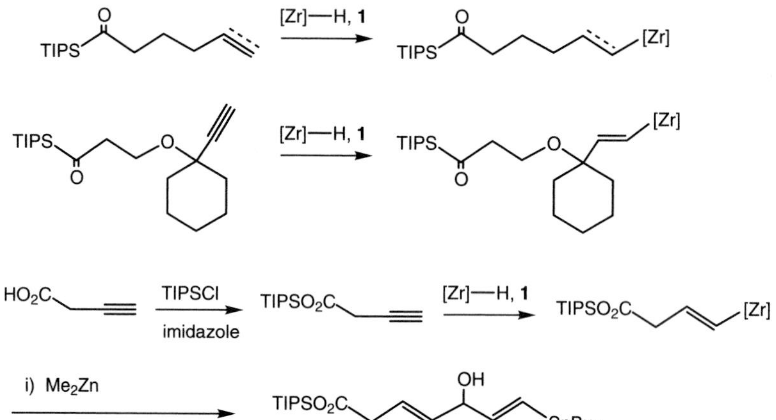

Scheme 8-27 Hydrozirconation with protected carbonyl functions

The rate of hydrozirconation of a terminal olefin is much faster than the reduction of the oxirane ring. This chemoselectivity was used in the preparation of various cycloalkylmethanols by hydrozirconation of alkenyloxiranes (Scheme 8-28) [217].

As described in the previous section for unsaturated functional groups, Schwartz's reagent (**1**) and most zirconocene(IV) hydrides readily deprotonate acidic moieties [183, 218].

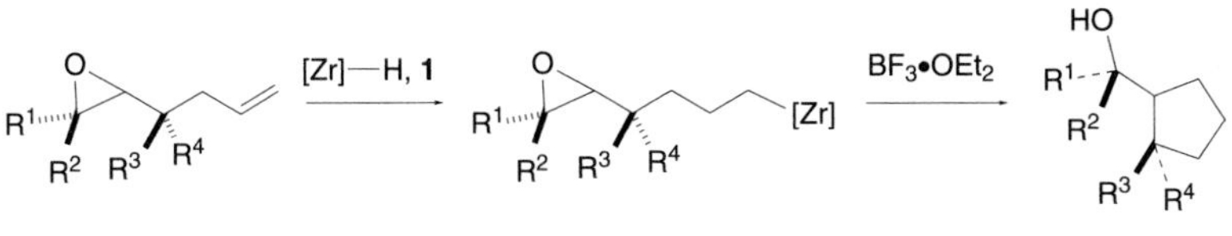

Scheme 8-28 Hydrozirconation of alkenyloxiranes

Dienols with 2 equiv. of **1** gave the corresponding bimetallic alkoxide organozirconocene complexes; however, protolysis allowed recovery of the alcohol functionality (Scheme 8-29) [107]. Alcohols can also be easily converted to ethers. Alkyl, aryl, silyl [85, 112, 183, 210] and THP [17, 153, 211, 219] ethers are stable under hydrozirconation conditions; side products were observed only with the trimethylsilyl group [220, 221].

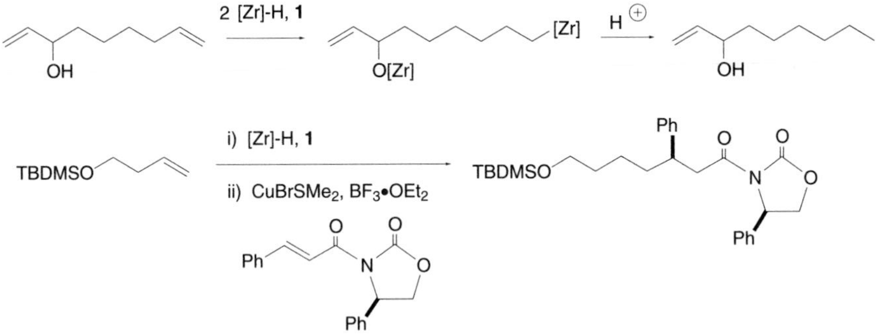

Scheme 8-29 Hydrozirconation of alkenyl (protected)alcohols compounds

Little is known about the tolerance of **1** with unsaturated (poly)halogen compounds. Hydrozirconation of chloroalkenes can lead to competitive cyclization, and simple reduction of both C=C and C–Cl bonds [98, 222]. However addition of **1** to an alkenyl- or propargyl bromide led to the expected product as the sole product of the reaction in excellent yield (Scheme 8-30) [134, 223].

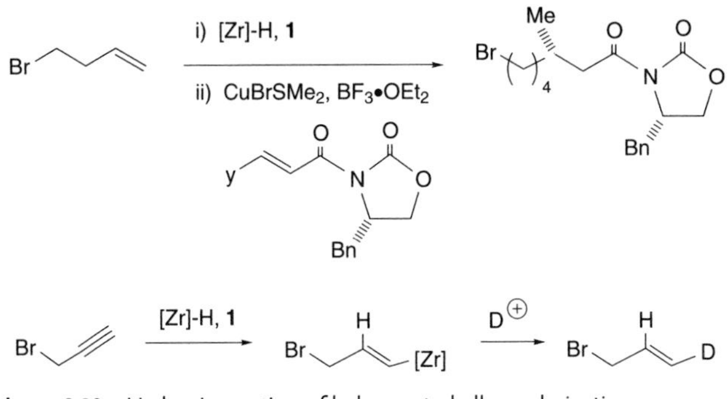

Scheme 8-30 Hydrozirconation of halogenated alkene derivations

8.6

Bimetallic Transition Metal-Zirconocene Complexes from Zirconium Hydrides

The bis-zirconocene complex $Cp_2ClZrCH_2CH_2ZrCp_2Cl$ has been isolated upon double hydrozirconation of acetylene with **1** [102]. Recently, the preparation of a heterogeneous bis-zirconocene catalyst was succesfully achieved from zirconocene dichloride complexes containing alkenyl or alkynyl substituents [224].

$Cp(PMe_3)_2RuC{\equiv}CH$ and the corresponding vinyl compound react with **1** to give the expected hydrozirconated complexes [225, 226]. Hydrozirconation reactions were also observed with the vinyl and acetylenic ferrocenyl complexes [227]. In marked contrast, the iron complex $Cp^*(dppe)Fe\text{-}C{\equiv}CH$ reacts anomalously with Schwartz's reagent to form acetylide C_2-bridged heterodinuclear complexes (Scheme 8-31) [228].

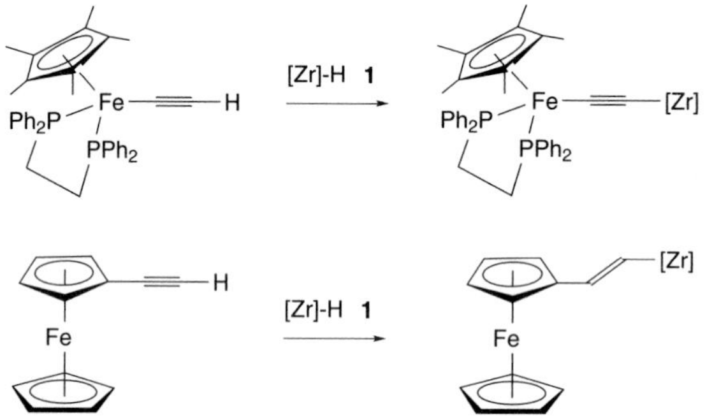

Scheme 8-31 Transition metalterminal alkynes with $[Cp_2Zr(H)Cl]_n$ **1**

CO_2-Bridged bimetallic zirconocene complexes have been formed from **1** and metallocarboxylic acids [229]. Reaction of **1** with metal enolates $Cp(CO)_3WCHR'$ COX (X = OEt, Me, Ph) gives $Cp(CO)_3WCH(R')CH(R)OZrCp_2(Cl)$. The structure for R' = H and R = Me was solved by an X-ray analysis and the chemical reactivity of these organometallic products have been studied [230].

Note that reaction of the vinyl chloride complex $Tp'Rh(CNCH_2{}^tBu)(CH{=}CH_2)Cl$ with $[Cp_2ZrH_2]_2$ did not give the hydrozirconation reaction on the vinyl group but led quantitatively to the formation of the vinyl hydride complex $Tp'Rh(CNCH_2{}^tBu)$ $(CH{=}CH_2)H$ [231].

8.7

Zirconocene Hydrides and Related Derivatives as Catalyst

Hydride complexes of Group IV metallocenes have been implicated as catalysts and as important intermediates in olefin hydrogenation and polymerization reactions [232, 233].

Schwartz's reagent (**1**) is an excellent catalyst for pinacolborane hydroboration of alkynes (Scheme 8-32) [234, 235].

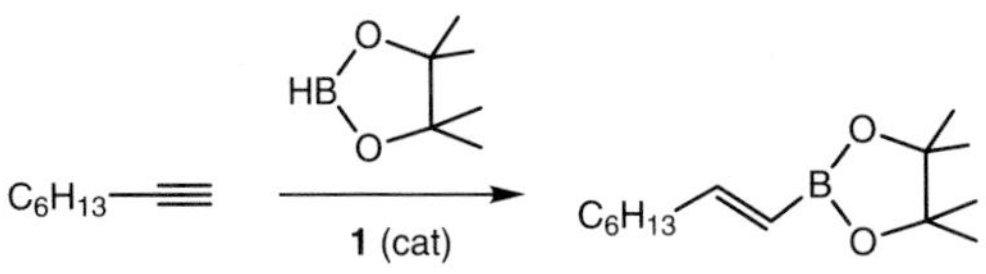

Scheme 8-32 [Cp$_2$Zr(H)Cl]$_n$ 1-catalyzed hydroboration

The selective dimerization of aldehydes to esters [236] as well as the cross-condensation of cycloalkanones with aldehydes and primary alcohols [237] were found to proceed very efficiently under mild conditions with **1** as catalyst (Scheme 8-33).

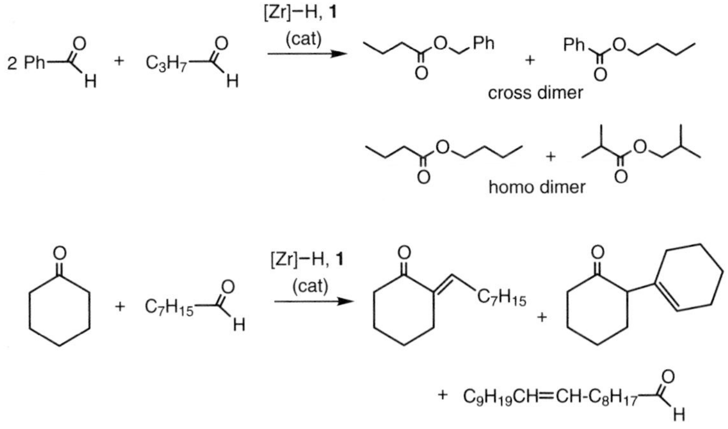

Scheme 8-33 [Cp$_2$Zr(H)Cl]$_n$ 1-catalyzed cross-condensation reactions

Some hydrometalation reactions have been shown to be catalyzed by zirconocene. For instance, Cp$_2$ZrCl$_2$-catalyzed hydroaluminations of alkenes [238] and alkynes [239] with ^{i}Bu$_3$Al have been observed (Scheme 8-34). With alkyl-substituted internal alkynes the process is complicated by double bond migration, and with terminal alkynes double hydrometalation is observed. The reaction with ^{n}Pr$_3$Al and Cp$_2$ZrCl$_2$ gives simultaneously hydrometalation and C–H activation. Cp$_2$ZrCl$_2$/^{n}BuLi-catalyzed hydrosilation of acyclic alkenes [64, 240] was also reported to involve hydrogen transfer *via* hydrozirconation.

An unexpected *cis,cis* to *trans,trans* isomerization of a disilanyl analog of 1,5-cyclooctadiene was observed with **1** as the catalyst [241].

8.8
Hydrozirconation: Conclusion and Perspectives

Since the initial reports of Schwartz's reagent ([Cp$_2$Zr(H)Cl]$_n$, **1**) over 30 years ago there has been explosive growth of zirconocene chemistry. Hydrozirconation is an efficient method of hydrometalation because of (i) the mild nature of the conditions involved, (ii) the excellent regio- and stereo-chemical control of hydrozirconation, (iii) the one-pot nature of the procedure, and (iv) the price of Zr, which is one of the

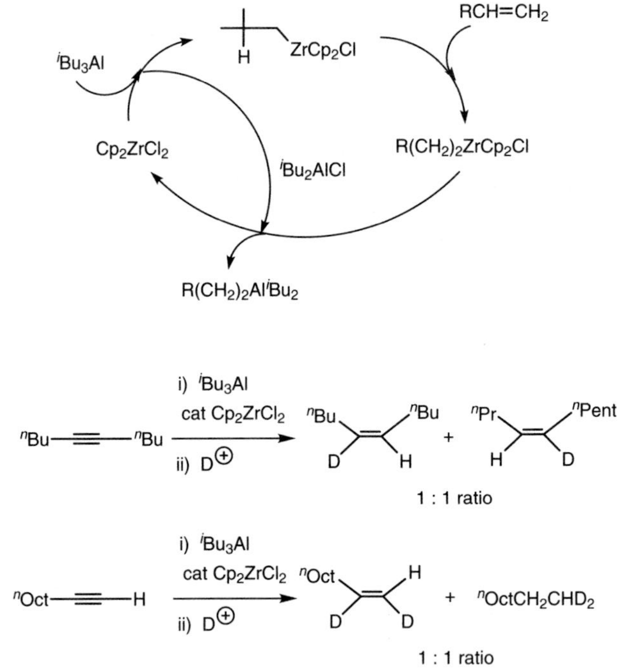

Scheme 8-34 Cp$_2$Zr(H)Cl$_2$-catalyzed hydroalumination with ^{i}Bu$_3$Al of alkenes and alkynes

least expensive transition metals (about \$11 per mole for 98% pure ZrCl$_4$ – comparable with the costs of Fe, Cu, Mn, and Ti).

In terms of scope and chemoselectivity, hydrozirconation takes its place between hydroboration and hydroalumination. However, the synthetic applications of organozirconocene complexes have been considerably expanded over these last few decades, and it can be expected that they will become more and more attractive in the future. Beside the direct substitution sequences, indirect reaction pathways involving transmetalation or activation by ligand abstraction have been successfully applied in a number of cross-coupling and C–C bond-forming reactions.

It is important to remember that the yield and reproducibility of hydrozirconations and the subsequent transformations depend on the presence of ionic impurities in **1** which often hamper the desired transformations. For example, *in situ* preparations of **1** with LiEt$_3$BH and ^{t}BuMgCl were not appropriate for hydrozirconation/copper-catalyzed conjugate addition sequences; otherwise preformed **1** is well adapted for these protocols.

The facile isomerization of alkyl and vinyl zirconocenes complicates the stereocontrol in the transformations of internal acyclic alkenes and allenes, and so further studies are certainly needed to circumvent this problem.

Enantioselective hydrozirconation using, for example, chiral ansa-metallocenes [242] remain largely unexplored.

Developments in the area of polymer-supported zirconocene hydride reagents are just now being investigated [243] and will be further explored.

Abbreviations

Bn	benzyl	TBDPS	*tert*-butyldiphenylsilyl
Cp	cyclopentadienyl	Tf	trifluoromethylsulfonyl
Cp*	cyclopentamethyldienyl	THF	tetrahydrofuran
EBTHI	ethylene-1,2-bis(η^5-tetrahydroindenyl)	THI	η^5-tetrahydroindenyl
		THP	tetrahydropyranyl
rt	room temperature	TIPS	triisopropylsilyl
TBAF	tetrabutylammonium fluoride	TMS	trimethylsilyl
		Tp'	hydridotris(3,5-dimethylpyrazolyl)-borate
TBDMS	*tert*-butyldimethylsilyl		

References

1 E. NEGISHI, J.-L. MONTCHAMP in *Metallocenes, Synthesis, Reactivity, Applications*, A. TOGNI, R. L. HALTERMAN (eds.), Wiley-VCH, Weinheim, **1998**, Vol. 1, p. 241.

2 P. WIPF, H. JAHN, *Tetrahedron*, **1996**, *52*, 12853.

3 J. M. TAKACS in *Comprehensive Organometallic Chemistry II*, E. W. ABEL, F. G. A. STONE, G. WILKINSON, L. S. HEGEDUS (eds.), Pergamon Press, Oxford, **1995**, Vol. 4, p. 39.

4 E. NEGISHI, T. TAKAHASHI, *Synthesis*, **1988**, 1. E. NEGISHI, T. TAKAHASHI, *Aldrichim. Acta*, **1985**, *18*, 31.

5 J. SCHWARTZ, J. A. LABINGER, *Angew. Chem. Int. Ed. Engl.*, **1976**, *15*, 333.

6 J. P. MAJORAL, A. IGAU, *Coord. Chem. Rev.*, **1998**, *176*, 1.

7 P. WIPF, W. XU, H. TAKAHASHI, H. JAHN, P. D. G. COISH, *Pure Appl. Chem.*, **1997**, *69*, 639.

8 P. BINGER, S. PODUBRIN, in *Comprehensive Organometallic Chemistry II*, E. W. ABEL, F. G. A. STONE, G. WILKINSON, M. F. LAPPERT (eds.), Pergamon Press, Oxford, **1995**, Vol. 4, p. 439. R. D. BROENE, in *Comprehensive Organometallic Chemistry II*, E. W. ABEL, F. G. A. STONE, G. WILKINSON, L. S. HEGEDUS (eds.), Pergamon Press, Oxford, **1995**, Vol. 12, p. 323. S. L. BUCHWALD, R. D. BROENE, in *Comprehensive Organometallic Chemistry II*, E. W. ABEL, F. G. A. STONE, G. WILKINSON, L. S. HEGEDUS (eds.), Pergamon Press, Oxford, **1995**, Vol. 12, p. 721.

9 R. D. BROENE, S. L. BUCHWALD, *Science*, **1993**, *261*, 1696.

10 E. NEGISHI, in *Comprehensive Organic Chemistry*, B. M. TROST, I. FLEMING, L. A. PAQUETTE, (eds.), Pergamon Press, Oxford, **1991**, Vol. 5, p. 1163. C. FERRERI, G. PALUMBO, R. CAPUTO in *Comprehensive Organic Chemistry*, B. M. TROST, I. FLEMING, S. L. SCHREIBER (eds.), Pergamon Press, Oxford, **1991**, Vol. 1, p. 162.

11 P. C. WAILES, H. WEIGOLD, A.P. BELL, *J. Organometal. Chem.* **1971**, *27*, 373; **1972**, *43*, C32.

12 P. WIPF, W. XU, *J. Org. Chem.* **1993**, *58*, 825.

13 H. MAETA, K. SUZUKI, *Tetrahedron Lett.* **1993**, *34*, 341.

14 H. MAETA, T. HASHIMOTO, T. HASEGAWA, K. SUZUKI, *Tetrahedron Lett.* **1992**, *33*, 5965.

15 M. J. LOOTS, J. SCHWARTZ, *J. Am. Chem. Soc.* **1977**, *99*, 8045.

16 E. NEGISHI, D. E. VAN HORN, *J. Am. Chem. Soc.* **1977**, *99*, 3168.

17 NEGISHI, E.; TAKAHASHI, T.; BABA, S.; VAN HORN, D. E.; OKUKADO, N. *J. Am. Chem. Soc.* **1987**, *109*, 2393-401.

18 E. NEGISHI, N. OKUKADO, A. O. KING, D. E. VAN HORN, B. I. SPIEGEL, *J. Am. Chem. Soc.* **1978**, *100*, 2254.

19 D. B. CARR, J. SCHWARTZ, *J. Am. Chem. Soc.* **1979**, *101*, 3521.

20 D. B. CARR, J. SCHWARTZ, *J. Am. Chem. Soc.* **1977**, *99*, 638.

21 P. WIPF, S. RIBE, *J. Org. Chem.* **1998**, *63*, 6454.

22 P. WIPF, W. XU, *Org. Synth.* **1997**, *74*, 205.

23 S. PEREIRA, B. ZHENG, M. SREBNIK, *J. Org. Chem.* **1995**, *60*, 6260.

24 B. H. LIPSHUTZ, A. BHANDARI, C. LINDS-LEY, R. KEIL, M. R. WOOD, *Pure Appl. Chem.* **1994**, *66*, 1493.

25 P. WIPF, *Synthesis*, **1993**, 537.

26 K. SONOGASHIRA, in *Metal-catalyzed Cross-coupling Reactions*, F. DIEDERICH, P. STANG (eds.), Wiley-VCH, Weinheim, **1998**, p. 223. I. MAREK, J. F. NORMANT, in *Metal-catalyzed Cross-coupling Reactions*, F. DIEDERICH, P. STANG (eds.), Wiley-VCH, Weinheim, **1998**, p. 271.

27 J. P. MAJORAL, M. ZABLOCKA, A. IGAU, N. CÉNAC, *Chem. Ber.*, **1996**, *129*, 879.

28 J. A. LABINGER, in *Comprehensive Organic Synthesis*, Vol. 8, B. M. TROST, 1. FLEMING (eds.), Pergamon Press, Oxford, **1991**, Vol. 8, p. 667.

29 For an overview of organozirconium and -hafnium chemistry, see: P. C. WAILES, R. S. P. COUTTS, H. WEIGOLD, *Organometallic Chemistry of Titanium, Zirconium, and Hafnium*, Academic Press, New York, **1974**, p. 302. D. J. CARDIN, M. F. LAPPERT, C. L. RASTON *Chemistry of Organozirconium and Hafnium Compounds*, John Wiley & Sons, New York, **1986**, p. 451.

30 For recent symposiums on zirconocene chemistry, see: E. NEGISHI, *Recent Advances in the Chemistry of Zirconocene and Related Compounds*, Tetrahedron Symposia-in-print No. 57, *Tetrahedron* **1995**, *51* (special issue). R. F. JORDAN *Metallocene and Single Site Olefin Catalysis, J. Mol. Catal.* **1998**, *128* (special issue) and references cited therein. R. F. JORDAN, A. S. GURAM, in *Comprehensive Organometallic Chemistry II*, E. W. ABEL, F. G. A. STONE, G. WILKINSON, M. F. LAPPERT (eds.), Pergamon Press, Oxford, **1995**, Vol. 4, p. 589.

31 S. L. BUCHWALD, S. J. LaMAIRE, R. B. NIELSEN, B. T. WATSON, S. M. KING, *Org. Synth.* **1993**, *71*, 77. S. L. BUCHWALD, S. J. LaMAIRE, R. B. NIELSEN, B. T. WATSON, S. M. KING, *Tetrahedron Lett.* **1987**, *28*, 3895.

32 P. C. WAILES, H. WEIGOLD, *Inorg. Synth.* **1979**, *19*, 223.

33 T. GIBSON, L. TULICH, *J. Org. Chem.* **1981**, *46*, 1821.

34 C. E. TUCKER, B. GREVE, W. KLEIN, P. KNOCHEL, *Organometallics* **1994**, *13*, 94.

35 A. MARAVAL, A. IGAU, J. P. MAJORAL, unpublished results.

36 Y. HUANG, B. WAN, *Huaxue Xuebao* **1982**, *40*, 217.

37 J. ENDO, N. KOGA, K. MOROKUMA, *Organometallics* **1993**, *12*, 2777. N. KOGA, K. MOROKUMA, *Chem. Rev.* **1991**, *91*, 823.

38 G. ERKER, R. SCHLUND, C. KRUEGER, *Organometallics* **1989**, *8*, 2349.

39 J. ALVHAELL, S. GRONOWITZ, A. HALLBERG, *Chem. Scr.* **1988**, *28*, 285.

40 H. YASUDA, K. NAGASUNA, M. AKITA, K. LEE, A. NAKAMURA, *Organometallics* **1984**, *3*, 1470.

41 U. ANNBY, S. GRONOWITZ, A. HALLBERG, *Acta Chem. Scand.* **1990**, *44*, 862.

42 G. L. HILLHOUSE, J. E. BERCAW, *J. Am. Chem. Soc.* **1984**, *106*, 5472.

43 P. T. WOLCZANSKI, R. S. THELKEL, J. E. BERCAW, *J. Am. Chem. Soc.* **1979**, *101*, 218.

44 P. T. WOLCZANSKI, J. E. BERCAW, *J. Am. Chem. Soc.* **1979**, *101*, 6450.

45 S. NIELSEN-MARCH, R. J. CROWTE, P. G. EDWARDS, *J. Chem. Soc., Chem. Commun.* **1992**, 699. J. HO, D. W. STEPHAN, *Organometallics* **1991**, *10*, 3001.

46 R. F. JORDAN, C. S. BAJUR, D. E. DASHER, A. L. RHEINGOLD, *Organometallics* **1987**, *6*, 1041.

47 C. J. HARLAN, S. G. BOTT, A. R. BARRON, *J. Chem. Soc., Dalton Trans.* **1997**, 637.

48 Y. RAOULT, R. CHOUKROUN, C. BLANDY, *Organometallics* **1992**, *11*, 2443. R. CHOUKROUN, D. GERVAIS, *J. Chem. Soc., Chem. Commun.* **1985**, 224.

49 N. S. HUESGEN, G. A. LUINSTRA, *Inorg. Chim. Acta* **1997**, *259*, 185.

50 G. A. LUINSTRA, U. RIEF, M. H. PROSENC, *Organometallics* **1995**, *14*, 1551.

51 P. C. WAILES, H. WEIGOLD, A.P. BELL, *J. Organometal. Chem.* **1970**, *24*, 405.

52 D. G. BICKLEY, H. NGUYEN, P. BOUGEARD, B. G. SAYER, R. C. BURNS, M. J. McGLINCHEY, *J. Organomet. Chem.* **1983**, *246*, 257-68.

53 P. T. WOLCZANSKI, J. E. BERCAW, *Organometallics* **1982**, *1*, 793.

54 J. M. MANRIQUEZ, D. R. McALISTER, R. D. SANNER, J. E. BERCAW, *J. Am. Chem. Soc.* **1978**, *100*, 2716.

55 P. J. Chirik, M. W. Day, J. E. Bercaw, *Organometallics* **1999**, *18*, 1873.

56 T. Cuenca, M. Galakhovi, E. Royo, P. Royo, *J. Organomet. Chem.* **1996**, *10*, 150.

57 H. Lee, P. J. Desrosiers, I. Guzei, A. L. Rheingold, G. Parkin, *J. Am. Chem. Soc.* **1998**, *120*, 3255.

58 S. Couturier, G. Tainturier, B. Gautheron, *J. Organomet. Chem.* **1980**, *195*, 291.

59 S. B. Jones, J. L. Petersen, *Inorg. Chem.* **1981**, *20*, 2889.

60 A. M. Larsonneur, R. Choukroun, J. Jaud, *Organometallics* **1993**, *12*, 3216.

61 R. Choukroun, F. Dahan, A. M. Larsonneur, E. Samuel, J. Petersen, P. Meunier, C. Sornay, *Organometallics* **1991**, *10*, 374.

62 R. B. Grossman, R. A. Doyle, S. L. Buchwald, *Organometallics* **1991**, *10*, 1501.

63 Y. Mu, C. Aitken, B. Cote, J. F. Harrod, E. Samuel, *Can. J. Chem.* **1991**, *69*, 264.

64 T. Takahashi, M. Hasegawa, M., N. Suzuki, M. Saburi, C. J. Rosset, P. E. Fanwick, E. Negishi, *J. Am. Chem. Soc.* **1991**, *113*, 8564.

65 K. I. Gell, J. Schwartz, *J. Am. Chem. Soc.* **1981**, *103*, 2687. K. I. Gell, J. Schwartz, *J. Am. Chem. Soc.* **1978**, *100*, 3246. K. I. Gell, J. Schwartz, *J. Organomet. Chem.* **1978**, *162*, C11.

66 D. R. McAllister, D. K. Erwin, J. E. Bercaw, *J. Am. Chem. Soc.* **1978**, *100*, 5966.

67 E. Negishi, J. A. Miller, T. Yoshida, *Tetrahedron Lett.* **1984**, *25*, 3407.

68 E. M. Zippi, H. Andres, H. Morimoto, P. G. Williams, *Synth. Commun.* **1994**, *24*, 1037.

69 B. H. Lipshutz, R. Keil, E. L. Ellsworth, *Tetrahedron Lett.* **1990**, *31*, 7257.

70 C. Xu, E. Negishi, *Tetrahedron Lett.* **1999**, *40*, 431.

71 D. R. Swanson, T. Nguyen, Y. Noda, E. Negishi, *J. Org. Chem.* **1991**, *56*, 2590.

72 H. Makabe, E. Negishi, *Eur. J. Org. Chem.* **1999**, 969.

73 P. Wipf, H. Takahashi, N. Zhuang, *Pure Appl. Chem.* **1998**, *70*, 1077.

74 G. J. Lynch, Monsanto Co., United State Patent US 4147709 A.

75 K. I. Gell, B. Posin, J. Schwartz, G. M. Williams, *J. Am. Chem. Soc.* **1982**, *104*, 1846.

76 E. Hey-Hawkins, *Chem. Rev.* **1994**, *94*, 1661.

77 A. J. Hoskin, D. W. Stephan, *Organometallics* **1999**, *18*, 2479. M. C. Fermin, D. W. Stephan, *J. Am. Chem. Soc.* **1995**, *117*, 12645.

78 H. Lee, T. Hascall, P. J. Desrosiers, G. Parkin, *J. Am. Chem. Soc.* **1998**, *120*, 5830.

79 L. Jia, X. Yang, C. L. Stern, T. J. Marks, *Organometallics* **1997**, *16*, 842. Y. W. Alelyunas, N. C. Baenziger, P. K. Bradley, R. F. Jordan, *Organometallics* **1994**, *13*, 148.

80 D. W. Hart, J. Schwartz, *J. Am. Chem. Soc.* **1974**, *96*, 8115.

81 U. Annby, S. Karlsson, S. Gronowitz, A. Hallberg, J. Alvhaell, R. Svenson, *Acta Chem. Scand.* **1993**, *47*, 425.

82 U. Annby, J. Alvhaell, S. Gronowitz, A. Hallberg, *J. Organomet. Chem.* **1989**, *377*, 75.

83 U. Annby, S. Gronowitz, A. Hallberg, *J. Organomet. Chem.* **1989**, *365*, 233.

84 P. Wipf, W. Xu, *J. Org. Chem.* **1996**, *61*, 6556.

85 L. Crombie, A. J. W. Hobbs, M. A. Horsham, R. J. Blade, *Tetrahedron Lett.* **1987**, *28*, 4875.

86 S. L. Buchwald, S. J. LaMaire, *Tetrahedron Lett.* **1987**, *28*, 295.

87 S. P. Nohan, R. Lopez de la Vega, S. L. Mukerjee, A. A. Gonzalez, K. Zhang, C. D. Hoff, *Polyhedron* **1988**, *7*, 1491.

88 P. J. Chirik, M. W. Day, J. A. Labinger, J. E. Bercaw, *J. Am. Chem. Soc.* **1999**, *121*, 10308.

89 M.-H. Prosenc, H.-H. Brintzinger, *Organometallics* **1997**, *16*, 3889.

90 V. Huc, P. Boussaguet, P. Mazerolles, *J. Organomet. Chem.* **1996**, *521*, 253.

91 M. Kosugi, T. Tanji, Y. Tanaka, A. Yoshida, K. Fugami, M. Kameyama, T. Migita, *J. Organomet. Chem.* **1996**, *508*, 255.

92 E. VEDEJS, A. R. HAIGHT, W. O. MOSS, *J. Am. Chem. Soc.* **1992**, *114*, 6556.

93 F. MOULINES, L. DJAKOVITCH, J. L. FIL-LAUT, D. ASTRUC, *Synlett* **1992**, 57.

94 X. ZHOU, S. R. STOBART, R. A. GOSSAGE, *Inorg. Chem.* **1997**, *36*, 3745.

95 S. KARLSSON, A. HALLBERG, S. GRONOWITZ, *J. Organomet. Chem.* **1992**, *430*, 53.

96 S. KARLSSON, A. HALLBERG, S. GRONOWITZ, *J. Organomet. Chem.* **1991**, *403*, 133.

97 J. E. NELSON, J. E. BERCAW, J. A. LABINGER, *Organometallics* **1989**, *8*, 2484.

98 T. GIBSON, *Organometallics* **1987**, *6*, 918.

99 Y. LIU, Q. GUO, X. LEI, *Youji Huaxue* **1984**, 33.

100 A. R. DIAS, J. A. MARTINHO SIMOES, *Polyhedron* **1988**, *7*, 1531.

101 U. ANNBY, S. GRONOWITZ, A. HALLBERG, *J. Organomet. Chem.* **1989**, *368*, 295.

102 G. ERKER, K. KROPP, J. L. ATWOOK, W. E. HUNTER, *Organometallics* **1983**, *2*, 1555.

103 C. A. BERTELO, J. SCHWARTZ, *J. Am. Chem. Soc.* **1976**, *98*, 262.

104 J. ALVHAELL, S. GRONOWITZ, A. HALL-BERG, *Chem. Scr.* **1985**, *25*, 393.

105 J. ALVHAELL, S. GRONOWITZ, A. HALL-BERG, R. SVENSON, *Chem. Scr.* **1984**, *24*, 170.

106 J. ALVHAELL, S. GRONOWITZ, A. HALL-BERG, R. SVENSON, *JAOCS, J. Am. Oil Chem. Soc.* **1984**, *61*, 430.

107 R. SVENSON, S. GRONOWITZ, *Chem. Scr.* **1982**, *19*, 149.

108 M. ZABLOCKA, F. BOUTONNET, A. IGAU, F. DAHAN, J. P. MAJORAL, K. M. PIETRUSIEWICZ, *Angew. Chem. Int. Ed. Engl.* **1993**, *32*, 1735.

109 F. BOUTONNET, M. ZABLOCKA, A. IGAU, J. P. MAJORAL, J JAUD, K. M. PIETRUSIEWICZ, *J. Chem. Soc., Chem. Commun.* **1993**, 1487. M. ZABLOCKA, A. IGAU, J. P. MAJORAL, K. M. PIETRUSIEWICZ, *Organometallics* **1993**, *12*, 603.

110 S. L. BUCHWALD, R. B. NIELSEN, J. C. DEWAN, *Organometallics* **1988**, *7*, 2324.

111 S. KARLSSON, A. HALLBERG, S. GRONOWITZ, *J. Am. Oil Chem. Soc.* **1989**, *66*, 1815.

112 U. ANNBY, S. GRONOWITZ, A. HALLBERG, *Chem. Scr.* **1987**, *27*, 445.

113 J. ALVHAELL, S. GRONOWITZ, A. HALL-BERG, R. SVENSON, *Chem. Scr.* **1984**, *23*, 207.

114 P. WIPF, J. H. SMITROVICH, *J. Org. Chem.* **1991**, *56*, 6494.

115 N. CENAC, M. ZABLOCKA, A. IGAU, J. P. MAJORAL, A. SKOWRONSKA, *J. Org. Chem.* **1996**, *61*, 796.

116 N. CENAC, M. ZABLOCKA, A. IGAU, G. COMMENGES, J. P. MAJORAL, A. SKOWRONSKA, *Organometallics* **1996**, *15*, 1208.

117 N. CENAC, M. ZABLOCKA, A. IGAU, J. P. MAJORAL, K. M. PIETRUSIEWICZ, *Organometallics* **1994**, *13*, 5166.

118 B. ZHENG, L. DELOUX, S. PEREIRA, E. SKRZYPCKAK-JANKUN, B. V. CHEESMAN, M. SABAT, M. SREBNIK, *Appl. Organomet. Chem.* **1996**, *10*, 267.

119 S. PEREIRA, M. SREBNIK, *J. Org. Chem.* **1995**, *60*, 4316.

120 B. ZHENG, M. SREBNIK, *J. Organomet. Chem.* **1994**, *474*, 49.

121 S.-J. EWA, B. V. CHEESMAN, B. ZHENG, R. M. LEMERT, S. ASTHANA, M. SREBNIK, *J. Chem. Soc., Chem. Commun.* **1994**, 127.

122 S. PEREIRA, M. SREBNIK, *Tetrahedron Lett.* **1995**, *36*, 1805.

123 B. ZHENG, M. SREBNIK, *J. Org. Chem.* **1995**, *60*, 486.

124 B. ZHENG, M. SREBNIK, *Tetrahedron Lett.* **1994**, *35*, 1145.

125 B. ZHENG, M. SREBNIK, *Tetrahedron Lett.* **1993**, *34*, 4133-6.

126 C. E. TUCKER, P. J. KNOCHEL, *J. Am. Chem. Soc.* **1991**, *113*, 9888

127 K. SUZUKI, T. IMAI, S. YAMANOI, M. CHINO, T. MATSUMOTO, *Angew. Chem., Int. Ed. Engl.* **1997**, *36*, 2469.

128 M. CHINO, G. H. LIANG, T. MATSUMO-TO, K. SUZUKI, *Chem. Lett.* **1996**, 231.

129 K. SUZUKI, T. HASEGAWA, T. IMAI, H. MAETA, S. OHBA, *Tetrahedron* **1995**, *51*, 4482.

130 M. CHINO, T. MATSUMOTO, K SUZUKI, *Synlett* **1994**, 359.

131 H. MAETA, T. HASEGAWA, K. SUZUKI, *Synlett* **1993**, 341.

132 S. YAMANOI, T. MATSUMOTO, K. SUZUKI, *Tetrahedron Lett.* **1999**, *40*, 2793.

133 S. Yamanoi, T. Imai, T. Matsumoto, K. Suzuki, *Tetrahedron Lett.* **1997**, *38*, 3031.

134 D. Orain, J.-C. Guillemin, *J. Org. Chem.* **1999**, *64*, 3563.

135 T. Matsushima, M. Mori, B.-Z. Zheng, H. Maeda, N. Nakajima, J.-I. Uenishi, O. Yonemitsu, *Chem. Pharm. Bull.* **1999**, *47*, 308-321.

136 S. Chen, K. D. Janda, *Tetrahedron Lett.* **1998**, *39*, 3943.

137 P. Wipf, P. D. G. Coish, *Tetrahedron Lett.* **1997**, *38*, 5073.

138 B. H. Lipshutz, G. Bulow, R. F. Lowe, K. L. Stevens, *Tetrahedron* **1996**, *52*, 7265.

139 T. Mandai, T. Matsumoto, M. Kawada, J. Tsuji, *J. Org. Chem.* **1992**, *57*, 6090.

140 J. E. Ireland, T. K. Highsmich, L. D. Gegnas, J. L. Gleason, *J. Org. Chem.* **1992**, *57*, 5060.

141 A. B. Smith, S.-Y. Chen, F. C. Nelson, J. M. Reichert, B. A. Salvatore, *J. Am. Chem. Soc.* **1995**, *117*, 12013.

142 E. J. Corey, E. J. Trybulski, L. S. Melvin, K. C. Jr.; Nicolaou, J. A. Secrist, R. Lett, P. W. Sheldrake, J. R. Falck, D. J. Brunelle, *J. Am. Chem. Soc.* **1978**, *100*, 4618.

143 M. J. Dabdoub, M. L. Begnini, T. M. Cassol, P. G. Jr. Guerrero, C. C. Silveira, *Tetrahedron Lett.* **1995**, *36*, 7623.

144 B. H. Lipshutz, K. Kato, *Tetrahedron Lett.* **1991**, *32*, 5647.

145 J. A. Labinger, D. W. Hart, W. E. Seibert III, J. Schwartz, *J. Am. Chem. Soc.* **1975**, *97*, 3851.

146 X. Huang, X.-H. Xu, W.-X. Zheng, *Synth. Commun.* **1999**, *29*, 2399.

147 X. Huang, J.-H. Wang, *Synlett* **1999**, 569.

148 D.-H. Duan, X. Huang, *Synlett* **1999**, 317.

149 X. Huang, X.-H. Xu, *J. Chem. Soc., Perkin Trans. 1* **1998**, 3321.

150 P. Zhong, X. Huang, Z.-X. Xiong, *Synlett* **1999**, 721.

151 M. Virgili, M. A. Pericas, A. Moyano, A. Riera, *Tetrahedron* **1997**, *53*, 13427.

152 M. Virgili, A. Moyano, M. A. Pericas, A. Riera, *Tetrahedron Lett.* **1997**, *38*, 6921.

153 P. Vincent, J. P. Beaucourt, L. Pichat, J. Balzarini, E. De Clercq, *Nucleosides Nucleotides* **1985**, *4*, 447.

154 M. J. Dabdoub, M. L. Begnini, P. G. Jr. Guerrero, *Tetrahedron* **1998**, *54*, 2371.

155 X. Huang, L.-S. Zhu, *J. Organomet. Chem.* **1996**, *523*, 9.

156 L.-S. Zhu, Z.-Z. Huang, X. Huang, *Tetrahedron* **1996**, *52*, 9819.

157 L.-S. Zhu, Z.-Z. Huang, X. Huang, *J. Chem. Res., Synop.* **1996**, 112.

158 I. Hyla-Kryspin, R. Gleiter, C. Krueger, R. Zwettler, G. Erker, *Organometallics* **1990**, *9*, 517.

159 A. Igau, J. P. Majoral, unpublished results.

160 M. J. Dabdoub, A. C. M. Baroni, *J. Org. Chem.* **2000**, *65*, 54.

161 D. W. Hart, T. F. Blackburn, J. Schwartz, *J. Am. Chem. Soc.* **1975**, *97*, 679.

162 J. Schwartz, *Pure Appl. Chem.* **1980**, *52*, 753.

163 J. Dupont, A. J. Donato, *Tetrahedron: Asymmetry* **1998**, *9*, 949.

164 J. S. Panek, T. Hu, *J. Org. Chem.* **1997**, *62*, 4912.

165 D. Romo, D. D. Johnson, L. Plamondon, T. Miwa, S. L. Schreiber, *J. Org. Chem.* **1992**, *57*, 5060

166 J. W. Sung, C. W. Lee, D. Y. Oh, *Tetrahedron Lett.* **1995**, *36*, 1503.

167 Park, C. P.; Sung, J. W.; Oh, D. Y. *Synlett* **1999**, 1055.

168 X. Huang, P. Zhong, *J. Chem. Soc., Perkin Trans. 1* **1999**, 1543.

169 Y. Ma, X. Huang, *Synth. Commun.* **1999**, *29*, 429.

170 B. H. Lipshutz, R. Keil, J. C. Barton, *Tetrahedron Lett.* **1992**, *33*, 5861.

171 X.-H. Xu, W.-X. Zheng, X. Huang, *Synth. Commun.* **1998**, *28*, 4165.

172 M.; Sabat, *J. Mol. Struct* **1996**, *374*, 291.

173 L. Deloux, M. Srebnik, *J. Org. Chem.* **1994**, *59*, 6871.

174 L. Deloux, E. Skrzypczak-Jankun, B. V. Cheesman, M. Srebnik, M. Sabat, *J. Am. Chem. Soc.* **1994**, *116*, 10302.

175 L. Deloux, M. Srebnik, M. Sabat, *J. Org. Chem.* **1995**, *60*, 3276.

176 X. Huang, J.-H. Wang, D.-Y. Yang, *J. Chem. Soc., Perkin Trans. 1* **1999**, 673.

177 B. ZHENG, M. SREBNIK, *Tetrahedron Lett.* **1994**, *35*, 6247.

178 A. IGAU, J. A. GLADYSZ, *Organometallics* **1991**, *10*, 2327.

179 F. H. ELSNER, H. G. WOO, T. D. TILLEY, *J. Am. Chem. Soc.* **1988**, *110*, 313.

180 M. C. BARDEN, J. SCHWARTZ, *J. Org. Chem.* **1995**, *60*, 5963.

181 T. NAKANO, Y. KINO, Y. ISHII, M. OGAWA, *Technol. Rep. Kansai Univ* **1988**, *30*, 81.

182 E. CESAROTTI, A. CHIESA, S. MAFFI, R. UGO, *Inorg. Chim. Acta* **1982**, *64*, L207.

183 S. COUTURIER, B. GAUTHERON, *J. Organomet. Chem.* **1978**, *157*, C61.

184 J. WANG, W. QI, Y. ZHANG, Y. XU, *Youji Huaxue* **1985**, 317.

185 J. WANG, Y. XU, Y. ZHANG, W. QI, *Youji Huaxue* **1986**, 138.

186 B. H. LIPSHUTZ, C. LINDSLEY, A. BHANDARI, A. *Tetrahedron Lett.* **1994**, *35*, 4669.

187 J. WANG, Y. ZHANG, Y. XU, G. BAI, *Youji Huaxue* **1989**, *9*, 41.

188 J. WANG, Y. ZHANG, Y. XU, Z. LIN, *Gaodeng Xuexiao Huaxue Xuebao* **1989**, *10*, 263.

189 W. FROEMBERG, G. ERKER, *J. Organomet. Chem.* **1985**, *280*, 343.

190 G. ERKER, W. FROEMBERG, J. L. ATWOOD, W. E. HUNTER, *Angew. Chem.* **1984**, *96*, 72.

191 P. ETIEVANT, G. TAINTURIER, B. GAUTHERON, *C. R. Acad. Sci., Ser. C* **1976**, *283*, 233.

192 A. MARAVAL, B. DONNADIEU, A. IGAU, J.-P. MAJORAL, *Organometallics* **1999**, *18*, 3138.

193 A. MARAVAL, B. DONNADIEU, A. IGAU, J.-P. MAJORAL, unpublished results.

194 F. BOUTONNET, N. DUFOUR, T. STRAW, A. IGAU, J.-P. MAJORAL, *J. Organometallics* **1991**, *10*, 3939.

195 A. MARAVAL, B. DONNADIEU, A. IGAU, J.-P. MAJORAL, unpublished results.

196 D. J. A. SCHEDLER, J. LI, B. GANEM, *J. Org. Chem.* **1996**, *61*, 4115.

197 NG, K. S.; LAYCOCK, D. E.; ALPER, H. *J. Org. Chem.* **1981**, *46*, 2899.

198 S. GAMBAROTTA, C. FLORIANI, A. CHIESI-VILLA, C. GUASTINI, *Inorg. Chem.* **1983**, *22*, 2029.

199 S. GAMBAROTTA, M. BASSO-BERT, C. FLORIANI, C. GUASTINI, *J. Chem. Soc., Chem. Commun.* **1982**, 374.

200 G. ERKER, U. HOFFMANN, R. ZWETTLER, P. BETZ, C. KRUEGER, *Angew. Chem.* **1989**, *101*, 644.

201 S. GAMBAROTTA, C. FLORIANI, A. CHIESI-VILLA, C. GUASTINI, *J. Am. Chem. Soc.* **1983**, *105*, 1690.

202 G. FACHINETTI, C. FLORIANI, A. ROSELLI, S. PUCCI, *J. Chem. Soc., Chem. Commun.* **1978**, 269.

203 W. FROEMBERG, G. ERKER, *J. Organomet. Chem.* **1985**, *280*, 355.

204 S. GAMBAROTTA, S. STROLOGO, C. FLORIANI, A. CHIESI-VILLA, C. GUASTINI, *J. Am. Chem. Soc.* **1985**, *107*, 6278.

205 M. WANG, S. LU, M. BEI, H. GUO, Z. JIN, *J. Organomet. Chem.* **1993**, *447*, 227.

206 A. GUDIMA, A. IGAU, B. DONNADIEU, J.-P. MAJORAL, *J. Org. Chem.* **1996**, *61*, 9585.

207 N. DUFOUR, A.-M. CAMINADE, M. BASSO-BERT, A. IGAU, J.-P. MAJORAL, *Organometallics* **1992**, *11*, 1131.

208 G. ERKER, P. CZISCH, R. MYNOTT, Y. H. TSAY, C. KRUEGER, *Organometallics* **1985**, *4*, 1310.

209 D. MAENNIG, H. NOETH, M. SCHWARTZ, S. WEBER, U. WIETELMANN, *Angew. Chem.* **1985**, *97*, 979.

210 M. D. FRYZUK, G. S. BATES, C. STONE, *J. Org. Chem.* **1991**, *56*, 7201.

211 P. VINCENT, J. P. BEAUCOURT, L. PICHAT, *Tetrahedron Lett.* **1982**, *23*, 63.

212 P. WIPF, W. XU, *Synlett* **1992**, 718.

213 B. H. LIPSHUTZ, R. KEIL, *J. Am. Chem. Soc.* **1992**, *114*, 7919.

214 J. R. HAUSKE, P. DORFF, S. JULIN, G. MARTINELLI, J. BUSSOLARI, *Tetrahedron Lett.* **1992**, *33*, 3715.

215 M. D. FRYZUK, G. S. BATES, C. STONE, *Tetrahedron Lett.* **1986**, *27*, 1537.

216 J. SCHWARTZ, M. J. LOOTS, H. KOSUGI, *J. Am. Chem. Soc.* **1980**, *102*, 1333.

217 S. HARADA, N. KOWASE, N. TABUCHI, T. TAGUCHI, Y. DOBASHI, A. DOBASHI, Y. HANZAWA, *Tetrahedron* **1998**, *54*, 753.

218 M. T. ASHBY, S. S. ALGUINDIGUE, M. A. KHAN, *Inorg. Chim. Acta* **1998**, *270*, 227.

219 G. A. TOLSTIKOV, M. S. MIFTAKHOV, N. A. DANILOVA, F. Z. GALIN, *Zh. Org. Khim.* **1983**, *19*, 1857.

220 E. Uhlig, B. Buerglen, C. Krueger, P. Betz, *J. Organomet. Chem.* **1990**, *382*, 77.

221 B. Buerglen, E. Uhlig, *Z. Chem.* **1988**, *28*, 408.

222 W. Tam, M. F. Rettig, *J. Organomet. Chem.* **1976**, *108*, C1.

223 P. Wipf, H. Takahashi, *Chem. Commun.* **1996**, 2675.

224 B. Peifer, W. Milius, H. G. Alt, *J. Organomet. Chem.* **1998**, *553*, 205.

225 F. R. Lemke, R. M. Bullock, *Organometallics* **1992**, *11*, 4261.

226 F. R. Lemke, D. J. Szalda, R. M. Bullock, *J. Am. Chem. Soc.* **1991**, *113*, 8466.

227 P. Meunier, I. Ouattara, B. Gautheron, J. Tirouflet, D. Camboli, J. Besancon, *Eur. J. Med. Chem.* **1991**, *26*, 351.

228 X. Gu, M. B. Sponsler, *Organometallics* **1998**, *17*, 5920-5923.

229 D. H. Gibson, J. M. Mehta, B. A. Sleadd, M. S. Mashuta, J. F. Richardson, *Organometallics* **1995**, *14*, 4886.

230 E. N. Jacobsen, M. K. Trost, R. G. Bergman, *J. Am. Chem. Soc.* **1986**, *108*, 8092.

231 D. H. Duan, X. Huang, *Synlett* **1999**, 317.

232 D. R. Kelsey, D. L. Jr. Handlin, M. Narayana, B. M. Scardino, *J. Polym. Sci., Part A: Polym. Chem.* **1997**, *35*, 3027.

233 T. Imori, V. Lu, H. Cai, T. D. Tilley, *J. Am. Chem. Soc.* **1995**, *117*, 9931.

234 S. Pereira, M. Srebnik, *J. Am. Chem. Soc.* **1996**, *118*, 909.

235 S. Pereira, M. Srebnik, *Organometallics* **1995**, *14*, 3127.

236 K. Morita, Y. Nishiyama, Y. Ishii, *Organometallics* **1993**, *12*, 3748.

237 T. Nakano, S. Irifune, S. Umano, A. Inada, Y. Ishii, M. Ogawa, *J. Org. Chem.* **1987**, *52*, 2239.

238 E. Negishi, T. Yoshida, *Tetrahedron* **1980**, *21*, 1501.

239 E. Negishi, D. Y. Kondakov, D. Choueiry, K. Kasai, T. Takahashi, *J. Am. Chem. Soc.* **1996**, *118*, 9577.

240 J. Y. Corey, X. H. Zhu, *Organometallics* **1992**, *11*, 672-83.

241 L. Zhang, D.-Y. Jung, E. R. Bittner, M. S. Sommer, E. L. Dias, T. R. J. Lee, *Org. Chem.* **1998**, *63*, 8624.

242 R. L. Alterman, in *Metallocenes, Synthesis, Reactivity, Applications*, A. Togni, R. L. Halterman (eds.), Wiley-VCH, Weinheim, **1998**, Vol. 1, p. 455.

243 U. Blaschke, G. Erker, M. Nissinen, E. Wegelius, R. Froehlich, *Organometallics* **1999**, *18*, 1224.

Index

T